Conversion Factors

Length

1 m = 39.37 in. = 3.281 ft

1 in. = 2.54 cm

1 km = 0.621 mi

1 mi = 5280 ft = 1.609 km

1 lightyear = 9.461×10^{15} m

1 angstrom (Å) = 10^{-10} m

Mass

1 kg = 10^3 g = 6.85×10^{-2} slug

1 slug = 14.59 kg

1 u = 1.66×10^{-27} kg

Time

1 min = 60 s

1 h = 3600 s

1 day = 8.64×10^4 s

1 year = 365.242 days = 3.156×10^7 s

Volume

1 liter = 1000 cm^3 = 3.531×10^{-2} ft^3

1 ft^3 = 2.832×10^{-2} m^3

1 gallon = 3.786 liter = 231 in.^{3}

Angle

180° = π rad

1 rad = 57.30°

1° = 60 min = 1.745×10^{-2} rad

Speed

1 km/h = 0.278 m/s = 0.621 mi/h

1 m/s = 2.237 mi/h = 3.281 ft/s

1 mi/h = 1.61 km/h = 0.447 m/s = 1.47 ft/s

Force

1 N = 0.2248 lb = 10^5 dynes

1 lb = 4.448 N

1 dyne = 10^{-5} N = 2.248×10^{-6} lb

Work and energy

1 J = 10^7 erg = 0.738 ft·lb = 0.239 cal

1 cal = 4.186 J

1 ft·lb = 1.356 J

1 Btu = 1.054×10^3 J = 252 cal

1 J = 6.24×10^{18} eV

1 eV = 1.602×10^{-19} J

1 kWh = 3.60×10^6 J

Pressure

1 atm = 1.013×10^5 N/m^2 (or Pa) = 14.70 lb/in.^{2}

1 Pa = 1 N/m^2 = 1.45×10^{-4} lb/in.^{2}

1 lb/in.^{2} = 6.895×10^3 N/m^2

Power

1 hp = 550 ft·lb/s = 0.746 kW

1 W = 1 J/s = 0.738 ft·lb/s

1 Btu/h = 0.293 W

COLLEGE PHYSICS

Fifth Edition

Volume 1

CO

PHYS

Fifth Edition

Volume 1

Raymond A. Serway

North Carolina State University

Jerry S. Faughn

SAUNDERS GOLDEN SUNBURST SERIES

Saunders College Publishing

Harcourt Brace College Publishers

FORT WORTH   PHILADELPHIA   SAN DIEGO   NEW YORK   ORLANDO   AUSTIN
SAN ANTONIO   TORONTO   MONTREAL   LONDON   SYDNEY   TOKYO

Requests for permission to make copies of any part of the work should be mailed to: Permissions Department, Harcourt Brace & Company, 6277 Sea Harbor Drive, Orlando, Florida 32887-6777.

Publisher: Emily Barrosse
Publisher: John Vondeling
Product Manager: Nick Agnew
Developmental Editor: Susan Dust Pashos
Project Editor: Elizabeth Ahrens
Production Manager: Charlene Catlett Squibb
Art Director: Carol Bleistine
Text and Cover Designer: Ruth A. Hoover

Cover Credit: © 1995 John Terence Turner/FPG International LLC

Frontispiece Credit: Snowboarder in air, © Colin Samuels/Photonica

Printed in the United States of America

COLLEGE PHYSICS, Fifth Edition: Volume 1
0-03-022507-8

Library of Congress Catalog Card Number: 97-81032

8901234567   048   10 9876543 21

Preface

College Physics is written for a one-year course in introductory physics. The course is usually taken by students majoring in biology; the health professions; environmental, earth, and social sciences; and technical fields such as architecture. The mathematical techniques used in the book include algebra, geometry, and trigonometry, but not calculus.

The main objectives of this introductory book are twofold: to provide the student with a clear and logical presentation of the basic concepts and principles of physics, and to strengthen an understanding of the concepts and principles through a broad range of interesting applications to the real world. In order to meet these objectives, emphasis is placed on sound physical arguments and discussions of everyday experiences and observations. At the same time, we have attempted to motivate the student through practical examples that demonstrate the role of physics in other disciplines.

This book, which covers the standard topics in classical physics and 20th century physics, is divided into six parts. Part 1 (Chapters 1–9) deals with Newtonian mechanics and the physics of fluids; Part 2 (Chapters 10–12) is concerned with heat and thermodynamics; Part 3 (Chapters 13–14) covers wave motion and sound; Part 4 (Chapters 15–21) is concerned with electricity and magnetism; Part 5 (Chapters 22–25) treats the properties of light and the field of geometric and wave optics; and Part 6 (Chapters 26–30) represents an introduction to special relativity, quantum physics, and atomic and nuclear physics.

© *Index Stock Photography, Inc.*

CHANGES TO THE FIFTH EDITION

A number of changes and improvements have been made in preparing the fifth edition of this text. Many changes were in response to comments and suggestions submitted by reviewers of the manuscript and instructors using the fourth edition. The new features added to this edition are based on current trends in science education. The following describe the major changes in the fifth edition.

Conceptual Thinking A concerted effort was made to place more emphasis on critical thinking and teaching physical concepts. This was accomplished through the use of approximately 200 conceptual examples, called **Thinking Physics,** and over 400 **Conceptual Questions.** Many of the **Thinking Physics** examples provide students with a means of reviewing concepts presented in that section. Some examples demonstrate the connection between the concepts presented in that chapter and other scientific disciplines. These examples also serve as models for students when assigned the task of responding to the **Conceptual Questions** presented at the end of each chapter. The answers to the odd-numbered **Conceptual Questions** are contained in the answer section at the end of the textbook.

Applications We have expanded the number of applications of physics throughout the text. Some applications are developed within the text proper, while many are in the form of new examples called **Applying Physics.** Because the examples chosen should be familiar to most students, they should help motivate the student and show the relevance of physical concepts as they apply to real-world problems. Applications in the text proper are identified by special marginal notes of the form

APPLICATION

Connecting Your Stereo
Speakers.

A complete list of the applications can be found on page xvii.

A complete list of the applications can be found on page xvii.

QUICKLAB

Place a cup of water on a sheet of paper with one side of the paper overhanging the table that supports them. Remove the sheet of paper from underneath the cup by *quickly* pulling on the sheet. (Until you become good at this, you might want to use a plastic cup rather than a glass cup in case of an accident.) Explain why this is possible.

QuickLabs   This new feature encourages students to perform simple experiments on their own so that they will be more engaged in the learning process. Most **QuickLab** experiments can be performed with low-cost items such as string, rubber bands, a ruler, tape, and balloons. In most cases, students are simply asked to observe the outcome of an experiment and explain their results in terms of what they have learned in that chapter. Where appropriate, students are also asked to obtain some data and graph their results.

Multiple-Choice Questions   We have included a number of multiple-choice questions at the end of each chapter, with answers provided in the answer section at the back of the book. These questions are intended to give students practice in this form of testing, commonly used in the MCAT exam and in many large enrollment courses.

Problems and Conceptual Questions   A substantial revision of the end-of-chapter problems and conceptual questions was made in this fifth edition. Most of the new problems that have been added are intermediate in level (as identified by their blue problem numbers). All problems have been carefully worded and have been checked for clarity and accuracy. Some problem sets begin with a Review Problem that requires the use of concepts learned in previous chapters. Solutions to approximately 20 percent of the end-of-chapter problems are included in the *Study Guide and Student Solutions Manual.* These problems are identified by boxes around their numbers. A smaller subset of solutions has been posted on the World Wide Web (**http://www.saunderscollege.com**) at a site accessible to students and instructors using *College Physics;* these problems are identified with the label WEB. See below for a complete description of other features of the problem set.

Chapter-Opening Photos and Puzzlers   All chapter-opening photographs now include puzzlers in the form of a question to the student. The student should be able to answer the puzzler after completing that chapter.

Other Content Changes   The material entitled "For Further Study" in the fourth edition for Chapters 7, 9, and 15 was modified and moved from the back of the book to the main text in the form of optional sections. Some features found in the fourth edition were either deleted or reduced in length to allow for the new features described earlier. The biographical sketches of important scientists were replaced by short descriptions of their work and placed in margins. Most essays by guest authors were omitted from the text. However, those essays that pertain to the health sciences were retained in this edition. The lists of suggested additional readings were removed from the text, but will be included in the *Instructor's Manual.*

FEATURES

Most instructors would agree that the textbook assigned in a course should be the student's major guide for understanding and learning the subject matter. With this in mind, we have retained or added many pedagogical features in the textbook that should enhance its usefulness to both the student and instructor. These are as follows:

Style We have attempted to write the book in a style that is clear, relaxed, and pleasing to the reader. New terms are carefully defined, and we have tried to avoid jargon. At the same time, we have attempted to keep the presentation accurate and precise.

Organization The book is divided into six parts: mechanics, thermodynamics, vibrations and wave motion, electricity and magnetism, light and optics, and modern physics. Each part includes an overview of the subject matter to be covered in that part and some historical perspectives.

Introductory Chapter The introductory chapter, which "sets the stage" for the text, discusses the building blocks of matter, the units of physical quantities, order-of-magnitude calculations, dimensional analysis, significant figures, and mathematical notation.

Units The International System of Units (SI) is used throughout the book. The British engineering system of units is used only to a limited extent in the problem sets of the early chapters on mechanics.

Previews Most chapters begin with a chapter preview, which includes a brief discussion of the chapter objectives and content.

Equations, Marginal Notes, and Important Statements Important equations are highlighted with a light gold screen, and marginal notes are often used to describe their meaning. Marginal notes are also used to locate specific definitions and important statements in the text. Most important statements and definitions are highlighted with a light blue screen or are set in **boldface type** for added emphasis and ease of review.

Problem-Solving Strategies General strategies and suggestions are included for solving the types of problems featured in both the worked examples and end-of-chapter problems. This feature is intended to help students identify the essential steps in solving problems and increase their skills as problem solvers. This feature is highlighted by a light blue screen to help students locate the strategies quickly.

Physics in Action This boxed material focuses on photographs of interesting demonstrations and phenomena in physics, accompanied by detailed explanations. The material can also serve as a source of information for initiating classroom discussions.

Topics Relevant to the Life Sciences and Other Disciplines Many chapters include sections which are intended to expose the student to various practical and interesting applications of physical principles. Those topics dealing with applications of physics to the life sciences are identified by the DNA icon.

Worked Examples A large number of worked examples, including many new ones, are presented as an aid in understanding and/or reinforcing physical concepts. In many cases, these examples serve as models for solving end-of-chapter problems. The examples are set off from the text for ease of location and are given titles to describe their content.

Telegraph Colour Library/FPG

Many examples include a **Reasoning** section to illustrate the underlying concepts and methodology used in arriving at a correct solution. This will help students understand the logic behind the solution and the advantage of using a particular approach to solve the problem. The solution answer is highlighted with a light blue screen. Many worked examples are followed immediately by exercises with answers. These exercises represent extensions of the worked examples and are intended to sharpen students' problem-solving skills and test their understanding of concepts. Students who work through these exercises on a regular basis should find the end-of-chapter problems less intimidating.

NASA

Illustrations and Photographs
The text material, worked examples, and end-of-chapter questions and problems are accompanied by numerous figures, photographs, and tables. Full color is used to add clarity to the figures and to make the visual presentation as realistic and pleasing as possible. Three-dimensional effects are rendered where appropriate. Vectors are color coded, and curves in *xy*-plots are drawn in color. Color photographs have been carefully selected, and their accompanying captions have been written to serve as an added instructional tool. A complete description of the pedagogical use of color appears on the inside front cover.

Summaries
Each chapter contains a summary which reviews the important concepts and equations discussed in that chapter.

Conceptual Questions
A set of conceptual questions is provided at the end of each chapter. The **Thinking Physics** and **Applying Physics** examples presented in the text should serve as models for students when conceptual questions are assigned or used in tests. The questions all provide the student with a means of self-testing the concepts presented in the chapter. Some conceptual questions are appropriate for initiating classroom discussions. Answers to all odd-numbered conceptual questions are located in the answer section at the end of the book.

Multiple-Choice Questions
A set of multiple-choice questions is included at the end of each chapter. The answers to these questions are given at the end of the book.

End-of-Chapter Problems
An extensive set of problems is included at the end of each chapter. Answers to odd-numbered problems are given at the end of the book. For the convenience of both the student and instructor, about two thirds of the problems are keyed to specific sections of the chapter. The remaining problems, labeled "Additional Problems," are not keyed to specific sections. There are three levels of problems according to their difficulty. Straightforward problems are numbered in black, intermediate-level problems are numbered in blue, and the most challenging problems are numbered in magenta. Those problems that are accompanied by an Interactive Physics™ 3.5 computer simulation (see below) are labeled with the Interactive Physics™ icon (📖), and those with a focus on the life sciences are identified by the DNA icon (🧬). Solutions for problems marked WEB may be found on our Web Site: **http://www.saunderscollege.com,** provided the proper instructor or student password is known.

Appendices
Several appendices are provided at the end of the book. The appendix material includes a review of mathematical techniques used in the book, such as scientific notation, algebra, geometry, and trigonometry. Reference to these appendices is made as needed throughout the book. Most of the mathematical review sections include worked examples and exercises with answers. Some appendices contain useful tables that supplement textual information. For easy reference, the front endpapers contain a chart explaining the use of color throughout the book and a list of frequently used conversion factors.

SAUNDERS CORE CONCEPTS IN COLLEGE PHYSICS CD-ROM

The *Saunders Core Concepts in College Physics* CD-ROM developed by Archipelago Productions applies the power of multimedia to the introductory physics course, offering full-motion animation and video, engaging interactive graphics, clear and concise text, and guiding narration. *Saunders Core Concepts in College Physics* focuses on those concepts students typically find most difficult in the course, drawing from topics in mechanics, thermodynamics, electric fields, magnetic fields, and optics. The animations and graphics are presented to aid the student in developing accurate conceptual models of difficult topics—topics often hard to explain with words or chalkboard illustrations.

The CD-ROM also presents a step-by-step exploration of problem-solving strategies and provides animations of problems in order to promote conceptual understanding and to sharpen problem-solving skills.

STUDENT ANCILLARIES

Saunders College Publishing offers several items to supplement and enhance the classroom experience. These ancillaries will allow instructors to customize the textbook to their students' needs and their own style of instruction. One or more of these ancillaries may be shrink-wrapped with the text at a reduced price:

Study Guide and Student Solutions Manual by John R. Gordon, Charles Teague, and Raymond A. Serway. The manual features detailed solutions to 20 percent of the end-of-chapter problems from the text. These are indicated in the text with boxed problem numbers. The manual also features a skills section, important notes from key sections of the text, and a list of important equations and concepts.

Pocket Guide to Accompany College Physics by V. Gordon Lind. This 5″ by 7″ paperback is a section-by-section capsule of the textbook that provides a handy guide for looking up important concepts, equations, and problem-solving hints.

Interactive Physics™ Simulations by Raymond A. Serway and Knowledge Revolution. Many physical situations discussed in the textbook involving the motion of objects are brought to life with computer simulations developed using the highly acclaimed Interactive Physics™ program by Knowledge Revolution. Approximately 110 simulations, most keyed to specific worked examples or end-of-chapter problems, are available on a dual-format CD-ROM also containing a Textbook version (version 3.5) of the program. (This Textbook version will enable students to print out their work, but not to save or to export data.) Other simulations portray material that is difficult to display in a printed textbook, such as the tracking of a specific point on a rotating object, or complement concepts or applications discussed in the text. A complete list of Interactive Physics™ simulations is found on pages xix–xx.

Physics Laboratory Manual, second edition by David Loyd. Updated and redesigned, this manual supplements the learning of basic physical principles while introducing laboratory procedures and equipment. Each chapter contains a pre-laboratory assignment, objectives, an equipment list, the theory behind the experiment, step-by-step experimental

Richard Megna/Fundamental Photographs, NYC

procedures, and questions. A laboratory report form is provided for each experiment so that students can record data and make calculations. Students are encouraged to apply statistical analysis to their data in order to develop their ability to judge the validity of their results.

So You Want to Take Physics: A Preparatory Course by Rodney Cole. This introductory-level book covers numerous physical principles and is ideal for strengthening mathematical skills essential to the study of physics.

Practice Problems with Solutions After consulting with their instructor, students may wish to purchase an additional set of skill-building practice problems with solutions in print form. These problems will provide extra practice in unit conversions and other mathematical basics for students who need to build confidence in solving physics problems.

Student Web Site Students will find an abundance of material at **http://www.saunderscollege.com/physics/college,** including a sampling of solutions from the Study Guide/Student Solutions Manual, a sample of material in the Pocket Guide, and a glossary that includes over 300 physics terms. Students will also be able to explore our Practice Exercises and Testing area to access extra problems, including multiple-choice and interactive-simulation problems.

© R. Folwell/SPL/Photo Researchers, Inc.

INSTRUCTOR ANCILLARIES

Instructor's Manual with Complete Solutions by Charles Teague and Jerry S. Faughn. This manual consists of complete solutions to all the problems in the text, as well as answers to even-numbered problems and conceptual questions. Instructors will also find a list of suggested readings from journals and other resources.

Printed Test Bank by Robert Beichner. The printed test bank contains approximately 2500 problems and questions (both open-ended and multiple-choice). The format allows instructors to duplicate pages for distribution to students.

Computerized Test Bank Available for the IBM and Macintosh computers, the test bank contains approximately 2500 open-ended and multiple-choice problems and questions, representing every chapter of the text. The test bank allows instructors to customize tests by rearranging, editing, or adding questions. The questions are graded in level of difficulty for the instructor's convenience; the program places the answers to all questions on a separate grading key.

Overhead Transparency Acetates The collection of transparencies consists of approximately 250 full-color figures from the text to enhance lectures; they feature large print for easy viewing in the classroom.

Interactive Physics™ *Simulations* by Raymond A. Serway and Knowledge Revolution. Instructors who wish to take full advantage of the power of the Interactive Physics™ program are encouraged to purchase a site license or lab pack of the program directly from Knowledge Revolution by calling (800) 766-6615 or through their Web site at **http://www.krev.com.** The simulations keyed to specific worked examples and end-of-chapter problems in *College Physics* are downloadable from the Saunders College Publishing Web site for instructors adopting *College Physics,* or may be obtained through your local sales representative.

Practice Problems with Solutions   About 400 problems that do not appear in the text are available with full solutions in a printed version and on the MediaActive™ Instructor's Resource CD-ROM. These can be used for examinations, homework assignments, or student practice and drill exercises. This problems booklet is also available at a low cost to the student.

Instructor's Manual for Physics Laboratory Manual, second edition by David Loyd. Each chapter contains a discussion of the experiment, teaching hints, answers to selected questions from the student laboratory manual, and a post-laboratory quiz with short answers and essay questions. The author has also included a list of the suppliers of scientific equipment and a summary of the equipment needed for all the experiments in the manual.

Physics Demonstration Videotapes   A unique collection of 70 physics demonstrations is provided on videotape to supplement classroom presentations and to help motivate students.

Instructor's Web Site   At the instructor's area at **http://www.saunderscollege.com/ physics,** instructors will find (among other resources) an overhead transparency listing; a chapter-by-chapter guide to relevant experiments in *Physics Laboratory Manual,* 2nd edition by David Loyd; a correlation guide between chapters in *College Physics,* 5th edition and the *Saunders Core Concepts in College Physics* CD-ROM; and a quizzing administration system.

Instructor's Resource CD-ROM   Instructor's Resource CD-ROMs have been created to provide our textbook users with an exciting new tool for classroom presentation. The IRCD accompanying *College Physics,* 5th edition contains a collection of graphics files of line art from the textbook. These files can be opened directly or can be imported into a variety of presentation packages. The labels for each piece of art have been enlarged and made bold for classroom viewing. The IRCD will also contain electronic files of the Instructor's Manual and Test Bank, plus the Practice Problems with Solutions.

Saunders College Publishing may provide complementary instructional aids and supplements or supplement packages to those adopters qualified under our adoption policy. Please contact your sales representative for more information. If as an adopter or potential user you receive supplements you do not need, please return them to your sales representative or send them to

> Attn: Returns Department
> Troy Warehouse
> 465 South Lincoln Drive
> Troy, MO 63379

TEACHING OPTIONS

College Physics, 5th edition, contains more than enough material for a one-year course in introductory physics. This serves two purposes. First, it gives the instructor more flexibility in choosing topics for a specific course. Second, the book becomes more useful as a resource for students. On the average, it should be possible to cover about one chapter each week for a class that meets three hours per week. Many sections containing interesting applications are labeled as being optional material. Those optional sections dealing with applications of physics to the life sciences are identified with the DNA icon (). Instructors are encouraged to cover those optional sections which best match their students' interests. We offer the following suggestions for shorter courses for those instructors who choose to move at a slower pace through the year:

Option A: If you choose to place more emphasis on contemporary topics in physics, you should consider omitting all or parts of Chapter 8 (Rotational Equilibrium and Rotational Dynamics), Chapter 21 (Alternating Current Circuits and Electromagnetic Waves), and Chapter 25 (Optical Instruments).

Option B: If you choose to place more emphasis on classical physics, you could omit all or parts of Part 6 of the textbook, which deals with special relativity and other topics in 20th century physics.

Ben Rose/The Image Bank

ACKNOWLEDGMENTS

In preparing the fifth edition of this textbook, we have been guided by the expertise of many people who have reviewed manuscript and/or provided pre-revision suggestions. We wish to acknowledge the following reviewers and express our sincere appreciation for their helpful suggestions, criticisms, and encouragement:

Paul D. Beale, *University of Colorado at Boulder*

David H. Bennum, *University of Nevada at Reno*

Neal M. Cason, *University of Notre Dame*

Steven D. Davis, *University of Arkansas at Little Rock*

John DeFord, *University of Utah*

Chris J. DeMarco, *Jackson Community College*

Tom French, *Montgomery County Community College*

Grant Hart, *Brigham Young University*

John Ho, *State University of New York at Buffalo*

Murshed Hossain, *Rowan University*

Ivan Kramer, *University of Maryland, Baltimore County*

David Markowitz, *University of Connecticut*

Steven Morris, *Los Angeles Harbor College*

Martin Nikolo, *Saint Louis University*

Ed Oberhofer, *University of North Carolina at Charlotte*

David G. Onn, *University of Delaware*

John Simon, *University of Toledo*

We thank William G. Buckman of Western Kentucky University and David Griffing of Miami University for their interesting essays that follow Chapters 9 and 11.

We are especially grateful to John W. Jewett, Jr. of the California State Polytechnic University at Pomona for his exceptional contribution of chapter opener puzzlers, new applications, and many of the **Thinking Physics** and **Applying Physics** examples. We thank Ed Oberhofer for his contribution of new problems, including all of the Review Problems in this edition. Charles Teague carefully compiled the problems manuscript, in addition to selecting and editing solutions to appear in the *Study Guide and Student Solutions Manual*. We thank Ralph McGrew of Broome Community College and John Jewett for their painstaking accuracy checking of the text. John R. Gordon put together the many features of the *Study Guide and Student Solutions Manual* and oversaw its composition and layout by Michael Rudmin. We are grateful to our friend and colleague Robert J. Beichner for his suggestions and encouragement regarding the **QuickLabs** and other features of the text.

We thank Linda Miller for her painstaking work with a word processor in typing and checking the fifth edition of the manuscript. Dena Digilio-Betz is to be recognized for locating and suggesting many excellent photographs for this edition.

We are extremely grateful to the publishing team at Saunders College Publishing for their expertise and outstanding work in all aspects of this project. Susan Pashos carefully analyzed the reviews of the fifth edition, provided important guidance in all aspects of the project, and made many important improvements in the final manuscript. Beth Ahrens managed the book through all production stages and was invaluable in her help with the revised illustrations. We appreciate the fine work of Alexandra Buczek for managing the development of the ancillary materials, Christine Benedetto for developing the Web site, and Carol Bleistine for managing the overall art and design program. We also recognize important contributions by Nick Agnew, Ruth Hoover, Sally Kusch, and Charlene Squibb. We thank John Vondeling for his continued friendship, enthusiasm, and sense of humor, even under a very tight schedule.

Finally, we dedicate this book to our wives and children, for their love, support, and long-term sacrifices.

Raymond A. Serway
Chapel Hill, North Carolina

Jerry S. Faughn
Richmond, Kentucky

Applications

Although physics is relevant to so much in our modern lives, this may not be obvious to students in an introductory course. In this fifth edition of *College Physics*, we have attempted to make the relevance of physics to everyday life more obvious by pointing out specific applications with a new design feature in the form of a marginal note. Additionally, some of these applications we have marked pertain to the life sciences and are identified with the DNA icon ( ). The list below is not intended to be a complete listing of all the applications of the principles of physics found in the textbook. Many other applications are to be found within the text and especially in the worked examples, Conceptual Questions, and end-of-chapter problems.

APPLICATION

Directs you to sections discussing applied principles of physics.

⊞ Interactive Physics™ Simulations

The following worked examples, end-of-chapter problems, and other items have simulations written for use with Interactive Physics™ by Knowledge Revolution. A dual-platform (IBM and Macintosh-compatible) CD-ROM is available at a greatly reduced price, containing the Textbook version 3.5 of the program and the simulations files. If the CD-ROM is not found on the inside back cover of your text, and you wish to order a copy, please phone (800)782-4479. Instructors who have purchased a site license or lab pack directly from Knowledge Revolution may download these simulations from the Saunders College Publishing instructor's Web site (**http://www.saunderscollege.com**) provided the proper instructor password is known.

To the Student

We feel it is appropriate to offer some words of advice that should be of benefit to you, the student. Before doing so, we shall assume that you have read the preface, which describes the various features of the text that will help you through the course.

HOW TO STUDY

Very often we are asked "How should I study physics and prepare for examinations?" There is no simple answer to this question, but we would like to offer some suggestions based on our own experiences in learning and teaching over the years.

First and foremost, maintain a positive attitude toward the subject matter, keeping in mind that physics is the most fundamental of all natural sciences. Other science courses that follow will use the same physical principles, so it is important that you understand and be able to apply the various concepts and theories discussed in the text.

CONCEPTS AND PRINCIPLES

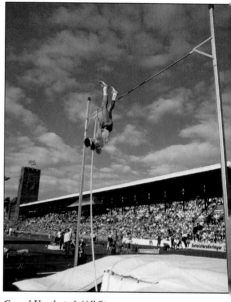

Gerard Vandystadt/All Sport

It is essential that you understand the basic concepts and principles *before* attempting to solve assigned problems. This is best accomplished through a careful reading of the textbook before attending your lecture on that material. In the process, it is useful to jot down certain points that are not clear to you. Take careful notes in class, and then ask questions pertaining to those ideas that require clarification. Keep in mind that few people are able to absorb the full meaning of scientific material after one reading. Several readings of the text and notes may be necessary. Your lectures and laboratory work should supplement the text and clarify some of the more difficult material. You should reduce memorization of material to a minimum. Memorizing passages from a text, equations, and derivations does not necessarily mean you understand the material. Your understanding of the material will be enhanced through a combination of efficient study habits, discussions with other students and instructors, and your ability to solve the problems in the text. Ask questions whenever you feel it is necessary. If you are reluctant to ask questions in class, seek private consultation or initiate discussions with your classmates. Many individuals are able to speed up the learning process when the subject is discussed on a one-to-one basis.

STUDY SCHEDULE

It is important to set up a regular study schedule, preferably on a daily basis. Be sure to read the syllabus for the course and adhere to the schedule set by your instructor. The lectures will be much more meaningful if you have read the corresponding textual material *before* attending the lecture. As a general rule, you should devote about two hours of study time for every hour in class. If you are having trouble with the course, seek the advice of the instructor or students who have already taken the course. You may find it necessary to seek further instruction from experienced students. Very often, instructors offer review sessions in addition to regular class periods. It is important that you avoid the practice of delaying study until a day or two before an exam. More often than not, this will lead to disastrous results. Rather than an all-night study session before an exam, it is better to briefly review the basic concepts and equations, followed by a good night's rest. The Pocket Guide that accompanies the text provides formulas and helpful hints and can be a useful review tool before exams. If you feel you need additional help in understanding the concepts, preparing for exams, or in problem solving, we suggest that you acquire a copy of the student study guide that accompanies the text; it should be available at your college bookstore.

USE THE FEATURES

You should make *full* use of the various features of the text discussed in the preface. For example, marginal notes are useful for locating and describing important equations, while important statements and definitions are highlighted with a light blue screen. Many useful tables are contained in the appendices, but most are incorporated into the text where they are used most often. Appendix A is a convenient review of mathematical techniques. Answers to odd-numbered problems are given at the end of the text, along with answers to the odd-numbered Conceptual Questions and all of the Multiple-Choice Questions. Exercises (with answers), which follow some worked examples, represent extensions of those examples, and in most cases you are expected to perform a simple calculation. Their purpose is to test your problem-solving skills as you read through the text. Problem-solving strategies are included in selected chapters throughout the text to give you additional information to help you solve the problems. An overview of the entire text is given in the table of contents, while the index will enable you to locate specific material quickly. Footnotes are sometimes used to supplement the discussion or to cite other references on the subject. Your instructor can suggest additional readings for you.

After reading a chapter, you should be able to define any new quantities introduced in that chapter and to discuss the principles and assumptions that were used to arrive at certain key relations. The chapter summaries and the review sections of the study guide should help you in this regard. In some cases, it will be necessary to refer to the index of the text to locate certain topics. You should be able to correctly associate with each physical quantity a symbol used to represent that quantity and the unit in which the quantity is specified. Furthermore, you should be able to express each important relation in a concise and accurate prose statement.

THINKING PHYSICS AND APPLYING PHYSICS

These new features are intended to develop your conceptual understanding of physics. Answers ("Explanation") are provided immediately for you because these non-mathematical examples often extend or further develop the chapter content. They may, for instance, illuminate the connection between physics and everyday life. You should attempt to answer the Conceptual Questions at the end of each chapter using the model for your answers provided by the Explanations in **Thinking Physics** and **Applying Physics.**

PROBLEM SOLVING

R. P. Feynman, Nobel laureate in physics, once said, "You do not know anything until you have practiced." In keeping with this statement, we strongly advise that you develop the skills necessary to solve a wide range of problems. Your ability to solve problems will be one

Heinz Fischer/The Image Bank

of the main tests of your knowledge of physics, and therefore you should try to solve as many problems as possible. It is essential that you understand basic concepts and principles before attempting to solve problems. It is good practice to try to find alternative solutions to the same problem. For example, problems in mechanics can be solved using Newton's laws, but very often an alternative method using energy considerations is more direct. You should not deceive yourself into thinking you understand the problem after seeing its solution in class. You must be able to solve the problem and similar problems on your own.

The method of solving problems should be carefully planned. A systematic plan is especially important when a problem involves several concepts. First, read the problem several times until you are confident you understand what is being asked. Look for any key words that will help you interpret the problem, and perhaps allow you to make certain assumptions. Your ability to interpret the question properly is an integral part of problem solving. You should acquire the habit of writing down the information given in a problem, and decide what quantities need to be found. You might want to construct a table listing quantities given and quantities to be found. This procedure is sometimes used in the worked examples of the text. After you have decided on the method you feel is appropriate for the situation, proceed with your solution. General problem-solving strategies of this type are included in the text and are highlighted by a light blue screen.

We often find that students fail to recognize the limitations of certain formulas or physical laws in a particular situation. It is very important that you understand and remember the assumptions underlying a particular theory or formalism. For example, we shall find that certain equations in kinematics apply only to an object moving with constant acceleration. They are not valid for situations in which the acceleration is not constant, as in the cases of the motion of an object connected to a spring and the motion of an object through a fluid.

Telegraph Colour Library/FPG

EXPERIMENTS AND "QUICKLABS"

Physics is a science based upon experimental observations. In view of this fact, we recommend that you make every effort to supplement the text through various types of "hands-on" experiments, either in the laboratory or at home, especially using the **QuickLabs** in this book. Many simple experiments can be used to test ideas and models discussed in class or in the text. For example, an object swinging on the end of a long string together with a wristwatch can be used to investigate pendulum motion; an object attached to the end of a vertical spring or a rubber band can be used to determine the nature of restoring forces and to investigate periodic motion; collisions between equal masses can be observed while playing billiards; an approximate value for the acceleration due to gravity can be obtained by dropping an object from a known height and simply measuring the time of its fall with a stopwatch; traveling waves can be investigated with the aid of a stretched rope or the common "Slinky" toy (a stretched spring); a pair of Polaroid sunglasses and some discarded lenses or a magnifying glass can be used to perform various experiments in optics. The list is endless. When physical models are not available, be imaginative and try to develop models of your own.

NEW MEDIA

We strongly encourage you to use one or more of the multimedia products that accompany this textbook. You may purchase a Textbook version of the Interactive Physics™ program by Knowledge Revolution, containing simulations keyed to selected worked examples and end-of-chapter problems dealing with mechanics and wave motion. Alternatively or in addition, you may purchase the *Saunders Core Concepts in College Physics* CD-ROM developed by Archipelago Productions for a complete multimedia presentation of selected topics in mechanics, thermodynamics, electromagnetism, and optics. It is far easier to understand physics if you see it in action, and these new materials will enable you to be a part of that action.

CLOSING COMMENTS

Someone once said that there are only two professions in which people truly enjoy what they are doing: professional sports and physics. Although this statement is most likely an exaggeration, both professions are truly exciting and stretch your skills to the limit. It is our sincere hope that you too will find physics exciting and that you will benefit from this experience, regardless of your chosen profession.

Welcome to the exciting world of physics.

To see a World in a Grain of Sand
And a Heaven in a Wild Flower,
Hold infinity in the palm of your hand
And Eternity in an hour.

W. Blake, "Auguries of Innocence"

Contents Overview

Volume 2

Contents

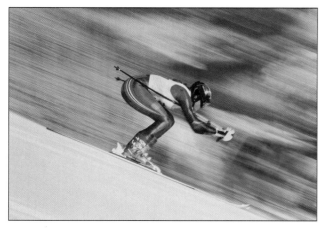

Jean Y. Rusniewski/Tony Stone Images

Stephen Krasemann/Tony Stone Images

2   Thermodynamics   *311*

Tom Mareschel/The Image Bank

3 Vibrations and Wave Motion   403

COLLEGE PHYSICS

Fifth Edition

Volume 1

Mechanics

Physics, the most fundamental science, is concerned with the basic principles of the Universe. It is one of the foundations on which the other physical sciences — astronomy, chemistry, and geology — are based. The beauty of physics lies in the simplicity of its fundamental theories and in the way just a small number of basic concepts, equations, and assumptions can alter and expand our view of the world.

The myriad physical phenomena in our world are parts of one or more of the following five areas of physics:

- Mechanics, which is concerned with the effects of forces on material objects
- Thermodynamics, which deals with heat, temperature, and the behavior of large numbers of particles
- Electromagnetism, which deals with charges, currents, and electromagnetic fields
- Relativity, a theory that describes particles moving at any speed, and connects space and time
- Quantum mechanics, a theory dealing with the behavior of particles at the submicroscopic level as well as the macroscopic world

The first part of this textbook addresses mechanics, sometimes referred to as classical mechanics or Newtonian mechanics. This is an appropriate subject with which to begin an introductory book because many of the basic principles of mechanical systems can be used later to describe such natural phenomena as waves and heat transfer. Furthermore, the laws of conservation of energy and momentum to be introduced in our study of mechanics retain their importance in the fundamental theories that follow, including the theories of modern physics.

The first serious attempts to develop a theory of motion were made by the Greek astronomers and philosophers. Although they devised a complex model to describe the motions of heavenly bodies, their model lacked correlation between such motions and the motions of objects on Earth. The study of mechanics was enhanced by a number of careful astronomical investigations by Copernicus, Brahe, and Kepler in the 16th century. In the 16th and 17th centuries, Galileo attempted to relate the motions of falling bodies and projectiles to the motions of planetary bodies, and Sevin and Hooke studied forces and their relationship to motion. A major development in the theory of mechanics was provided by Newton in 1687 when he published his *Principia*. Newton's elegant theory, which remained unchallenged for more than 200 years, was based on his hypothesis of universal gravitation together with contributions made by Galileo and others.

Today, mechanics is of vital importance to students from all disciplines. It is highly successful in describing the motions of material bodies, such as planets, rockets, and baseballs. In the first nine chapters of this book, we shall describe the laws of mechanics and examine a wide range of phenomena that can be understood through these laws.

1

Introduction

▲ **PHYSICS PUZZLER**

Shown here is an eclipse over Stonehenge. Stonehenge is a circle of stone built by the ancient Britons on Salisbury Point, England. Its orientation marks the seasonal rising and setting points of the Sun. During the summer solstice, where would the builders of Stonehenge place a marker stone to locate the rising Sun from the center of the stone circle? *(Fred Espenak/Science Photo Library/ Photo Researchers, Inc.)*

The goal of physics is to provide an understanding of nature by developing theories based on experiments. The theories are usually expressed in mathematical form. Fortunately, it is possible to explain the behavior of a variety of physical systems with a limited number of fundamental laws.

Scientists continually work at improving our comprehension of fundamental laws, and new discoveries are made every day. In many research areas, a great deal of overlap occurs among physics, chemistry, and biology. The numerous recent technological advances are results of the efforts of many scientists, engineers, and technicians. Some of the most notable include unmanned space missions and manned Moon landings, microcircuitry and high-speed computers, and sophisticated imaging techniques used in scientific research and medicine. The impacts of such developments on our society have indeed been great, and it is likely that future discoveries will be exciting, challenging, and of great benefit to humanity.

Because following chapters will be concerned with the laws of physics, we must begin by clearly defining the basic quantities involved in these laws. For example, such physical quantities as force, velocity, volume, and acceleration can be described in terms of more fundamental quantities. In the next several chapters we shall encounter three basic quantities: **length** (L), **mass** (M), and **time** (T). In later chapters we will need to add two other standard units to our list, for temperature (the kelvin) and for electric current (the ampere). In our study of mechanics, however, we shall be concerned only with the units of length, mass, and time.

(Top) The National Standard Kilogram No. 20, an accurate copy of the International Standard Kilogram kept at Sèvres, France, is kept under a double bell jar in a vault at the National Institute of Standards and Technology. *(Bottom)* The primary frequency standard (an atomic clock) at the National Institute of Standards and Technology. This device keeps time with an accuracy of about one millionth of a second per year. *(Courtesy of National Institute of Standards and Technology, U.S. Department of Commerce)*

1.1 STANDARDS OF LENGTH, MASS, AND TIME

If we are to report the result of a measurement of a certain quantity to someone who wishes to reproduce this measurement, a unit for the quantity must be defined. For example, if someone familiar with our system of measurement and weights reports that a wall is 2.0 meters high and our fundamental unit of length is defined to be 1.0 meter, we know that the height of the wall is twice the fundamental unit of length. Likewise, if we are told that a person has a mass of 75 kilograms and our fundamental unit of mass is defined as 1.0 kilogram, then that person has a mass 75 times as great as the fundamental unit of mass. In 1960, an international committee agreed on a system of standards and designations for these fundamental quantities, called the **SI system** (Système International) of units. Its units of length, mass, and time are the meter, kilogram, and second.

Length

In 1799, the legal standard of length in France became the meter, defined as one ten-millionth of the distance from the Equator to the North Pole. As recently as 1960, the official length of the meter was the distance between two lines on a specific bar of platinum-iridium alloy stored under controlled conditions. This standard was abandoned for several reasons, a principal one being that the limited accuracy with which the separation between the lines can be determined does not meet the current requirements of science and technology. Then the meter was defined as 1 650 763.73 wavelengths of orange-red light emitted from a krypton-86 lamp. In October 1983, this definition was abandoned also and **the meter was redefined as the distance traveled by light in vacuum during a time interval**

◀ Definition of the meter

of 1/299 792 458 second. This latest definition establishes that the speed of light is 299 792 458 meters per second.

Mass

Definition of the kilogram ▶

The SI unit of mass, the kilogram, is defined as the mass of a specific platinum-iridium alloy cylinder kept at the International Bureau of Weights and Measures at Sèvres, France. As we shall see in Chapter 4, mass is a quantity used to measure the resistance to a change in state of motion of an object. It is more difficult to cause a change in the state of motion of an object with a large mass than an object with a small mass.

Time

Before 1960, the time standard was defined in terms of the average length of a solar day in the year 1900. (A solar day is the time interval between successive appearances of the Sun at the highest point it reaches in the sky each day.) The basic unit of time, the second, was defined to be $(1/60)(1/60)(1/24) = 1/86\ 400$ of the average solar day. In 1967, the second was redefined to take advantage of the high precision attainable with a device known as an atomic clock, which uses the characteristic frequency of the light emitted from the cesium-133 atom as its "reference clock." **The second is now defined as 9 192 631 700 times the period of oscillation of radiation from the cesium atom.**

Magnified view of the gears in a Swiss watch. *(Dale Boyer/Tony Stone Images)*

Definition of the second ▶

Approximate Values for Length, Mass, and Time Intervals

Approximate values of some lengths, masses, and time intervals are presented in Tables 1.1, 1.2, and 1.3, respectively. Note the wide ranges of values. Study these tables and get a feel for what is meant by a kilogram of mass (this book has a mass

TABLE 1.2
Approximate Values of Some Masses

	Mass (kg)
Universe	10^{52}
Milky Way galaxy	7×10^{41}
Sun	2×10^{30}
Earth	6×10^{24}
Moon	7×10^{22}
Shark	1×10^{2}
Human	7×10^{1}
Frog	1×10^{-1}
Mosquito	1×10^{-5}
Bacterium	1×10^{-15}
Hydrogen atom	2×10^{-27}
Electron	9×10^{-31}

TABLE 1.1   Approximate Values of Some Measured Lengths

	Length (m)
Distance from Earth to most remote known quasar	1×10^{26}
Distance from Earth to most remote known normal galaxies	4×10^{25}
Distance from Earth to nearest large galaxy (M31 in Andromeda)	2×10^{22}
Distance from Earth to nearest star (Proxima Centauri)	4×10^{16}
One lightyear	9×10^{15}
Mean orbit radius of the Earth about the Sun	2×10^{11}
Mean distance from Earth to Moon	4×10^{8}
Mean radius of the Earth	6×10^{6}
Typical altitude of a satellite orbiting Earth	2×10^{5}
Length of a football field	9×10^{1}
Length of a housefly	5×10^{-3}
Size of smallest dust particles	1×10^{-4}
Size of cells of most living organisms	1×10^{-5}
Diameter of a hydrogen atom	1×10^{-10}
Diameter of an atomic nucleus	1×10^{-14}
Diameter of a proton	1×10^{-15}

TABLE 1.3 Approximate Values of Some Time Intervals

	Time Interval (s)
Age of the Universe	5×10^{17}
Age of the Earth	1×10^{17}
Average age of a college student	6×10^{8}
One year	3×10^{7}
One day (time required for one revolution of Earth about its axis)	9×10^{4}
Time between normal heartbeats	8×10^{-1}
Period[a] of audible sound waves	1×10^{-3}
Period of typical radio waves	1×10^{-6}
Period of vibration of an atom in a solid	1×10^{-13}
Period of visible light waves	2×10^{-15}
Duration of a nuclear collision	1×10^{-22}
Time required for light to cross a proton	3×10^{-24}

[a] A *period* is defined as the time required for one complete vibration.

TABLE 1.4
Some Prefixes for Powers of Ten Used with Metric Units

Power	Prefix	Abbreviation
10^{-18}	atto-	a
10^{-15}	femto-	f
10^{-12}	pico-	p
10^{-9}	nano-	n
10^{-6}	micro-	μ
10^{-3}	milli-	m
10^{-2}	centi-	c
10^{-1}	deci-	d
10^{1}	deka-	da
10^{3}	kilo-	k
10^{6}	mega-	M
10^{9}	giga-	G
10^{12}	tera-	T
10^{15}	peta-	P
10^{18}	exa-	E

of about 2 kilograms) or a time interval of 10^{10} seconds (one year is about 3×10^{7} seconds). Study Appendix A if you need to learn or review the notation for powers of 10, such as the expression of the number 50 000 in the form 5×10^{4}.

Systems of units commonly used are the SI system, in which the units of length, mass, and time are the meter (m), kilogram (kg), and second (s), respectively; the cgs or Gaussian system, in which the units of length, mass, and time are the centimeter (cm), gram (g), and second, respectively; and the British engineering system (sometimes called the conventional system), in which the units of length, mass, and time are the foot (ft), slug, and second, respectively. Throughout most of this book we shall use SI units, because they are almost universally accepted in science and industry. We shall make limited use of British engineering units in the study of mechanics.

Some of the most frequently used metric prefixes representing powers of 10 and their abbreviations are listed in Table 1.4. For example, 10^{-3} m is equivalent to 1 millimeter (mm), and 10^{3} m is 1 kilometer (km). Likewise, 1 kg is equal to 10^{3} g, and 1 megavolt (MV) is 10^{6} volts (V).

1.2 THE BUILDING BLOCKS OF MATTER

A 1-kg cube of solid gold has a length of about 3.73 cm on a side. Is this cube nothing but wall-to-wall gold, with no empty space? If the cube is cut in half, the two resulting pieces still retain their chemical identity as solid gold. But what if the pieces of the cube are cut again and again, indefinitely? Will the smaller and smaller pieces always be the same substance, gold? Questions such as these can be traced back to early Greek philosophers. Two of them — Leucippus and Democritus — could not accept the idea that such cutting could go on forever. They speculated that the process ultimately must end when it produces a particle that can no longer be cut. In Greek, *atomos* means "not sliceable." From this comes our English word *atom*, once believed to be the smallest, ultimate particle of matter. Elementary-

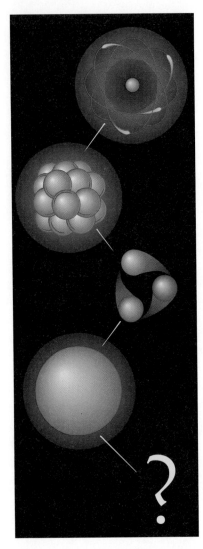

Figure 1.1 Distances at the frontier of nuclear physics are astonishingly short. An atom is so small that a single-file line of 250 000 of them would fit within the thickness of aluminum foil. The nucleus at the atom's center is a cluster of nucleons, each 100 000 times smaller than the atom itself. The three quarks inside each nucleon are smaller still. *(Courtesy of SURA, Inc.)*

particle physicists still engage in speculation and experimentation concerning the ultimate building blocks of matter.

Let us review briefly what is known about the ultimate structure of the world around us. It is useful to view the atom as a miniature Solar System with a dense, positively charged nucleus occupying the position of the Sun and negatively charged electrons orbiting like the planets. This model of the atom enables us to understand certain properties of the simpler atoms, such as hydrogen, but fails to explain many fine details of atomic structure.

Following the discovery of the nucleus in the early 1900s, the question arose: Does it have structure? That is, is the nucleus a single particle or a collection of particles? The exact composition of the nucleus has not been defined completely even today, but by the early 1930s a model evolved that helps us understand how the nucleus behaves. Scientists determined that occupying the nucleus are two basic entities, protons and neutrons. The *proton* is nature's fundamental carrier of positive charge, and the number of protons in a nucleus determines what element the material is. For instance, one proton in the nucleus means that the atom is an atom of hydrogen, regardless of how many neutrons may be present; two protons mean an atom of helium.

The existence of *neutrons* was verified conclusively in 1932. A neutron has no charge and a mass about equal to that of a proton. One of its primary purposes is to act as a "glue" to hold the nucleus together. If neutrons were not present, the repulsive electrical force between the positively charged particles would cause the nucleus to fly apart.

But is this where the breaking down stops? As we shall explore more carefully in Chapter 30, even more elementary building blocks than protons and neutrons exist. Protons, neutrons, and a zoo of other exotic particles are now thought to be composed of six particles called **quarks,** which have been given the names *up, down, strange, charmed, bottom,* and *top* (Fig. 1.1). The up, charmed, and top quarks have charges of + 2/3 that of the proton, whereas the down, strange, and bottom quarks have charges of − 1/3 that of the proton. The proton consists of two up quarks and one down quark, which one can easily show leads to the correct charge for the proton. Likewise, the neutron is composed of two down quarks and one up quark, giving a net charge of zero.

1.3   DIMENSIONAL ANALYSIS

The word *dimension* has a special meaning in physics. It usually denotes the physical nature of a quantity. Whether the separation between two points is measured in units of feet or meters or furlongs, it is a distance. We say its dimension is *length*.

The symbols that will be used in this section to specify length, mass, and time are L, M, and T, respectively. We shall often use brackets [ ] to denote the dimensions of a physical quantity. For example, in this notation the dimensions of velocity, v, are written $[v] = L/T$, and the dimensions of area, A, are $[A] = L^2$. The dimensions of area, volume, velocity, and acceleration are listed in Table 1.5, along with their units in the three common systems. The dimensions of other quantities, such as force and energy, will be described later as they are introduced.

In many situations, you may have to derive or check a specific formula. Although you may have forgotten the details of the derivation, you can use a powerful procedure called **dimensional analysis.** Dimensional analysis makes use of the fact

TABLE 1.5 Dimensions and Some Units of Area, Volume, Velocity, and Acceleration

System	Area (L^2)	Volume (L^3)	Velocity (L/T)	Acceleration (L/T^2)
SI	m^2	m^3	m/s	m/s^2
cgs	cm^2	cm^3	cm/s	cm/s^2
British engineering (conventional)	ft^2	ft^3	ft/s	ft/s^2

that **dimensions can be treated as algebraic quantities:** That is, quantities can be added or subtracted only if they have the same dimensions. Furthermore, the terms on both sides of an equation must have the same dimensions. By following these simple rules, you can use dimensional analysis to help determine whether or not an expression has the correct form. The relationship can be correct only if the dimensions on both sides of the equation are the same.

To illustrate this procedure, suppose you wish to derive a formula for the distance x traveled by a car in a time t if the car starts from rest and moves with constant acceleration a. In Chapter 2 we shall find that the correct expression for this special case is $x = \frac{1}{2}at^2$. Let us check the validity of this expression from a dimensional analysis approach.

The quantity x on the left side has the dimension length. In order for the equation to be dimensionally correct, the quantity on the right side must also have the dimension length. We can perform a dimensional check by substituting the basic units for acceleration, L/T^2, and time, T, into the equation. That is, the dimensional form of the equation $x = \frac{1}{2}at^2$ can be written

$$L = \frac{L}{T^2} \cdot T^2 = L$$

The units of time cancel as shown, leaving the unit of length. Note that the factor $\frac{1}{2}$ was ignored because it has no dimensions. Dimensional analysis cannot give us numerical factors; this is its characteristic limitation.

EXAMPLE 1.1 Analysis of an Equation

Show that the expression $v = v_0 + at$ is dimensionally correct, where v and v_0 represent velocities, a is acceleration, and t is a time interval.

Solution Because

$$[v] = [v_0] = \frac{L}{T}$$

and the dimensions of acceleration are L/T^2, the dimensions of at are

$$[at] = \frac{L}{T^2}(T) = \frac{L}{T}$$

and the expression is dimensionally correct. However, if the expression were given as $v = v_0 + at^2$, it would be dimensionally *incorrect*. Try it and see!

1.4   SIGNIFICANT FIGURES

When measurements are performed on certain quantities, the measured values are known only to within the limits of the experimental uncertainty. The value of the uncertainty can depend on factors such as the quality of the apparatus, the skill of the experimenter, and the number of measurements performed.

Suppose that in a laboratory experiment we are asked to measure the area of a rectangular plate with a meter stick. Let us assume that the accuracy to which we can measure a particular dimension of the plate is ±0.1 cm. If the length of the plate is measured to be 16.3 cm, we can claim only that its length lies somewhere between 16.2 cm and 16.4 cm. In this case, we say that the measured value has three significant figures. Likewise, if the plate's width is measured to be 4.5 cm, the actual value lies between 4.4 cm and 4.6 cm. This measured value has only two significant figures. Note that the significant figures include the first estimated digit. Thus, we could write the measured values as 16.3 ± 0.1 cm and 4.5 ± 0.1 cm.

Suppose now that we would like to find the area of the plate by multiplying the two measured values together. If we were to claim that the area is (16.3 cm)(4.5 cm) = 73.35 cm^2, our answer would be unjustifiable, because it contains four significant figures, which is greater than the number of significant figures in either of the measured lengths. A good rule of thumb for determining the number of significant figures that can be claimed is as follows. **When multiplying several quantities, the number of significant figures in the final answer is the same as the number of significant figures in the *least accurate* of the quantities being multiplied, where *least accurate* means *having the lowest number of significant figures*. The same rule applies to division.**

Applying this rule to the preceding multiplication example, we see that the answer for the area can have only two significant figures because the dimension 4.5 cm has only two significant figures. Thus, we can only claim the area to be 73 cm^2, realizing that the value can range between (16.2 cm)(4.4 cm) = 71 cm^2 and (16.4 cm)(4.6 cm) = 75 cm^2.

Zeros may or may not be significant figures. Those used to position the decimal point in such numbers as 0.03 and 0.0075 are not significant. Thus there are one and two significant figures, respectively, in these two values. When the positioning of zeros comes after other digits, however, there is the possibility of misinterpretation. For example, suppose the mass of an object is given as 1500 g. This value is ambiguous because we do not know whether the last two zeros are being used to locate the decimal point or whether they represent significant figures in the measurement. In order to remove this ambiguity, it is common to use scientific notation to indicate the number of significant figures. In this case, we would express the mass as 1.5×10^3 g if there are two significant figures in the measured value, 1.50×10^3 g if there are three significant figures, and 1.500×10^3 g if there are four. Likewise, 0.000 15 should be expressed in scientific notation as 1.5×10^{-4} if it has two significant figures or as 1.50×10^{-4} if it has three significant figures. The three zeros between the decimal point and the digit 1 in the number 0.000 15 are not counted as significant figures because they are present only to locate the decimal point. In general, a **significant figure** is a reliably known digit (other than a zero used to locate the decimal point).

For addition and subtraction, the number of decimal places must be considered when you are determining how many significant figures to report. **When numbers**

This sundial is located in Majorca, Spain. The principle of the sundial is as follows. If the number 12 points to true north, the shadow gives the correct standard time. *(David R. Frazier/Photo Researchers, Inc.)*

are added or subtracted, the number of decimal places in the result should equal the smallest number of decimal places of any term in the sum. For example, if we wish to compute $123 + 5.35$, the answer is 128 and not 128.35. If we compute the sum $1.0001 + 0.0003 = 1.0004$, the result has the correct number of decimal places; consequently it has five significant figures even though one of the terms in the sum, 0.0003, has only one significant figure. Likewise, if we perform the subtraction $1.002 - 0.998 = 0.004$, the result has only one significant figure even though one term has four significant figures and the other has three. In this book, **most of the numerical examples and end-of-chapter problems will yield answers having either two or three significant figures.**

EXAMPLE 1.2 Installing a Carpet

A carpet is to be installed in a room whose length is measured to be 12.71 m (four significant figures) and whose width is measured to be 3.46 m (three significant figures). Find the area of the room.

Solution If you multiply 12.71 m by 3.46 m on your calculator, you will get the answer 43.9766 m². How many of these numbers should you claim? Our rule of thumb for multiplication says that you can claim only the number of significant figures in the least accurate of the quantities being measured. In this example, we have three significant figures in our less accurate measurement, so we should express our final answer as 44.0 m². Note that in the answer given, we used a general rule for rounding off numbers, which states that the last digit retained is to be increased by 1 if the first digit dropped was equal to 5 or greater. Furthermore, if the last digit is 5, the result should be rounded to the nearest even number. (This helps avoid accumulation of errors.)

1.5 CONVERSION OF UNITS

Sometimes it is necessary to convert units from one system to another. Conversion factors between the SI and conventional systems for units of length are as follows:

$$1 \text{ mile} = 1609 \text{ m} = 1.609 \text{ km} \qquad 1 \text{ ft} = 0.3048 \text{ m} = 30.48 \text{ cm}$$

$$1 \text{ m} = 39.37 \text{ in.} = 3.281 \text{ ft} \qquad 1 \text{ in.} = 0.0254 \text{ m} = 2.54 \text{ cm}$$

A more extensive list of conversion factors can be found on the inside of the front cover.

Units can be treated as algebraic quantities that can cancel each other. For example, suppose we wish to convert 15.0 in. to centimeters. Because 1 in. = 2.54 cm, we find that

$$15.0 \text{ in.} = 15.0 \text{ in.} \times 2.54 \frac{\text{cm}}{\text{in.}} = 38.1 \text{ cm}$$

This road sign near Raleigh, North Carolina, shows distances in miles and kilometers. How accurate are the conversions? *(Billy E. Barnes/Stock Boston)*

EXAMPLE 1.3   Pull Over, Buddy!

If a car is traveling at a speed of 28.0 m/s, is it exceeding the speed limit of 55.0 mi/h?

Solution   We can perform the conversion from meters per second to miles per hour in two steps—meters to miles and then seconds to hours. Converting the length units, we find that

$$28.0 \text{ m/s} = \left(28.0 \,\frac{\cancel{m}}{s}\right)\left(\frac{1 \text{ mi}}{1609 \,\cancel{m}}\right) = 1.74 \times 10^{-2} \text{ mi/s}$$

Now we finish by converting seconds to hours, as follows:

$$1.74 \times 10^{-2} \text{ mi/s} = \left(1.74 \times 10^{-2} \,\frac{\text{mi}}{\cancel{s}}\right)\left(60 \,\frac{\cancel{s}}{\cancel{min}}\right)\left(60 \,\frac{\cancel{min}}{h}\right) = 62.6 \text{ mi/h}$$

The car should slow down because it is exceeding the speed limit.

    An alternative approach is to use the single conversion relationship 1 m/s = 2.237 mi/h:

$$28.0 \text{ m/s} = \left(28.0 \,\frac{\cancel{m}}{\cancel{s}}\right)\left(\frac{2.237 \text{ mi/h}}{1 \,\cancel{m/s}}\right) = 62.6 \text{ mi/h}$$

Exercise   Use the speedometer pictured in Figure 1.2 to see whether the foregoing result is in agreement with the speedometer dial (calibrated in miles per hour and kilometers per hour).

Figure 1.2   (Example 1.3) This car speedometer gives readings in both miles per hour and kilometers per hour. See if you can confirm the conversions for a few readings on the dial. *(Paul Silverman, Fundamental Photographs)*

1.6   ORDER-OF-MAGNITUDE CALCULATIONS

Often it is useful to estimate an answer to a problem in which little information is given. This answer can then be used to determine whether or not a more precise calculation is necessary. Such an approximation is usually based on certain assumptions, which must be modified if greater precision is needed. Sometimes it is necessary to know a quantity only within a factor of 10. In such a case we refer to the **order of magnitude** of the quantity, by which we mean the power of ten that is closest to the actual value of the quantity. For example, the mass of a person might be 75 kg ≈ 10^2 kg. We would say that the person's mass is *on the order of* 10^2 kg. Usually, when an order-of-magnitude calculation is made, the results are reliable to within a factor of 10. If a quantity increases in value by three orders of magnitude, this means that its value increases by a factor of $10^3 = 1000$.

EXAMPLE 1.4 French Fries to the Moon?

McDonald's sells about 250 million packages of French fries per year. If these fries were placed end to end, how far would they reach?

Solution This is a relatively simple estimation problem to give you some idea of how to approach more complicated problems. First, we must guess the length of an average French fry. A reasonable guess is three inches. Of course, some are longer and some are shorter, but three inches seems to be a reasonable estimate of the length of a typical fry. Next, we estimate that the number of fries in a typical package is 30. Multiplying the length of an average fry by the total number of fries gives us

$$(30 \text{ fries/package})\ (250 \times 10^6 \text{ packages})\ (3 \text{ in./fry}) \approx 2 \times 10^{10} \text{ in.}$$

which is about 5×10^8 m. Because the Earth-Moon distance is about 4×10^8 m, the chain of French fries would reach the Moon. Our estimate of 5×10^8 m should be accurate to within a factor of ten. That is, our stretched-out length of fries will be longer than about 5×10^7 m but shorter than 5×10^9 m.

EXAMPLE 1.5 How Much Gasoline Do We Use?

Estimate the number of gallons of gasoline used by all cars in the United States each year.

Solution Because there are about 240 million people in the United States, an estimate of the number of cars in the country is 60 million (assuming one car and four people per family). We also estimate that the average distance traveled per car per year is 10 000 miles. If we assume a gas mileage of 20 mi/gal, corresponding to a gasoline consumption of 0.05 gal/mi, each car uses about 500 gal/year. Multiplying this by the total number of cars in the United States gives an estimated total consumption of 3×10^{10} gal. This corresponds to a yearly consumer expenditure of more than $30 billion! This is probably a low estimate because we have not accounted for commercial consumption and families with two or more vehicles.

EXAMPLE 1.6 The High Price of Gold

All the gold that has ever been mined could be melted and molded into a cube approximately 20 m on a side. Use the fact that 1 kg of gold occupies a cube 3.73 cm on a side and the fact that gold sells for about $12 000 per kg to estimate the value of all the gold existing on this planet.

Solution The volume of a block 20 m on a side is $(20 \text{ m})(20 \text{ m})(20 \text{ m}) = 8\,000 \text{ m}^3$, and the volume of 1 kg of gold is

$$(3.73 \times 10^{-2} \text{ m})(3.73 \times 10^{-2} \text{ m})(3.73 \times 10^{-2} \text{ m}) \approx 5 \times 10^{-5} \text{ m}^3$$

Thus, our 8 000 m³ block of gold would have a mass of

$$\frac{8\,000 \text{ m}^3}{5 \times 10^{-5} \text{ m}^3/\text{kg}} = 1.5 \times 10^8 \text{ kg}$$

At \$12 000 per kg, this mass of gold would cost about $\$2 \times 10^{12}$.

1.7   MATHEMATICAL NOTATION

Many mathematical symbols will be used throughout this book. You are no doubt familiar with some, such as the symbol = to denote the equality of two quantities.

The symbol $\propto$ denotes a proportionality. For example, $y \propto x^2$ means that y is proportional to the square of x.

The symbol $<$ means *is less than,* and $>$ means *is greater than.* For example, $x > y$ means x is greater than y.

The symbol $\ll$ means *is much less than,* and $\gg$ means *is much greater than.*

The symbol $\approx$ indicates that two quantities are *approximately equal* to each other.

The symbol $\equiv$ means *is defined as.* This is a stronger statement than a simple =.

It is convenient to use the notation Δx (read as "delta x") to indicate the *change in the quantity x.* (Note that Δx does not mean "the product of Δ and x.") For example, suppose that a person out for a morning stroll starts measuring her distance away from home when she is 10 m from her doorway. The person then moves along a straight-line path and stops strolling 50 m from the door. Her change in position during the walk is $\Delta x = 50 \text{ m} - 10 \text{ m} = 40 \text{ m}$ or, in symbolic form,

$$\Delta x = x_f - x_i$$

In this equation x_f is the *final position* and x_i is the *initial position.*

We shall often have occasion to add several quantities. A useful abbreviation for representing such a sum is the Greek letter Σ (capital sigma). Suppose we wish to add a set of five numbers represented by $x_1, x_2, x_3, x_4,$ and x_5. In the abbreviated notation, we would write the sum as

$$x_1 + x_2 + x_3 + x_4 + x_5 = \sum_{i=1}^{5} x_i$$

where the subscript i on x represents any one of the numbers in the set. For example, if there are five masses in a system, $m_1, m_2, m_3, m_4,$ and m_5, the total mass of the system $M = m_1 + m_2 + m_3 + m_4 + m_5$ could be expressed as

$$M = \sum_{i=1}^{5} m_i$$

Finally, the magnitude of a quantity x, written $|x|$, is simply the absolute value of that quantity. The sign of $|x|$ is always positive, regardless of the sign of x. For example, if $x = -5$, $|x| = 5$; if $x = 8$, $|x| = 8$.

1.8 COORDINATE SYSTEMS AND FRAMES OF REFERENCE

Many aspects of physics deal in some way with locations in space. For example, the mathematical description of the motion of an object requires a method for describing the position of the object. Thus, it is fitting that we first discuss how to describe the position of a point in space. This is done by means of coordinates. A point on a line can be located with one coordinate; a point in a plane is located with two coordinates; and three coordinates are required to locate a point in space.

A coordinate system used to specify locations in space consists of

- A fixed reference point O, called the *origin*
- A set of specified axes or directions with an appropriate scale and labels on the axes
- Instructions that tell us how to label a point in space relative to the origin and axes

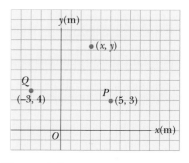

Figure 1.3 Designation of points in a two-dimensional Cartesian coordinate system. Every point is labeled with coordinates (x, y).

One convenient coordinate system that we shall use frequently is the **Cartesian coordinate system,** sometimes called the **rectangular coordinate system.** Such a system in two dimensions is illustrated in Figure 1.3. An arbitrary point in this system is labeled with the coordinates (x, y). For example, the point P in the figure has coordinates $(5, 3)$. This notation means that if we start at the origin (O), we can reach P by moving 5 meters to the right of the origin and then 3 meters above the origin. (Positive x is usually selected to the right of the origin and positive y upward from the origin. Negative x is usually to the left of the origin and negative y downward from the origin. This convention is not an absolute necessity, however. There may be instances in which you would like to take positive x to the left of the origin and negative x to the right. Feel free to do so.) In the same way, the point Q has coordinates $(-3, 4)$, which corresponds to going 3 meters to the left of the origin and 4 meters above the origin.

Sometimes it is more convenient to locate a point in space by its **plane polar coordinates** (r, θ), as in Figure 1.4. In this coordinate system, an origin and a reference line are selected, as shown. A point is then specified by the distance r from the origin to the point and by the angle θ between a line from the origin to the point and the reference line. (Frequently the reference line is selected to be the positive x axis of a Cartesian coordinate system, although this is not necessary.) The angle θ is considered positive when measured counterclockwise from the reference line and negative when measured clockwise. For example, if a point is specified by the polar coordinates 3 m and 60°, one locates this point by moving out 3 m from the origin at an angle of 60° above (or counterclockwise from) the reference line. A point specified by polar coordinates 3 m and $-60°$ is located 3 m out from the origin and 60° below (or clockwise from) the reference line.

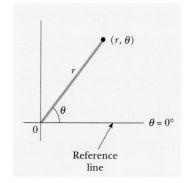

Figure 1.4 A polar coordinate system.

1.9 TRIGONOMETRY

The portion of mathematics that is based on the special properties of a right triangle is called trigonometry. Many of the concepts of this branch of mathematics are of utmost importance in the study of physics. We will now review

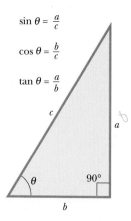

Figure 1.5 Certain trigonometric functions for a right triangle.

some of the more inherent fundamental definitions and ideas that you will need to know.

Consider the right triangle shown in Figure 1.5, where side a is opposite the angle θ, side b is adjacent to the angle θ, and side c is the hypotenuse of the triangle. The basic trigonometric functions defined by such a triangle are the ratios of the lengths of certain sides of the triangle. Specifically, the most important relationships are called the sine (sin), cosine (cos), and tangent (tan) functions. In terms of the angle θ, these functions are[1]

$$\sin\theta = \frac{\text{side opposite }\theta}{\text{hypotenuse}} = \frac{a}{c}$$

$$\cos\theta = \frac{\text{side adjacent to }\theta}{\text{hypotenuse}} = \frac{b}{c} \qquad [1.1]$$

$$\tan\theta = \frac{\text{side opposite }\theta}{\text{side adjacent to }\theta} = \frac{a}{b}$$

For example, if the angle θ is equal to 30°, it is found that the ratio of a to c is always 0.50; that is, sin 30° = 0.50. Note that the sin, cos, and tan are quantities without units, because each represents the ratio of two lengths.

Another important relationship, called the **Pythagorean theorem,** exists between the lengths of the sides of a right triangle; it is

$$c^2 = a^2 + b^2 \qquad [1.2]$$

Finally, it will often be necessary to find the value of **inverse** relationships. For example, suppose you know the value of the sine of an angle, but you need to know the value of the angle itself. The inverse sine function may be expressed as $\sin^{-1}(0.866)$, which is a shorthand way of asking the question, "What angle has a sine of 0.866?" Punching a couple of buttons on your calculator reveals that this angle is 60°. Try it for yourself and show that $\tan^{-1}(0.400) = 21.8°$.

The following examples will give you a little practice in working with these trigonometric principles in practical situations.

EXAMPLE 1.7 How High Is the Building?

A person attempts to measure the height of a building by walking out a distance of 46.0 m from its base and shining a flashlight beam toward its top. He finds that when the beam is elevated at an angle of 39.0° with respect to the horizontal, as shown in Figure 1.6, the beam just strikes the top of the building. Find the height of the building and the distance the flashlight beam has to travel before it strikes the top of the building.

---

[1] In order to recall these trigonometric relationships, consider using the following memory device. Recall the mnemonic *SOHCAHTOA* whose letters stand for *Sine = Opposite/Hypotenuse, Cosine = Adjacent/Hypotenuse,* and *Tangent = Opposite/Adjacent.* (Thanks go to Professor Don Chodrow for pointing this out.)

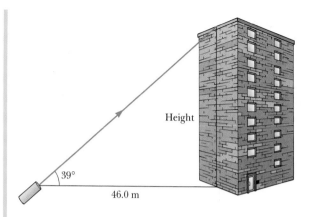

Figure 1.6 (Example 1.7)

Solution We know the length of the adjacent side of the triangle shown in the figure, and we know the value of the angle. Thus, we can use the definition of the tangent function to find the height of the building, which is the opposite side of the triangle.

$$\tan 39.0° = \frac{\text{height of building}}{46.0 \text{ m}}$$

or

Height of building $= (\tan 39.0°)(46.0 \text{ m}) = (0.810)(46.0 \text{ m}) = \boxed{37.3 \text{ m}}$

Now that we know the lengths of both the adjacent side and the opposite side of the triangle, we can find the length of the hypotenuse, which is the distance, c, that the beam travels before it strikes the top of the building.

$$c = \sqrt{a^2 + b^2} = \sqrt{(37.3 \text{ m})^2 + (46.0 \text{ m})^2} = \boxed{59.2 \text{ m}}$$

Exercise Try using a different method to find the distance traveled by the light. Would another trigonometric function, such as the sin or cos, give you the answer? Try it.

EXAMPLE 1.8 Things to Do When the CB Goes Bad

A truck driver moves up a straight mountain highway, as shown in Figure 1.7. Elevation markers at the beginning and ending points of the trip show that he has risen vertically 0.530 km, and the mileage indicator on the truck shows that he has traveled a total distance of 3.00 km during the ascent. Find the angle of incline of the hill.

Figure 1.7 (Example 1.8)

Solution   This is an example in which we must use an inverse trigonometric relationship. From the sine of the angle, defined as the opposite side over the hypotenuse, we find

$$\sin \theta = \frac{0.530}{3.00} = 0.177$$

To find θ, we use the inverse sine relationship:

$$\theta = \sin^{-1}(0.177) = \boxed{10.2°}$$

1.10   PROBLEM-SOLVING STRATEGY

Most courses in general physics require the student to learn the skills of problem solving, and examinations usually include problems that test such skills. This brief section presents some useful suggestions that will help increase your accuracy in solving problems, enhance your understanding of physical concepts, eliminate initial panic or lack of direction in approaching a problem, and organize your work. One way to help accomplish these goals is to adopt a problem-solving strategy. Many chapters in this book will include a section labeled "Problem-Solving Strategy" that should help you through the rough spots.

The following steps are commonly used to develop a problem-solving strategy:

1. Read the problem carefully at least twice. Be sure you understand the nature of the problem before proceeding further.
2. Draw a suitable diagram with appropriate labels and coordinate axes, if needed.
3. Imagine a movie, running in your mind, of what happens in the problem.
4. As you examine what is being asked in the problem, identify the basic physical principle or principles that are involved, listing the knowns and unknowns.
5. Select a basic relationship or derive an equation that can be used to find the unknown, and symbolically solve the equation for the unknown.
6. Substitute the given values with the appropriate units into the equation.
7. Obtain a numerical value with units for the unknown. You can have confidence in your result if the following questions can be properly answered: Do the units match? Is the answer reasonable? Is the plus or minus sign proper or meaningful?

One of the purposes of this strategy is to promote accuracy. Properly drawn diagrams can eliminate many sign errors. Diagrams also help to isolate the physical principles of the problem. Symbolic solutions and carefully labeled knowns and unknowns will help eliminate other careless errors. The use of symbolic solutions should help you think in terms of the physics of the problem. A check of units at the end of the problem can indicate a possible algebraic error. The physical layout and organization of your problem will make the final product more understandable and easier to follow. Once you have developed an organized system for examining problems and extracting relevant information, you will become a more confident problem solver.

The following example illustrates this procedure.

A GUIDE FOR
PROBLEM SOLVING

Read Problem

↓

Draw Diagram

↓

Identify Data

↓

Choose Equation(s)

↓

Solve Equation(s)

↓

Evaluate and Check Answer

EXAMPLE 1.9 A Round Trip by Air

An airplane travels 450 km due east and then travels an unknown distance due north. Finally, it returns to its starting point by traveling a distance of 525 km. How far did the airplane travel in the northerly direction?

Solution After reading the problem carefully, draw a diagram of the situation, as in Figure 1.8. Because the path of the airplane forms a right triangle, you can use the Pythagorean theorem, which relates the three sides of the triangle:

$$c^2 = a^2 + b^2$$

$a = 450$ km
$c = 525$ km
$b = ?$

Figure 1.8 (Example 1.9)

In this problem, the hypotenuse c and the side a are known quantities. We must solve for the unknown quantity, b. Subtracting a^2 from both sides of the preceding equation gives

$$b^2 = c^2 - a^2$$

We now take the square root of both sides to find the solution for b:

$$b = \sqrt{c^2 - a^2}$$

Note that the negative solution has been disregarded because it is not physically meaningful.

Finally, we substitute the given values for the quantities c and a to find a numerical value for b:

$$b = \sqrt{(525 \text{ km})^2 - (450 \text{ km})^2} = \boxed{270 \text{ km}}$$

SUMMARY

The physical quantities we shall encounter in our study of mechanics can be expressed in terms of three fundamental quantities, length, mass, and time—which have the units meters (m), kilograms (kg), and seconds (s), respectively, in the SI system. It is often helpful to use dimensional analysis to check equations and to assist in deriving equations.

When inserting numerical factors into equations, you must be sure that the dimensions of these factors are consistent throughout the equation. In many cases it will be necessary to use the table of conversion factors on the inside front cover to convert from one system of units to another.

Often it is useful to estimate an answer to a problem in which little information is given. Such estimates are called **order-of-magnitude calculations.**

The three most basic trigonometric functions for a right triangle are the sine, cosine, and tangent, defined as

$$\sin \theta = \frac{\text{side opposite } \theta}{\text{hypotenuse}}$$

$$\cos \theta = \frac{\text{side adjacent to } \theta}{\text{hypotenuse}}$$

$$\tan \theta = \frac{\text{side opposite } \theta}{\text{side adjacent to } \theta}$$

The **Pythagorean theorem** is an important relationship between the lengths of the sides of a right triangle:

$$c^2 = a^2 + b^2$$

where c is the hypotenuse of the triangle and a and b are its other two sides.

MULTIPLE-CHOICE QUESTIONS

1. A fathom is a unit of length about 6 ft long commonly used in measuring depths of water. Because a fathom is about the length of two outspread arms, it should not be surprising that it derives its name from the Viking word for *embrace*. Assuming that the distance to the Moon is about 4×10^8 m, the distance to the Moon in fathoms is about
   (a) 4×10^8 fathoms   (b) 2×10^8 fathoms
   (c) 3×10^8 fathoms   (d) 8×10^8 fathoms

2. Suppose that two quantities, A and B, have different dimensions. Determine which of the following arithmetic operations *could* be physically meaningful.
   (a) $A + B$ (b) $B - A$ (c) $A - B$ (d) A/B

3. Estimate the number of breaths taken during an average life span of 70 years. The answer is approximately
   (a) 3×10^7 breaths   (b) 3×10^8 breaths
   (c) 3×10^9 breaths

4. Which of the following relationships is dimensionally consistent with an expression yielding a value for acceleration? Acceleration has the units of distance divided by time squared. In these equations, x is a distance, t is time, and v is velocity with units of distance divided by time.
   (a) v/t^2 (b) v/x^2 (c) v^2/t (d) v^2/x

5. Use the rules for using significant figures to find the answer to the following addition problem:
   $21.4 + 15 + 17.17 + 4.003$
   (a) 57.573   (b) 57.57   (c) 57.6   (d) 58

6. A sphere has a surface area of 100 m². A second sphere has a radius twice that of the first sphere. What is the surface area of the second sphere? (*Hint:* You do not need to find the radius of the first or the second sphere.)
   (a) 50 m²   (b) 200 m²   (c) 157 m²
   (d) 400 m²   (e) 800 m²

7. A sphere has a volume of 100 m³. A second sphere has a radius twice that of the first sphere. The volume of the second sphere is
   (a) 33.3 m³   (b) 300 m³   (c) 400 m³
   (d) 800 m³   (e) No answer is correct.

CONCEPTUAL QUESTIONS

1. Estimate the order of magnitude of the length, in meters, of each of the following: (a) a mouse, (b) a pool cue, (c) a basketball court, (d) an elephant, (e) a city block.
2. Estimate the number of times your heart beats in a month.
3. Estimate your age in seconds.
4. Estimate the number of atoms in 1 cm³ of a solid. (Note that the diameter of an atom is about 10^{-10} m.)
5. The height of a horse is sometimes given in units of "hands." Why is this a poor standard of length?
6. How many of the lengths or time intervals given in Tables 1.2 and 1.3 could you verify using only equipment found in a typical dormitory room?
7. An ancient unit of length called the *cubit* was equal to six palms, where a palm was the width of the four fingers of an open hand. Noah's ark was 300 cubits long, 50 cubits wide, and 30 cubits high. Estimate the volume of the ark in cubic feet. Also, estimate the volume of a typical home and compare it to the volume of the ark in cubic meters.
8. If an equation is dimensionally correct, does this mean that the equation must be true?
9. An automobile tire is rated to last for 50 000 miles. Estimate the number of revolutions the tire will make in its lifetime.
10. Figure Q1.10 is a photograph showing unit conversions on the labels of some grocery-store items. Check the accuracy of these conversions. Are the manufacturers using significant figures correctly?

Figure Q1.10   *(Courtesy of Henry Leap and Jim Lehman)*

PROBLEMS

1, 2, 3 = straightforward, intermediate, challenging   ☐ = full solution available in Study Guide/Student Solutions Manual
WEB = solution posted at **http://www.saunderscollege.com/physics/college**   = biomedical application   = Interactive Physics

Section 1.3 Dimensional Analysis

1. A shape that covers an area A and has a uniform height h has a volume given by $V = Ah$. (a) Show that $V = Ah$ is dimensionally correct. (b) Show that the volumes of a cylinder and of a rectangular box can be written in the form $V = Ah$, identifying A in each case. (Note that A, sometimes called the "footprint" of the object, can have any shape and the height can be replaced by average thickness in general.)
2. In a desperate attempt to come up with an equation to use during an examination, a student tries $v^2 = ax$. Use dimensional analysis to determine whether this equation might be valid.

3. The period of a simple pendulum, defined as the time for one complete oscillation, is measured in time units and is given by

$$T = 2\pi\sqrt{\frac{\ell}{g}}$$

where ℓ is the length of the pendulum and g is the acceleration due to gravity, in units of length divided by time squared. Show that this equation is dimensionally consistent.
4. (a) One of the fundamental laws of motion states that the acceleration of an object is directly proportional to the resultant force on it and inversely proportional to its

mass. From this statement, determine the dimensions of force. (b) The newton is the SI unit of force. According to the results for (a), how can you express a force having units of newtons using the fundamental units of mass, length, and time?

5. (a) Suppose that the displacement of an object is related to the time according to the expression $x = Bt^2$. What are the dimensions of B? (b) A displacement is related to the time as $x = A \sin(2\pi ft)$, where A and f are constants. Find the dimensions of A. (*Hint:* A trigonometric function appearing in an equation must be dimensionless.)

Section 1.4   Significant Figures

6. The value of the speed of light is now known to be $2.99\,792\,458 \times 10^8$ m/s. Express the speed of light to (a) three significant figures, (b) five significant figures, and (c) seven significant figures.

7. How many significant figures are there in (a) 78.9 ± 0.2, (b) 3.788×10^9, (c) 2.46×10^{-6}, (d) 0.0032?

8. Carry out the following arithmetic operations: (a) the sum of the numbers 756, 37.2, 0.83, and 2.5; (b) the product 3.2×3.563; (c) the product $5.67 \times \pi$.

9. (a) Using your calculator find, in scientific notation with appropriate rounding, the value of $(2.437 \times 10^4)(6.5211 \times 10^9)/(5.37 \times 10^4)$. (b) Find the value of $(3.14159 \times 10^2)(27.01 \times 10^4)/(1234 \times 10^6)$.

10. Calculate (a) the circumference of a circle of radius 3.5 cm and (b) the area of a circle of radius 4.65 cm.

11. A farmer measures the distance around a rectangular field. The length of each long side of the rectangle is found to be 38.44 m, and the length of each short side is found to be 19.5 m. What is the total distance around the field?

12. A fisherman catches two sturgeon. The smaller of the two has a measured length of 93.46 cm (two decimal places, four significant figures), and the larger fish has a measured length of 135.3 cm (one decimal place, four significant figures). What is the total length of fish caught for the day?

Section 1.5   Conversion of Units

13. Estimate the distance to the nearest star, in feet, using the data in Table 1.1 and the appropriate conversion factors.

14. Estimate the age of the Earth in years, using the data in Table 1.3 and the appropriate conversion factors.

15. The speed of light is about 3.00×10^8 m/s. Convert this to miles per hour.

16. (a) Find a conversion factor to convert from miles per hour to kilometers per hour. (b) For a while, federal law mandated that the maximum highway speed would be 55 mi/h. Use the conversion factor from part (a) to find the speed in kilometers per hour. (c) The maximum highway speed has been raised to 65 mi/h in some places. In kilometers per hour, how much of an increase is this over the 55-mi/h limit?

17. A painter is to cover the walls in a room that is 8.0 ft high and 12.0 ft along a side. What surface area, in square meters, must she cover?

18. A house is 50.0 ft long and 26.0 ft wide, and has 8.0-ft-high ceilings. What is the volume of the interior of the house in cubic meters and in cubic centimeters?

19. Assume that 2.0 million stone blocks that average 2.5 tons each are used to build a pyramid. Find the weight of this pyramid in pounds and in newtons (see Problem 4).

20. A billionaire offers to give you $1 billion if you can count it out using only $1 bills. Should you accept her offer? Assume that you can count at an average rate of one bill every second, and be sure to allow for the fact that you need about 8 hours a day for sleeping and eating.

21. The amount of water in reservoirs is often measured in acre-ft. One acre-ft is a volume that covers an area of one acre to a depth of one foot. An acre is an area of $43\,560$ ft^2. Find the volume in SI units of a reservoir containing 25.0 acre-ft of water.

22. The base of a pyramid covers an area of 13.0 acres (1 acre = $43\,560$ ft^2) and has a height of 481 ft (Fig. P1.22). If the volume of a pyramid is given by the expression $V = (1/3)bh$, where b is the area of the base and h is the height, find the volume of this pyramid in cubic meters.

Figure P1.22

WEB 23. A quart-container of ice cream is to be made in the form of a cube. What should be the length of a side, in centimeters? (Use the conversion 1 gallon = 3.786 liter.)

24. One cubic meter (1.00 m^3) of aluminum has a mass of 2.70×10^3 kg, and 1.00 m^3 of iron has a mass of 7.86×10^3 kg. Find the radius of an aluminum sphere whose mass is the same as that of an iron sphere of ra-

dius 2.00 cm. (*Note:* Density is defined as the mass of an object divided by its volume.)

25. One cubic centimeter (1.0 cm³) of water has a mass of 1.0×10^{-3} kg. (a) Determine the mass of 1.0 m³ of water. (b) Assuming biological substances are 98% water, estimate the masses of a cell with a diameter of 1.0 μm, a human kidney, and a fly. Assume a kidney is roughly a sphere with a radius of 4.0 cm, and a fly is roughly a cylinder 4.0 mm long and 2.0 mm in diameter. (See the note in Problem 24.)

Section 1.6 Order-of-Magnitude Calculations

Note: In developing answers to the problems in this section, you should state your important assumptions, including the numerical values assigned to parameters used in the solution. Because only order-of-magnitude results are expected, do not be surprised if your results differ from those given in the answer section.

26. A hamburger chain advertises that it has sold more than 50 billion hamburgers. Estimate how many pounds of hamburger meat must have been used by the restaurant chain and how many head of cattle were required to furnish the meat.

27. Imagine that you are the equipment manager of a professional baseball team. One of your jobs is to keep baseballs on hand for games. Balls are sometimes lost when players hit them into the stands as either home runs or foul balls. Estimate how many baseballs you have to buy per season in order to make up for such losses. Assume your team plays an 81-game home schedule in a season.

28. Estimate the number of piano tuners living in New York City. This question was raised by the physicist Enrico Fermi, who was well known for making order-of-magnitude calculations.

Section 1.8 Coordinate Systems and Frames of Reference

29. A point is located in a polar coordinate system by the coordinates $r = 2.5$ m and $\theta = 35°$. Find the x and y coordinates of this point, assuming the two coordinate systems have the same origin.

30. A certain corner of a room is selected as the origin of a rectangular coordinate system. If a fly is crawling on an adjacent wall at a point having coordinates (2.0, 1.0), where the units are meters, what is the distance of the fly from the corner of the room?

31. Express the location of the fly in Problem 30 in polar coordinates.

32. Two points in a rectangular coordinate system have the coordinates (5.0, 3.0) and (−3.0, 4.0), where the units are centimeters. Determine the distance between these points.

Section 1.9 Trigonometry

33. For the triangle shown in Figure P1.33, what are (a) the length of the unknown side, (b) the tangent of θ, and (c) the sin of ϕ?

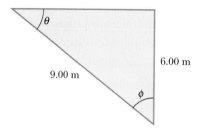

Figure P1.33

WEB 34. A right triangle has a hypotenuse of length 3.00 m, and one of its angles is 30.0°. What are the lengths of (a) the side opposite the 30.0° angle and (b) the side adjacent to the 30.0° angle?

35. In Figure P1.35, find (a) the side opposite θ, (b) the side adjacent to ϕ, (c) cos θ, (d) sin ϕ, and (e) tan ϕ.

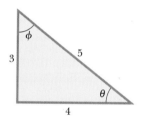

Figure P1.35

36. In a certain right triangle, the two sides that are perpendicular to each other are 5.00 m and 7.00 m long. What is the length of the third side of the triangle?

37. In Problem 36, what is the tangent of the angle for which 5.00 m is the opposite side?

ADDITIONAL PROBLEMS

38. Soft drinks are commonly sold in aluminum containers. Estimate the number of such containers thrown away each year by U.S. consumers. Approximately how many tons of aluminum does this represent?

39. The radius of the planet Saturn is 5.85×10^7 m, and its mass is 5.68×10^{26} kg (Fig. P1.39). (a) Find the density of Saturn (its mass divided by its volume) in grams per cubic centimeter. (The volume of a sphere is given by $(4/3)\pi r^3$.) (b) Find the surface area of Saturn in square feet. (The surface area of a sphere is given by $4\pi r^2$.)

Figure P1.39   A view of Saturn from *Voyager 2.*
(Courtesy of NASA)

40. The consumption of natural gas by a company satisfies the empirical equation $V = 1.5t + 0.0080t^2$, where V is the volume in millions of cubic feet and t the time in months. Express this equation in cubic feet and seconds. Put the proper units on the coefficients. Assume a month is 30 days.

WEB 41. The displacement of an object moving under uniform acceleration is some function of time and the acceleration. Suppose we write this displacement as $s = k \, a^m t^n$, where k is a dimensionless constant. Show by dimensional analysis that this expression is satisfied if $m = 1$ and $n = 2$. Can this analysis give the value of k?

42. Sphere 1 has surface area A_1 and volume V_1 and sphere 2 has surface area A_2 and volume V_2. If the radius of sphere 2 is double the radius of sphere 1, what is the ratio of **(a)** the areas A_2/A_1, and **(b)** the volumes V_2/V_1?

43. Compute the order of magnitude of the mass of **(a)** a bathtub filled with water, and **(b)** a bathtub filled with pennies. In your solution list the quantities you estimate and the value you estimate for each.

44. One gallon of paint (volume $= 3.78 \times 10^{-3} \text{ m}^3$) covers an area of 25.0 m^2. What is the thickness of the paint on the wall?

45. The radius r of a circle inscribed in any triangle whose sides are a, b, and c is given by

$$r = \left[\frac{(s - a)(s - b)(s - c)}{s} \right]^{1/2}$$

where s is an abbreviation for $(a + b + c)/2$. Check this formula for dimensional consistency.

46. Newton's law of universal gravitation is given by

$$F = G \frac{Mm}{r^2}$$

Here F is the force of gravity, M and m are masses, and r is a length. Force has the units $\text{kg} \cdot \text{m/s}^2$. What are the SI units of the proportionality constant G?

47. **(a)** How many seconds are there in a year? **(b)** If one micrometeorite (a sphere with a diameter of $1.0 \times 10^{-6} \text{ m}$) struck each square meter of the Moon each second, how many years would it take to cover the Moon to a depth of 1.0 m? (*Hint:* Consider a cubic box, 1.0 m on a side, on the Moon, and find how long it would take to fill the box.)

WEB 48. You can obtain a rough estimate of the size of a molecule by the following simple experiment. Let a droplet of oil spread out on a smooth water surface. The resulting oil slick will be approximately one molecule thick. Given an oil droplet of mass $9.00 \times 10^{-7} \text{ kg}$ and density 918 kg/m^3 that spreads out into a circle of radius 41.8 cm on the water surface, what is the diameter of an oil molecule?

49. At the time of this book's printing, the U.S. national debt is about \$4 trillion. **(a)** If this debt were paid at a rate of \$1000 per second, how long (in years) would it take to pay off the debt, assuming no interest were charged? **(b)** A dollar bill is about 15.5 cm long. If 4 trillion dollar bills were laid end to end around the Earth's Equator, how many times would they circle the Earth? Take the radius of the Earth to be about 6378 km. (*Note:* Before doing any of the calculations, try to guess at the answers. You may be surprised.)

Motion in One Dimension

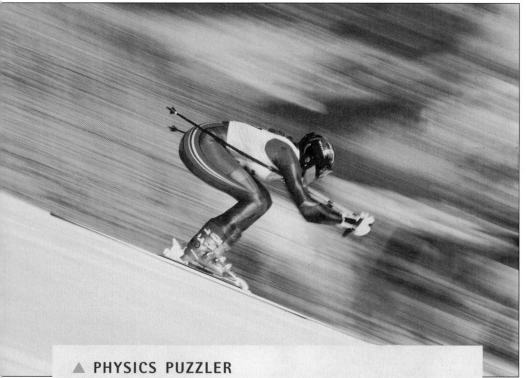

▲ PHYSICS PUZZLER

Downhill skiers can reach speeds of greater than 100 km/h. If the skier starts from rest at the top of the slope and moves in a straight line, can you estimate the time it takes the skier to reach this speed? *(Jean Y. Ruszniewski/Tony Stone Images)*

People have always been concerned with some form of motion. For example, in today's world you consider motion when you describe how fast your new car will go or how much pickup it has. Likewise, prehistoric cave dwellers must have, in their own way, pondered motion as they devised methods for capturing a rapidly moving antelope. The branch of physics concerned with the study of the motion of an object and the relationship of this motion to such physical concepts as force and mass is called **dynamics.** The part of dynamics that describes motion without regard to its causes is called **kinematics.** In this chapter we shall focus on kinematics and on one-dimensional motion — that is, motion along a straight line. We shall start by discussing displacement, velocity, and acceleration. These concepts will enable us to study the motion of objects undergoing constant acceleration. In Chapter 3 we shall discuss the motions of objects in two dimensions.

The first recorded evidence of the study of mechanics can be traced to the peoples of ancient Sumerian and Egyptian civilizations, whose primary concern was to understand the motions of heavenly bodies. The most systematic and detailed early studies of the heavens were conducted by the Greeks from about 300 B.C. to A.D. 300. Ancient scientists and lay people regarded the Earth as the center of the Universe. This geocentric model was accepted by such notables as Aristotle (384–322 B.C.) and Claudius Ptolemy (about A.D. 140). Largely because of the authority of Aristotle, the geocentric model became the accepted theory of the Universe until the 17th century.

About 250 B.C. the Greek philosopher Aristarchus worked out the details of a model of the Solar System based on a moving Earth. He proposed that the sky appears to turn westward because the Earth is really turning eastward. This model was not given much consideration because it was believed that if the Earth turned, it would set up a great wind as it moved through the air. The critics could not see that the Earth was carrying the air and everything else with it as it rotated.

The Polish astronomer Nicolaus Copernicus (1473–1543) is credited with initiating the revolution that finally replaced the geocentric model. In his system, called the *heliocentric model,* the Earth and the other planets revolve in circular orbits around the Sun.

This early knowledge formed the foundation for the work of Galileo Galilei (1564–1642). Galileo stands out as perhaps the dominant facilitator of the entrance of physics into the modern era. In 1609 he became one of the first to make astronomical observations with a telescope. He observed mountains on the Moon, the larger satellites of Jupiter, the rings of Saturn, spots on the Sun, and the phases of Venus. His observations convinced him of the correctness of the Copernican theory. Galileo's work with motion is particularly well known, and because of his leadership, experimentation has become an important part of our search for knowledge.

2.1 DISPLACEMENT

In order to describe the motion of an object, one must be able to specify its position at all times using some convenient coordinate system and a specified origin. For example, consider a sprinter moving along the x axis from an initial position, x_i, to some final position, x_f, as in Figure 2.1. **The displacement of an object, defined as its *change in position*, is given by the difference between its final and initial coordinates, or $x_f - x_i$.** As mentioned in Chapter 1, we use the Greek letter delta

Figure 2.1 A sprinter moving along the x axis from x_i to x_f undergoes a displacement of $\Delta x = x_f - x_i$.

(Δ) to denote a change in a quantity. Therefore, we write the displacement, or change in position, of the object as

$$\Delta x \equiv x_f - x_i \qquad\qquad [2.1]$$

◀ Definition of displacement

From this definition, we see that Δx is positive if x_f is greater than x_i and negative if x_f is less than x_i. If the sprinter moves from an initial position of $x_i = 3$ m to a final position of $x_f = 15$ m, his displacement is $\Delta x = 15$ m $- 3$ m $= 12$ m.

Displacement is an example of a vector quantity. Many physical quantities in this book, including displacement, velocity, and acceleration, are vectors. In general, **a vector is a physical quantity that requires the specification of both direction and magnitude.** By contrast, **a scalar is a quantity that has magnitude and no direction.** Scalar quantities, such as mass and temperature, can be specified by numbers with appropriate units.

◀ A vector has both direction and magnitude.

We shall usually designate vector quantities with boldface type. For example, **v** denotes a velocity vector, and **a** denotes an acceleration vector. In this chapter, which deals with one-dimensional motion, it will not be necessary to use this notation. The reason is that in one-dimensional motion there are only two directions in which an object can move, and these two directions are easily specified by plus and minus signs. For example, if a truck moves from an initial position of 10 m to a final position of 80 m, as in Figure 2.2a, its displacement is

$$\Delta x = x_f - x_i = 80 \text{ m} - 10 \text{ m} = +70 \text{ m}$$

In this case, the displacement has a magnitude of 70 m and is directed in the positive x direction, as indicated by the plus sign of the result. (Sometimes the plus sign is omitted, but a result of 70 m for the displacement is understood to be the same as $+70$ m.) The displacement vector is usually represented by an arrow, as in Figure 2.2a. The length of the arrow represents the magnitude of the displacement, and the head of the arrow indicates its direction.

Now suppose the truck moves to the left from an initial position of 80 m to a final position of 20 m, as in Figure 2.2b. In this situation, the displacement is

$$\Delta x = x_f - x_i = 20 \text{ m} - 80 \text{ m} = -60 \text{ m}$$

The minus sign in this result indicates that the displacement is in the negative x direction. Likewise, the arrow representing the displacement vector is to the left, as in Figure 2.2b.

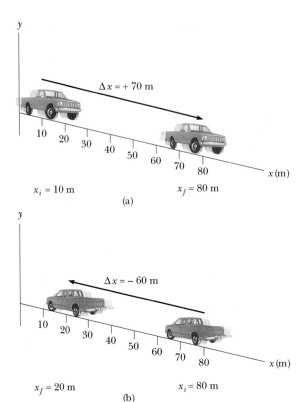

Figure 2.2   (a) A truck moving to the right from $x_i = 10$ m to $x_f = 80$ m undergoes a displacement $\Delta x = +70$ m. (b) A truck moving to the left from $x_i = 80$ m to $x_f = 20$ m undergoes a displacement $\Delta x = -60$ m.

2.2   AVERAGE VELOCITY

The motion of an object is completely known if its position is known at all times. Consider a truck moving along a highway (the x axis), as in Figure 2.2. Let the truck's position be x_i at some time t_i, and let its position be x_f at time t_f. (The indices i and f refer to the initial and final locations, respectively.) In the time interval $\Delta t = t_f - t_i$, the displacement of the truck is $\Delta x = x_f - x_i$.

▶ Definition of average velocity   **The average velocity, $\overline{v}$, is defined as the displacement, Δx, divided by the time interval during which the displacement occurred:**

$$\overline{v} \equiv \frac{\Delta x}{\Delta t} = \frac{x_f - x_i}{t_f - t_i} \qquad [2.2]$$

The average velocity of an object can be either positive or negative, depending on the sign of the displacement. (The time interval, Δt, is always positive.) For example, if we select a coordinate system so that a car moves from a position 100 m from the origin to a position 50 m from that origin in a time interval of 2 s, the x component of the average velocity is -25 m/s. In this case, the minus sign indicates motion to the left. If an object moves in the negative x direction during some time interval, its velocity in the x direction during that interval is necessarily negative.

In order to understand the vector nature of velocity, consider the following situation. Suppose a friend tells you that she will be taking a trip in her car and will travel at a constant rate of 55 mi/h in a straight line for 1 h. If she starts from her home, where will she be at the end of the trip? Obviously, you cannot answer this question because she did not specify the direction of her trip. All you can say is that she will be located 55 mi from her starting point. However, if she tells you that she will be driving at the rate of 55 mi/h directly northward, then her final location will be known exactly. The velocity of an object is known only if its direction and its magnitude (speed) are specified. In general, **any physical quantity that is a vector must be characterized by both a magnitude and a direction.**

As an example of the use of Equation 2.2, suppose that the truck in Figure 2.2a moves 100 m to the right in 5.0 s. The substitution of these values into Equation 2.2 gives the average velocity in this time interval as

$$\bar{v} = \frac{\Delta x}{\Delta t} = \frac{100 \text{ m}}{5.0 \text{ s}} = +20 \text{ m/s}$$

Note that the units of average velocity are units of length divided by units of time. These are meters per second (m/s) in SI. Other units for velocity might be feet per second (ft/s) in the conventional system or centimeters per second (cm/s) if we decided to measure distances in centimeters.

Let us assume we are watching a drag race from the Goodyear blimp. In one run we see a car follow the straight-line path from P to Q shown in Figure 2.3 during the time interval Δt, and in a second run a car follows the curved path during the same time interval. If you examine the definition of average velocity (Eq. 2.2) carefully, you will see that the two cars had the same average velocity. This is because they had the same displacement ($x_f - x_i$) during the same time interval (Δt), as indicated in Figure 2.3.

Figure 2.4 shows the unusual path of a confused football player. He receives a kickoff at his own goal, runs downfield to within inches of a touchdown, and then reverses direction to race backward until he is tackled at the exact location where he first caught the ball. What is his average velocity in the x direction (along the field) during this run? From the definition of average velocity, we see that the football player's average velocity is zero because his displacement is zero. In other words, because x_i and x_f have the same value, $\bar{v} = \Delta x/\Delta t = 0$. From this we see that displacement should not be confused with distance traveled. The confused football player clearly traveled a great distance, almost 200 yards; yet his displacement is zero.

Figure 2.3 A drag race viewed from a blimp. One car follows the red straight-line path from P to Q, and a second car follows the blue curved path.

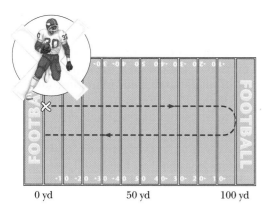

0 yd 50 yd 100 yd

Figure 2.4 The path followed by a confused football player.

Figure 2.5 Position-time graph for an object moving along the x axis. The average velocity $\bar{v}$ in the time interval $\Delta t = t_f - t_i$ is the slope of the blue straight line connecting the points P and Q.

Graphical Interpretation of Velocity

Figure 2.5 is a graph of the motion of an object moving along a straight-line path from the position x_i at time t_i to the position x_f at time t_f. (Note that the motion is along a straight line, yet the position-time graph is not a straight line. Why?) The straight line connecting points P and Q provides us with a geometric interpretation of average velocity. The slope of the line is Δx $(= x_f - x_i)$ divided by the time interval for the motion, Δt $(= t_f - t_i)$. Therefore, **the average velocity of an object during the time interval t_i to t_f is equal to the slope of the straight line joining the initial and final points on a graph of the position of the object plotted versus time.**

2.3 INSTANTANEOUS VELOCITY

Imagine that you take a trip in your car along a perfectly straight highway. At the end of your journey, it is a relatively simple task to calculate your average velocity. The car's odometer gives you the distance traveled, a compass can give you direction, and a watch can supply the time interval. However, such a calculation would omit a great deal of information about what actually occurred on your trip. For example, if your calculation indicated that your average velocity was 55 mi/h, this would not necessarily mean that your velocity at every instant was 55 mi/h. Instead, you might have traveled at 55 mi/h for a short distance, stopped for lunch and then made up some time by traveling at 70 mi/h, paused again as a police officer wrote out a ticket, and then raced forward again when the coast was clear.

▶ Definition of instantaneous velocity

More precisely, **the instantaneous velocity, v, is defined as the limit of the average velocity as the time interval Δt becomes infinitesimally short.** In mathematical language this is written

$$v \equiv \lim_{\Delta t \to 0} \frac{\Delta x}{\Delta t} \qquad [2.3]$$

The notation $\lim \Delta t \to 0$ means that the ratio $\Delta x/\Delta t$ is to be evaluated as the time interval Δt approaches zero.

To better understand the meaning of instantaneous velocity as expressed by Equation 2.3, consider the data in Table 2.1. Assume you have been observing a runner racing along a track. One second after starting into motion, the runner has moved to a position 1.00 m from the starting point; at $t = 1.50$ s, the runner is 2.25 m from the starting point; and so on. After collecting these data, suppose you wish to determine the velocity of the runner at the time $t = 1.00$ s. Table 2.2 presents some of the calculations you might perform to determine the velocity in question. Let us use $t = 1.00$ s as our initial time, and first consider the top row of the table. For the observed portion of the run (from 1.00 s to 3.00 s), the time interval $\Delta t = 2.00$ s and the displacement $\Delta x = +8.00$ m. Thus, the average velocity in this interval is $\Delta x/\Delta t = +4.00$ m/s. This gives only a rough approximation of the instantaneous velocity at $t = 1.00$ s. According to Equation 2.3, we can find increasingly more nearly correct answers by letting the time interval become smaller and smaller. Therefore, consider the second entry in Table 2.2, from 1.00 s to 2.00 s, corresponding to a time interval of $\Delta t = 1.00$ s. In this interval, $\Delta x = +3.00$ m and

TABLE 2.1
Positions of a Runner at Specific Instants of Time

t (s)	x (m)
1.00	1.00
1.01	1.02
1.10	1.21
1.20	1.44
1.50	2.25
2.00	4.00
3.00	9.00

TABLE 2.2 Calculated Values of the Time Intervals, Displacements, and Average Velocities for the Runner of Table 2.1

Time Interval (s)	Δt (s)	Δx (m)	$\bar{v}$ (m/s)
1 to 3.00	2.00	8.00	4.00
1 to 2.00	1.00	3.00	3.00
1 to 1.50	0.50	1.25	2.5
1 to 1.20	0.20	0.44	2.2
1 to 1.10	0.10	0.21	2.1
1 to 1.01	0.01	0.02	2

so the average velocity is $+3.00$ m/s. This is closer to the correct answer because the time interval is smaller. Now let us consider a very short time interval indicated by the last entry in Table 2.2. In this case, $\Delta t = 0.01$ s and the displacement $\Delta x = +0.02$ m. Thus, the average velocity in this interval is $+2$ m/s. We could improve the reliability of our calculation by allowing the time interval to become even smaller, but we can state with some degree of confidence that the instantaneous velocity of the runner was $+2$ m/s at the time $t = 1.00$ s.

In day-to-day usage, the terms *speed* and *velocity* are interchangeable. In physics, however, there is a clear distinction between these two quantities. **The instantaneous speed of an object, which is a scalar quantity, is defined as the magnitude of the instantaneous velocity.** Hence, by definition, speed can never be negative.

At appropriate places within each chapter we will include interludes that we hope will motivate you to think critically about the material you have been reading. These interesting interludes, called "Thinking Physics" or "Applying Physics," are conceptual rather than mathematical and should serve as models in responding to the conceptual questions at the end of each chapter.

Applying Physics 1

Does an automobile speedometer measure average or instantaneous speed?

Explanation Based on the fact that changes in the speed of the automobile are reflected in changes in the speedometer reading, we might be tempted to say that a speedometer measures instantaneous speed. We would be wrong, however. The reading on a speedometer is related to the rotation of the wheels. Often, a rotating magnet, whose rotation is related to that of the wheels, is used in a speedometer. The reading is based on the *time interval* between rotations of the magnet. Thus, the reading is an average speed over this time interval. This time interval is generally very short, so that the average speed is close to the instantaneous speed. As long as the speed changes are gradual, this approximation is very good. If the automobile brakes to a quick stop, however, the speedometer may not be able to keep up with the rapid changes in speed, and the measured average speed over the time interval may be very different than the instantaneous speed at the end of the interval.

APPLICATION

Automobile Speedometers.

Figure 2.6 Geometric construction for obtaining the instantaneous velocity from the x versus t curve. The instantaneous velocity at P is defined as the slope of the line tangent to the curve at P.

Graphical Interpretation of Instantaneous Velocity

Figure 2.6 is a graph, similar to Figure 2.5, for the position of an object versus time. To find the instantaneous velocity of the object at point P, we must find the average velocity during an infinitesimally short time interval. This means that point Q on the curve must be brought closer and closer to point P until the two points nearly overlap each other. From this construction we see that the line joining P and Q is approaching the line tangent to the curve at point P. **The slope of the line tangent to the position-time curve at P is defined to be the instantaneous velocity at that time.**

EXAMPLE 2.1   A Toy Train

A toy train moves slowly along a straight portion of track according to the graph of position versus time in Figure 2.7. Find (a) the average velocity for the total trip, (b) the average velocity during the first 4.0 s of motion, (c) the average velocity during the next 4.0 s of motion, (d) the instantaneous velocity at t = 2.0 s, and (e) the instantaneous velocity at t = 5.0 s.

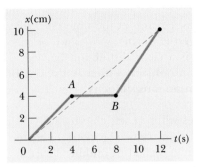

Figure 2.7   (Example 2.1)

Solution   (a) The slope of the line joining the starting point and end point on the graph (the dashed line) provides the average velocity for the total trip. A measurement of the slope gives

$$\bar{v} = \frac{\Delta x}{\Delta t} = \frac{10 \text{ cm}}{12 \text{ s}} = \quad +0.83 \text{ cm/s}$$

(b) The slope of the line joining the starting point to the point at t = 4.0 s on the curve gives us the average velocity during the first 4 s:

$$\bar{v} = \frac{\Delta x}{\Delta t} = \frac{4.0 \text{ cm}}{4.0 \text{ s}} = \quad +1.0 \text{ cm/s}$$

(c) Following the same procedure for the next 4.0-s interval, we see that the slope of the line between points A and B is zero. During this time interval, the train has remained at the same location, 4.0 cm from the starting point.

(d) A line drawn tangent to the curve at the point corresponding to t = 2.0 s has the same slope as the line in part (b). Thus, the instantaneous velocity at this time is +1.0 cm/s. This has to be true because the graph indicates that during the first 4.0 s of motion the train covers equal distances in equal intervals of time. In other words, the train moves at a constant velocity during the first 4.0 s. Under these conditions the average velocity and the instantaneous velocity are identical at all times.

(e) At $t = 5.0$ s, the slope of the position-time curve is zero. Therefore, the instantaneous velocity is zero at this instant. In fact, the train is at rest during the entire time interval between 4.0 s and 8.0 s.

2.4   ACCELERATION

As you travel from place to place in your car, you normally do not travel long distances at a constant velocity. The velocity of the car increases when you step on the gas and decreases when you apply the brakes. Furthermore, the velocity changes when you round a curve, altering your direction of motion. When the velocity of an object changes with time, the object is said to undergo an *acceleration*. However, we need a more precise definition of acceleration.

Average Acceleration

Suppose a car moves along a straight highway as in Figure 2.8. At time t_i it has a velocity of v_i, and at time t_f its velocity is v_f. **The average acceleration during this time interval is defined as the change in velocity divided by the time interval during which this change occurs:**

$$\bar{a} \equiv \frac{\Delta v}{\Delta t} = \frac{v_f - v_i}{t_f - t_i} \qquad [2.4]$$

For example, suppose the car shown in Figure 2.8 accelerates from an initial velocity of $v_i = +10$ m/s to a final velocity of $v_f = +30$ m/s in a time interval of 2.0 s. (Note that both velocities are toward the right, the direction selected as the positive direction.) These values can be inserted into Equation 2.4 to give the average acceleration:

$$\bar{a} = \frac{30 \text{ m/s} - 10 \text{ m/s}}{2.0 \text{ s}} = +10 \text{ m/s}^2$$

Acceleration is a vector quantity having dimensions of length divided by the square of time. Some common units of acceleration are meters per second per second [(m/s)/s, which is usually written m/s²] and feet per second per second (ft/s²). The acceleration calculated in the preceding case was $+10$ m/s². This notation means that, on the average, the car accelerates in the positive x direction such that its velocity increases at a rate of 10 m/s every second.

As a second example, consider the car pictured in Figure 2.9. In this case, the velocity of the car has changed from an initial value of $+30$ m/s to a final value of $+10$ m/s in a time interval of 2.0 s. The average acceleration of the car during this time interval is

$$\bar{a} = \frac{10 \text{ m/s} - 30 \text{ m/s}}{2.0 \text{ s}} = -10 \text{ m/s}^2$$

Figure 2.8   A car moving to the right accelerates from a velocity of v_i to a velocity of v_f in the time interval $\Delta t = t_f - t_i$.

◀ Definition of average acceleration

Figure 2.9   The velocity of the car decreases from $+30$ m/s to $+10$ m/s in a time interval of 2.0 s.

The minus sign indicates that the acceleration vector is in the negative *x* direction (to the left). For the case of motion in a straight line, the direction of the velocity of an object and the direction of its acceleration are related as follows. **When the object's velocity and acceleration are in the same direction, the speed of the object increases with time.** (The first case we cited demonstrates this situation.) **When the object's velocity and acceleration are in opposite directions, the speed of the object decreases with time.** (Decreases in speed are sometimes called *decelerations*.)

To clarify this point, consider the following situation. Suppose the velocity of a car changes from − 10 m/s to − 30 m/s in a time interval of 2.0 s. The minus signs indicate that the velocity of the car is in the negative *x* direction. The average acceleration of the car in this time interval is

$$\bar{a} = \frac{-30 \text{ m/s} - (-10 \text{ m/s})}{2.0 \text{ s}} = -10 \text{ m/s}^2$$

The minus sign indicates that the acceleration vector is also in the negative *x* direction. Because the velocity and acceleration vectors are in the same direction, the speed of the car must increase as the car moves to the left.

Instantaneous Acceleration

In some situations, the value of the average acceleration differs in different time intervals. It is useful, therefore, to define instantaneous acceleration. This concept is analogous to the definition of instantaneous velocity discussed in Section 2.3. In mathematical terms, **the instantaneous acceleration, *a*, is defined as the limit of the average acceleration as the time interval Δt goes to zero.** That is,

▶ Definition of instantaneous acceleration

$$a \equiv \lim_{\Delta t \to 0} \frac{\Delta v}{\Delta t} \qquad [2.5]$$

Here again, the notation $\Delta t \to 0$ means that the ratio $\Delta v / \Delta t$ is to be evaluated as the time interval Δt approaches zero.

Figure 2.10 is useful for understanding the concept of instantaneous acceleration. It plots the velocity of an object versus time. This could represent, for example, the motion of a car along a busy street. The average acceleration of the car between times t_i and t_f can be found by determining the slope of the line joining points *P* and *Q*. If we imagine that point *Q* is brought closer and closer to point *P*, the value that we find for the average acceleration between these points approaches the value of the acceleration of the car at point *P*. The instantaneous acceleration at point *P*, for example, is the slope of the graph at the point. That is, **the instantaneous acceleration of an object at a certain time equals the slope of the velocity-time graph at that instant of time.** From now on we shall use the term *acceleration* to mean "instantaneous acceleration."

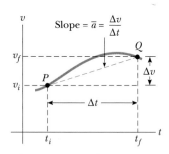

Figure 2.10 Velocity-time graph for an object moving in a straight line. The slope of the blue line connecting points *P* and *Q* is defined as the average acceleration in the time interval $\Delta t = t_f - t_i$.

EXAMPLE 2.2 A Fly Ball Is Caught

A baseball player moves in a straight-line path in order to catch a fly ball hit to the outfield. His velocity as a function of time is shown in Figure 2.11. Find his instantaneous acceleration at points *A*, *B*, and *C* on the curve.

Solution The instantaneous acceleration at any time is the slope of the velocity-time curve at that instant. The slope of the curve at point *A* is

$$a = \frac{\Delta v}{\Delta t} = \frac{4 \text{ m/s}}{2 \text{ s}} = +2 \text{ m/s}^2$$

At point *B*, the slope of the line is zero and hence the instantaneous acceleration at this time is also zero. Note that even though the instantaneous acceleration has dropped to zero at this instant, the velocity of the player is not zero. (In general, if $a = 0$, the velocity is constant in time but *not necessarily* zero.) Instead, the player is running at a constant velocity of $+4$ m/s. A value of zero for an instantaneous acceleration means that at a particular instant the velocity of the object is not changing.

Finally, at point *C* we calculate the slope of the velocity-time curve and find that the instantaneous acceleration is

$$a = \frac{\Delta v}{\Delta t} = \frac{-2 \text{ m/s}}{1 \text{ s}} = -2 \text{ m/s}^2$$

For this situation, the minus sign and the fact that the velocities are positive indicate that the speed of the player is decreasing as he approaches the location where he will attempt to catch the baseball.

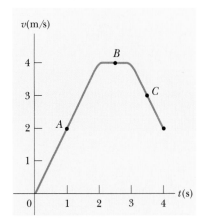

Figure 2.11 (Example 2.2)

Thinking Physics 1

Parts (a), (b), and (c) of Figure 2.12 represent three graphs of the velocities of different objects moving in straight-line paths as functions of time. The possible accelerations of each object as functions of time are shown in parts (d), (e), and (f). Match each velocity-time graph with the acceleration-time graph that best describes the motion.

Explanation The velocity-time graph (a) has a constant slope, indicating a constant acceleration, which is represented by acceleration-time graph (e).

Graph (b) represents an object that has a constantly increasing speed, but the speed is not increasing at a uniform rate. Thus, the acceleration must be increasing, and the acceleration-time graph that best indicates this is (d).

Graph (c) depicts an object that first has a velocity that increases at a constant rate, which means its acceleration is constant. The motion then changes to one at constant speed, indicating that the acceleration of the object becomes zero. Thus, the best match to this situation is graph (f).

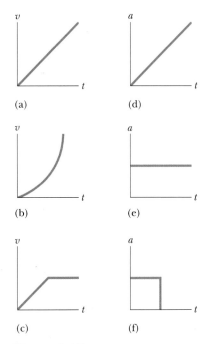

Figure 2.12 (Thinking Physics 1) Match each velocity-time graph to its corresponding acceleration-time graph.

2.5   MOTION DIAGRAMS

The concepts of velocity and acceleration are often confused with each other, but in fact they are quite different quantities. It is instructive to make use of motion diagrams to describe the velocity and acceleration vectors as time progresses while an object is in motion. In order not to confuse these two vector quantities, we use red for velocity vectors and violet for acceleration vectors as in Figure 2.13.[1] The vectors are sketched at several instants during the motion of the object, and the time intervals between adjacent positions are assumed to be equal.

A stroboscopic photograph of a moving object shows several images of the object, each image taken as the strobe light flashes. Figure 2.13 represents three sets of strobe photographs of cars moving along a straight roadway, from left to right. The time intervals between flashes of the stroboscope are equal in each diagram. Let us describe the motion of the car in each diagram.

In Figure 2.13a, the images of the car are equally spaced — the car moves the same distance in each time interval. Thus, the car moves with *constant positive velocity* and has zero acceleration.

In Figure 2.13b, the images of the car become farther apart as time progresses. In this case, the velocity vector increases in time because the car's displacement between adjacent positions increases as time progresses. Thus, the car is moving with a *positive velocity,* and a *positive acceleration.*

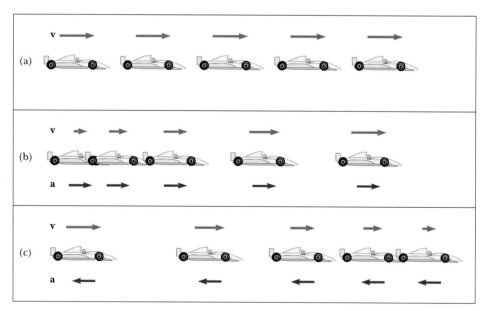

Figure 2.13   (a) Motion diagram for a car moving at constant velocity (zero acceleration). (b) Motion diagram for a car whose constant acceleration is in the direction of its velocity. The velocity vector at each instant is indicated by a red arrow, and the constant acceleration vector by a violet arrow. (c) Motion diagram for a car whose constant acceleration is in the direction *opposite* the velocity at each instant.

---

[1] Items marked with the icon are accompanied by Interactive Physics 4.0 simulations on disk.

In Figure 2.13c, the car slows as it moves to the right because its displacement between adjacent positions decreases as time progresses. In this case, the car moves initially to the right with a constant negative acceleration. The velocity vector decreases in time and eventually reaches zero. (This type of motion is exhibited by a car that skids to a stop after applying its brakes.) From this diagram we see that the acceleration and velocity vectors are *not* in the same direction. The car is moving with a *positive velocity,* but with a *negative acceleration.*

You should be able to construct motion diagrams for a car that moves initially to the left with a constant positive or negative acceleration. You should also construct appropriate motion diagrams after completing the mathematical solutions to kinematic problems, to see if your answers are consistent with the diagrams.

2.6 ONE-DIMENSIONAL MOTION WITH CONSTANT ACCELERATION

Most of the applications in this book that deal with mechanics will be concerned with objects moving with *constant acceleration.* This type of motion is important because it applies to many objects in nature. For example, an object in free fall near the Earth's surface moves in the vertical direction with constant acceleration, assuming that air resistance can be neglected. When an object moves with constant acceleration, *the average acceleration equals the instantaneous acceleration.* Consequently, the velocity increases or decreases at the same rate throughout the motion.

Because the average acceleration equals the instantaneous acceleration when a is constant, we can eliminate the bar used to denote average values from our defining equation for acceleration. That is, because $\bar{a} = a$, we can write Equation 2.4 as

$$a = \frac{v_f - v_i}{t_f - t_i}$$

For convenience, let $t_i = 0$ and t_f be any arbitrary time t. Also, we shall let $v_i = v_0$ (the initial velocity at $t = 0$) and $v_f = v$ (the velocity at any arbitrary time t). With this notation, we can express the acceleration as

$$a = \frac{v - v_0}{t}$$

or

$$v = v_0 + at \qquad \text{(for constant } a\text{)} \qquad [2.6]$$

One of the features of one-dimensional motion with constant acceleration is the manner in which the initial, final, and average velocities are related. Because the velocity is increasing or decreasing *uniformly* with time, we can express the **average velocity** in any time interval as the arithmetic average of the initial velocity, v_0, and the final velocity, v:

$$\bar{v} = \frac{v_0 + v}{2} \qquad \text{(for constant } a\text{)} \qquad [2.7]$$

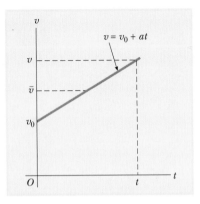

Figure 2.14 The velocity varies linearly with time for a particle moving with constant acceleration. The average velocity is the mean value of the initial and final velocities.

Note that this expression is valid only when the acceleration is constant — that is, when the velocity changes uniformly with time. The graphical interpretation of $\bar{v}$ is shown in Figure 2.14. As you can see, the velocity varies linearly with time according to Equation 2.6.

We can now use this result along with the defining equation for average velocity, Equation 2.2, to obtain an expression for the displacement of an object as a function of time. Again we choose $t_i = 0$, and for convenience we shall also assume that we have selected our coordinate system so that the initial position of the object under consideration is at the origin; that is, $x_i = 0$. This gives

$$x = \bar{v}t = \left(\frac{v_0 + v}{2}\right) t$$

$$x = \tfrac{1}{2}(v + v_0)t \qquad [2.8]$$

We can obtain another useful expression for displacement by substituting the equation for v (Eq. 2.6) into Equation 2.8:

$$x = \tfrac{1}{2}(v_0 + at + v_0)t \qquad [2.9]$$

$$x = v_0 t + \tfrac{1}{2}at^2 \qquad \text{(for constant } a\text{)}$$

Figure 2.15 shows a plot of x versus t for this equation. For example, you should be able to show that the area under the curve shown in Figure 2.14 is equal to $v_0 t + \tfrac{1}{2}at^2$, which is equal to the displacement. In fact, the area under *any* curve of v versus t can be shown to be equal to the displacement of the object.

Finally, we can obtain an expression that does not contain time by substituting the value of t from Equation 2.6 into Equation 2.8. This gives

$$x = \tfrac{1}{2}(v + v_0)\left(\frac{v - v_0}{a}\right) = \frac{v^2 - v_0^2}{2a}$$

$$v^2 = v_0^2 + 2ax \qquad \text{(for constant } a\text{)} \qquad [2.10]$$

Figure 2.15 The displacement versus time for an object moving with constant acceleration. Note that the displacement varies as t^2. The constantly increasing slope indicates that the velocity increases with time.

TABLE 2.3 Equations for Motion in a Straight Line Under Constant Acceleration

Equation	Information Given by Equation
$v = v_0 + at$	Velocity as a function of time
$x = \tfrac{1}{2}(v_0 + v)t$	Displacement as a function of velocity
$x = v_0 t + \tfrac{1}{2}at^2$	Displacement as a function of time
$v^2 = v_0^2 + 2ax$	Velocity as a function of displacement

Note: Motion is along the x axis. At $t = 0$, the particle is at the origin ($x_0 = 0$) and its velocity is v_0.

Equations 2.6 through 2.10 may be used to solve any problem in one-dimensional motion with constant acceleration. The four equations that are used most often are listed in Table 2.3 for convenience.

The best way to gain confidence in the use of these equations is to work a number of problems. Many times you will discover that there is more than one method for solving a given problem.

Thinking Physics 2

Consider the following motions of an object in one dimension: (a) A ball is thrown directly upward, rises to a highest point and falls back into the thrower's hand. (b) A race car starts from rest and speeds up to 100 m/s. (c) A *Voyager* spacecraft drifts through space at constant velocity. Are there any points in the motion of these particles at which the average velocity (over the entire interval) and the instantaneous velocity (at an instant of time within the interval) are the same? If so, identify the point(s).

Explanation   (a) The average velocity for the thrown ball is zero—the ball returns to the starting point at the end of the time interval. There is one point at which the instantaneous velocity is zero—at the top of the motion. (b) The average velocity for the motion of the race car cannot be evaluated unambiguously with the information given, but it must be some value between 0 and 100 m/s. Because the car will have every instantaneous velocity between 0 and 100 m/s at some time during the interval, there must be some instant at which the instantaneous velocity is equal to the average velocity. (c) Because the instantaneous velocity of the spacecraft is constant, its instantaneous velocity over *any* time interval and its average velocity at *any* time are the same.

Problem-Solving Strategy

Accelerated Motion

The following procedure is recommended for solving problems involving accelerated motion:

1. Make sure all the units in the problem are consistent. That is, if distances are measured in meters, be sure that velocities have units of meters per second and accelerations have units of meters per second per second.
2. Chose a coordinate system.
3. Make a list of all the quantities given in the problem and a separate list of those to be determined.
4. Select from the list of kinematic equations the one or ones that will enable you to determine the unknowns.
5. Construct an appropriate motion diagram, and check to see if your answers are consistent with the diagram.

EXAMPLE 2.3   The Indianapolis 500

A race car starting from rest accelerates at a rate of 5.00 m/s². What is the velocity of the car after it has traveled 100 ft?

Reasoning and Solution   You should refer to the preceding Problem-Solving Strategy section to see how the steps indicated there are applied to this example. (Step 1) Be sure that the units you use are consistent. The units in this problem are *not* consistent. If we choose to leave the distance in feet, we must change the length dimension of the units of acceleration from meters to feet. Alternatively, we can leave the units of acceleration as they are and convert the distance traveled to meters. Let's do the latter. The table of conversion factors on the inside front cover gives 1 m = 3.281 ft, so we find that 1 ft = 0.305 m and, thus, 100 ft = 30.5 m.

(Step 2) You must choose a coordinate system. A convenient one for this problem is shown in Figure 2.16. The origin of the coordinate system is at the initial location of the car, and the positive direction is to the right. Using this convention, we require that velocities, accelerations, and displacements to the right are positive, and vice versa.

Figure 2.16   (Example 2.3)

(Step 3) Make a list of the quantities given in the problem and a separate list of those to be determined:

Given	To Be Determined
$v_0 = 0$	v
$a = +5.00$ m/s²	
$x = +30.5$ m	

(Step 4) Select from the kinematic equations (Table 2.3) those that will allow you to determine the unknowns. In the present case, the equation

$$v^2 = v_0{}^2 + 2ax$$

is our best choice because it will give us a value for v directly:

$$v^2 = (0)^2 + 2(5.00 \text{ m/s}^2)(30.5 \text{ m}) = 305 \text{ m}^2/\text{s}^2$$

from which

$$v = \sqrt{305 \text{ m}^2/\text{s}^2} = \pm 17.5 \text{ m/s}$$

Because the car is moving to the right, we choose $+17.5$ m/s as the correct solution for v. Alternatively, the problem can be solved by using $x = v_0 t + \frac{1}{2}at^2$ to find t and then using the expression $v = v_0 + at$ to find v. Try it!

EXAMPLE 2.4   Watch Out for the Speed Limit

A car traveling at a constant speed of 30.0 m/s ($\approx$67 mi/h) passes a trooper hidden behind a billboard. One second after the speeding car passes the billboard, the trooper sets off in chase with a constant acceleration of 3.00 m/s². How long does it take the trooper to overtake the speeding car?

Reasoning   To solve this problem algebraically, we write an expression for the position of each vehicle as a function of time. It is convenient to choose the origin at the position of the billboard and take $t = 0$ as the time the trooper begins moving. At that instant, the speeding car has already traveled a distance of 30.0 m, because it travels at a constant speed of 30.0 m/s. Thus, the initial position of the speeding car is $x_0 = 30.0$ m.

Solution   Because the car moves with constant speed, its acceleration is zero, and applying Equation 2.9 gives

$$x_C = 30.0 \text{ m} + (30.0 \text{ m/s})t$$

Note that at $t = 0$, this expression does give the car's correct initial position, $x_C = x_0 = 30.0$ m.

For the trooper, who starts from the origin at $t = 0$, we have $x_0 = 0$, $v_0 = 0$, and $a = 3.00$ m/s². Hence, the position of the trooper as a function of time is

$$x_T = \tfrac{1}{2}at^2 = \tfrac{1}{2}(3.00 \text{ m/s}^2)t^2$$

The trooper overtakes the car at the instant that $x_T = x_C$, or

$$\tfrac{1}{2}(3.00 \text{ m/s}^2)t^2 = 30.0 \text{ m} + (30.0 \text{ m/s})t$$

This gives the quadratic equation

$$1.50t^2 - 30.0t - 30.0 = 0$$

whose positive solution is $t = 21.0$ s. Note that in this time interval, the trooper travels a distance of about 660 m.

Frequently in this book, an example will be followed by an exercise whose purpose is to test your understanding of the example. The answer will be provided immediately after the exercise, when appropriate. Here is your first exercise, which relates to Example 2.4.

Exercise   This problem also can be easily solved graphically. On the *same* graph, plot the position versus time for each vehicle, and from the intersection of the two curves determine the time at which the trooper overtakes the speeding car.

2.7   FREELY FALLING OBJECTS

It is now well known that all objects dropped near the surface of the Earth in the absence of air resistance fall toward the Earth with the same constant acceleration. It was not until about 1600 that this conclusion was accepted. Prior to that time,

Galileo Galilei, Italian physicist and astronomer (1564–1642)

Galileo formulated the laws that govern the motion of objects in free fall. He also investigated the motion of an object on an inclined plane, established the concept of relative motion, invented the thermometer, and discovered that the motion of a swinging pendulum could be used to measure time intervals. After designing and constructing his own telescope, he discovered four of Jupiter's moons, found that the Moon's surface is rough, discovered sunspots and the phases of Venus, and showed that the Milky Way consists of an enormous number of stars. Galileo publicly defended Nicolaus Copernicus's assertion that the Sun is at the center of the Universe (the heliocentric system). He published *Dialogue Concerning Two New World Systems* to support the Copernican model, a view which the Church declared to be heretical. After being taken to Rome in 1633 on a charge of heresy, he was sentenced to life imprisonment, and later was confined to his villa at Arcetri, near Florence, where he died in 1642. *(North Wind)*

the teachings of the great philosopher Aristotle (384–322 B.C.) had held that heavier objects fell faster than lighter ones.

It was Galileo who originated our present-day ideas concerning falling objects. There is a legend that he discovered the law of falling objects by observing that two different weights dropped simultaneously from the Leaning Tower of Pisa hit the ground at approximately the same time. Although it is doubtful that this particular experiment was carried out, it is well established that Galileo performed many systematic experiments on objects moving on inclined planes. In his experiments he rolled balls down a slight incline and measured the distances they covered in successive time intervals. The purpose of the incline was to reduce the acceleration and enable Galileo to make accurate measurements of the time intervals. (Some people refer to this experiment as "diluting gravity.") By gradually increasing the slope of the incline, he was finally able to draw conclusions about freely falling objects, because a falling ball is equivalent to a ball falling down a vertical incline. Galileo's achievements in the science of mechanics paved the way for Newton in his development of the laws of motion, which we shall study in Chapter 4.

You might want to try the following experiment. Simultaneously drop a coin and a crumpled-up piece of paper from the same height. If the effects of air friction are negligible, both will have the same motion and hit the floor at the same time. In the idealized case, in which air resistance is absent, such motion is referred to as *free fall*. If this same experiment could be conducted in a vacuum, where air friction is zero, the paper and coin would fall with the same acceleration, regardless of the shape of the paper. (See the photographs in Physics in Action.) On August 2, 1971, such a demonstration was conducted on the Moon by astronaut David Scott. He simultaneously released a hammer and a feather, and they fell with the same acceleration to the lunar surface. This demonstration surely would have pleased Galileo!

We shall denote the free-fall acceleration with the symbol **g.** The magnitude of **g** decreases with increasing altitude. Furthermore, slight variations in the magnitude of **g** occur with latitude. However, at the surface of the Earth the magnitude of **g** is approximately 9.8 m/s², or 980 cm/s², or 32 ft/s². Unless stated otherwise, we shall use the value 9.80 m/s² when doing calculations. Furthermore, we shall assume that the vector **g** is directed downward toward the center of the Earth.

When we use the expression *freely falling object* we do not necessarily mean an object dropped from rest. **A freely falling object is an object moving under the influence of gravity only, regardless of its initial motion. Objects thrown upward or downward and those released from rest are all falling freely once they are released! Once they are in free fall, all objects have an acceleration downward, which is the free-fall acceleration g.**

If we neglect air resistance and assume that the free-fall acceleration does not vary with altitude over short vertical distances, then the motion of a freely falling object is equivalent to motion in one dimension under constant acceleration. Therefore, the equations developed in Section 2.6 for objects moving with constant acceleration can be applied. The only modification that we need to make in these equations for freely falling objects is to note that the motion is in the vertical direction (the y direction) rather than the horizontal (x) direction, and that the acceleration is downward and has a magnitude of 9.80 m/s². Thus, we always take $a = -g = -9.80$ m/s², where the minus sign means that the acceleration of a freely falling object is downward. In Chapter 7 we shall study how to deal with variations in g with altitude.

PHYSICS IN ACTION

Freely Falling Objects

On the left is a multiflash photograph of a falling billiard ball. As the ball falls, the spacing between successive images increases, indicating that the ball accelerates downward. The motion diagram shows that the ball's velocity (red arrows) increases with time, and its acceleration (violet arrows) remains constant.

The middle photograph shows a multiflash image of two freely falling balls released simultaneously. The ball on the left side is solid, and the ball on the right is a hollow Ping-Pong ball. The time interval between flashes is (1/30) s and the scale is in centimeters. Can you determine **g** from these data? Note that the effect of air resistance is greater for the smaller, hollow ball on the right, as indicated by the smaller separation between successive images.

In the photograph on the right, a feather and an apple are released from rest in a 4-ft vacuum chamber. The trap door that released the two objects was opened with an electronic switch at the same instant the camera shutter was opened. The two objects fell at approximately the same rate, as indicated by the horizontal alignment of the multiple images. When air resistance is negligible, all objects fall with the same acceleration, regardless of their masses.

(Richard Megna 1990, Fundamental Photographs)

(Education Development Center, Newton, MA)

(© James Sugar, Black Star)

Applying Physics 2

A skydiver jumps out of a helicopter. A few seconds later, another skydiver jumps out, so that they both fall along the same vertical line. Ignore air resistance, so that both skydivers fall with the same acceleration. Does the vertical distance between them stay the same? Does the difference in their velocities stay the same? If they were connected

APPLICATION

Skydiving.

QUICKLAB

Reaction Time

To measure your reaction time, have a friend hold a ruler vertically between your index finger and thumb as shown. Note the position of the ruler with respect to your index finger. Your friend must release the ruler and you must catch it (without moving your hand downward) after it falls through some distance that you measure from the new position of the ruler. The ruler, a freely falling object, falls through a distance $d = \frac{1}{2}gt^2$, where t is the reaction time and $g = 9.80$ m/s². Repeat this measurement of d five times, average your results, and calculate an average value of t. Now measure your friend's reaction time using the same procedure. Compare your results. For most people, the reaction time is at best about 0.2 s.

Friend's hand

Ruler

Your hand

by a long bungee cord, would the tension in the cord become larger, smaller, or stay the same?

Explanation   At any given instant of time, the velocities of the jumpers are definitely different, because one had a head start. In a time interval after this instant, however, each jumper increases his or her velocity by the same amount, because they have the same acceleration. Thus, the difference in velocities remains the same. The first jumper will always be moving with a higher velocity than the second. Thus, in a given time interval, the first jumper will cover more distance than the second. Thus, the separation distance between them increases. As a result, once the distance between the divers reaches the value at which the bungee cord becomes straight, the tension in the bungee cord will increase as the jumpers move apart. Of course, the bungee cord will pull downward on the second jumper and upward on the first, resulting in a force in addition to gravity that will affect their motion. If the bungee cord is robust enough, this will result in the separation distance of the divers becoming smaller as the cord pulls them together.

EXAMPLE 2.5   Look Out Below!

A golf ball is released from rest at the top of a very tall building. Neglecting air resistance, calculate the position and velocity of the ball after 1.00, 2.00, and 3.00 s.

Solution   We choose our coordinates so that the starting point of the ball is at the origin ($y_0 = 0$ at $t = 0$) and remember that we have defined y to be positive upward. Because $v_0 = 0$, and $a = -g = -9.80$ m/s², Equations 2.6 and 2.9 become

$$v = at = (-9.80 \text{ m/s}^2)t$$

$$y = \tfrac{1}{2}at^2 = \tfrac{1}{2}(-9.80 \text{ m/s}^2)t^2$$

where t is in seconds, v is in meters per second, and y is in meters. These expressions give the velocity and displacement at any time t after the ball is released. Therefore, at $t = 1.00$ s,

$$v = (-9.80 \text{ m/s}^2)(1.00 \text{ s}) = \boxed{-9.80 \text{ m/s}}$$

$$y = \tfrac{1}{2}(-9.80 \text{ m/s}^2)(1.00 \text{ s})^2 = \boxed{-4.90 \text{ m}}$$

Likewise, at $t = 2.00$ s, we find that $v = -19.6$ m/s and $y = -19.6$ m. Finally, at $t = 3.00$ s, $v = -29.4$ m/s and $y = -44.1$ m. The minus sign for v indicates that the velocity vector is directed downward, and the minus sign for y indicates a position below the original position.

Exercise   Calculate the position and velocity of the ball after 4.00 s.

Answer -78.4 m, -39.2 m/s.

 ## EXAMPLE 2.6   Not a Bad Throw for a Rookie!

A stone is thrown from the top of a building with an initial velocity of 20.0 m/s straight upward. The building is 50.0 m high, and the stone just misses the edge of the roof on its way down, as in Figure 2.17. Determine (a) the time needed for the stone to reach its maximum height, (b) the maximum height, (c) the time needed for the stone to return to the level of the thrower, (d) the velocity of the stone at this instant, and (e) the velocity and position of the stone at $t = 5.00$ s.

Solution (a) To find the time necessary for the stone to reach the maximum height, use Equation 2.6, $v = v_0 + at$, noting that $v = 0$ at maximum height:

$$20.0 \text{ m/s} + (-9.80 \text{ m/s}^2)t = 0$$

$$t = \frac{20.0 \text{ m/s}}{9.80 \text{ m/s}^2} = \boxed{2.04 \text{ s}}$$

(b) This value of time can be substituted into Equation 2.9, $y = v_0 t + \frac{1}{2}at^2$, to give the maximum height as measured from the position of the thrower:

$$y_{max} = (20.0 \text{ m/s})(2.04 \text{ s}) + \tfrac{1}{2}(-9.80 \text{ m/s}^2)(2.04 \text{ s})^2 = \boxed{20.4 \text{ m}}$$

(c) When the stone is back at the height of the thrower, the y coordinate is zero. From the expression $y = v_0 t + \frac{1}{2}at^2$ (Eq. 2.9), with $y = 0$, we obtain the expression

$$0 = 20.0t - 4.90t^2$$

This is a quadratic equation and has two solutions for t. The equation can be factored to give

$$t(20.0 - 4.90t) = 0$$

One solution is $t = 0$, corresponding to the time at which the stone starts its motion. The other solution is $t = 4.08$ s, which is the solution we are after.

(d) The value for t found in (c) can be inserted into $v = v_0 + at$ (Eq. 2.6) to give

$$v = 20.0 \text{ m/s} + (-9.80 \text{ m/s}^2)(4.08 \text{ s}) = \boxed{-20.0 \text{ m/s}}$$

Note that the velocity of the stone when it arrives back at its original height is equal in magnitude to the stone's initial velocity but opposite in direction. This indicates that the motion is symmetric.

(e) From $v = v_0 + at$ (Eq. 2.6), the velocity after 5.00 s is

$$v = 20.0 \text{ m/s} + (-9.80 \text{ m/s}^2)(5.00 \text{ s}) = \boxed{-29.0 \text{ m/s}}$$

We can use $y = v_0 t + \frac{1}{2}at^2$ (Eq. 2.9) to find the position of the particle at $t = 5.00$ s:

$$y = (20.0 \text{ m/s})(5.00 \text{ s}) + \tfrac{1}{2}(-9.80 \text{ m/s}^2)(5.00 \text{ s})^2 = \boxed{-22.5 \text{ m}}$$

Exercise Find the velocity of the stone just before it hits the ground.

Answer -37.0 m/s.

Figure 2.17 (Example 2.6) Position and velocity versus time for a freely falling object thrown initially upward with a velocity of $v_0 = +20.0$ m/s.

SUMMARY

The **displacement** of an object moving along the x axis is defined as the change in position of the object, and is given by

$$\Delta x \equiv x_f - x_i \qquad [2.1]$$

where x_i is the initial position of the object and x_f is its final position.

The **average velocity** of an object moving along the x axis during some time interval is equal to the displacement of the object, Δx, divided by the time interval, Δt, during which the displacement occurred.

$$\bar{v} \equiv \frac{\Delta x}{\Delta t} = \frac{x_f - x_i}{t_f - t_i} \qquad \text{[2.2]}$$

The average velocity is equal to the slope of the straight line joining the initial and final points on a graph of the position of the object versus time.

The slope of the line tangent to the position-time curve at some point is equal to the instantaneous velocity at that time. The **instantaneous speed** of an object is defined as the magnitude of the instantaneous velocity.

The **average acceleration** of an object during some time interval is defined as the change in velocity, Δv, divided by the time interval, Δt, during which the change occurred:

$$\bar{a} \equiv \frac{\Delta v}{\Delta t} = \frac{v_f - v_i}{t_f - t_i} \qquad \text{[2.4]}$$

The **instantaneous acceleration** of an object at a certain time equals the slope of a velocity-time graph at that instant.

A **vector** quantity is any quantity that is characterized by both a magnitude and a direction. In contrast, a **scalar** quantity such as mass has a magnitude only. Displacement, velocity, and acceleration are all vector quantities. In the case of motion along a straight line, we use algebraic signs to describe the directions of the vectors. For example, an object moving to the right with a speed of 8 m/s would have a velocity of $+8$ m/s, assuming the positive direction has been chosen to point to the right. If the object were moving to the left, its velocity would be -8 m/s.

The equations that describe the motion of an object moving with constant acceleration along the x axis are

$$v = v_0 + at \qquad \text{[2.6]}$$

$$x = v_0 t + \tfrac{1}{2} a t^2 \qquad \text{[2.9]}$$

$$v^2 = v_0^2 + 2ax \qquad \text{[2.10]}$$

An object falling in the presence of the Earth's gravity experiences a free-fall acceleration directed toward the center of the Earth. If air friction is neglected and if the altitude of the falling object is small compared with the Earth's radius, then one can assume that the free-fall acceleration, **g,** is constant over the range of motion, where g is equal to 9.80 m/s², or 32.0 ft/s². Assuming that the positive direction for y is chosen to be upward, the acceleration is $-g$ (downward) and the equations describing the motion of the falling object are the same as the foregoing equations, with the substitutions $x \rightarrow y$ and $a \rightarrow -g$.

MULTIPLE-CHOICE QUESTIONS

1. People become uncomfortable in an elevator if it accelerates downward at a rate such that it attains a speed of about 6 m/s after ten stories (about 30 m). What is this acceleration?
(a) 9.8 m/s²   (b) 0.33 m/s²   (c) 0.6 m/s²
(d) 1.2 m/s²   (e) 2 m/s²

2. Races are timed to 1/1000 of a second. What distance could a roller blader moving at a speed of 10 m/s travel in that period of time?
(a) 10 mm   (b) 10 cm   (c) 10 m   (d) 1 mm   (e) 1 km

3. A rock is thrown downward from the top of a 40-m tower with an initial speed of 12 m/s. Assuming negligi-

ble air resistance, what is the speed of the rock just before hitting the ground?
(a) 28 m/s (b) 30 m/s (c) 56 m/s (d) 784 m/s

4. An x versus t graph is drawn for a ball moving in one direction. The graph starts at the origin, increases with a positive slope for five seconds, and then levels off to a zero slope for 2 s. We can be sure that
(a) the speed is decreasing during the first 5 seconds.
(b) the acceleration is constant throughout the complete motion.

(c) the speed is zero between $t = 5$ s and $t = 7$ s.
(d) the displacement is zero throughout the motion.
(e) more than one of the above is true.

5. A ball is thrown straight up. For which situation are both the instantaneous velocity and the acceleration zero?
(a) on the way up
(b) at the top of the flight path
(c) on the way down
(d) none of the above

CONCEPTUAL QUESTIONS

1. If a car is traveling eastward, can its acceleration be westward? Explain.

2. The speed of sound in air is 331 m/s. During the next thunderstorm, try to estimate your distance from a lightning bolt by measuring the time lag between the flash and the thunderclap. You can ignore the time it takes for the light flash to reach you. Why?

3. Can an object having constant acceleration ever stop and *stay* stopped?

4. If the average velocity of an object is zero in some time interval, what can you say about the displacement of the object for that interval?

5. A child throws a marble into the air with an initial speed v_0. Another child drops a ball at the same instant. Compare the accelerations of the two objects while they are in flight.

6. The strobe photographs in Figure Q2.6 were taken of a disk moving from left to right under different conditions, and the time interval between images is constant. Taking the direction to the right to be positive, describe the motion of the disk in each case. For which case is
(a) the acceleration positive? (b) the acceleration negative? (c) the velocity constant?

7. Can the instantaneous velocity of an object at an instant of time ever be greater in magnitude than the average velocity over a time interval containing the instant? Can it ever be less?

8. Car A, traveling from New York to Miami, has a speed of 25 m/s. Car B, traveling from New York to Chicago, also has a speed of 25 m/s. Are their velocities equal? Explain.

9. A ball is thrown vertically upward. (a) What are its velocity and acceleration when it reaches its maximum altitude? (b) What is its acceleration just before it hits the ground?

10. A rule of thumb for driving is that a separation of one car length for each 10 mi/h of speed should be maintained between moving vehicles. Assuming a constant reaction time, discuss the relevance of this rule for (a) mo-

(a)

(b)

(c)

Figure Q2.6 *(Courtesy David Rogers)*

tion with constant velocity and (b) motion with constant acceleration.

11. Two cars are moving in the same direction in parallel lanes along a highway. At some instant, the velocity of car A exceeds the velocity of car B. Does this mean that the acceleration of A is greater than that of B? Explain.

12. Consider the following combinations of signs and values for velocity and acceleration of a particle with respect to a one-dimensional x-axis:

	Velocity	Acceleration		Velocity	Acceleration
a.	Positive	Positive	e.	Negative	Negative
b.	Positive	Negative	f.	Negative	Zero
c.	Positive	Zero	g.	Zero	Positive
d.	Negative	Positive	h.	Zero	Negative

Figure Q2.15

Describe what a particle is doing in each case, and give a real-life example for an automobile on an east–west one-dimensional axis, with east considered the positive direction.

13. A student at the top of a building of height h throws one ball upward with a speed of v_0 and then throws a second ball downward with the same initial speed, v_0. How do the final velocities of the balls compare when they reach the ground?

14. You drop a ball from a window on an upper floor of a building. It strikes the ground with velocity v. You now repeat the drop, but you have a friend down on the street who throws another ball upward at velocity v. Your friend throws the ball upward at exactly the same time that you drop yours from the window. At some location, the balls pass each other. Is this location *at the* halfway point between window and ground, *above* this point, or *below* this point?

15. A pebble is dropped into a water well, and the splash is heard 16 s later, as illustrated in the cartoon strip shown in Figure Q2.15. Estimate the distance from the rim of the well to the water's surface.

16. A ball rolls in a straight line along the horizontal direction. Using motion diagrams (or multiflash photographs) as in Figure 2.13, describe the velocity and acceleration of the ball for each of the following situations: **(a)** The ball moves to the right at a constant speed. **(b)** The ball moves from right to left and continually slows down. **(c)** The ball moves from right to left and continually speeds up. **(d)** The ball moves to the right, first speeding up at a constant rate, and then slowing down at a constant rate.

PROBLEMS

1, 2, 3 = straightforward, intermediate, challenging   ☐ = full solution available in Study Guide/Student Solutions Manual
WEB = solution posted at **http://www.saunderscollege.com/physics/college**   🖼 = biomedical application   🖼 = Interactive Physics

Section 2.2 Average Velocity

Section 2.3 Instantaneous Velocity

1. A person travels by car from one city to another with different constant speeds between pairs of cities. She drives for 30.0 min at 80.0 km/h, 12.0 min at 100 km/h, and 45.0 min at 40.0 km/h, and spends 15.0 min eating lunch and buying gas. **(a)** Determine the average speed for the trip. **(b)** Determine the distance between the initial and final cities along this route.

2. An athlete swims the length of a 50.0-m pool in 20.0 s and makes the return trip to the starting position in 22.0 s. Determine her average velocities in **(a)** the first half of the swim, **(b)** the second half of the swim, and **(c)** the round trip.

3. Two boats start together and race across a 60 km wide lake and back. Boat A goes across at 60 km/h and returns at 60 km/h. Boat B goes across at 30 km/h and its crew, realizing how far behind it is getting, returns at 90 km/h. Turn-around times are negligible, and the boat that completes the round-trip first wins. **(a)** Which boat wins and by how much? (Or is it a tie?) **(b)** What is the average velocity of the winning boat?

4. A certain bacterium swims with a speed of 3.5 μm/s. How long would it take this bacterium to swim across a petri dish having a diameter of 8.4 cm?

5. The Olympic record for the marathon is 2h, 9 min, 21 s. The marathon distance is 26 mi, 385 yd. Determine the average speed (in miles per hour) of this record.

6. A tennis player moves in a straight-line path as shown in Figure P2.6. Find her average velocities in the time intervals (a) 0 to 1.0 s, (b) 0 to 4.0 s, (c) 1.0 s to 5.0 s, and (d) 0 to 5.0 s.

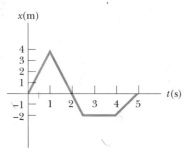

Figure P2.6 (Problems 6 and 15)

7. Two cars travel in the same direction along a straight highway, one at a constant speed of 55 mi/h and the other at 70 mi/h. (a) Assuming that they start at the same point, how much sooner does the faster car arrive at a destination 10 mi away? (b) How far must the faster car travel before it has a 15-min lead on the slower car?

8. If the average speed of an orbiting space shuttle is 19 800 mi/h, determine the time required for it to circle the Earth. Make sure you consider the fact that the shuttle is orbiting about 200 mi above the Earth's surface, and assume that the Earth's radius is 3963 miles.

9. A person takes a trip, driving with a constant speed of 89.5 km/h except for a 22.0-min rest stop. If the person's average speed is 77.8 km/h, how much time is spent on the trip and how far does the person travel?

10. A tortoise can run with a speed of 10.0 cm/s, and a hare can run 20 times as fast. In a race, they both start at the same time, but the hare stops to rest for 2.0 minutes. The tortoise wins by a shell (20 cm). (a) How long does the race take? (b) What is the length of the race?

WEB 11. In order to qualify for the finals in a racing event, a race car must achieve an average speed of 250 km/h on a track with a total length of 1600 m. If a particular car covers the first half of the track at an average speed of 230 km/h, what minimum average speed must it have in the second half of the event in order to qualify?

12. Runner A is initially 4.0 mi west of a flagpole and is running with a constant velocity of 6.0 mi/h due east. Runner B is initially 3.0 mi east of the flagpole and is running with a constant velocity of 5.0 mi/h due west. How far are the runners from the flagpole when they meet?

13. A runner moves so that his positions at certain times are given by the data in the following table. Use these data to construct a table like Table 2.2 in the text. From your table, find (a) the average velocity during the complete interval and (b) the instantaneous velocity at $t = 2.00$ s.

t(s)	x(m)	t(s)	x(m)
2.00	5.66	2.50	6.32
2.01	5.674	3.00	6.92
2.20	5.93	4.00	8.00

14. A race car moves such that its position fits the relationship

$$x = (5.0 \text{ m/s})t + (0.75 \text{ m/s}^3)t^3$$

where x is measured in meters and t in seconds. (a) Plot a graph of position versus time. (b) Determine the instantaneous velocity at $t = 4.0$ s, using time intervals of 0.40 s, 0.20 s, and 0.10 s. (c) Compare the average velocity during the first 4.0 s with the results of (b).

15. Find the instantaneous velocities of the tennis player of Figure P2.6 at (a) 0.50 s, (b) 2.0 s, (c) 3.0 s, (d) 4.5 s.

Section 2.4 Acceleration

16. A car traveling initially at $+7.0$ m/s accelerates at the rate of $+0.80$ m/s^2 for an interval of 2.0 s. What is its velocity at the end of the acceleration?

17. A car traveling in a straight line has a velocity of $+5.0$ m/s at some instant. After 4.0 s, its velocity is $+8.0$ m/s. What is its average acceleration in this time interval?

18. A tennis ball with a speed of 10.0 m/s is thrown perpendicularly at a wall. After striking the wall, the ball rebounds in the opposite direction with a speed of 8.0 m/s. If the ball is in contact with the wall for 0.012 s, what is the average acceleration of the ball while it is in contact with the wall?

WEB 19. A certain car is capable of accelerating at a rate of $+0.60$ m/s^2. How long does it take for this car to go from a speed of 55 mi/h to a speed of 60 mi/h?

20. The velocity-versus-time graph for an object moving along a straight path is shown in Figure P2.20.

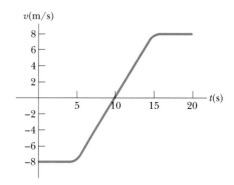

Figure P2.20

(a) Find the average accelerations of this object during the time intervals 0 to 5.0 s, 5.0 s to 15 s, and 0 to 20 s. (b) Find the instantaneous accelerations at 2.0 s, 10 s, and 18 s.

21. The engine of a model rocket accelerates the rocket vertically upward for 2.0 s as follows: at $t = 0$, its speed is zero; at $t = 1.0$ s, its speed is 5.0 m/s; at $t = 2.0$ s, its speed is 16 m/s. Plot a velocity-time graph for this motion, and from it determine (a) the average acceleration during the 2.0-s interval and (b) the instantaneous acceleration at $t = 1.5$ s.

Section 2.6   One-Dimensional Motion with Constant Acceleration

22. Jules Verne in 1865 proposed sending men to the Moon by firing a space capsule from a 220-m-long cannon with final velocity of 10.97 km/s. What would have been the unrealistically large acceleration experienced by the space travelers during launch? Compare your answer with the free-fall acceleration, 9.80 m/s^2.

23. A racing car reaches a speed of 40 m/s. At this instant, it begins a uniform negative acceleration, using a parachute and a braking system, and comes to rest 5.0 s later. (a) Determine the acceleration of the car. (b) How far does the car travel after acceleration starts?

24. A jet plane lands with a velocity of $+100$ m/s and can accelerate at a maximum rate of -5.0 m/s^2 as it comes to rest. (a) From the instant it touches the runway, what is the minimum time needed before it can come to rest? (b) Can this plane land on a small island airport where the runway is 0.80 km long?

25. A truck on a straight road starts from rest accelerating at 2.0 m/s^2 until it reaches a speed of 20 m/s. Then the truck travels for 20 s at constant speed until the brakes are applied, stopping the truck in a uniform manner in an additional 5.0 s. (a) How long is the truck in motion? (b) What is the average velocity of the truck for the motion described?

26. A Cessna aircraft has a lift-off speed of 120 km/h. (a) What minimum constant acceleration does this require if the aircraft is to be airborne after a takeoff run of 240 m? (b) How long does it take the aircraft to become airborne?

27. An electron moving in a straight line has an initial speed of 3.0×10^5 m/s. It undergoes an acceleration of 8.0×10^{14} m/s^2. (a) How long will it take to reach a speed of 5.4×10^5 m/s? (b) How far will it have traveled in this time?

28. A driver in a car traveling at a speed of 60 mi/h sees a deer 100 m away on the road. Calculate the minimum constant acceleration that is necessary for the car to stop without hitting the deer (assuming that the deer does not move in the meantime).

29. A speedboat increases in speed uniformly from 20 m/s to 30 m/s in a distance of 200 m. Find (a) the magnitude of its acceleration and (b) the time it takes the boat to travel this distance.

30. Two cars are traveling along a straight line in the same direction, the lead car at 25 m/s and the other car at 30 m/s. At the moment the cars are 40 m apart, the lead driver applies the brakes, causing her car to have an acceleration of -2.0 m/s^2. (a) How long does it take for the lead car to stop? (b) Assuming that the chasing car brakes at the same time as the lead car, what must be the chasing car's minimum negative acceleration so as not to hit the lead car? (c) How long does it take for the chasing car to stop?

31. A record of travel along a straight path is as follows.
    1. Start from rest with constant acceleration of 2.77 m/s^2 for 15.0 s
    2. Constant velocity for the next 2.05 min
    3. Constant negative acceleration -9.47 m/s^2 for 4.39 s
    (a) What was the total displacement for the complete trip?
    (b) What were the average speeds for legs 1, 2, and 3 of the trip as well as for the complete trip?

32. A train is traveling down a straight track at 20 m/s when the engineer applies the brakes, resulting in an acceleration of -1.0 m/s^2 as long as the train is in motion. How far does the train move during a 40 s time interval starting at the instant the brakes are applied?

33. A car accelerates uniformly from rest to a speed of 40.0 mi/h in 12.0 s. (a) Find the distance the car travels during this time and (b) the constant acceleration of the car.

34. A car starts from rest and travels for 5.0 s with a uniform acceleration of $+1.5$ m/s^2. The driver then applies the brakes, causing a uniform acceleration of -2.0 m/s^2. If the brakes are applied for 3.0 s, how fast is the car going at the end of the braking period, and how far has it gone?

35. A train 400 m long is moving on a straight track with a speed of 82.4 km/h. The engineer applies the brakes at a crossing, and later the last car passes the crossing with a speed of 16.4 km/h. Assuming constant acceleration, determine how long the train blocked the crossing. Disregard the width of the crossing.

36. A hockey player is standing on his skates on a frozen pond when an opposing player, moving with a uniform speed of 12 m/s, skates by with the puck. After 3.0 s, the first player makes up his mind to chase his opponent. If he accelerates uniformly at 4.0 m/s^2, (a) how long does it take him to catch his opponent, and (b) how far has he traveled in this time? (Assume the player with the puck remains in motion at constant speed.)

Section 2.7 Freely Falling Objects

37. A ball is thrown vertically upward with a speed of 25.0 m/s. **(a)** How high does it rise? **(b)** How long does it take to reach its highest point? **(c)** How long does it take to hit the ground after it reaches its highest point? **(d)** What is its speed when it returns to the level from which it started?

38. A peregrine falcon dives at a pigeon. The falcon starts downward from rest and falls with free-fall acceleration. If the pigeon is 76.0 m below the initial position of the falcon, how long does it take the falcon to reach the pigeon? Assume that the pigeon remains at rest.

39. A small mailbag is released from a helicopter that is descending steadily at 1.50 m/s. After 2.00 s, **(a)** what is the speed of the mailbag, and **(b)** how far is it below the helicopter? **(c)** What are your answers to parts **(a)** and **(b)** if the helicopter is rising steadily at 1.50 m/s?

40. A rocket moves upward, starting from rest with an acceleration of +29.4 m/s^2 for 4.00 s. It runs out of fuel at the end of this 4.00 s and continues to move upward for a while. How high does it rise above its original starting point?

41. A ball thrown vertically upward is caught by the thrower after 2.00 s. Find **(a)** the initial velocity of the ball and **(b)** the maximum height it reaches.

42. A model rocket is launched straight upward with an initial speed of 50.0 m/s. It accelerates with a constant upward acceleration of 2.00 m/s^2 until its engines stop at an altitude of 150 m. **(a)** What is the maximum height reached by the rocket? **(b)** How long after lift-off does the rocket reach its maximum height? **(c)** How long is the rocket in the air?

43. A parachutist with a camera, both descending at a speed of 10 m/s, releases that camera at an altitude of 50 m. **(a)** How long does it take the camera to reach the ground? **(b)** What is the velocity of the camera just before it hits the ground?

ADDITIONAL PROBLEMS

44. The total area under the velocity versus time curve shown in Figure P2.44 represents the distance traveled. Break this area into a rectangular area A_1 and a triangular area A_2. Compute the areas A_1 and A_2 and compare their sum to the result given in Equation 2.8.

45. A bullet is fired through a board 10.0 cm thick in such a way that the bullet's line of motion is perpendicular to the face of the board. If the initial speed of the bullet is 400 m/s and it emerges from the other side of the board with a speed of 300 m/s, find **(a)** the acceleration of the bullet as it passes through the board and **(b)** the total time the bullet is in contact with the board.

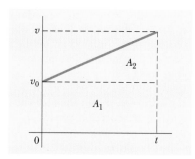

Figure P2.44

46. A ball is thrown upward from the ground with an initial speed of 25 m/s; at the same instant, a ball is dropped from a building 15 m high. After how long will the balls be at the same height?

47. A ranger in a national park is driving at 35.0 mi/h when a deer jumps into the road 200 ft ahead of the vehicle. After a reaction time of t, the ranger applies the brakes to produce an acceleration of $a = -9.00$ ft/s^2. What is the maximum reaction time allowed if she is to avoid hitting the deer?

48. Two students are on a balcony 19.6 m above the street. One student throws a ball vertically downward at 14.7 m/s; at the same instant the other student throws a ball vertically upward at the same speed. The second ball just misses the balcony on the way down. **(a)** What is the difference in their time in air? **(b)** What is the velocity of each ball as it strikes the ground? **(c)** How far apart are the balls 0.800 s after they are thrown?

49. The driver of a car slams on the brakes as a truck approaches the car head-on while slowing uniformly with acceleration −5.60 m/s^2. The driver of the car is frozen with horror for 4.20 s while the truck makes skid marks 62.4 m long. With what speed does the truck strike the car?

50. A young woman named Kathy Kool buys a sports car that can accelerate at the rate of 4.90 m/s^2. She decides to test the car by dragging with another speedster, Stan Speedy. Both start from rest, but experienced Stan leaves the starting line 1.00 s before Kathy. If Stan moves with a constant acceleration of 3.50 m/s^2 and Kathy maintains an acceleration of 4.90 m/s^2, find **(a)** the time it takes Kathy to overtake Stan, **(b)** the distance she travels before she catches him, and **(c)** the speeds of both cars at the instant she overtakes him.

51. A mountain climber stands at the top of a 50.0-m cliff that overhangs a calm pool of water. He throws two stones vertically downward 1.0 s apart and observes that they cause a single splash. The first stone has an initial velocity of −2.00 m/s. **(a)** How long after release of the first stone will the two stones hit the water?

(b) What initial velocity must the second stone have if they are to hit simultaneously? **(c)** What will the velocity of each stone be at the instant they hit the water?

52. In order to pass a physical education class at a university, a student must run 1.0 mi in 12 min. After running for 10 min, she still has 500 yd to go. If her maximum acceleration is 0.15 m/s², can she make it? If the answer is no, determine what acceleration she would need to be successful.

53. In Mostar, Bosnia, the ultimate test of a young man's courage once was to jump off a 400-year-old bridge (now destroyed) into the River Neretva, 23 m below the bridge. **(a)** How long did the jump last? **(b)** How fast was the diver traveling on impact with the river? **(c)** If the speed of sound in air is 340 m/s, how long after the diver took off did a spectator on the bridge hear the splash?

54. One swimmer in a relay race has a 0.50-s lead and is swimming at a constant speed of 4.0 m/s. He has 50 m to swim before reaching the end of the pool. A second swimmer moves in the same direction as the leader. What constant speed must the second swimmer have in order to catch up to the leader at the end of the pool?

55. A person sees a lightning bolt pass close to an airplane that is flying in the distance. The person hears thunder 5.0 s after seeing the bolt, and sees the airplane overhead 10 s after hearing the thunder. The speed of sound in air is 1100 ft/s. **(a)** Find the distance of the airplane from the person at the instant of the bolt. (Neglect the time it takes the light to travel from the bolt to the eye.) **(b)** Assuming that the plane travels with a constant speed toward the person, find the velocity of the airplane. **(c)** Look up the speed of light in air, and defend the approximation used in (a).

56. A person walks first at a constant speed of 5.00 m/s along a straight line from Point *A* to Point *B* and then back along the line from *B* to *A* at a constant speed of 3.00 m/s. What is **(a)** her average speed over the entire trip? **(b)** her average velocity over the entire trip?

57. A hard rubber ball, released at chest height, falls to the pavement and bounces back to nearly the same height. When it is in contact with the pavement, the lower side of the ball is temporarily flattened. Before this dent in the ball pops out, suppose that its maximum depth is on the order of one centimeter. Estimate the maximum acceleration of the ball. State your assumptions, the quantities you estimate, and the values you estimate for them.

58. A ball is thrown directly downward, with an initial speed of 8.00 m/s, from a height of 30.0 m. After what interval does the ball strike the ground?

59. A hot air balloon is traveling vertically upward at a constant speed of 5.00 m/s. When it is 21.0 m above the ground, a package is released from the balloon. **(a)** For how long after being released is the package in the air? **(b)** What is the velocity of the package just before impact with the ground? **(c)** Repeat (a) and (b) for the case of the balloon descending at 5.00 m/s.

WEB 60. A stunt woman sitting on a tree limb wishes to drop vertically onto a horse galloping under the tree. The speed of the horse is 10.0 m/s, and the woman is initially 3.00 m above the level of the saddle. **(a)** What must be the horizontal distance between the saddle and limb when the woman makes her move? **(b)** How long is she in the air?

The Wizard of Id   **by Parker and Hart**

Vectors and Two-Dimensional Motion

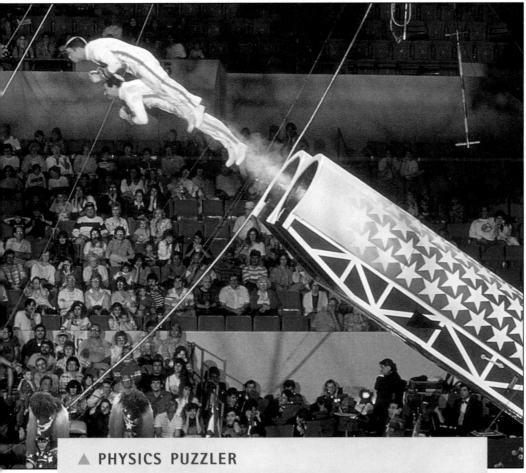

▲ PHYSICS PUZZLER

The circus stuntmen being shot out of cannons are human projectiles. Neglecting air resistance, they move in parabolic paths until they land in a stategically placed net. What initial condition(s) will determine where the catching net should be placed? *(Ringling Brothers Circus)*

I n our discussion of one-dimensional motion in Chapter 2, we used the concept of vectors only to a limited extent. As we progress in our study of motion, the ability to manipulate vector quantities will become increasingly important. As a result, much of this chapter will be devoted to techniques for adding vectors, subtracting them, and so forth. We will then apply our newfound skills to a special case of two-dimensional motion—projectiles. We shall also see that an understanding of vector manipulation is necessary in order to work with and understand relative motion.

3.1   VECTORS AND SCALARS REVISITED

Each of the physical quantities we shall encounter in this book can be categorized as either a *scalar* or a *vector quantity*. A scalar is a quantity that can be completely specified by its magnitude with appropriate units; that is, a **scalar** has only magnitude and no direction. A **vector** is a physical quantity that requires the specification of both direction and magnitude.

Temperature is an example of a scalar quantity. If someone tells you that the temperature of an object is $-5°C$, that information completely specifies the temperature of the object; no direction is required. Other examples of scalars are masses, time intervals, and the number of pages in this textbook. Scalar quantities can be manipulated using the rules of ordinary arithmetic. For example, if you have 2 liters of water in a container (where 1 liter is defined to be $1000\ cm^3$) and you add 3 more liters, by ordinary arithmetic the amount of water in the container is then 5 liters.

An example of a vector quantity is force. If you are told that someone is going to exert a force of 10 lb on an object, that is not enough information to let you

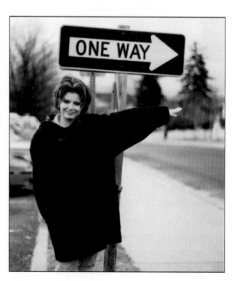

(Left) The number of apples in the basket is one example of a scalar quantity. Can you think of other examples? *(Superstock)*   *(Right)* Jennifer pointing in the right direction tells us to travel 5 blocks to the north to reach the courthouse. A vector is a physical quantity that must be specified by both magnitude and direction. *(Raymond A. Serway)*

know what will happen to the object. The effect of a force of 10 lb exerted horizontally is different from the effect of a force of 10 lb exerted vertically upward or downward. In other words, you need to know the direction of the force as well as its magnitude.

Velocity is another example of a vector quantity. If we wish to describe the velocity of a moving vehicle, we must specify both its speed (say, 30 m/s) and the direction in which the vehicle is moving (say, northeast). Other examples of vector quantities include displacement and acceleration, which were defined in Chapter 2.

3.2 SOME PROPERTIES OF VECTORS

Equality of Two Vectors. Two vectors, **A** and **B,** are defined as equal if they have the same magnitude and the same direction. This property allows us to translate a vector parallel to itself in a diagram without affecting the vector. In fact, for most purposes, any vector can be moved parallel to itself without being affected.

Adding Vectors. When two or more vectors are added together, they must all have the same units. For example, it would be meaningless to add a velocity vector to a displacement vector, because they are different physical quantities. Scalars also obey this rule. For example, it would be meaningless to add temperatures and areas.

When a vector quantity is handwritten, it is often represented with an arrow over the letter ($\vec{A}$). In this book, a vector quantity will be represented by boldface type (for example, **A**). The magnitude of a vector such as **A** will be represented by italic type such as A. Likewise, italic type will be used to represent scalars.

The procedures for adding vectors rely on geometric methods. (Later we shall develop an algebraic technique for adding vectors that is much more convenient and will be used throughout the remainder of the text.) To add vector **B** to vector **A,** first draw **A** on a piece of graph paper to some scale such as 1 cm = 1 m. Vector **A** must be drawn so that its direction is specified relative to a coordinate system. Then draw vector **B** to the same scale and with the tail of **B** starting at the tip of **A,** as in Figure 3.1a. Vector **B** must be drawn along the direction that makes the proper angle relative to vector **A.** The resultant vector, **R,** given by **R** = **A** + **B,** is

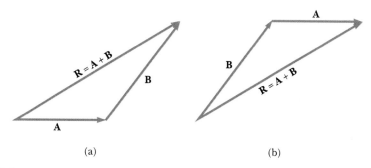

(a)   (b)

Figure 3.1 (a) When vector **B** is added to vector **A,** the vector sum, **R,** is the vector that runs from the tail of **A** to the tip of **B.** (b) Here the resultant runs from the tail of **B** to the tip of **A.** These constructions prove that **A** + **B** = **B** + **A.**

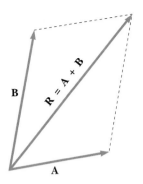

Figure 3.2 In this construction, the resultant **R** is the diagonal of a parallelogram with sides **A** and **B.**

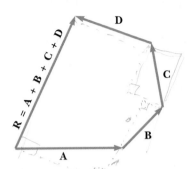

Figure 3.3 A geometric construction for summing four vectors. The resultant vector, **R,** is the vector that completes the polygon.

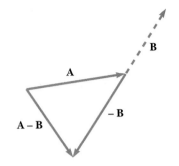

Figure 3.4 This construction shows how to subtract vector **B** from vector **A.** The vector −**B** has the same magnitude as the vector **B** but points in the opposite direction.

the vector drawn from the tail of **A** to the tip of **B**. This is known as the *triangle method of addition*.

When two vectors are added, the sum is independent of the order of the addition. That is, **A** + **B** = **B** + **A.** This can be seen from the geometric construction in Figure 3.1b.

An alternative graphical procedure for adding two vectors, known as the *parallelogram rule of addition,* is shown in Figure 3.2. In this construction, the tails of vectors **A** and **B** are joined together, and the resultant vector, **R,** is the diagonal of the parallelogram formed with **A** and **B** as its sides.

This same general approach can also be used to add more than two vectors, as is done in Figure 3.3 for four vectors. The resultant vector sum, **R** = **A** + **B** + **C** + **D,** is the vector drawn from the tail of the first vector to the tip of the last vector. Again, the order in which vectors are added is unimportant.

Negative of a Vector. The negative of the vector **A** is defined as the vector that when added to **A** gives zero for the vector sum. This means that **A** and −**A** have the same magnitude but opposite directions.

Subtraction of Vectors. Vector subtraction makes use of the definition of the negative of a vector. We define the operation **A** − **B** as vector −**B** added to vector **A:**

$$\mathbf{A} - \mathbf{B} = \mathbf{A} + (-\mathbf{B}) \tag{3.1}$$

Thus, vector subtraction is really a special case of vector addition. The geometric construction for subtracting two vectors is shown in Figure 3.4.

Multiplication and Division of Vectors by Scalars. The multiplication or division of a vector by a scalar gives a vector. For example, if a vector, **A,** is multiplied by the scalar number 3, the result, written 3**A,** is a vector with a magnitude three times that of the original vector **A,** pointing in the same direction as **A.** On the other hand, if we multiply vector **A** by the scalar −3, the result is a vector with a magnitude three times that of **A,** pointing in the direction opposite **A** (because of the negative sign).

EXAMPLE 3.1 Taking a Trip

A car travels 20.0 km due north and then 35.0 km in a direction 60.0° west of north, as in Figure 3.5. Find the magnitude and direction of the car's resultant displacement.

Solution The problem can be solved geometrically using graph paper and a protractor, as shown in Figure 3.5. The resultant displacement, **R,** is the sum of the two individual displacements, **A** and **B.**

The length of **R,** drawn to the same scale as **A** and **B,** indicates that the displacement of the car is 48.2 km, and a measurement of the angle β shows that the displacement is approximately 38.9° west of north.

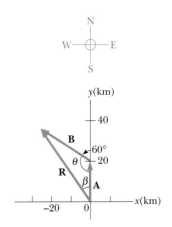

Figure 3.5 (Example 3.1) A graphical method for finding the resultant displacement vector **R = A + B.**

3.3 COMPONENTS OF A VECTOR

One method of adding vectors makes use of the projections of a vector along the axes of a rectangular coordinate system. These projections are called **components.** Any vector can be completely described by its components.

Consider a vector, **A,** in a rectangular coordinate system, as shown in Figure 3.6. Note that **A** can be expressed as the sum of two vectors, $\mathbf{A}_x$ parallel to the x axis, and $\mathbf{A}_y$ parallel to the y axis. That is,

$$\mathbf{A} = \mathbf{A}_x + \mathbf{A}_y$$

$\mathbf{A}_x$ and $\mathbf{A}_y$ are the component vectors of **A.** The projection of **A** along the x axis, A_x, is called the x component of **A,** and the projection of **A** along the y axis, A_y, is called the y component of **A.** These components can be either positive or negative numbers with units. From the definitions of sine and cosine of an angle, we see that $\cos \theta = A_x/A$ and $\sin \theta = A_y/A$. Hence, the magnitudes of the components of **A** are

$$A_x = A \cos \theta$$

$$A_y = A \sin \theta$$

[3.2]

These components form two sides of a right triangle, the hypotenuse of which has the magnitude A. Thus, it follows that **A**'s magnitude and direction are related to its components through the Pythagorean theorem and the definition of the tangent:

$$A = \sqrt{A_x^2 + A_y^2}$$

[3.3]

$$\tan \theta = \frac{A_y}{A_x}$$

[3.4]

To solve for the angle θ, we can write Equation 3.4 in the form

$$\theta = \tan^{-1}\left(\frac{A_y}{A_x}\right)$$

If a coordinate system other than the one shown in Figure 3.6 is chosen, the components of the vector must be modified accordingly. In many applications it is

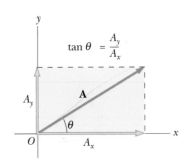

Figure 3.6 Any vector **A** lying in the xy plane can be represented by its rectangular components, A_x and A_y.

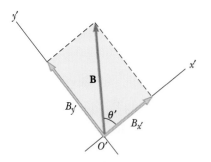

Figure 3.7 The components of vector **B** in a tilted coordinate system.

more convenient to express the components of a vector in a coordinate system having axes that are not horizontal and vertical but are still perpendicular to each other. Suppose a vector, **B,** makes an angle of θ' with the x' axis defined in Figure 3.7. The rectangular components of **B** along the axes of Figure 3.7 are given by $B_{x'} = B \cos \theta'$ and $B_{y'} = B \sin \theta'$, as in Equation 3.2. The magnitude and direction of **B** are obtained from expressions equivalent to Equations 3.3 and 3.4. Thus, we can express the components of a vector in any coordinate system that is convenient for the situation.

Thinking Physics 1

You may have asked someone directions to a destination in a city and have been told something like, "Walk 3 blocks east and then 5 blocks south." If so, are you experienced with vector components?

Explanation Yes, you are! Although you may not have thought of vector component language when you heard these directions, this is exactly what the directions represent. The perpendicular streets of the city reflect an x–y coordinate system—we can assign the x axis to the east–west streets and the y axis to the north–south streets. Thus, the comment of the person giving you directions can be translated as, "Undergo a displacement vector that has an x component of $+3$ blocks and a y component of -5 blocks." You would arrive at the same destination by undergoing the y-component first, followed by the x component, demonstrating the commutative law of addition.

Thinking Physics 2

Consider your commute to work or school in the morning. Which is larger, the distance you traveled or the magnitude of the displacement vector?

Explanation The distance traveled, unless you have a very unusual commute, *must* be larger than the magnitude of the displacement vector. The distance includes all of the twists and turns that you made in following the roads from home to work or school. However, the magnitude of the displacement vector is the length of a straight line from your home to work or school. This is often described informally as "the distance as the crow flies." The only way that the distance could be the same as the magnitude of the displacement vector is if your commute is a perfect straight line, which is unlikely! The distance could *never* be less than the magnitude of the displacement vector, because the shortest distance between two points is a straight line.

Adding Vectors

In order to add vectors algebraically rather than graphically, the following procedure is used. First, find the components of all the vectors in some coordinate system that is appropriate for the problem. Next, add all the x components to find the resultant component in the x direction. Similarly, add all the y components to find the resultant component in the y direction. Because the resultant x and y components are at right angles to each other, you can use the Pythagorean theorem to determine the magnitude of the resultant vector. Finally, use a suitable trigonometric function and the components of the resultant vector to find the angle that the resultant vector makes with the *x* axis. These steps are summarized as a Problem-Solving Strategy and are illustrated in the worked examples that follow.

Problem-Solving Strategy

Adding Vectors

When two or more vectors are to be added, the following steps are used.

1. Select a coordinate system.
2. Draw a sketch of the vectors to be added (or subtracted), with a label on each vector.
3. Find the *x* and *y* components of all vectors.
4. Find the resultant components (the algebraic sum of the components) in both the *x* and *y* directions.
5. Use the Pythagorean theorem to find the magnitude of the resultant vector.
6. Use a suitable trigonometric function to find the angle the resultant vector makes with the *x* axis.

EXAMPLE 3.2 Help Is on the Way!

Find the horizontal and vertical components of the 100-m displacement of a superhero who flies from the top of a tall building along the path shown in Figure 3.8a.

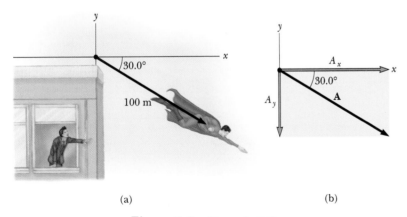

(a)　　　　　　　　　　　　　　　　(b)

Figure 3.8 (Example 3.2)

Solution The triangle formed by the displacement and its components is shown in Figure 3.8b. Because $A = 100$ m and $\theta = -30.0°$ (θ is negative because it is measured clockwise from the *x* axis), we have

$$A_y = A \sin \theta = (100 \text{ m}) \sin(-30.0°) = -50.0 \text{ m}$$

Note that $\sin(-\theta) = -\sin \theta$. The negative sign for A_y reflects the fact that displacement in the *y* direction is *downward* from the origin.

The *x* component of displacement is

$$A_x = A \cos \theta = (100 \text{ m}) \cos(-30.0°) = \boxed{+86.6 \text{ m}}$$

Note that $\cos(-\theta) = \cos \theta$. Also, from an inspection of the figure, you should be able to see that A_x is positive in this case.

EXAMPLE 3.3 Taking a Hike

A hiker begins a trip by first walking 25.0 km due southeast from her base camp. On the second day she walks 40.0 km in a direction 60.0° north of east, at which point she discovers a forest ranger's tower.
(a) Determine the components of the hiker's displacements in the first and second days.

Solution   If we denote the displacement vectors on the first and second days by **A** and **B,** respectively, and use the camp as the origin of coordinates, we get the vectors shown in Figure 3.9. Displacement **A** has a magnitude of 25.0 km and is 45.0° south of east. Its components are

$$A_x = A \cos(-45.0°) = (25.0 \text{ km})(0.707) = \boxed{17.7 \text{ km}}$$

$$A_y = A \sin(-45.0°) = -(25.0 \text{ km})(0.707) = \boxed{-17.7 \text{ km}}$$

The negative value of A_y indicates that the *y* coordinate decreased in this displacement. The signs of A_x and A_y are also evident from Figure 3.9.

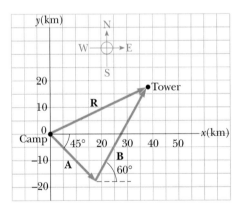

Figure 3.9   (Example 3.3)

The second displacement, **B,** has a magnitude of 40.0 km and is 60.0° north of east. Its components are

$$B_x = B \cos 60.0° = (40.0 \text{ km})(0.500) = \boxed{20.0 \text{ km}}$$

$$B_y = B \sin 60.0° = (40.0 \text{ km})(0.866) = \boxed{34.6 \text{ km}}$$

(b) Determine the components of the hiker's total displacement for the trip.

Solution The resultant displacement for the trip, $\mathbf{R} = \mathbf{A} + \mathbf{B}$, has components given by

$$R_x = A_x + B_x = 17.7 \text{ km} + 20.0 \text{ km} = \boxed{37.7 \text{ km}}$$

$$R_y = A_y + B_y = -17.7 \text{ km} + 34.6 \text{ km} = \boxed{16.9 \text{ km}}$$

Exercise Determine the magnitude and direction of the total displacement.

Answer 41.3 km, 24.1° north of east from the base camp.

<h2>3.4 VELOCITY AND ACCELERATION IN TWO DIMENSIONS</h2>

In our discussion of one-dimensional motion in Chapter 2, the vector nature of displacement, velocity, and acceleration was taken into account through the use of positive and negative signs. To completely describe the motion of an object in two or three dimensions, we must make use of vectors.

Consider an object moving through space as shown in Figure 3.10. When the object is at some point P at time t_i, its position is described by the position vector $\mathbf{r}_i$, drawn from the origin to P. Likewise, when the object has moved to some other point, Q, at time t_f, its position vector is $\mathbf{r}_f$. As you can see from the vector diagram in Figure 3.10, the final position vector is the sum of the initial position vector and $\Delta\mathbf{r}$. Because $\mathbf{r}_f = \mathbf{r}_i + \Delta\mathbf{r}$, the displacement of the object is defined as the change in the position vector:

$$\Delta\mathbf{r} \equiv \mathbf{r}_f - \mathbf{r}_i \qquad [3.5]$$

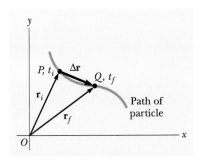

Figure 3.10 An object moving along some curved path between points P and Q. The displacement vector, $\Delta\mathbf{r}$, is the difference in the position vectors. That is, $\Delta\mathbf{r} = \mathbf{r}_f - \mathbf{r}_i$.

The average velocity of a particle during the time interval Δt is the ratio of the displacement to the time interval for this displacement.

$$\overline{\mathbf{v}} \equiv \frac{\Delta\mathbf{r}}{\Delta t} \qquad [3.6]$$

◀ Average velocity

Because the displacement is a vector and the time interval is a scalar, we conclude that the average velocity is a *vector* quantity directed along $\Delta\mathbf{r}$.

The instantaneous velocity, $\mathbf{v}$, is defined as the limit of the average velocity, $\Delta\mathbf{r}/\Delta t$, as Δt goes to zero:

$$\mathbf{v} \equiv \lim_{\Delta t \to 0} \frac{\Delta\mathbf{r}}{\Delta t} \qquad [3.7]$$

◀ Instantaneous velocity

The direction of the instantaneous velocity vector is along a line that is tangent to the path of the particle and in the direction of motion.

The average acceleration of an object whose velocity changes by $\Delta\mathbf{v}$ in the time interval Δt is a vector defined as the ratio $\Delta\mathbf{v}/\Delta t$.

$$\overline{\mathbf{a}} \equiv \frac{\Delta\mathbf{v}}{\Delta t} \qquad [3.8]$$

◀ Average acceleration

Instantaneous acceleration ▷   **The instantaneous acceleration vector, a, is defined as the limit of the average acceleration vector as Δt goes to zero.**

It is important to recognize that a particle can accelerate in several ways. First, the magnitude of the velocity vector (the speed) may change with time. Second, a particle accelerates when the direction of the velocity vector changes with time (makes a curved path) even though the speed is constant. Third, acceleration may be due to changes in both the magnitude and the direction of the velocity vector.

Thinking Physics 3

The gas pedal in an automobile is called the *accelerator*. Are there any other controls that could also be considered to be an accelerator?

Explanation   The gas pedal is called the accelerator because the common usage of the word *acceleration* refers to an *increase in speed*. The scientific definition, however, is that **acceleration occurs whenever the velocity changes in any way.** Thus, the *brake pedal* can also be considered to be an accelerator, because it causes the car to slow down. The *steering wheel* is also an accelerator, because it changes the direction of the velocity vector.

3.5   PROJECTILE MOTION

In the situations we considered in Chapter 2, objects moved along straight-line paths, such as the *x* axis. Now let us look at some cases in which an object moves in a plane. By this we mean that the object may move in both the *x* and *y* directions simultaneously or move in two dimensions. The particular form of two-dimensional

Projectile motion ▷   motion we shall concentrate on is called **projectile motion.** Anyone who has observed a baseball in motion (or, for that matter, any object thrown into the air) has observed projectile motion. It is surprisingly simple to analyze if the following three assumptions are made.

1. The free-fall acceleration, **g,** has a magnitude of 9.80 m/s², is constant over the range of motion, and is directed downward.
2. The effect of air resistance is negligible.
3. The rotation of the Earth does not affect the motion.

With these assumptions, we shall find that the path of a projectile is curved as shown in Figure 3.11. Such a curve is called a *parabola*.

Let us choose our coordinate system so that the *y* direction is vertical and positive upward. In this case, **the acceleration in the *y* direction is $-g$, just as in free fall, and the acceleration in the *x* direction is 0 (because air resistance is neglected).** Furthermore, let us assume that at $t = 0$, the projectile leaves the origin

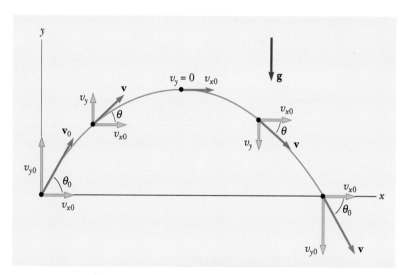

Figure 3.11 The parabolic trajectory of a particle that leaves the origin with a velocity of $\mathbf{v}_0$. Note that the velocity, $\mathbf{v}$, changes with time. However, the x component of the velocity, v_x, remains constant in time. Also, $v_y = 0$ at the peak, but the acceleration is always equal to the free-fall acceleration and acts vertically downward.

with a velocity of $\mathbf{v}_0$, as in Figure 3.11. If the velocity vector makes an angle of θ_0 with the horizontal where θ_0 is called the projection angle, then from the definitions of cosine and sine functions and Figure 3.11, we have

$$v_{x0} = v_0 \cos \theta_0 \qquad \text{and} \qquad v_{y0} = v_0 \sin \theta_0$$

In order to analyze projectile motion, we shall separate the motion into two parts, the x (or horizontal) motion and the y (or vertical) motion, and solve each part separately. We shall look first at the x motion. As noted before, motion in the x direction occurs with $a_x = 0$. This means that **the velocity component along the x direction remains constant.** Thus, if the initial value of the velocity component in the x direction is $v_{x0} = v_0 \cos \theta_0$, this is also the value of the velocity at any later time. That is,

$$v_x = v_{x0} = v_0 \cos \theta_0 = \text{constant} \qquad \text{[3.9]}$$

Equation 3.9 can be substituted into the defining equation for velocity (Eq. 2.2) to give us an expression for the horizontal position of the projectile as a function of time:

$$x = v_{x0}t = (v_0 \cos \theta_0)t \qquad \text{[3.10]}$$

These equations tell us all we need to know about the motion in the x direction. Let us now consider the y motion. Because it has constant acceleration, the equations developed in Section 2.6 can be used. In these equations, v_{y0} shall denote the initial velocity in the y direction and $-g$ the free-fall acceleration. The negative sign

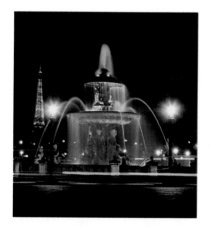

An evening view of an illuminated water fountain in la Place de la Concorde, Paris. The individual water streams follow parabolic trajectories. The horizontal range and maximum height of a given stream of water depend on the elevation angle of its initial velocity as well as its initial speed. (© *The Telegraph Colour Library/ FPG*)

QUICKLAB

For this investigation, you need to be outdoors with a small ball such as a tennis ball and a wristwatch with a second hand. Throw the ball upward as hard as you can and determine the initial speed of your throw and the approximate maximum height of the ball using only your wristwatch.

APPLICATION

Trajectory of a Baseball.

for g indicates that the positive direction for the vertical motion is assumed to be upward. With this choice of signs, we have

$$v_y = v_{y0} - gt \qquad [3.11]$$

$$y = v_{y0}t - \tfrac{1}{2}gt^2 \qquad [3.12]$$

$$v_y^2 = v_{y0}^2 - 2gy \qquad [3.13]$$

where $v_{y0} = v_0 \sin \theta_0$. The speed, v, of the projectile at any instant can be calculated from the components of velocity at that instant, using the Pythagorean theorem:

$$v = \sqrt{v_x^2 + v_y^2}$$

Applying Physics 1

A home run is hit in a baseball game. The ball is hit from home plate into the center field stands, along a long parabolic path. What is the acceleration of the ball (a) while it is rising, (b) at the highest point of the trajectory, and (c) while it is descending after reaching the highest point? Ignore air resistance.

Explanation The answers to all three parts are the same—the acceleration is that due to gravity, 9.8 m/s^2, because the force of gravity is pulling downward on the ball during the entire motion. During the rising part of the trajectory, the downward acceleration results in the decreasing positive values of the vertical component of the velocity of the ball. During the falling part of the trajectory, the downward acceleration results in the increasing negative values of the vertical component of the velocity. Many people have trouble with the topmost point, claiming that the acceleration of the ball at the highest point is zero. This interpretation arises from a confusion between zero vertical velocity and zero acceleration. Even though the ball has momentarily come to rest (vertically; it is still moving horizontally), it is still accelerating, because the force of gravity is still acting. If the ball were to come to rest and experience zero acceleration, then the velocity would not change—the ball would remain suspended at the highest point! We do not see that happening, because the acceleration is indeed not equal to zero.

Before we examine some numerical examples dealing with projectile motion, let us pause to summarize what we have learned so far about this kind of motion:

- Provided air resistance is negligible, the horizontal component of velocity, v_x, remains constant because there is no horizontal component of acceleration.
- The vertical component of acceleration is equal to the free-fall acceleration, g.
- The vertical component of velocity, v_y, and the displacement in the y direction are identical to those of a freely falling body.
- Projectile motion can be described as a superposition of the two motions in the x and y directions.

Problem-Solving Strategy

Projectile Motion

We suggest that you use the following approach to solving projectile motion problems:

1. Select a coordinate system.
2. Resolve the initial velocity vector into x and y components.
3. Treat the horizontal motion and the vertical motion independently.
4. Follow the techniques for solving problems with constant velocity to analyze the horizontal motion of the projectile.
5. Follow the techniques for solving problems with constant acceleration to analyze the vertical motion of the projectile.

EXAMPLE 3.4 The Stranded Explorers

An Alaskan rescue plane drops a package of emergency rations to a stranded party of explorers, as shown in Figure 3.12. The plane is traveling horizontally at 40.0 m/s at a height of 100 m above the ground.
(a) Where does the package strike the ground relative to the point at which it was released?

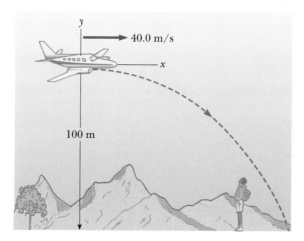

Figure 3.12 (Example 3.4) From the point of view of an observer on the ground, a package released from the rescue plane travels along the path shown.

Solution The coordinate system for this problem is selected as shown in Figure 3.12, with the positive x direction to the right and the positive y direction upward.

Consider first the horizontal motion of the package. The only equation available to us is

$$x = v_{x0}t$$

The initial x component of the package velocity is the same as the velocity of the plane when the package was released, 40.0 m/s. Thus, we have

$$x = (40.0 \text{ m/s})t$$

If we know t, the length of time the package is in the air, we can determine x, the distance traveled by the package along the horizontal. To find t, we move to the equations for the vertical motion of the package. We know that at the instant the package hits the ground, its y coordinate is -100 m. We also know that the initial velocity of the package in the vertical direction, v_{y0}, is zero because the package was released with only a horizontal component of velocity. From Equation 3.12 we have

$$y = -\tfrac{1}{2}gt^2$$

$$-100 \text{ m} = -\tfrac{1}{2}(9.80 \text{ m/s}^2)t^2$$

$$t^2 = 20.4 \text{ s}^2$$

$$t = 4.51 \text{ s}$$

This value for the time of flight substituted into the equation for the x coordinate gives

$$x = (40.0 \text{ m/s})(4.51 \text{ s}) = \boxed{180 \text{ m}}$$

(b) What are the horizontal and vertical components of the velocity of the package just before it hits the ground?

Solution   We already know the horizontal component of the velocity of the package just before it hits, because the velocity in the horizontal direction remains constant at 40.0 m/s throughout the flight.

The vertical component of the velocity just before the package hits the ground may be found by using Equation 3.11, with $v_{y0} = 0$:

$$v_y = v_{y0} - gt = 0 - (9.80 \text{ m/s}^2)(4.51 \text{ s}) = \boxed{-44.1 \text{ m/s}}$$

EXAMPLE 3.5   The Long Jump

APPLICATION

Long Jumping.

A long jumper leaves the ground at an angle of 20.0° to the horizontal and at a speed of 11.0 m/s, as in Figure 3.13.
(a) How far does he jump? (Assume that the motion of the long jumper is equivalent to that of a particle — in other words, disregard the motion of the jumper's arms and legs).

Solution   His horizontal motion is described by using Equation 3.10:

$$x = (v_0 \cos \theta_0)t = (11.0 \text{ m/s})(\cos 20.0°)t$$

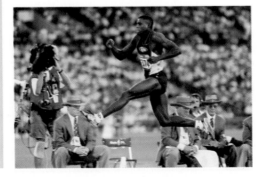

Figure 3.13   (Example 3.5) Carl Lewis sets a record long jump of 8.50 m in the 1996 Olympics held in Atlanta, Georgia. (*J. O. Atlanta 96/ Gamma*)

The value of x can be found if t, the total duration of the jump, is known. We can find t with $v_y = v_0 \sin \theta_0 - gt$ (Eq. 3.11) by noting that at the top of the jump the vertical component of velocity goes to zero:

$$v_y = v_0 \sin \theta_0 - gt$$

$$0 = (11.0 \text{ m/s}) \sin 20.0° - (9.80 \text{ m/s}^2) t_1$$

$$t_1 = 0.384 \text{ s}$$

Note that t_1 is the time interval to the *top* of the jump. Because of the symmetry of the vertical motion, an identical time interval passes before the jumper returns to the ground. Therefore, the *total time* in the air is $t = 2t_1 = 0.768$ s, and the distance jumped is

$$x = (11.0 \text{ m/s}) (\cos 20.0°) (0.768 \text{ s}) = \boxed{7.94 \text{ m}}$$

(b) What is the maximum height reached?

Solution The maximum height reached is found using $y = (v_0 \sin \theta_0) t - \frac{1}{2} g t^2$ (Eq. 3.12) with $t = t_1 = 0.384$ s.

$$y_{max} = (11.0 \text{ m/s}) (\sin 20.0°) (0.384 \text{ s}) - \frac{1}{2} (9.80 \text{ m/s}^2) (0.384 \text{ s})^2 = \boxed{0.722 \text{ m}}$$

Although the assumption that the motion of the long jumper is that of a projectile is an oversimplification of the situation, the values obtained are reasonable.

EXAMPLE 3.6 That's Quite an Arm

A stone is thrown upward from the top of a building at an angle of 30.0° to the horizontal and with an initial speed of 20.0 m/s, as in Figure 3.14. The height of the building is 45.0 m.
(a) How long is the stone "in flight"?

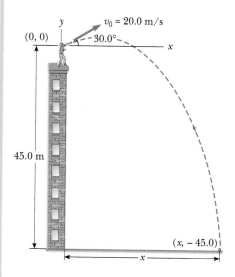

Figure 3.14 (Example 3.6)

Solution The initial x and y components of the velocity are

$$v_{x0} = v_0 \cos \theta_0 = (20.0 \text{ m/s}) (\cos 30.0°) = +17.3 \text{ m/s}$$

$$v_{y0} = v_0 \sin \theta_0 = (20.0 \text{ m/s}) (\sin 30.0°) = +10.0 \text{ m/s}$$

To find t, we can use $y = v_{y0}t - \frac{1}{2}gt^2$ (Eq. 3.12) with $y = -45.0$ m and $v_{y0} = 10.0$ m/s. (We have chosen the point of release as the origin, as shown in Fig. 3.14.)

$$-45.0 \text{ m} = (10.0 \text{ m/s})t - \tfrac{1}{2}(9.80 \text{ m/s}^2)t^2$$

Solving the quadratic equation for t (see Appendix A) gives, for the positive root, the value $t = 4.22$ s. Does the negative root have any physical meaning? (Why not think of another way of finding t from the information given?)

(b) What is the speed of the stone just before it strikes the ground?

Solution The y component of the velocity just before the stone strikes the ground can be obtained using Equation 3.11 with $t = 4.22$ s:

$$v_y = v_{y0} - gt = 10.0 \text{ m/s} - (9.80 \text{ m/s}^2)(4.22 \text{ s}) = -31.4 \text{ m/s}$$

Because $v_x = v_{x0} = 17.3$ m/s, the required speed is given by

$$v = \sqrt{v_x^2 + v_y^2} = \sqrt{(17.3)^2 + (-31.4)^2} \text{ m/s} = \quad 35.9 \text{ m/s}$$

Exercise Where does the stone strike the ground?

Answer 73.0 m from the base of the building.

3.6   RELATIVE VELOCITY

Observers in different frames of reference may measure different displacements or velocities for an object in motion. That is, two observers moving with respect to each other would generally not agree on the outcome of a measurement.

For example, if two cars were moving in the same direction with speeds of 50 mi/h and 60 mi/h, a passenger in the slower car would measure the speed of the faster car relative to the slower car as 10 mi/h. Of course, a stationary observer would measure the speed of the faster car as 60 mi/h. This simple example demonstrates that velocity measurements differ in different frames of reference.

We solved the previous problem with a minimum of thought and effort, but you will encounter many situations in which a more systematic method for attacking such problems is beneficial. To develop this method, let us write down all the information we are given and that which we want to know in the form of velocities with subscripts appended. We have

$\mathbf{v}_{se} = +50$ mi/h   (The subscript "se" means the velocity of the *slower* car with respect to the *Earth*.)

$\mathbf{v}_{fe} = +60$ mi/h   (The subscript "fe" means the velocity of the *faster* car with respect to the *Earth*.)

We want to know $\mathbf{v}_{fs}$, which is the velocity of the *faster* car with respect to the *slower* car. To find this, we write an equation for $\mathbf{v}_{fs}$ in terms of the other velocities. For this case, we have

$$\mathbf{v}_{\boxed{fs}} = \mathbf{v}_{\boxed{f}e} + \mathbf{v}_{e\boxed{s}} \qquad\qquad [3.14]$$

Note the pattern of the subscripts in this expression. The first subscript f on the left side of the equation is also the first subscript on the right side of the equation.

PHYSICS IN ACTION

Parabolic Paths

On the left is a multiflash photograph of a popular lecture demonstration in which a projectile is fired at a target that is being held by a magnet in the device at the top right of the photograph. The conditions of the experiment are that the gun is aimed at the target and the projectile leaves the gun at the instant the target is released from rest. Under these conditions the projectile will hit the target, independent of the initial speed of the projectile. The reason is that they both experience the same downward acceleration and hence the velocities of the projectile and target change by the same amount in the same time interval. Note that the velocity of the projectile (red arrows) changes in direction and magnitude, and its downward acceleration (violet arrows) remains constant.

On the right, a welder cuts holes through a heavy metal construction beam with a hot torch. The sparks generated in the process follow parabolic paths.

(Central Scientific Company)

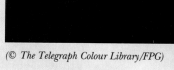

(© The Telegraph Colour Library/FPG)

The second subscript s on the left side of the equation is also the last subscript on the right side of the equation. Boxes are used in the subscripts to emphasize this pattern. The boldface notation is, of course, indicative of the fact that velocity is a vector quantity. As we shall see in the following examples, this vector nature of the velocity is of paramount importance in certain instances.

We know that $\mathbf{v}_{es} = -\mathbf{v}_{se}$, so

$$\mathbf{v}_{fs} = +60 \text{ mi/h} - 50 \text{ mi/h} = +10 \text{ mi/h}$$

There is no general equation to memorize in order to work relative velocity problems; instead, you should develop the necessary equations on your own by following the technique already demonstrated for writing subscripts. We suggest that you practice this technique to work the examples that follow.

EXAMPLE 3.7　Where Does the Ball Land?

A passenger at the rear of a train, traveling at 15 m/s relative to the Earth, throws a baseball with a speed of 15 m/s in the direction opposite the motion of the train. What is the velocity of the baseball relative to the Earth?

Solution　We first write down our knowns and unknowns with appropriate subscripts:

$$\mathbf{v}_{te} = +15 \text{ m/s} \quad \text{(velocity of the } train \text{ relative to the } Earth\text{)}$$

$$\mathbf{v}_{bt} = -15 \text{ m/s} \quad \text{(velocity of the } baseball \text{ relative to the } train\text{)}$$

We need $\mathbf{v}_{be}$, the velocity of the *baseball* relative to the *Earth*. Following the strategy for watching the subscripts while we write down the equation for our unknown, we have

$$\mathbf{v}_{\boxed{be}} = \mathbf{v}_{\boxed{b}t} + \mathbf{v}_{t\boxed{e}}$$

$$\mathbf{v}_{\boxed{be}} = -15 \text{ m/s} + 15 \text{ m/s} = \boxed{0}$$

EXAMPLE 3.8　Crossing a River

A boat heading due north crosses a wide river with a velocity of 10.0 km/h relative to the water. The river has a uniform velocity of 5.00 km/h due east. Determine the velocity of the boat with respect to an observer on the riverbank.

Solution　We have

$$\mathbf{v}_{br} = 10.0 \text{ km/h due north} \quad \text{(velocity of the } boat \text{ with respect to the } river\text{)}$$

$$\mathbf{v}_{re} = 5.00 \text{ km/h due east} \quad \text{(velocity of the } river \text{ with respect to the } Earth\text{)}$$

and we want, $\mathbf{v}_{be}$, the velocity of the *boat* with respect to the *Earth*. Our equation becomes

$$\mathbf{v}_{\boxed{be}} = \mathbf{v}_{\boxed{b}r} + \mathbf{v}_{r\boxed{e}}$$

Because the velocities are not along the same direction as in preceding examples, the terms in the equation must be manipulated as vector quantities, which are shown in Figure 3.15. The quantity $\mathbf{v}_{br}$ is due north, $\mathbf{v}_{re}$ is due east, and the vector sum of the two, $\mathbf{v}_{be}$, is at the angle θ, as defined in Figure 3.15. Thus, we see that the velocity of the boat with respect to the Earth can be found from the Pythagorean theorem as

$$v_{be} = \sqrt{(v_{br})^2 + (v_{re})^2} = \sqrt{(10.0 \text{ km/h})^2 + (5.00 \text{ km/h})^2} = \boxed{11.2 \text{ km/h}}$$

and the direction of $\mathbf{v}_{be}$ is

$$\theta = \tan^{-1}\left(\frac{v_{re}}{v_{br}}\right) = \tan^{-1}\frac{5.00}{10.0} = 26.6°$$

Therefore, the boat travels at a speed of 11.2 km/h in the direction 63.4° north of east with respect to the Earth.

Figure 3.15　(Example 3.8)

EXAMPLE 3.9 Which Way Should the Boat Head?

If the boat of the preceding example travels with the same speed of 10.0 km/h relative to the water and is to travel due north, as in Figure 3.16, in what direction should it travel?

Solution We know

$$\mathbf{v}_{br} = \text{velocity of the } boat \text{ with respect to the } river$$

$$\mathbf{v}_{re} = \text{velocity of the } river \text{ with respect to the } Earth$$

and we need $\mathbf{v}_{be}$, the velocity of the *boat* with respect to the *Earth*. We have

$$\mathbf{v}_{\boxed{be}} = \mathbf{v}_{\boxed{b}r} + \mathbf{v}_{r\boxed{e}}$$

The relationship among these three vectors is shown in Figure 3.16; it agrees with our intuitive guess that the boat must head upstream in order to be pushed directly northward across the water. The speed v_{be} can be found from the Pythagorean theorem:

$$v_{be} = \sqrt{(v_{br})^2 - (v_{re})^2} = \sqrt{(10.0 \text{ km/h})^2 - (5.00 \text{ km/h})^2} = 8.66 \text{ km/h}$$

and the direction of v_{be} is

$$\theta = \tan^{-1} \frac{v_{re}}{v_{be}} = \tan^{-1}\left(\frac{5.00}{8.66}\right) = 30°$$

where θ is west of north.

Figure 3.16 (Example 3.9)

SUMMARY

Two vectors, **A** and **B,** can be added geometrically by either the triangle method or the parallelogram rule. In the **triangle method,** the two vectors are drawn to scale, on graph paper, so that the tail of the second vector starts at the tip of the first. The resultant vector is the vector drawn from the tail of the first to the tip of the second. In the **parallelogram method,** the vectors are drawn to scale on graph paper with the tails of the two vectors joined. The resultant vector is the diagonal of the parallelogram formed with the two vectors as its sides.

The negative of a vector **A** is a vector with the same magnitude as **A** but pointing in the opposite direction.

The x component of a vector is equivalent to its projection along the x axis of a coordinate system. Likewise, the y component is the projection of the vector along the y axis of this coordinate system. The resultant of two or more vectors can be found mathematically by resolving all vectors into their x and y components, finding the resultant x and y components, and then using the Pythagorean theorem to find the resultant vector. The angle of the resultant vector with respect to the x axis can be found by use of a suitable trigonometric function.

An object moving above the surface of the Earth in both the x and y directions simultaneously is said to be undergoing **projectile motion.** The object moves along the horizontal (x) direction so that its velocity in this direction, v_x, is a constant, and the object also moves in the vertical (y) direction with a constant downward free-fall acceleration of magnitude $g = 9.80$ m/s^2. The equations describing the motion of a projectile are

$$x = v_{x0}t \qquad [3.10]$$

$$v_y = v_{y0} - gt \qquad [3.11]$$

$$y = v_{y0}t - \tfrac{1}{2}gt^2 \qquad [3.12]$$

$$v_y^2 = v_{y0}^2 - 2gy \qquad [3.13]$$

where $v_{y0} = v_0 \sin \theta_0$ is the initial vertical component of the velocity and $v_{x0} = v_0 \cos \theta_0$ is the initial horizontal component of the velocity.

Observations made in different frames of reference can be related to one another through the techniques of the transformation of relative velocities. There is no general equation for you to memorize in order to work relative velocity problems. Instead, you should follow the technique for writing subscripts on each velocity under consideration as described in the text.

MULTIPLE-CHOICE QUESTIONS

1. A 6-km displacement and an 8-km displacement are at right angles. The resultant displacement is found to be 10 km due east. Which of the following is a possible set of directions for the 6 km and 8 km displacements?
   (a) The 6 km is north of east and the 8 km is south of east.
   (b) The 6 km is north of west and the 8 km is south of west.
   (c) The 6 km is north of east and the 8 km is north of west.
   (d) The 6 km is south of west and the 8 km is south of east.

2. Ball A weighs twice as much as ball B. A is dropped straight down from the roof of a building and at the same time B is thrown off the building horizontally at a high speed. Neglect air resistance and determine which statement below is true.
   (a) A hits the ground before B.
   (b) B hits the ground before A.
   (c) They both hit at the same time.
   (d) There is insufficient information in this problem to enable one to determine an answer.

3. A tennis ball is given an initial horizontal velocity on the surface of a table. Which of the diagrams that follow best describes the motion of the ball as it rolls off the end of the table?

(a)   (b)

(c)   (d)

4. The record for a ski jump is 180 m set in 1989. Assume the jumper comes off the end of the ski jump horizontally and falls 90 m vertically before contacting the ground. What was the initial horizontal speed of the jumper?
   (a) 18.4 m/s   (b) 4.28 m/s   (c) 42 m/s
   (d) 84 m/s   (e) 360 m/s

5. It is said that in his youth George Washington threw a silver dollar across the Potomac River. Assuming the river was 75 m wide and the coin was thrown at a 45° angle, what minimum initial speed was required to get the coin across the river?
(a) 7 m/s   (b) 14 m/s   (c) 21 m/s   (d) 27 m/s

CONCEPTUAL QUESTIONS

1. (a) Can an object accelerate if its speed is constant?
(b) Can an object accelerate if its velocity is constant?

2. If **B** is added to **A,** under what conditions does the resultant vector have a magnitude equal to $A + B$? Under what conditions is the resultant vector equal to zero?

3. (a) Can an object have a constant velocity and varying speed? (b) Can an object have a constant speed and varying velocity? Give examples.

4. Two vectors have unequal magnitudes. Can their sum be zero? Explain.

5. Suppose that you are running at constant speed. You wish to throw a ball such that you will catch it as it comes back down. How should you throw the ball?

6. Can a vector have a component greater than its magnitude?

7. Vector **A** lies in the xy plane. For what orientations will both of its rectangular components be negative? For what orientations will its components have opposite signs?

8. Under what circumstances would a vector have components that are equal in magnitude?

9. As a projectile moves in its path, is there any point along the path where the velocity and acceleration vectors are (a) perpendicular to each other? (b) parallel to each other?

10. A rock is dropped at the same instant that a ball, at the same initial elevation, is thrown horizontally. Which will have the greater speed when it reaches ground level?

11. Explain whether or not the following particles have an acceleration: (a) a particle moving in a straight line with constant speed and (b) a particle moving around a curve with constant speed.

12. Correct the following statement: "The racing car rounds the turn at a constant velocity of 90 miles per hour."

13. A spacecraft drifts through space at a constant velocity. Suddenly a gas leak in the side of the spacecraft causes a constant acceleration of the spacecraft in a direction perpendicular to the initial velocity. The orientation of the spacecraft does not change, so that the acceleration remains perpendicular to the original direction of the velocity. What is the shape of the path followed by the spacecraft in this situation?

14. A ball is projected horizontally from the top of a building. One second later another ball is projected horizontally from the same point with the same velocity. At what point in the motion will the balls be closest to each other? Will the first ball always be traveling faster than the second ball? What will be the time difference between when the balls hit the ground? Can the horizontal projection velocity of the second ball be changed so that the balls arrive at the ground at the same time?

15. Two projectiles are thrown with the same magnitude of initial velocity, one at an angle θ with respect to the level ground and the other at angle $90° - \theta$. Both projectiles will strike the ground at the same distance from the projection point. Will both projectiles be in the air for the same time interval?

16. A baseball is thrown such that its initial x and y components of velocity are known. Neglecting air resistance, describe how you would calculate, at the instant the ball reaches the top of its trajectory, (a) its coordinates, (b) its velocity, and (c) its acceleration. How would these results change if air resistance were taken into account?

17. A projectile is fired at some angle to the horizontal with some initial speed v_0, and air resistance is neglected. Is the projectile a freely falling body? What is its acceleration in the vertical direction? What is its acceleration in the horizontal direction?

18. The yellow ball in Figure Q3.18 is projected horizontally with some initial velocity. At the instant the yellow ball is projected, the red ball is released from the same height

Figure Q3.18   (© *Richard Megna 1990, Fundamental Photographs*)

and falls to the floor. Explain why the two balls hit the floor simultaneously, even though the yellow ball has an initial velocity.

19. Describe how a driver can steer a car traveling at constant speed so that (a) the acceleration is zero or (b) the magnitude of the acceleration remains constant.

20. A ball is thrown upward in the air by a passenger on a train that is moving with constant velocity. (a) Describe the path of the ball as seen by the passenger. Describe the path as seen by a stationary observer outside the train. (b) How would these observations change if the train were accelerating along the track?

PROBLEMS

1, 2, 3 = straightforward, intermediate, challenging   □ = full solution available in Study Guide/Student Solutions Manual
WEB = solution posted at **http://www.saunderscollege.com/physics/college**   🖾 = biomedical application   🖳 = Interactive Physics

Review Problem

When the triangular plot of land shown in the figure is surveyed, it is found that the side from *A* to *B* has a length of 311.4 ft and runs due east from point *A*. The side from *B* to *C* has a length of 260.8 ft and is in a direction 40.0° west of north. (a) When the side from *C* to *A* is surveyed, what is the expected length and direction for this side? (*Hint:* First draw a line from point *C* perpendicular to *AB*, giving you two right triangles to consider.) (b) Find the area of the plot of land in acres (1 acre = 43 560 ft²).

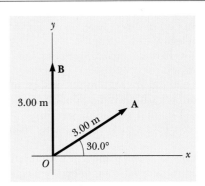

Figure P3.3

5. A man lost in a maze makes three consecutive displacements so that at the end of the walk he is right back where he started. The first displacement is 8.00 m westward, and the second is 13.0 m northward. Find the magnitude and direction of the third displacement, using the graphical method.

6. A jogger runs 100 m due west, then changes direction for the second leg of the run. At the end of the run, she is 175 m away from the starting point at an angle of 15.0° north of west. What were the direction and length of her second displacement? Use graphical techniques.

Section 3.2   Some Properties of Vectors

1. A dog searching for a bone walks 3.50 m south, then 8.20 m at an angle 30.0° north of east, and finally 15.0 m west. Find the dog's resultant displacement vector using graphical techniques.

2. A roller coaster moves 200 ft horizontally, then rises 135 ft at an angle of 30.0° above the horizontal. It then travels 135 ft at an angle of 40.0° below the horizontal. Use graphical techniques to find its displacement from the starting point at the end of this movement.

3. Each of the displacement vectors **A** and **B** shown in Figure P3.3 has a magnitude of 3.0 m. Graphically find (a) **A** + **B**, (b) **A** − **B**, (c) **B** − **A**, and (d) **A** − 2**B**.

4. Vector **A** is 3.00 units in length and points along the positive *x* axis. Vector **B** is 4.00 units in length and points along the negative *y* axis. Use graphical methods to find the magnitude and direction of the vectors (a) **A** + **B** and (b) **A** − **B**.

Section 3.3   Components of a Vector

7. A golfer takes two putts to get his ball into the hole once he is on the green. The first putt displaces the ball 6.00 m east, and the second, 5.40 m south. What displacement would have been needed to get the ball into the hole on the first putt?

8. A person walks 25.0° north of east for 3.10 km. How far would a person walk due north and due east to arrive at the same location?

WEB 9. A girl delivering newspapers covers her route by traveling 3.00 blocks west, 4.00 blocks north, then 6.00 blocks east. (a) What is her resultant displacement? (b) What is the total distance she travels?

10. A quarterback takes the ball from the line of scrimmage, runs backward for 10.0 yards, then runs sideways parallel to the line of scrimmage for 15.0 yards. At this point, he throws a 50.0-yard forward pass straight downfield, perpendicular to the line of scrimmage. What is the magnitude of the football's resultant displacement?

11. The eye of a hurricane passes over Grand Bahama Island. It is moving in a direction 60.0° north of west with a speed of 41.0 km/h. Three hours later, the course of the hurricane suddenly shifts due north, and its speed slows to 25.0 km/h. How far from Grand Bahama is the hurricane 4.50 h after it passes over the island?

12. A small map shows Atlanta to be 730 miles in a direction of 5° north of east from Dallas. The same map shows that Chicago is 560 miles in a direction of 21° west of north from Atlanta. Assume a flat Earth and use this information to find the displacement from Dallas to Chicago.

13. A commuter airplane starts from an airport and takes the route shown in Figure P3.13. It first flies to city A located at 175 km in a direction 30.0° north of east. Next, it flies 150 km 20.0° west of north to city B. Finally, it flies 190 km due west to city C. Find the location of city C relative to the location of the starting point.

Figure P3.13

14. Two people pull on a stubborn mule, as seen from a helicopter in Figure P3.14. Find (a) the single force that is equivalent to the two forces shown, and (b) the force that a third person would have to exert on the mule to make the net force equal to zero.

15. A man pushing a mop across a floor causes it to undergo two displacements. The first has a magnitude of 150 cm and makes an angle of 120° with the positive x axis. The resultant displacement has a magnitude of 140 cm and is directed at an angle of 35.0° to the positive x axis. Find the magnitude and direction of the second displacement.

Section 3.4 Velocity and Acceleration in Two Dimensions

Section 3.5 Projectile Motion

16. The fastest recorded pitch in major-league baseball, thrown by Nolan Ryan in 1974, was clocked at

Figure P3.14

100.8 mi/h. If a pitch were thrown horizontally with this velocity, how far would the ball fall vertically by the time it reached home plate 60.0 ft away?

WEB 17. Tom the cat is chasing Jerry the mouse across a table surface 1.5 m above the floor. Jerry steps out of the way at the last second, and Tom slides off the edge of the table at a speed of 5.0 m/s. Where will Tom strike the floor, and what velocity components will he have just before he hits?

18. A student stands at the edge of a cliff and throws a stone horizontally over the edge with a speed of 18.0 m/s. The cliff is 50.0 m above a flat horizontal beach, as shown in Figure P3.18. How long after being released does the stone strike the beach below the cliff? With what speed and angle of impact does it land?

19. A brick is thrown upward from the top of a building at an angle of 25° to the horizontal and with an initial speed of 15 m/s. If the brick is in flight for 3.0 s, how tall is the building?

20. A place kicker must kick a football from a point 36.0 m (about 40.0 yd) from the goal, and the ball must clear the crossbar, which is 3.05 m high. When kicked, the ball leaves the ground with a speed of 20.0 m/s at an angle of 53.0° to the horizontal. (a) By how much does the ball clear or fall short of clearing the crossbar? (b) Does the ball approach the crossbar while still rising or while falling?

21. A car is parked on a cliff overlooking the ocean on an incline that makes an angle of 24.0° below the horizontal. The negligent driver leaves the car in neutral, and

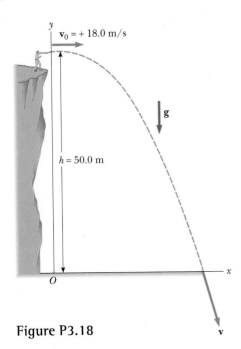

$v_0 = +18.0$ m/s

g

$h = 50.0$ m

O

x

v

Figure P3.18

the emergency brakes are defective. The car rolls from rest down the incline with a constant acceleration of 4.00 m/s² for a distance of 50.0 m to the edge of the cliff. The cliff is 30.0 m above the ocean. Find (a) the car's position relative to the base of the cliff when the car lands in the ocean, and (b) the length of time the car is in the air.

22. A firefighter, 50.0 m away from a burning building, directs a stream of water from a ground level fire hose at an angle of 30.0° above the horizontal. If the speed of the stream as it leaves the hose is 40.0 m/s, at what height will the stream of water strike the building?

23. A projectile is launched with an initial speed of 60 m/s at an angle of 30° above the horizontal. The projectile lands on a hillside 4.0 s later. Neglect air friction. (a) What is the projectile's velocity at the highest point of its trajectory? (b) What is the straight-line distance from where the projectile was launched to where it hits?

Section 3.6   Relative Velocity

24. A jet airliner moving initially at 300 mph due east enters a region where the wind is blowing at 100 mph in a direction 30.0° north of east. What is the new velocity of the aircraft relative to the ground?

25. A boat moves through the water of a river at 10 m/s relative to the water, regardless of the boat's direction. If the water in the river is flowing at 1.5 m/s, how long does it take the boat to make a round trip consisting of a 300-m displacement downstream followed by a 300-m displacement upstream?

26. A river flows due east at 1.50 m/s. A boat crosses the river from the south shore to the north shore by maintaining a constant velocity of 10.0 m/s due north relative to the water. (a) What is the velocity of the boat relative to shore? (b) If the river is 300 m wide, how far downstream has the boat moved by the time it reaches the north shore?

WEB 27. A rowboat crosses a river with a velocity of 3.30 mi/h at an angle 62.5° north of west relative to the water. The river is 0.505 mi wide and carries an eastward current of 1.25 mi/h. How far upstream is the boat when it reaches the opposite shore?

28. The pilot of an aircraft wishes to fly due west in a 50.0-km/h wind blowing toward the south. If the speed of the aircraft relative to the air is 200 km/h, (a) in what direction should the aircraft head, and (b) what will be its speed relative to the ground?

29. How long does it take an automobile traveling in the left lane at 60 km/h to overtake (become even with) another car that is traveling in the right lane at 40 km/h, when the cars' front bumpers are initially 100 m apart?

30. A science student is riding on a flatcar of a train traveling along a straight horizontal track at a constant speed of 10.0 m/s. The student throws a ball along a path that she judges to make an initial angle of 60.0° with the horizontal and to be in line with the track. The student's professor, who is standing on the ground nearby, observes the ball to rise vertically. How high does the ball rise?

ADDITIONAL PROBLEMS

31. A person walks half way around the circumference of a circular path of radius 5.00 m. (a) Find the magnitude of the displacement vector. (b) How far did the person walk? (c) What is the magnitude of the displacement if the circle is completed?

32. A particle undergoes two displacements. The first has a magnitude of 150 cm and makes an angle of 120.0° with the positive x axis. The *resultant* of the two displacements is 140 cm directed at an angle of 35.0° to the positive x axis. Find the magnitude and direction of the second displacement.

33. Figure P3.33 is a multiflash photograph of two golf balls released simultaneously. The time interval between flashes is 0.033 s, and the white parallel lines were placed 15 cm apart. (a) Find the speed at which the right ball was projected, and (b) show that both balls should be expected to reach the floor simultaneously.

34. A ball is projected horizontally from the edge of a table that is 1.00 m high, and it strikes the floor at a point 1.20 m from the base of the table. (a) What is the initial speed of the ball? (b) How high is the ball above the floor when its velocity vector makes a 45.0° angle with the horizontal?

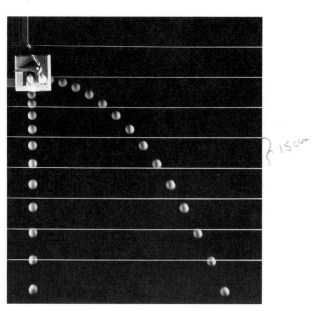

Figure P3.33   *(Educational Development Center)*

35. A car travels due east with a speed of 50.0 km/h. Rain is falling vertically with respect to the Earth. The traces of the rain on the side windows of the car make an angle of 60.0° with the vertical. Find the velocity of the rain with respect to (a) the car and (b) the Earth.

36. Towns A and B in Figure P3.36 are 80.0 km apart. A couple arranges to drive from town A and meet a couple driving from town B at the lake, L. The two couples leave simultaneously and drive for 2.50 h in the direc-

tions shown. Car 1 has a speed of 90.0 km/h. If the cars arrive simultaneously at the lake, what is the speed of car 2?

37. A rocket is launched at an angle of 53.0° above the horizontal with an initial speed of 100 m/s. It moves for 3.00 s along its initial line of motion with an acceleration of 30.0 m/s². At this time its engines fail and the rocket proceeds to move as a free body. Find (a) the maximum altitude reached by the rocket, (b) its total time of flight, and (c) its horizontal range.

38. Two paddlers in identical canoes exert the same effort paddling and hence maintain the same speed relative to the water. One paddles directly upstream (and moves upstream), whereas the other paddles directly downstream. If downstream is the positive direction, an observer on shore determines the velocities of the two canoes to be −1.2 m/s and +2.9 m/s, respectively. (a) What is the speed of the water relative to shore? (b) What is the speed of each canoe relative to the water?

39. If a person can jump a maximum horizontal distance (by using a 45° projection angle) of 3.0 m on the Earth, what would be his maximum range on the Moon, where the free-fall acceleration is g/6 and g = 9.80 m/s²? Repeat for Mars, where the acceleration due to gravity is 0.38g.

WEB 40. A daredevil decides to jump a canyon of width 10 m. To do so, he drives a motorcycle up an incline sloped at an angle of 15°. What minimum speed must she have in order to clear the canyon?

41. The determined coyote is out once more to try to capture the elusive roadrunner. The coyote wears a new pair of Acme power roller skates, which provide a constant horizontal acceleration of 15 m/s², as shown in Figure P3.41. The coyote starts off at rest 70 m from the edge of a cliff at the instant the roadrunner zips by in the direction of the cliff. (a) If the roadrunner moves with constant speed, determine the minimum speed he

Figure P3.36

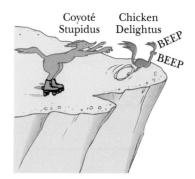

Coyoté   Chicken
Stupidus   Delightus

Figure P3.41

must have in order to reach the cliff before the coyote.
(b) If the cliff is 100 m above the base of a canyon, determine where the coyote lands in the canyon. (Assume
that his skates are still in operation when he is in
"flight" and that his horizontal component of acceleration remains constant at 15 m/s².)

42. An artillery shell is fired with an initial speed of
1.70×10^3 m/s (approximately five times the speed of
sound) at an angle of 55.0° above the horizontal and returns to its original level before impact. Neglecting air
resistance, find (a) the time it is in motion and (b) the
horizontal distance traveled.

43. A home run is hit in such a way that the baseball just
clears a wall 21 m high, located 130 m from home plate.
The ball is hit at an angle of 35° to the horizontal, and
air resistance is negligible. Find (a) the initial speed of
the ball, (b) the time it takes the ball to reach the wall,
and (c) the velocity components and the speed of the
ball when it reaches the wall. (Assume the ball is hit at a
height of 1.0 m above the ground.)

44. A ball is thrown straight upward and returns to the
thrower's hand after 3.00 s in the air. A second ball is
thrown at an angle of 30.0° with the horizontal. At what
speed must the second ball be thrown so that it reaches
the same height as the one thrown vertically?

45. A quarterback throws a football toward a receiver with
an initial speed of 20 m/s, at an angle of 30° above the
horizontal. At that instant, the receiver is 20 m from the
quarterback. In what direction and with what constant
speed should the receiver run in order to catch the football at the level at which it was thrown?

46. A 2.00-m-tall basketball player wants to make a goal
from 10.0 m from the basket, as in Figure P3.46. If he
shoots the ball at a 45.0° angle, at what initial speed
must he throw the basketball so that it goes through the
hoop without striking the backboard?

47. In a popular lecture demonstration, a projectile is fired
at a falling target as in Figure P3.47. The projectile
leaves the gun at the same instant that the target is
dropped from rest. Assuming that the gun is initially
aimed at the target, show that the projectile will hit the
target. (One restriction of this experiment is that the
projectile must reach the target before the target strikes
the floor.)

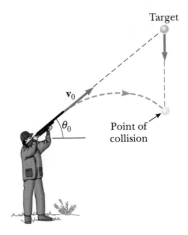

Figure P3.47

48. Figure P3.48 illustrates the difference in proportions between the male and female anatomies. The displace-
ments $\mathbf{d}_{1m}$ and $\mathbf{d}_{1f}$ from the bottom of the feet to the
navel have magnitudes of 104 cm and 84.0 cm, respec-
tively. The displacements $\mathbf{d}_{2m}$ and $\mathbf{d}_{2f}$ have magnitudes
of 50.0 cm and 43.0 cm, respectively. (a) Find the vector
sum of the displacements $\mathbf{d}_1$ and $\mathbf{d}_2$ in each case.
(b) The male figure is 180 cm tall, the female 168 cm.
Normalize the displacements of each figure to a common height of 200 cm, and re-form the vector sums as
in part (a). Then find the vector difference between the
two sums.

Figure P3.46

Figure P3.48

49. A girl can throw a ball a maximum horizontal distance of R on a level field. How far can she throw the same ball vertically upward? Assume that her muscles give the ball the same speed in each case. (Is this assumption valid?)

50. A projectile is fired with an initial speed of v_0 at an angle of θ_0 to the horizontal, as in Figure 3.11. When it reaches its peak, it has (x, y) coordinates given by $(R/2, h)$, and when it strikes the ground, its coordinates are $(R, 0)$, where R is called the horizontal range. **(a)** Show that it reaches a maximum height, h, given by

$$h = \frac{v_0^2 \sin^2 \theta_0}{2g}$$

(b) Show that its horizontal range is given by

$$R = \frac{v_0^2 \sin 2\theta_0}{g}$$

51. A hunter wishes to cross a river that is 1.5 km wide and flows with a speed of 5.0 km/h parallel to its banks. The hunter uses a small powerboat that moves at a maximum speed of 12 km/h with respect to the water. What is the minimum time necessary for crossing?

52. A skater (water spider) maintains an average position on the surface of a stream by darting upstream against the current, then drifting downstream with the current to its original position. The current in the stream is 0.500 m/s relative to the shore, and the skater darts upstream 0.560 m (relative to a spot on shore) in 0.800 s during the first part of its motion. Take upstream as the positive direction. **(a)** Determine the velocity of the skater relative to the water (i) during its dash upstream and (ii) during its drift downstream. **(b)** How far upstream relative to the water does the skater move during one cycle of this motion? **(c)** What is the average velocity of the skater relative to the water?

53. A daredevil is shot out of a cannon at 45.0° to the horizontal with an initial speed of 25.0 m/s. A net is positioned a horizontal distance of 50.0 m from the cannon. At what height above the cannon should the net be placed in order to catch the daredevil?

54. A dart gun is fired while being held horizontally at a height of 1.00 m above ground level and at rest relative to the ground. The dart from the gun travels a horizontal distance of 5.00 m. A college student holds the same gun in a horizontal position while sliding down a 45.0° incline at a constant speed of 2.00 m/s. How far will the dart travel if the student fires the gun when it is 1.00 m above the ground?

55. A student decides to measure the muzzle velocity of a pellet from her gun. She points the gun horizontally. She places a target on a vertical wall a distance x away from the gun. The pellet hits the target a vertical dis-

tance y below the gun. **(a)** Show that the position of the pellet when traveling through the air is given by $y = Ax^2$, where A is a constant. **(b)** Express the constant A in terms of the initial velocity and the free-fall acceleration. **(c)** If $x = 3.00$ m and $y = 0.210$ m, what is the initial speed of the pellet?

56. Indiana Jones is trapped in a maze. To find his way out, he walks 10.0 m, makes a 90.0° right turn, walks 5.00 m, makes another 90.0° right turn, and walks 7.00 m. What is his displacement from his initial position?

57. Instructions for finding a buried treasure include the following: Go 75 paces at 240°, turn to 135° and walk 125 paces, then travel 100 paces at 160°. Determine the resultant displacement from the starting point.

58. A jet airliner moving initially at 300 mph due east enters a region where the wind is blowing at 100 mph in a direction 30.0° north of east. What are the new velocity and direction of the aircraft?

59. One strategy in a snowball fight is to throw a snowball at a high angle over level ground. While your opponent is watching the first one, you throw a second snowball at a low angle timed to arrive before or at the same time as the first one. Assume both snowballs are thrown with a speed of 25.0 m/s. The first one is thrown at an angle of 70.0° with respect to the horizontal. **(a)** At what angle should the second snowball be thrown to arrive at the same point as the first? **(b)** How many seconds later should the second snowball be thrown after the first to arrive at the same time?

60. When baseball outfielders throw the ball, they usually allow it to take one bounce on the theory that the ball arrives sooner this way. Suppose that after the bounce the ball rebounds at the same angle θ as it had when released (Fig. P3.60) but loses half its speed. **(a)** Assuming the ball is always thrown with the same initial speed, at what angle θ should the ball be thrown in order to go the same distance D with one bounce (blue path) as one thrown upward at 45.0° with no bounce (green path)? **(b)** Determine the ratio of the times for the one-bounce and no-bounce throws.

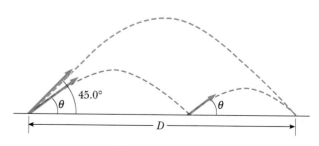

Figure P3.60

61. A projectile is fired up an incline (having an angle of ϕ) with an initial velocity of v_0 at an angle of θ_0 with respect to the horizontal ($\theta_0 > \phi$), as shown in Figure P3.61. Show that the projectile will travel a distance of d up the incline, where

$$d = \frac{2v_0^2 \cos(\theta_0)\ \sin(\theta_0 - \phi)}{g \cos^2(\phi)}$$

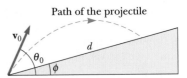

Figure P3.61

The Laws of Motion

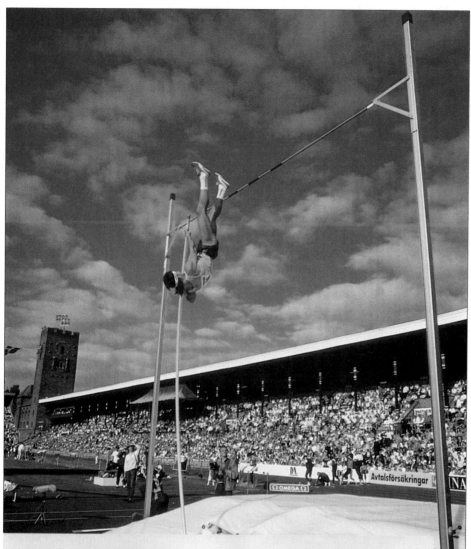

▲ PHYSICS PUZZLER

A pole vaulter falling from a height of about 6 m is protected from serious injury by landing on a thick soft pad. Why does the padding reduce the average force exerted on the vaulter? What information would you need to estimate the acceleration of the vaulter when he is in contact with the pad? *(Gerard Vandystadt/All Sport)*

Classical mechanics describes the relationship between the motion of an object and the forces acting on it. There are conditions under which classical mechanics does not apply or applies in only a limited way. Most often these conditions are encountered when dealing with objects whose size is comparable to that of atoms or smaller (about 10^{-10} m) and/or which move at speeds near the speed of light (3×10^8 m/s). Our study of relativity and quantum mechanics in later chapters will enable us to handle such situations. However, we can make accurate calculations and investigations in physics with the laws of classical mechanics discussed in this chapter.

We learn in this chapter that an object remains in motion with constant velocity if no net external force acts on it. We will also see that if a net external force is exerted on an object, the object accelerates in response to this net force. The acceleration can be determined if the net force acting on the object is known and if the object's mass is known.

4.1 THE CONCEPT OF FORCE

When we think of a force, we usually imagine a push or a pull exerted on some object. For instance, you exert a force on a ball when you throw it or kick it, and you exert a force on a chair when you sit down on it. What happens to an object when it is acted on by a force depends on the magnitude and the direction of the force. Force is a vector quantity; thus, we denote it with a directed arrow, just as we do velocity and acceleration.

If you pull on a spring, as in Figure 4.1a, the spring stretches. If a child pulls hard enough on a wagon, as in Figure 4.1b, the wagon moves. When a football is kicked, as in Figure 4.1c, it is deformed and set in motion. These are all examples of **contact forces,** so named because they result from physical contact between two objects.

Another class of forces does not involve physical contact between two objects. Early scientists, including Newton, were uneasy with the concept of forces that act between two disconnected objects. To overcome this conceptual difficulty, Michael Faraday (1791–1867) introduced the concept of a *field*. The corresponding forces are called **field forces.** According to this approach, when a mass, m, is placed at some point P near a second mass, M, we say that m interacts with M by virtue of the gravitational field that exists at P. Thus, the force of gravitational attraction between two objects, illustrated in Figure 4.1d, is an example of a field force. This force keeps objects bound to the Earth and gives rise to what we commonly call the *weight* of the object. The planets of our Solar System move under the actions of gravitational forces.

Another common example of a field force is the electric force that one electric charge exerts on another, as in Figure 4.1e. A third example is the force exerted by a bar magnet on a piece of iron, as in Figure 4.1f. The known fundamental forces in nature are all field forces. These are, in order of decreasing strength, (1) strong nuclear forces between subatomic particles, (2) electromagnetic forces between electric charges at rest or in motion, (3) weak nuclear forces, which arise in certain radioactive decay processes, and (4) gravitational attractions between objects. In classical physics we are concerned only with gravitational and electromagnetic forces.

Whenever a force is exerted on an object, its shape can change. For example, when you squeeze a rubber ball or strike a punching bag with your fist, the object

A football is set in motion by the contact force, **F,** on it due to the kicker's foot. The ball is deformed during the short time in contact with the foot. *(Ralph Cowan, Tony Stone Worldwide)*

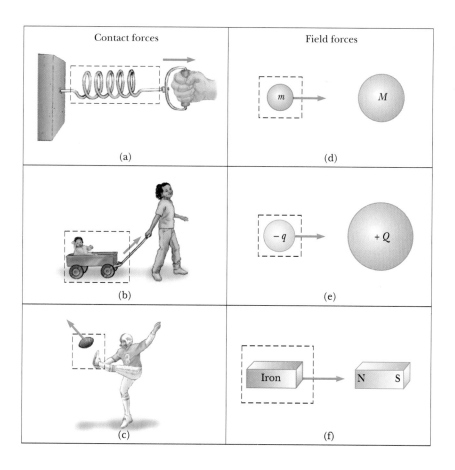

Figure 4.1 Examples of forces applied to various objects. In each case a force acts on the object surrounded by the dashed lines. Something in the environment external to the boxed area exerts this force.

deforms to some extent. Even more rigid objects, such as automobiles, are deformed under the action of external forces. Often the deformations are permanent, as in the case of a collision between vehicles.

4.2 NEWTON'S FIRST LAW

Consider the following simple experiment. Suppose a book is lying on a table. Obviously, the book remains at rest if left alone. Now imagine that you push the book with a horizontal force great enough to overcome the force of friction between book and table, so that the book is set in motion. If the magnitude of your applied force is equal to the magnitude of the friction force, the book moves with constant velocity. If the magnitude of your applied force exceeds the magnitude of the friction force, the book accelerates. If you stop applying the force, the book stops sliding after traveling a short distance because the force of friction retards its motion. Now imagine pushing the book across a smooth, waxed floor. The book again comes to rest once the force is no longer applied, but not as soon as before. Finally, imagine that the book is moving on a horizontal, frictionless surface. In this situation, the book continues to move in a straight line with constant velocity until it hits a wall or some other object.

Before about 1600, scientists felt that the natural state of matter was the state of rest. Galileo was the first to take a different approach. He devised thought

A multiflash exposure of a tennis ball being struck by a racket. The ball experiences a large contact force as it is struck, and a correspondingly large acceleration. *(Ben Rose/The Image Bank)*

experiments—such as an object moving on a frictionless surface, as just described—and concluded that it is not the nature of an object to *stop* once set in motion; rather, it is an object's nature to *resist acceleration*. This approach to motion was later formalized by Newton in a form that has come to be known as **Newton's first law of motion:**

Newton's first law ▷

> An object at rest remains at rest, and an object in motion continues in motion with constant velocity (that is, constant speed in a straight line), unless it experiences a net external force.

By external force we mean any force that results from the interaction between the object and its environment, such as the force exerted on an object when it is lifted. In simpler terms, Newton's first law says that **when the net external force on an object is zero, its acceleration is zero.** That is, when $\Sigma \mathbf{F} = 0$, then $\mathbf{a} = 0$.

Finally, consider a spaceship traveling in space, far from any planets or other matter. The spaceship requires a propulsion system to change its velocity. However, if the propulsion system is turned off when a velocity, **v,** is reached, the spaceship coasts in space with a constant velocity and the astronauts get a free ride.

Mass and Inertia

Imagine a bowling ball and a golf ball sitting side by side on the ground. Newton's first law tells us that both remain at rest as long as no net external force acts on them. Now imagine supplying a net force by striking each ball with a golf club.

An object at rest will remain at rest and an object in motion will continue in motion with constant velocity unless a net external force acts on the object. In this case, the wall of the building did not exert a large enough force on the moving train to stop it. *(Photo Roger Viollet, Mill Valley, CA, University Science Books, 1982)*

Both balls resist your attempt to change their states of motion. But you know from everyday experience that if the two are struck with equal force, the golf ball will travel much farther than the bowling ball. That is, the bowling ball is more successful in resisting your attempt to change the state of motion. The tendency of an object to resist any attempt to change its motion is called the **inertia** of the object.

Mass is a measurement of inertia, and the SI unit of mass is the kilogram. The greater the mass of a body, the less it accelerates under the action of an applied force. For example, if a given force acting on a 3-kg mass produces an acceleration of 4 m/s^2, the same force applied to a 6-kg mass will produce an acceleration of only 2 m/s^2. Mass is a scalar quantity that obeys the rules of ordinary arithmetic.

Inertia is the principle that underlies the operation of seat belts. In the event of an accident, the purpose of the seat belt is to hold the passenger firmly in place relative to the car, to prevent serious injury. Figure 4.2 illustrates how one type of shoulder harness operates. Under normal conditions, the ratchet turns freely to allow the harness to wind on or unwind from the pulleys as the passenger moves. If an accident occurs, the car undergoes a large negative acceleration and rapidly comes to rest. The large mass under the seat, because of its inertia, continues to slide forward along the tracks. The pin connection between the mass and the rod causes the rod to pivot about its center and engage the ratchet wheel. At this point the ratchet wheel locks in place, and the harness no longer unwinds.

A slight modification enables this device to also activate an air bag in the car. In this case, movement of the mass and pivoting of the rod open a valve on a cylinder that contains nitrogen under pressure. The nitrogen rushes into the air bag, causing it to expand rapidly so that it serves as a protective cushion for the passenger.

APPLICATION

Seat Belts and Air Bags.

Isaac Newton, English physicist and mathematician (1642–1727)

Newton was one of the most brilliant scientists in history. Before the age of 30, he formulated the basic concepts and laws of mechanics, discovered the law of universal gravitation, and invented the mathematical methods of calculus. As a consequence of his theories, Newton was able to explain the motions of the planets, the ebb and flow of the tides, and many special features of the motions of the Moon and the Earth. He also interpreted many fundamental observations concerning the nature of light. His contributions to physical theories dominated scientific thought for two centuries and remain important today. *(Giraudon/Art Resource)*

Figure 4.2 A mechanical arrangement for a safety belt.

4.3   NEWTON'S SECOND LAW

Newton's first law explains what happens to an object when the net force acting on it is zero: the object either remains at rest or moves in a straight line with constant speed. Newton's second law answers the question of what happens to an object that has a nonzero net force acting on it.

Imagine pushing a block of ice across a frictionless horizontal surface. When you exert some horizontal force on the block, it moves with an acceleration of, say, $2 \ \mathrm{m/s^2}$. If you apply a force twice as large, the acceleration doubles. Pushing three times as hard triples the acceleration, and so on. From such observations, we conclude that **the acceleration of an object is directly proportional to the net force acting on it.**

Common experience with pushing objects tells you that mass also affects acceleration. Imagine that you stack identical blocks of ice on top of each other while pushing the stack with constant force. If the force when you push one block produces an acceleration of $2 \ \mathrm{m/s^2}$, the acceleration will drop to half that value when two blocks are pushed, one third that initial value when three blocks are pushed, and so on. We conclude that **the acceleration of an object is inversely proportional to its mass.** These observations are summarized in **Newton's second law:**

Newton's second law ▶

> The acceleration of an object is directly proportional to the net force acting on it and inversely proportional to its mass.

In equation form, we can state Newton's second law as

$$\Sigma \mathbf{F} = m\mathbf{a} \qquad\qquad [4.1]$$

where **a** is the acceleration of the object, m is its mass, and $\Sigma \mathbf{F}$ represents the *vector sum of all external forces acting on the object.* (An external force is one that results from interaction between the object and its environment.) You should note that, because this is a vector equation, it is equivalent to the following three component equations:

$$\Sigma F_x = ma_x \qquad \Sigma F_y = ma_y \qquad \Sigma F_z = ma_z \qquad [4.2]$$

Units of Force and Mass

Definition of newton ▶

The SI unit of force is the **newton,** defined as the force that, when acting on a 1-kg mass, produces an acceleration of $1 \ \mathrm{m/s^2}$. From this definition and Newton's second law, we see that the newton can be expressed in terms of the fundamental units of mass, length, and time:

$$1 \ \mathrm{N} \equiv 1 \ \mathrm{kg \cdot m/s^2} \qquad\qquad [4.3]$$

Definition of dyne ▶

The unit of force in the cgs system is called the **dyne** and is defined as the force that, when acting on a 1-g mass, produces an acceleration equal to $1 \ \mathrm{cm/s^2}$:

$$1 \ \mathrm{dyne} \equiv 1 \ \mathrm{g \cdot cm/s^2} \qquad\qquad [4.4]$$

TABLE 4.1 Units of Mass, Acceleration, and Force

System	Mass	Acceleration	Force
SI	kg	m/s^2	N = kg · m/s^2
cgs	g	cm/s^2	dyne = g · cm/s^2

Note: 1 N = 10^5 dyne = 0.225 lb.

In the British engineering system the unit of force is the **pound.** The following conversion from pounds to newtons will be useful to you in many problems:

◀ Definition of pound

$$1 \text{ N} \equiv 0.225 \text{ lb} \qquad [4.5]$$

The units of mass, acceleration, and force in the SI and cgs system are summarized in Table 4.1.

Weight and the Gravitational Force

We are well aware that objects are attracted to the Earth. The force exerted by the Earth on an object is the gravitational force $\mathbf{F}_g$. This force is directed approximately toward the center of the Earth, and its magnitude varies with location. The magnitude of the gravitational force is called the **weight** of the object, $w = mg$, where m is the mass of the object.

We have seen that a freely falling object experiences an acceleration, $\mathbf{g}$, acting toward the center of the Earth. Applying Newton's second law to the freely falling object shown in Figure 4.3, we have $\mathbf{F} = m\mathbf{a}$. Because $\mathbf{F}_g = m\mathbf{g}$, it follows that $\mathbf{a} = \mathbf{g}$ and

$$w = mg \qquad [4.6]$$

Because it depends on g, weight varies with geographic location. Bodies weigh less at higher altitudes than at sea level because g decreases with increasing distance above the Earth's surface. Hence, weight, unlike mass, is not an inherent property of an object. For example, if an object has a mass of 70.0 kg, then its weight at a location where $g = 9.80$ m/s^2 is $mg = 686$ N. At the top of a mountain, where g might be 9.76 m/s^2, the object's weight would be 683 N. Therefore, if you want to lose weight quickly without going on a diet, move to the top of a mountain or weigh yourself at an altitude of 30 000 ft during a flight on a jet airplane.

Because $w = mg$, we can compare the masses of two objects by measuring their weights with a spring scale or balance. At a given location the ratio of the weights of the two objects equals the ratio of their masses.

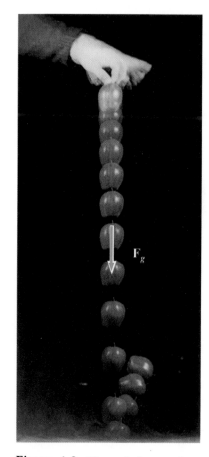

Figure 4.3 The only force acting on an object in free fall is the force of gravity, $\mathbf{F}_g$. The magnitude of the force of gravity is defined as the weight of the object, w. Newton's second law applied in the vertical direction gives $w = mg$.

Thinking Physics 1

Suppose you are talking by interplanetary telephone to your friend, who lives on the Moon. She tells you that she has just won 1 newton of gold in a contest. Excitedly, you tell her that you entered the Earth version of the same contest and also won 1 newton of gold! Who is richer?

Astronaut Edwin E. Aldrin, Jr., walking on the Moon after the Apollo 11 lunar landing. The weight of this astronaut on the Moon is less than it is on Earth, but his mass remains the same. *(NASA)*

Explanation Because the value of *g* on the Moon is smaller than on Earth by a factor of about 6, more mass of gold would be required on the Moon to represent a weight of 1 newton. Thus, your friend on the Moon is richer, by a factor of about 6!

4.4   NEWTON'S THIRD LAW

In Section 4.1 we found that a force is exerted on an object when that object comes into contact with some other object. For example, consider the task of driving a nail into a block of wood, as illustrated in Figure 4.4a. To accelerate the nail and drive it into the block, a net force must be supplied to the nail by the hammer. However, Newton recognized that a single isolated force (such as the force on the nail by the hammer) cannot exist. Instead, **forces in nature always exist in pairs.** According to Newton, the hammer exerts a force on the nail, and the nail exerts a force on the hammer. There is clearly a net force on the hammer, because it rapidly slows down after coming into contact with the nail.

Newton described this type of situation in terms of his **third law of motion:**

◀ Newton's third law

> If two objects interact, the force exerted on object 1 by object 2 is equal in magnitude but opposite in direction to the force exerted on object 2 by object 1.

This law, which is illustrated in Figure 4.4b, is equivalent to stating that **forces always occur in pairs,** or that **a single isolated force cannot exist.** The force that body 1 exerts on body 2 is sometimes called the *action force,* and the force of body 2 on body 1 is called the *reaction force.* In reality, either force can be labeled the action or reaction force. **The action force is equal in magnitude to the reaction force and opposite in direction. In all cases, the action and reaction forces act on different objects and must be of the same type.** For example, the force acting on a freely falling projectile is the force of the Earth on the projectile, $\mathbf{F}_g$, and the magnitude of this force is *mg.* The reaction to this force is the force of the projectile

(a)   (b)

Figure 4.4 Newton's third law. (a) The force exerted by the hammer on the nail is equal to and opposite the force exerted by the nail on the hammer. (b) The force exerted by object 1 on object 2 is equal to and opposite the force exerted by object 2 on object 1. *(John Gillmoure, The Stock Market)* on the Earth, $\mathbf{F}_g' = -\mathbf{F}_g$. The reaction force, $\mathbf{F}_g'$, must accelerate the Earth toward the projectile just as the action force, $\mathbf{F}_g$, accelerates the projectile toward the Earth. However, because the Earth has such a large mass, its acceleration due to this reaction force is negligibly small.

You directly experience the law if you slam your fist against a wall or kick a football with your bare foot. You should be able to identify the action and reaction forces in these cases.

As another application of Newton's third law, consider helicopters that you have seen. The most familiar helicopters have a large set of blades rotating in a horizontal plane above the body and another smaller set rotating in a vertical plane at the back. Other helicopters have two large sets of blades above the body rotating in opposite directions. Why do helicopters always have two sets of blades? The helicopter engine applies a force to the blades, causing them to change their rotational motion, but according to Newton's third law, the blades must exert an equal and opposite force on the helicopter. This force would cause the body of the helicopter to rotate in the direction opposite to the blades. A rotating helicopter would be impossible to control, so a second set of blades is used. The small blades in the back provide a force opposite to that tending to rotate the body of the helicopter, keeping the body oriented stably. In the case of two sets of large, counter-rotating blades, the engines apply forces in opposite directions, so that there is no net reaction force rotating the helicopter.

As we mentioned earlier, the Earth exerts a force $\mathbf{F}_g$ on any object. If the object is a TV at rest on a table, as in Figure 4.5a, the reaction force to $\mathbf{F}_g$ is the force the TV exerts on the Earth, $\mathbf{F}_g'$. The TV does not accelerate, because it is held up by the table. The table, therefore, exerts on the TV an upward action force, $\mathbf{n}$, called the **normal force.** This is the force that prevents the TV from falling through the

APPLICATION

Helicopter Flight.

(a)   (b)

Figure 4.5 When a TV set is sitting on a table, the forces acting on the set are the normal force, $\mathbf{n}$, and the force of gravity, $\mathbf{F}_g$, as illustrated in (b). The reaction to $\mathbf{n}$ is the force exerted by the TV set on the table, $\mathbf{n}'$. The reaction to $\mathbf{F}_g$ is the force exerted by the TV set on the Earth, $\mathbf{F}_g'$.

PHYSICS IN ACTION

Forces and Motion

On the left is an athlete running at sunset. The external forces acting on the athlete, as described by the blue vectors, are (a) the force exerted by the Earth, $\mathbf{F}_g$, (b) the force exerted by the ground, $\mathbf{F}$, and (c) the force of air resistance, $\mathbf{R}$.

On the right is a multiflash exposure of a tennis player executing a backhand stroke. The parabolic path of the ball before it is struck is visible, as is the follow-through of the swing. The force exerted by the racket on the ball is equal to and opposite the force exerted by the ball on the racket.

(Mitchell Funk/The Image Bank)

(© Zimmerman/FPG)

table; it can have any value needed, up to the point of breaking the table. The normal force balances the weight and provides equilibrium. The reaction to $\mathbf{n}$ is the force of the TV on the table, $\mathbf{n}'$. Therefore, we conclude that

$$\mathbf{F}_g = -\mathbf{F}_g' \qquad \text{and} \qquad \mathbf{n} = -\mathbf{n}'$$

The forces $\mathbf{n}$ and $\mathbf{n}'$ have the same magnitude, which is the same as $\mathbf{F}_g$ unless the table has broken. Note that the forces acting on the TV are $\mathbf{F}_g$ and $\mathbf{n}$, as shown in Figure 4.5b. The two reaction forces, $\mathbf{F}_g'$ and $\mathbf{n}'$ are exerted on objects other than the TV. Remember, the two forces in an action-reaction pair always act on two different objects.

From the second law, we see that, because the TV is in equilibrium ($\mathbf{a} = 0$), it follows that $F_g = n = mg$.

Applying Physics 1

APPLICATION

Colliding Vehicles.

If a small sports car collides head-on with a massive truck, which vehicle experiences the greater impact force? Which vehicle experiences the greater acceleration?

Explanation   The car and truck experience forces that are equal in magnitude but in opposite directions. A calibrated spring scale placed between the colliding vehicles reads the same whichever way it faces. Because the car has the smaller mass, it experiences much greater acceleration.

Thinking Physics 2

Consider a horse attempting to pull a wagon from rest. A student makes the following argument: If the horse pulls forward on the wagon, then Newton's third law holds that the wagon pulls back equally hard on the horse. Thus, the forces cancel and nothing happens.

Explanation   This is a common conundrum that has its basis in an incomplete understanding of Newton's third law. When applying the third law, it is important to recognize the objects on which the pair of forces act: "If object 1 exerts a force on object 2, then 2 exerts an equal and opposite force on 1." This form of the third law indicates clearly that the two forces act on *different* objects. In the statement of this example, we are equating the force of the horse *on the wagon* to the force of the wagon *on the horse*—the forces are acting on different objects. However, if you want to determine the motion of an object, you must consider *only the forces on that object*. The force that accelerates the horse and wagon (the object) is the force of the Earth on the horse's feet, which is the reaction to the force that the horse exerts backward on the Earth.

Thinking Physics 3

You have most likely had the experience of standing in an elevator that accelerates upward as it leaves to move toward a higher floor. In this case, you *feel* heavier. If you are standing on a bathroom scale at the time, it will *measure* a force larger than your weight. Thus, you have evidence that leads you to believe that you are heavier in this situation. *Are* you heavier?

Explanation   No, you are not. Your weight is unchanged. In order to provide the acceleration upward, the floor or the bathroom scale must apply an upward normal force larger than your weight. It is this larger force that you feel, which you interpret as feeling heavier. A bathroom scale reads this upward force, not your weight, so its reading also increases.

4.5   SOME APPLICATIONS OF NEWTON'S LAWS

This section applies Newton's laws to objects moving under the actions of constant external forces. We assume that objects behave as particles, and so we need not worry about rotational motion. We also neglect any friction effects. Finally, we neglect the masses of any ropes or strings involved; in these approximations, the magnitude of the force exerted along a rope (the tension) is the same at all points in the rope.

When we apply Newton's law to an object, we are interested only in those *external* forces that act *on the object*. For example, in Figure 4.4b, the only external forces acting on the TV are $\mathbf{n}$ and $\mathbf{F}_g$. The reactions to these forces, $\mathbf{n}'$ and $\mathbf{F}_g{}'$, act on the table and on the Earth, respectively, and do not appear in Newton's second law as applied to the TV.

Figure 4.6 (a) A crate being pulled to the right on a frictionless surface. (b) The free-body diagram that represents the external forces on the crate.

Consider a crate being pulled to the right on a frictionless, horizontal surface, as in Figure 4.6a. Suppose you are asked to find the acceleration of the crate and the force the surface exerts on it. The horizontal force applied to the crate acts through the rope. The force that the rope exerts on the crate is denoted by **T**. The magnitude of **T** is equal to the tension in the rope. A dashed circle is drawn around the crate in Figure 4.6a to remind you to isolate it from its surroundings.

Because we are interested only in the motion of the crate, we must be able to identify all external forces acting on it. These are illustrated in Figure 4.6b. In addition to the force **T,** the force diagram for the crate includes the force of gravity, F_g, and the normal force, **n**, exerted by the floor on the crate. Such a force diagram is referred to as a **free-body diagram.** The construction of a correct free-body diagram is an essential step in applying Newton's laws; its importance cannot be overemphasized. The *reactions* to the forces we have listed—namely, the force exerted by the rope on the hand doing the pulling, the force exerted by the crate on the Earth, and the force exerted by the crate on the floor—are not included in the free-body diagram because they act on other objects and not on the crate.

Now let us apply Newton's second law to the crate. First we must choose an appropriate coordinate system. In this case it is convenient to use the one shown in Figure 4.6b, with the x axis horizontal and the y axis vertical. We can apply Newton's second law in the x direction, y direction, or both, depending on what we are asked to find in a problem. In addition, we may be able to use the equations of motion for constant acceleration that are found in Chapter 2. However, you should use these equations only when the acceleration is constant.

Objects in Equilibrium and Newton's First Law

Objects that are either at rest or moving with constant velocity are said to be in equilibrium, and Newton's first law is a statement of one condition that must be true for equilibrium conditions to prevail. In equation form, this condition of equilibrium can be expressed as

$$\Sigma \mathbf{F} = 0 \qquad [4.7]$$

This statement signifies that the *vector* sum of all the forces (the net force) acting on an object in equilibrium is zero.

Usually, equilibrium problems are solved more easily if we work with Equation 4.7 in terms of the components of the external forces acting on an object. By this we mean that, in a two-dimensional problem, the sum of all the external forces in the x and y directions must separately equal zero; that is,

$$\Sigma F_x = 0 \qquad \text{and} \qquad \Sigma F_y = 0 \qquad [4.8]$$

This set of equations is often referred to as the **first condition for equilibrium.** We shall not consider three-dimensional problems in this book, but the extension of Equation 4.8 to a three-dimensional situation can be made by adding a third equation, $\Sigma F_z = 0$.

Problem-Solving Strategy

Objects in Equilibrium

The following procedure is recommended for problems involving objects in equilibrium:

1. Make a sketch of the situation described in the problem statement.
2. Draw a free-body diagram for the *isolated* object under consideration, and label all external forces acting on the object.
3. Resolve all forces into x and y components, choosing a convenient coordinate system.
4. Use the equations $\Sigma F_x = 0$ and $\Sigma F_y = 0$. Keep track of the signs of the various force components.
5. Application of Step 4 leads to a set of equations with several unknowns. Solve the simultaneous equations for the unknowns in terms of the known quantities.

EXAMPLE 4.1   A Traffic Light at Rest

A traffic light weighing 100 N hangs from a vertical cable tied to two other cables that are fastened to a support, as in Figure 4.7a. The upper cables make angles of 37.0° and 53.0° with the horizontal. Find the tension in each of the three cables.

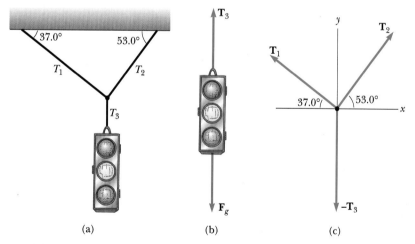

(a)   (b)   (c)

Figure 4.7   (Example 4.1) (a) A traffic light suspended by cables. (b) A free-body diagram for the traffic light. (c) A free-body diagram for the knot joining the cables.

Reasoning   We must construct two free-body diagrams in order to work this problem. The first of these is for the traffic light, shown in Figure 4.7b; the second is for the knot that holds the three cables together, as in Figure 4.7c. The knot is a convenient point to choose because all forces in question act at this point.

Solution   From the free-body diagram in Figure 4.7b, we see that $T_3 = w = 100$ N. Next, we choose the coordinate axes shown in Figure 4.7c and resolve the forces into their x and y components:

Force	x Component	y Component
$\mathbf{T}_1$	$-T_1 \cos 37.0°$	$T_1 \sin 37.0°$
$\mathbf{T}_2$	$T_2 \cos 53.0°$	$T_2 \sin 53.0°$
$\mathbf{T}_3$	0	-100 N

The first condition for equilibrium gives us the equations

$$(1) \qquad \Sigma F_x = T_2 \cos 53.0° - T_1 \cos 37.0° = 0$$

$$(2) \qquad \Sigma F_y = T_1 \sin 37.0° + T_2 \sin 53.0° - 100 \text{ N} = 0$$

From (1) we see that the horizontal components of $\mathbf{T}_1$ and $\mathbf{T}_2$ must be equal in magnitude, and from (2) we see that the sum of the vertical components of $\mathbf{T}_1$ and $\mathbf{T}_2$ must balance the force of gravity acting on the light. We can solve (1) for T_2 in terms of T_1 to give

$$T_2 = T_1 \left(\frac{\cos 37.0°}{\cos 53.0°} \right) = T_1 \left(\frac{0.799}{0.602} \right) = 1.33 T_1$$

This value for T_2 can be substituted into (2) to give

$$T_1 \sin 37.0° + (1.33 T_1)(\sin 53.0°) - 100 \text{ N} = 0$$

$$T_1 = \boxed{60.1 \text{ N}}$$

$$T_2 = 1.33 T_1 = 1.33(60.0 \text{ N}) = \boxed{79.9 \text{ N}}$$

Exercise   When will $T_1 = T_2$?

Answer   When the supporting cables make equal angles with the horizontal support.

EXAMPLE 4.2   Sled on a Frictionless Hill

A child holds a sled at rest on a frictionless, snow-covered hill, as shown in Figure 4.8a. If the sled weighs 77.0 N, find the force exerted on the rope by the child and the force exerted on the sled by the hill.

Reasoning   Figure 4.8b shows the forces acting on the sled and a convenient coordinate system to use for this type of problem. Note that $\mathbf{n}$, the force exerted on the sled by the ground, is perpendicular to the hill. The ground can exert a component of force along the incline only if there is friction between the sled and the hill. Because the sled is at rest, we are able to apply the first condition for equilibrium as $\Sigma F_x = 0$ and $\Sigma F_y = 0$.

Solution   Applying the first condition for equilibrium to the sled, we find that

$$\Sigma F_x = T - (77.0 \text{ N})(\sin 30.0°) = 0$$

$$T = \boxed{38.5 \text{ N}}$$

Figure 4.8 (Example 4.2) (a) A child holding a sled on a frictionless hill. (b) A free-body diagram for the sled.

$$\Sigma F_y = n - (77.0 \text{ N})(\cos 30.0°) = 0$$

$$n = \quad 66.7 \text{ N}$$

Note that n is *less* than the weight of the sled in this situation. This is so because the sled is on an incline and **n** is equal to and opposite the component of the force of gravity perpendicular to the incline.

Exercise What happens to the normal force as the angle of incline increases?

Answer It decreases.

Exercise When is the normal force equal to the weight of the sled?

Answer When the sled is on a horizontal surface and the applied force is either zero or along the horizontal.

Accelerating Objects and Newton's Second Law

In a situation in which a net force acts on an object, the object accelerates, and we use Newton's second law in order to analyze the motion. The representative examples and suggestions that follow should help you solve problems of this kind.

Problem-Solving Strategy

Newton's Second Law

The following procedure is recommended for working problems that involve the application of Newton's second law:

1. Draw a diagram of the system.
2. Isolate the object of interest whose motion is being analyzed. Draw a free-body diagram for this object, showing all external forces acting on it. For

systems containing more than one object, draw a *separate* diagram for each object.

3. Establish convenient coordinate axes for each object, and find the components of the forces along these axes. Apply Newton's second law in the *x* and *y* directions for each object.

4. Solve the component equations for the unknowns. Remember that, in order to obtain a complete solution, you must have as many independent equations as you have unknowns.

5. If necessary, use the equations of kinematics (motion with constant acceleration) from Chapter 2 to find all the unknowns.

Figure 4.9　(Example 4.3)

EXAMPLE 4.3　Moving a Crate

The combined weight of the crate and dolly in Figure 4.9 is 300 N. If the person pulls on the rope with a constant force of 20.0 N, what is the acceleration of the system (crate plus dolly), and how far will it move in 2.00 s? Assume that the system starts from rest and that there are no frictional forces opposing the motion of the system.

Reasoning　We can find the acceleration of the system from Newton's second law. Because the force exerted on the system is constant, its acceleration is constant. Therefore, we can apply the equations of motion with constant acceleration to find the distance traveled.

Solution　In order to apply Newton's second law to the system, we must first know the system's mass:

$$m = \frac{w}{g} = \frac{300 \text{ N}}{9.80 \text{ m/s}^2} = 30.6 \text{ kg}$$

Now we can find the acceleration of the system from the second law:

$$a_x = \frac{F_x}{m} = \frac{20.0 \text{ N}}{30.6 \text{ kg}} = 0.654 \text{ m/s}^2$$

Because the acceleration is constant, we can find the distance the system moves in 2.00 s using the relation $x = v_0 t + \frac{1}{2}at^2$ with $v_0 = 0$:

$$x = \tfrac{1}{2}at^2 = \tfrac{1}{2}(0.654 \text{ m/s}^2)(2.00 \text{ s})^2 = 1.31 \text{ m}$$

It is important to note that the constant applied force of 20.0 N is assumed to act on the system all during its motion. If the force were removed at some instant, the system would continue to move with constant velocity and hence zero acceleration.

 ## EXAMPLE 4.4　The Run-Away Car

A car of mass *m* is on an icy driveway inclined at an angle $\theta = 20.0°$, as in Figure 4.10a. Determine the acceleration of the car, assuming the incline is frictionless.

Reasoning　The free-body diagram for the car is shown in Figure 4.10b. The only forces on the car are the normal force, **n,** acting perpendicular to the driveway surface, and the force of gravity, $\mathbf{F}_g$, acting vertically downward. It is convenient to choose the coordinate axes with the *x* axis along the incline and the *y* axis perpendicular to it. Then we replace

(a)   (b)

Figure 4.10   (Example 4.4)

the force of gravity with a component of magnitude $mg \sin \theta$ along the positive x axis and a component of magnitude $mg \cos \theta$ in the negative y direction.

Solution   Applying Newton's second law in component form, with $a_y = 0$, gives

$$(1) \qquad \Sigma F_x = mg \sin \theta = ma_x$$

$$(2) \qquad \Sigma F_y = n - mg \cos \theta = 0$$

From (1) we see that the acceleration of the car along the driveway is provided by the component of the force of gravity directed down the incline:

$$(3) \qquad a_x = g \sin \theta$$

Note that the acceleration given by (3) is *constant* and *independent of the mass* of the car — it depends only on the angle of inclination and on g. In our example, $\theta = 20.0°$ and so, we find that

$$a_x = \quad 3.35 \text{ m/s}^2$$

Exercise   A compact car and a large luxury sedan are at rest at the top of an ice-covered (frictionless) driveway. If they both slide down the driveway, which reaches the bottom first?

Answer   Because the acceleration is independent of the mass, and thus is the same for both vehicles, they arrive at the bottom simultaneously.

Exercise   If the length of the driveway is 25.0 m and a car starts from rest at the top, how long does it take to travel to the bottom? What is the car's speed at the bottom?

Answer   3.86 s; 12.9 m/s

4.6   FORCE OF FRICTION

When a body is in motion either on a surface or through a viscous medium such as air or water, there is resistance to the motion because the body interacts with its surroundings. We call such resistance a **force of friction.** Forces of friction are very important in our everyday lives. They allow us to walk or run and are necessary for the motion of wheeled vehicles.

Consider a block on a horizontal table, as in Figure 4.11a. If we apply an external horizontal force **F** to the block, acting to the right, the block remains stationary if **F** is not too large. The force that counteracts **F** and keeps the block from moving acts to the left and is called the force of static friction, f_s. As long as the block is not moving, $f_s = F$. Thus, if **F** is increased, f_s also increases. Likewise, if **F** decreases, f_s also decreases. Experiments show that the frictional force arises from the nature of the two surfaces: Because of their roughness, contact is made only at a few points, as shown in the magnified view of the surface in Figure 4.11a. Actually, the frictional force is much more complicated than presented, because it ultimately involves the electrostatic force between atoms or molecules.

If we increase the magnitude of **F,** as in Figure 4.11b, the block eventually slips. When the block is on the verge of slipping, f_s is a maximum as shown in Figure

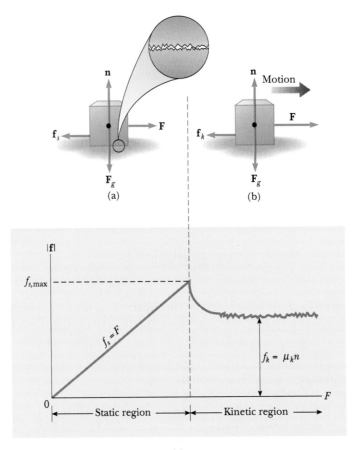

Figure 4.11   The direction of the force of friction, **f,** between a block and a horizontal surface is opposite the direction of an applied force, **F.** (a) The force of static friction equals the applied force. (b) When the magnitude of the applied force exceeds the maximum static friction force, $f_{s,max}$, the block accelerates to the right. (c) A graph of the magnitude of the frictional force versus the applied force. Note that $f_{s,max} > f_k$.

4.11c. When F exceeds $f_{s,max}$, the block moves and accelerates to the right. When the block is in motion, the frictional force becomes less than $f_{s,max}$ (Fig. 4.11c). We call the frictional force for an object in motion the **force of kinetic friction, f_k.** The unbalanced force in the x direction, $F - f_k$, produces an acceleration to the right. If $F = f_k$, the block moves to the right with constant speed. If the applied force is removed, then the frictional force acting to the left accelerates the block in the $-x$ direction and eventually brings it to rest.

Experimentally, one finds that, to a good approximation, both $f_{s,max}$ and f_k are proportional to the normal force acting on the block. The experimental observations can be summarized as follows:

- The magnitude of the force of static friction between any two surfaces in contact can have the values

$$f_s \leq \mu_s n \qquad [4.9]$$

where the dimensionless constant μ_s is called the **coefficient of static friction** and n is the magnitude of the normal force. The equality in Equation 4.9 holds when the block is on the verge of slipping, that is, when $f_s = f_{s,max} \equiv \mu_s n$. The inequality holds when the applied force is less than this value.

- The magnitude of the force of kinetic friction acting on an object is given by

$$f_k = \mu_k n \qquad [4.10]$$

where μ_k is the **coefficient of kinetic friction.**

- The values of μ_k and μ_s depend on the nature of the surfaces, but μ_k is generally less than μ_s. Table 4.2 lists some reported values.

- The coefficients of friction are nearly independent of the area of contact between the surfaces.

Finally, although the coefficient of kinetic friction varies with speed, we shall neglect any such variations. The approximate nature of the equations is easily demonstrated by trying to get an object to slide down an incline at constant speed.

TABLE 4.2 Coefficients of Friction [a]

	μ_s	μ_k
Steel on steel	0.74	0.57
Aluminum on steel	0.61	0.47
Copper on steel	0.53	0.36
Rubber on concrete	1.0	0.8
Wood on wood	0.25–0.5	0.2
Glass on glass	0.94	0.4
Waxed wood on wet snow	0.14	0.1
Waxed wood on dry snow	—	0.04
Metal on metal (lubricated)	0.15	0.06
Ice on ice	0.1	0.03
Teflon on Teflon	0.04	0.04
Synovial joints in humans	0.01	0.003

[a] All values are approximate.

QUICKLAB

Measuring μ

This is a simple method for measuring the coefficients of static and kinetic friction between an object and some surface. For this investigation, you will need a few coins, your book or some other flat surface that can be inclined, a protractor, and some double-stick tape. Place a coin at one edge of the book as it lies on a table, and lift that edge of the book until the coin just slips down the incline as in the figure. When this occurs, measure the inclined angle using your protractor. Repeat the measurement five times, and find the average value of this critical angle θ_C. The coefficient of static friction between the coin and book's surface is $\mu_s = \tan \theta_C$. (You should prove this as an exercise.) Calculate the average value of μ_s. To measure the coefficient of kinetic friction, find the angle θ_C' at which the coin moves down the incline with constant speed. This angle should be less than θ_C. Measure this new angle five times, and get its average value. Calculate the average value of μ_k using the fact that $\mu_k = \tan \theta_C'$, where $\theta_C' < \theta_C$. Repeat these measurement using two or three stacked coins, with double-stick tape between them. You should get the same results as with one coin. Why?

Coin

θ

Applying Physics 2

You are playing with your younger sister in the snow. She is sitting on a sled and asking you to slide her across a flat, horizontal field. You have a choice of pushing her from behind, by applying a force at 30° below the horizontal, or attaching a rope to the front of the sled and pulling with a force at 30° above the horizontal. Which would be easier for you and why?

Explanation   It is easier to attach the rope and pull. In this case, there is a component of your applied force that is upward. This reduces the normal force between the sled and the snow. In turn, this reduces the friction force between the sled and the snow, making it easier to move. If you push from behind, with a force with a downward component, the normal force is larger, the friction force is larger, and the sled is harder to move.

EXAMPLE 4.5   Moving into the Dormitory

At the beginning of a new school term, a student moves a box of books by attaching a rope to the box and pulling with a force of 90.0 N at an angle of 30.0°, as shown in Figure 4.12. The box of books has a mass of 20.0 kg, and the coefficient of kinetic friction between the bottom of the box and the floor is 0.500. Find the acceleration of the box.

Figure 4.12   (Example 4.5) A box of books being pulled to the right at an angle of 30.0°.

Reasoning   There are basically three steps required for the solution to this problem. (1) First find the normal force $\mathbf{n}$ by applying the first condition of equilibrium, $\Sigma F_y = 0$, in the vertical direction. (2) Calculate the force of kinetic friction on the box from $f_k = \mu_k n$. (3) Apply Newton's second law along the horizontal direction to find the acceleration of the box.

Solution   (Step 1) The box is not accelerating in the vertical direction, and so we find the normal force from $\Sigma F_y = 0$. The forces in the y direction are the force of gravity $\mathbf{F}_g$, the normal force, $\mathbf{n,}$ and the vertical component of the applied 90.0-N force. This vertical component has a magnitude equal to $(90.0 \text{ N})(\sin 30.0°)$. We find that

$$\Sigma F_y = n + (90.0 \text{ N})(\sin 30.0°) - (20.0 \text{ kg})(9.80 \text{ m/s}^2) = 0$$

$$n = 151 \text{ N}$$

The magnitude of the normal force is *not* equal to the weight of the box, because the vertical component of the 90.0-N force is helping to support some of that weight.

(Step 2) Knowing the normal force, we can find the force of kinetic friction:

$$f_k = \mu_k n = (0.500)(151 \text{ N}) = 75.5 \text{ N} \quad \text{(to the left)}$$

(Step 3) Finally, we determine the horizontal acceleration using Newton's second law:

$$\Sigma F_x = (90.0 \text{ N})(\cos 30.0°) - 75.5 \text{ N} = (20.0 \text{ kg})(a_x)$$

$$a_x = +0.122 \text{ m/s}^2$$

Exercise If the initial speed of the box is zero, what is its speed after it has traveled 2.00 m? How long does it take to pull it this distance?

Answer 0.699 m/s; 5.73 s

EXAMPLE 4.6 The Sliding Hockey Puck

A hockey puck is given an initial speed of 20.0 m/s on a frozen pond, as in Figure 4.13. The puck remains on the ice and slides 120 m before coming to rest. Determine the coefficient of kinetic friction between puck and ice.

Reasoning The puck slides to rest with a constant acceleration along the horizontal. Thus, we can use the kinematic equation $v^2 = v_0^2 + 2ax$ to find a. Newton's second law applied in the horizontal direction is $-f_k = -\mu_k n = ma$. To find μ_k, we first find the normal force **n** by applying $\Sigma F_y = 0$ in the vertical direction.

Solution With the final speed, v, equal to zero; the initial speed, $v_0 = 20.0$ m/s; and the displacement, $x = 120$ m:

$$v^2 = v_0^2 + 2ax$$
$$0 = (20.0 \text{ m/s})^2 + 2a(120 \text{ m})$$
$$a = -1.67 \text{ m/s}^2$$

The negative sign means that the acceleration is to the left, *opposite* the direction of the velocity.

The magnitude of the force of kinetic friction is found from $f_k = \mu_k n$, and n is found from $\Sigma F_y = 0$ as follows:

$$\Sigma F_y = n - F_g = 0$$
$$n = F_g = mg$$

Thus,

$$f_k = \mu_k n = \mu_k mg$$

Now we apply Newton's second law along the horizontal direction, taking the positive direction toward the right:

$$\Sigma F_x = -f_k = ma$$
$$-\mu_k mg = m(-1.67 \text{ m/s}^2)$$
$$\mu_k = \frac{1.67 \text{ m/s}^2}{9.80 \text{ m/s}^2} = 0.170$$

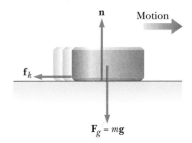

Figure 4.13 (Example 4.6) *After* the puck is given an initial velocity to the right, the external forces acting on it are the force of gravity, **F**$_g$, the normal force, **n**, and the force of kinetic friction, **f**$_k$.

EXAMPLE 4.7   Connected Objects

Two objects are connected by a light string that passes over a frictionless pulley, as in Figure 4.14a. The coefficient of sliding friction between the 4.00-kg object and the surface is 0.300. Find the acceleration of the two objects and the tension in the string.

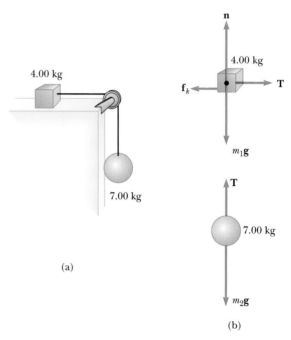

(a)

(b)

Figure 4.14   (Example 4.7) (a) Two objects connected by a light string that passes over a frictionless pulley. (b) Free-body diagrams for the objects.

Reasoning   Connected objects are handled by applying Newton's second law separately to each. The free-body diagrams for the block on the table and for the hanging block are shown in Figure 4.14b. We shall obtain two equations involving the unknowns T and a that can be solved simultaneously.

Solution   With the positive x direction to the right and the positive y direction upward, Newton's second law applied to the 4.00-kg object gives

$$\Sigma F_x = T - f_k = (4.00 \text{ kg})(a)$$

$$\Sigma F_y = n - (4.00 \text{ kg})(g) = 0$$

Because $f_k = \mu_k n$ and $n = (4.00 \text{ kg})g = 39.2 \text{ N}$, we have $f_k = (0.300)(39.2 \text{ N}) = 11.8 \text{ N}$. Therefore,

$$(1) \qquad T = f_k + (4.00 \text{ kg})(a) = 11.8 \text{ N} + (4.00 \text{ kg})(a)$$

Now we apply Newton's second law to the 7.00-kg object moving in the vertical direction, this time selecting downward as the positive direction:

$$\Sigma F_y = (7.00 \text{ kg})(g) - T = (7.00 \text{ kg})(a)$$

or

(2) $T = 68.6 \text{ N} - (7.00 \text{ kg})(a)$

Subtracting (1) from (2) eliminates T:

$56.8 \text{ N} - (11.0 \text{ kg})(a) = 0$

$a = \boxed{5.16 \text{ m/s}^2}$

When this value for the acceleration is substituted into (1), we get

$T = \boxed{32.4 \text{ N}}$

Friction and the Motion of a Car

Forces of friction are important in the analysis of the motion of cars and other wheeled vehicles. There are several types of friction forces to consider, the main ones being the force of friction between tires and road surface and the retarding force produced by air resistance.

As each of four wheels turns to propel a car forward, it exerts a backward force on the road through its tire. The reaction to this backward force is a forward force, **f,** exerted by the road on the tire (Fig. 4.15). If we assume that the same forward force **f** is exerted on each tire, the net forward force on the car is 4**f,** and the car's acceleration is therefore **a** = 4**f**/m.

When the car is in motion, we must also consider the force of air resistance, **R,** which acts in the direction opposite its velocity. The net force on the car is therefore 4**f** − **R,** and so the car's acceleration is **a** = (4**f** − **R**)/m. At normal driving speeds, the magnitude of **R** is proportional to the first power of the speed. That is, $R = bv$, where b is a constant. Thus, the force of air resistance increases with increasing speed. When R is equal to 4f the acceleration is zero, and the car moves at a constant speed.

A similar situation occurs when an object falls through air. When the upward force of air resistance balances the downward force of gravity, the net force on the object is zero and hence the object's acceleration is zero. Once this condition is reached, the object continues to move downward with some constant maximum speed called the **terminal speed.**

Figure 4.15 The horizontal forces acting on the car are the *forward* forces, **f,** exerted on each tire by the road and the force of air resistance, **R,** which acts *opposite* the car's velocity. (The wheels of the car exert a backward force on the road not shown in the diagram.)

Applying Physics 3

Consider a skydiver falling through air before reaching her terminal speed, as in Figure 4.16. As the speed of the skydiver increases, what happens to her acceleration? What is her acceleration once she reaches terminal speed?

Explanation   The forces exerted on the skydiver are the downward force of gravity, m**g,** and an upward force of air resistance, **R.** The magnitude of **R** is less than her weight before she reaches terminal speed. As her downward speed increases, the force of air resistance increases. The vector sum of the force of gravity and the force of air resistance gives a total force that decreases with time, so her acceleration decreases. Once the two forces balance each other so that the net force is zero, the acceleration is consequently zero, and she has reached terminal speed.

APPLICATION

Skydiving.

Figure 4.16   (Applying Physics 3) By spreading their arms and legs and by keeping their bodies parallel to the ground, skydivers experience maximum air resistance resulting in a minimum terminal speed. *(Guy Sauvage, Photo Researchers, Inc.)*

SUMMARY

Newton's first law states that an object at rest remains at rest, and an object in motion continues in motion with a constant velocity, unless it experiences a net external force.

The resistance of an object to a change in its state of motion is called **inertia.** Mass is the physical quantity used to measure inertia.

Newton's second law states that the resultant force acting on an object is equal to the product of the mass of the object and its acceleration:

$$\Sigma \mathbf{F} = m\mathbf{a} \qquad [4.1]$$

The magnitude of the force of gravity exerted on an object is called the **weight** of an object, denoted by w. The weight of an object of mass m is equal to the product mg:

$$w = mg \qquad [4.6]$$

Newton's third law states that, if two objects interact, the force exerted by object 1 on object 2 is equal in magnitude and opposite in direction to the force exerted by object 2 on object 1. Thus, an isolated force can never occur in nature.

An **object in equilibrium** has no net external force acting on it, and the first law, in component form, implies that $\Sigma F_x = 0$ and $\Sigma F_y = 0$.

The maximum force of static friction, $\mathbf{f}_{s,\max}$, between an object and a surface is proportional to the normal force acting on the object. This maximum force occurs when the object is on the verge of slipping. In general,

$$f_s \leq \mu_s n \qquad [4.9]$$

where μ_s is the **coefficient of static friction.** When an object slides over a surface, the direction of the force of kinetic friction, $\mathbf{f}_k$, is opposite the direction of the motion, and the magnitude is proportional to that of the normal force. The magnitude of $\mathbf{f}_k$ is

$$f_k = \mu_k n \qquad\qquad [4.10]$$

where μ_k is the **coefficient of kinetic friction.** In general, $\mu_k < \mu_s$.

MULTIPLE-CHOICE QUESTIONS

1. A block of mass 10 kg remains at rest on a hill inclined at an angle of 30° with the horizontal. Which one of the statements below is correct about the magnitude of the force of friction that acts on the mass?
   (a) It is larger than the weight of the block.
   (b) It is equal to the weight of the block.
   (c) It is greater than the component of the force of gravity down the plane.
   (d) It is less than the component of the force of gravity down the plane.
   (e) It is equal to the component of the force of gravity down the plane.

2. A hockey puck struck by a hockey stick is given an initial speed of 10 m/s. If the coefficient of kinetic friction between ice and puck is 0.10, how far will the puck slide before stopping?
   (a) 39 m (b) 45 m (c) 51 m (d) 57 m

3. A thrown brick hits a window, breaking the glass, and ends up on the floor inside the room. Even though the brick broke the glass, we know that
   (a) the force of the brick on the glass was bigger than the force of the glass on the brick.
   (b) the force of the brick on the glass was the same size as the force of the glass on the brick.
   (c) the force of the brick on the glass was less than the force of the glass on the brick.
   (d) the brick didn't slow down as it broke the glass.

4. A manager of a restaurant pushes horizontally with a force of magnitude 200 N on a box of meat. The box moves across the floor with a constant forward acceleration. Which statement is most true about the magnitude of the force of kinetic friction acting on the box?
   (a) It is greater than 200 N.
   (b) It is less than 200 N.
   (c) It is equal to 200 N.
   (d) None of the above are necessarily true.

5. In the absence of air resistance, the force on a football at the top of its path after having been kicked is/are
   (a) the force due to the horizontal motion of the football.
   (b) the force of gravity.
   (c) the force exerted on it by the kicker and the force of gravity.
   (d) the force exerted on it by the Earth and the force of gravity.

6. Four forces act on an object, given by $\mathbf{A}$ = 40 N east, $\mathbf{B}$ = 50 N north, $\mathbf{C}$ = 70 N west, and $\mathbf{D}$ = 90 N south. What is the magnitude of the net force on the object?
   (a) 50 N (b) 131 N (c) 170 N (d) 250 N
   (e) not enough information given

CONCEPTUAL QUESTIONS

1. (a) An object has only one force acting on it. Can it be at rest? Can it have an acceleration? (b) An object has zero acceleration. Does this mean that no forces act on it?

2. If an object is at rest, can we conclude that no external forces are acting on it?

3. Is it possible for an object to move if no net force acts on it?

4. If gold were sold by weight, would you rather buy it in Denver or in Death Valley? If it were sold by mass, in which of the two locations would you prefer to buy it? Why?

5. A passenger sitting in the rear of a bus claims that she was injured as the driver slammed on the brakes, causing a suitcase to come flying toward her from the front of the bus. If you were the judge in this case, what disposition would you make? Why?

6. A space explorer is moving through space far from any planet or star. She notices a large rock, taken as a specimen from an alien planet, floating around the cabin of

the ship. Should she push it gently or kick it toward the storage compartment? Why?

7. What force causes an automobile to move? A propeller-driven airplane? A rowboat?

8. Analyze the motion of a rock dropped in water in terms of its speed and acceleration as it falls. Assume that a resistive force is acting on the rock that increases as the velocity increases.

9. In the motion picture *It Happened One Night* (Columbia Pictures, 1934), Clark Gable is standing inside a stationary bus in front of Claudette Colbert, who is seated. The bus suddenly starts moving forward and Clark falls into Claudette's lap. Why did this happen?

10. A weightlifter stands on a bathroom scale. He pumps a barbell up and down. What happens to the reading on the bathroom scale as this is done? Suppose he is strong enough to actually *throw* the barbell upward. How does the reading on the scale vary now?

11. In a tug-of-war between two athletes, each pulls on the rope with a force of 200 N. What is the tension in the rope? If the rope does not move, what horizontal force does each athlete exert against the ground?

12. As a rocket is fired from a launching pad, its speed and acceleration increase with time as its engines continue to operate. Explain why this occurs even though the thrust of the engines remains constant.

13. Identify the action-reaction pairs in the following situations: a man takes a step; a snowball hits a girl in the back; a baseball player catches a ball; a gust of wind strikes a window.

14. The driver of a speeding empty truck slams on the brakes and skids to a stop through a distance d. (a) If the truck carried a load that doubled its mass, what would be the truck's "skidding distance"? (b) If the initial speed of the truck were halved, what would be the truck's "skidding distance"?

15. Suppose you are driving a car at a high speed. Why should you avoid slamming on your brakes when you want to stop in the shortest possible distance?

16. Suppose a truck loaded with sand accelerates along a highway. If the driving force on the truck remains constant, what happens to the truck's acceleration if its trailer leaks sand at a constant rate through a hole in its bottom?

17. A large crate is placed on the bed of a truck but not tied down. (a) As the truck accelerates forward, the crate remains at rest relative to the truck. What force causes the crate to accelerate forward? (b) If the driver slammed on the brakes, what could happen to the crate?

18. Describe a few examples in which the force of friction exerted on an object is in the direction of motion of the object.

PROBLEMS

1, **2**, **3** = straightforward, intermediate, challenging   ☐ = full solution available in Study Guide/Student Solutions Manual
WEB = solution posted at **http://www.saunderscollege.com/physics/college**   🖾 = biomedical application   🖼 = Interactive Physics

Section 4.1   The Concept of Force

Section 4.2   Newton's First Law

Section 4.3   Newton's Second Law

Section 4.4   Newton's Third Law

1. A 6.0-kg object undergoes an acceleration of 2.0 m/s^2. (a) What is the magnitude of the resultant force acting on it? (b) If this same force is applied to a 4.0-kg object, what acceleration is produced?

2. A football punter accelerates a football from rest to a speed of 10 m/s during the time in which his toe is in contact with the ball (about 0.20 s). If the football has a mass of 0.50 kg, what average force does the punter exert on the ball?

3. A bag of sugar weighs 5.00 lb on Earth. What should it weigh in newtons on the Moon, where the acceleration due to gravity is 1/6 that on Earth? Repeat for Jupiter,

where g is 2.64 times Earth gravity. Find the mass in kilograms at each of the three locations.

4. A freight train has a mass of 1.5×10^7 kg. If the locomotive can exert a constant pull of 7.5×10^5 N, how long does it take to increase the speed of the train from rest to 80 km/h?

5. The air exerts a forward force of 10 N on the propeller of a 0.20-kg model airplane. If the plane accelerates forward at 2.0 m/s^2, what is the magnitude of the resistive force exerted by the air on the airplane?

6. A 5.0-g bullet leaves the muzzle of a rifle with a speed of 320 m/s. What force (assumed constant) is exerted on the bullet while it is traveling down the 0.82-m-long barrel of the rifle?

7. A boat moves through the water with two forces acting on it. One is a 2000-N forward push by the motor, and the other is an 1800-N resistive force due to the water.
(a) What is the acceleration of the 1000-kg boat?
(b) If it starts from rest, how far will it move in 10.0 s?
(c) What will its velocity be at the end of this time?

Figure P4.8

8. Two forces are applied to a car in an effort to move it, as shown in Figure P4.8. **(a)** What is the resultant of these two forces? **(b)** If the car has a mass of 3000 kg, what acceleration does it have?
9. The force of the wind on the sails of a sailboat is 390 N north. The water exerts a force of 180 N east. If the boat (including crew) has a mass of 270 kg, what are the magnitude and direction of its acceleration?
10. After falling from rest at a height of 30 m, a 0.50-kg ball rebounds upward, reaching a height of 20 m. If the contact between ball and ground lasted 2.0 ms, what average force was exerted on the ball?

Section 4.5 Some Applications of Newton's Laws

11. Find the tension in each cable supporting the 600-N cat burglar in Figure P4.11.

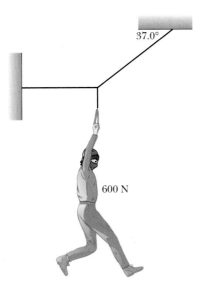

Figure P4.11

12. Find the tension in the two wires that support the 100-N light fixture in Figure P4.12.

Figure P4.12

WEB **13.** A 150-N bird feeder is supported by three cables, as shown in Figure P4.13. Find the tension in each cable.

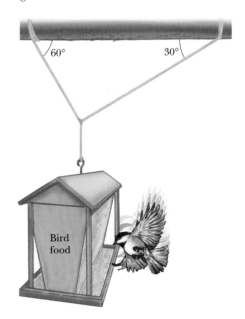

Figure P4.13

14. The leg and cast in Figure P4.14 weigh 220 N (w_1). Determine the weight w_2 and the angle α needed in order that there be no force exerted on the hip joint by the leg plus cast.

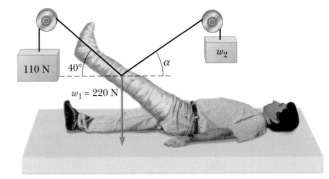

Figure P4.14

15. A block of mass $m = 2.0$ kg is held in equilibrium on an incline of angle $\theta = 60°$ by the horizontal force **F**, as shown in Figure P4.15. **(a)** Determine the value of F. **(b)** Determine the normal force exerted by the incline on the block (ignore friction).

Figure P4.15   (Problems 15 and 42)

16. Two persons are pulling a boat through the water, as in Figure P4.16. Each exerts a force of 600 N directed at a 30.0° angle relative to the forward motion of the boat. If the boat moves with constant velocity, find the resistive force, **F**, exerted on the boat by the water.

Figure P4.16   Two people pulling a boat.

17. A 5.0-kg bucket of water is raised from a well by a rope. If the upward acceleration of the bucket is 3.0 m/s², find the force exerted by the rope on the bucket.

18. A shopper in a supermarket pushes a loaded cart with a horizontal force of 10 N. The cart has a mass of 30 kg. **(a)** How far will it move in 3.0 s, starting from rest? (Ignore friction.) **(b)** How far will it move in 3.0 s if the shopper places her 30-N child in the cart before she begins to push it?

19. A 2000-kg car is slowed down uniformly from 20.0 m/s to 5.00 m/s in 4.00 s. What average force acted on the

car during this time, and how far did the car travel during the deceleration?

20. Two objects of masses 10.0 kg and 5.00 kg are connected by a light string that passes over a frictionless pulley as in Figure P4.20. The 5.00-kg object lies on a smooth incline of angle 40.0°. Find the acceleration of the 5.00 kg object and the tension in the string.

Figure P4.20

21. Assume that the three blocks in Figure P4.21 move on a frictionless surface and that a 42 N force acts as shown on the 3.0-kg block. Determine **(a)** the acceleration given this system, **(b)** the tension in the cord connecting the 3.0-kg and the 1.0-kg blocks, and **(c)** the force exerted on the 2.0-kg block by the 1.0-kg block.

Figure P4.21

22. A train has a mass of 5.22×10^6 kg and is moving at 90.0 km/h. The engineer applies the brakes, which results in a net backward force of 1.87×10^6 N on the train. The brakes are held on for 30.0 s. **(a)** What is the new speed of the train? **(b)** How far does it travel during this period?

23. A 2.0-kg mass starts from rest and slides down an inclined plane 80 cm long in 0.50 s. What *net force* is acting on the mass along the incline?

24. A 40.0-kg wagon is towed up a hill inclined at 18.5° with respect to the horizontal. The tow rope is parallel to the incline and has a tension of 140 N in it. Assume that the wagon starts from rest at the bottom of the hill, and neglect friction. How fast is the wagon going after moving 80 m up the hill?

25. A mass, $m_1 = 5.00$ kg, resting on a frictionless horizontal table is connected to a cable that passes over a pulley and then is fastened to a hanging mass, $m_2 = 10.0$ kg, as in Figure P4.25. Find the acceleration of each mass and the tension in the cable.

m_1

m_2

Figure P4.25 (Problems 25, 30, and 37)

26. Two blocks are fastened to the ceiling of an elevator, as in Figure P4.26. The elevator accelerates upward at 2.00 m/s². Find the tension in each rope.

A

B

10.0 kg   2.00 m/s²

C

D

10.0 kg

Figure P4.26

WEB 27. A 1000-kg car is pulling a 300-kg trailer. Together the car and trailer have an acceleration of 2.15 m/s² in the forward direction. Neglecting frictional forces on the trailer, determine (a) the net force on the car; (b) the net force on the trailer; (c) the force exerted on the car by the trailer; (d) the resultant force exerted on the road by the car.

28. Two masses of 3.00 kg and 5.00 kg are connected by a light string that passes over a frictionless pulley as in Figure P4.28. Determine (a) the tension in the string,

(b) the acceleration of each mass, and (c) the distance each mass will move in the first second of motion if both masses start from rest.

3.00 kg

5.00 kg

Figure P4.28

Section 4.6 Force of Friction

29. A dockworker loading crates on a ship finds that a 20-kg crate, initially at rest on a horizontal surface, requires a 75-N horizontal force to set it in motion. However, after the crate is in motion, a horizontal force of 60 N is required to keep it moving with a constant speed. Find the coefficients of static and kinetic friction between crate and floor.

30. In Figure P4.25, $m_1 = 10$ kg and $m_2 = 4.0$ kg. The coefficient of static friction between m_1 and the horizontal surface is 0.50 while the coefficient of kinetic friction is 0.30. (a) If the system is released from rest, what will its acceleration be? (b) If the system is set in motion with m_2 moving downward, what will be the acceleration of the system?

31. A 1000-N crate is being pushed across a level floor at a constant speed by a force **F** of 300 N at an angle of 20.0° below the horizontal as shown in Figure P4.31a. (a) What is the coefficient of kinetic friction between the crate and the floor? (b) If the 300-N force is instead pulling the block at an angle of 20.0° above the horizontal as shown in Figure P4.31b, what will be the acceleration of the crate? Assume that the coefficient of friction is the same as found in (a).

F

F

(a)

(b)

Figure P4.31

32. A hockey puck is hit on a frozen lake and starts moving with a speed of 12.0 m/s. Five seconds later, its speed is 6.00 m/s. **(a)** What is its average acceleration? **(b)** What is the average value of the coefficient of kinetic friction between puck and ice? **(c)** How far does the puck travel during this 5.00-s interval?

33. The coefficient of static friction between the 3.00-kg crate and the 35.0° incline of Figure P4.33 is 0.300. What minimum force **F** must be applied to the crate perpendicular to the incline to prevent the crate from sliding down the incline?

3.00 kg

35.0°

Figure P4.33

34. A box of books weighing 300 N is shoved across the floor of an apartment by a force of 400 N exerted downward at an angle of 35.2° below the horizontal. If the coefficient of kinetic friction between box and floor is 0.57, how long does it take to move the box 4.00 m, starting from rest?

35. An object falling under the pull of gravity experiences a frictional force of air resistance. The magnitude of this force is approximately proportional to the speed of the object, $f = bv$. Assume that $b = 15$ kg/s and $m = 50$ kg. **(a)** What is the terminal speed that the object reaches while falling? **(b)** Does your answer to part **(a)** depend on the initial speed of the object? Explain.

36. A student decides to move a box of books into her dormitory room by pulling on a rope attached to the box. She pulls with a force of 80.0 N at an angle of 25.0° above the horizontal. The box has a mass of 25.0 kg, and the coefficient of kinetic friction between box and floor is 0.300. **(a)** Find the acceleration of the box. **(b)** The student now starts moving the box up a 10.0° incline, keeping her 80.0 N force directed at 25.0° above the line of the incline. If the coefficient of friction is unchanged, what is the new acceleration of the box?

37. Masses $m_1 = 10.0$ kg and $m_2 = 5.00$ kg are connected by a light string that passes over a frictionless pulley as in Figure P4.25. If, when the system starts from rest, m_2 falls 1.00 m in 1.20 s, determine the coefficient of kinetic friction between m_1 and the table.

38. A car is traveling at 50.0 km/h on a flat highway. **(a)** If the coefficient of friction between road and tires on a

rainy day is 0.100, what is the minimum distance in which the car will stop? **(b)** What is the stopping distance when the surface is dry and the coefficient of friction is 0.600?

39. A box slides down a 30.0° ramp with an acceleration of 1.20 m/s². Determine the coefficient of kinetic friction between the box and the ramp.

40. Masses $m_1 = 4.00$ kg and $m_2 = 9.00$ kg are connected by a light string that passes over a frictionless pulley. As shown in Figure P4.40, m_1 is held at rest on the floor and m_2 rests on a fixed incline of $\theta = 40.0°$. The masses are released from rest, and m_2 slides 1.00 m down the incline in 4.00 s. Determine **(a)** the acceleration of each mass, **(b)** the tension in the string, and **(c)** the coefficient of kinetic friction between m_2 and the incline.

m_2

m_1

40.0°

Figure P4.40

41. Find the acceleration experienced by each of the two masses shown in Figure P4.41 if the coefficient of kinetic friction between the 7.00-kg mass and the plane is 0.250.

7.00 kg

12.0 kg

37.0°

Figure P4.41

42. A 2.00-kg block is held in equilibrium on an incline of angle $\theta = 60.0°$ by a horizontal force, **F**, applied in the direction shown in Figure P4.15. If the coefficient of static friction between block and incline is $\mu_s = 0.300$, determine **(a)** the minimum value of **F** and **(b)** the normal force of the incline on the block.

43. The person in Figure P4.43 weighs 170 lb. The crutches each make an angle of 22.0° with the vertical (as seen from the front). Half of the person's weight is sup-

ported by the crutches. The other half is supported by the vertical forces of the ground on his feet. Assuming the person is at rest and the force of the ground on the crutches acts along the crutches, determine **(a)** the smallest possible coefficient of friction between crutches and ground and **(b)** the magnitude of the compression force supported by each crutch.

Figure P4.43

44. A block of mass $m = 2.00$ kg rests on the left edge of a block of length $L = 3.00$ m and mass $M = 8.00$ kg. The coefficient of kinetic friction between the two blocks is $\mu_k = 0.300$, and the surface on which the 8.00-kg block rests is frictionless. A constant horizontal force of magnitude $F = 10.0$ N is applied to the 2.00-kg block, setting it in motion as shown in Figure P4.44a. **(a)** How long will

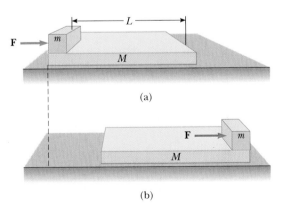

Figure P4.44

it take before this block makes it to the right side of the 8.00-kg block, as shown in Figure P4.44b? (*Note:* Both blocks are set in motion when **F** is applied.) **(b)** How far does the 8.00-kg block move in the process?

ADDITIONAL PROBLEMS

45. **(a)** What is the resultant force exerted by the two cables supporting the traffic light in Figure P4.45? **(b)** What is the weight of the light?

Figure P4.45

46. As a protest against the umpire's calls, a baseball pitcher throws a ball straight up into the air at a speed of 20.0 m/s. In the process, he moves his hand through a distance of 1.50 m. If the ball has a mass of 0.150 kg, find the average force he exerts on the ball to give it this upward speed.

47. A girl coasts down a hill on a sled, reaching a level surface at the bottom with a speed of 7.0 m/s. If the coefficient of friction between runners and snow is 0.050 and the girl and sled together weigh 600 N, how far does the sled travel on the level surface before coming to rest?

48. **(a)** What is the minimum force of friction required to hold the system of Figure P4.48 in equilibrium?

Figure P4.48

(b) What coefficient of static friction between the 100-N block and the table ensures equilibrium? (c) If the coefficient of kinetic friction between the 100-N block and the table is 0.250, what hanging weight should replace the 50.0-N weight to allow the system to move at a constant speed once it is set in motion?

WEB 49. A box rests on the back of a truck. The coefficient of static friction between box and truck bed is 0.300. (a) When the truck accelerates forward, what force accelerates the box? (b) Find the maximum acceleration the truck can have before the box slides.

50. A 4.00-kg block is pushed along the ceiling with a constant applied force of 85.0 N that acts at an angle of 55.0° with the horizontal, as in Figure P4.50. The block accelerates to the right at 6.00 m/s². Determine the coefficient of kinetic friction between block and ceiling.

85.0 N
55.0°

Figure P4.50

51. A 2.00-kg aluminum block and a 6.00-kg copper block are connected by a light string over a frictionless pulley. They are allowed to move on a fixed steel block-wedge (of angle $\theta = 30.0°$) as shown in Figure P4.51. Making use of Table 4.2, determine (a) the acceleration of the two blocks and (b) the tension in the string.

Aluminum
m_1
Copper
m_2
Steel
θ

Figure P4.51

52. Three masses are connected by light strings, as shown in Figure P4.52. The string connecting the 4.00-kg mass

and the 5.00-kg mass passes over a light frictionless pulley. Determine (a) the acceleration of each mass and (b) the tensions in the two strings.

4.00 kg
5.00 kg
3.00 kg

Figure P4.52

53. A 3.00-kg block starts from rest at the top of a 30.0° incline and slides 2.00 m down the incline in 1.50 s. Find (a) the acceleration of the block, (b) the coefficient of kinetic friction between it and the incline, (c) the frictional force acting on the block, and (d) the speed of the block after it has slid 2.00 m.

54. A 5.0-kg penguin sits on a 10-kg sled, as in Figure P4.54. A horizontal force of 45 N is applied to the sled, but the penguin attempts to impede the motion by holding onto a cord attached to a tree. The coefficient of kinetic friction between the sled and snow as well as that between the sled and the penguin is 0.20. (a) Draw a free-body diagram for the penguin and one for the sled, and identify the reaction force for each force you include. Determine (b) the tension in the cord and, (c) the acceleration of the sled.

45 N

Figure P4.54

55. Two blocks on a frictionless horizontal surface are connected by a light string as in Figure P4.55, where $m_1 = 10$ kg and $m_2 = 20$ kg. A force of 50 N is applied to the 20-kg block. **(a)** Determine the acceleration of each block and the tension in the string. **(b)** Repeat the problem for the case where the coefficient of kinetic friction between each block and the surface is 0.10.

Figure P4.55

56. In Figure P4.56, the coefficient of kinetic friction between the two blocks is 0.30. The table surface and the pulleys are frictionless. **(a)** Draw a free-body diagram for each block. **(b)** Determine the acceleration of each block. **(c)** Find the tensions in the strings.

Figure P4.56

57. Two people pull as hard as they can on ropes attached to a 200-kg boat. If they pull in the same direction, the boat has an acceleration of 1.52 m/s^2 to the right. If they pull in opposite directions, the boat has an acceleration of 0.518 m/s^2 to the left. What is the force exerted by each person on the boat? (Disregard any other forces on the boat.)

58. A 3.0-kg mass hangs at one end of a rope that is attached to a support on a railroad car. When the car accelerates to the right, the rope makes an angle of 4.0° with the vertical, as shown in Figure P4.58. Find the acceleration of the car.

Figure P4.58

59. The three blocks of masses 10.0 kg, 5.00 kg, and 3.00 kg are connected by light strings that pass over frictionless pulleys as shown in Figure P4.59. The acceleration of the 5.00-kg block is 2.00 m/s^2 to the left, and the surfaces are rough. Find **(a)** the tension in each string and **(b)** the coefficient of kinetic friction between blocks and surfaces. (Assume the same μ_k for both blocks in contact with surfaces.)

Figure P4.59

60. A sled weighing 60.0 N is pulled horizontally across snow and the coefficient of kinetic friction between sled and snow is 0.100. A penguin weighing 70.0 N rides on the sled. (See Fig. P4.60.) If the coefficient of static friction between penguin and sled is 0.700, find the maximum horizontal force that can be exerted on the sled before the penguin begins to slide off.

Figure P4.60

61. An 80-kg person escapes from a burning building by jumping from a window situated 30 m above a catching net. Assuming that air resistance exerts a 100-N force on the person as he falls, determine his velocity just before he hits the net.

62. The parachute on a race car of weight 8820 N opens at the end of a quarter-mile run when the car is traveling at 35 m/s. What total retarding force must be supplied by the parachute to stop the car in a distance of 1000 m?

WEB 63. On takeoff, the combined action of the engines and wings of an airplane exerts an 8000-N force on the plane, directed upward at an angle of 65.0° above the horizontal. The plane rises with constant velocity in the vertical direction while continuing to accelerate in the horizontal direction. (a) What is the weight of the plane? (b) What is its horizontal acceleration?

64. A van accelerates down a hill (Fig. P4.64), going from rest to 30.0 m/s in 6.00 s. During the acceleration, a toy (m = 0.100 kg) hangs by a string from the van's ceiling. The acceleration is such that the string remains perpendicular to the ceiling. Determine (a) the angle θ and (b) the tension in the string.

Figure P4.64

65. The board sandwiched between two other boards in Figure P4.65 weighs 95.5 N. If the coefficient of friction between the boards is 0.663, what must be the magnitude of the compression forces (assume horizontal) acting on both sides of the center board to keep it from slipping?

Figure P4.65

66. A 72-kg man stands on a spring scale in an elevator. Starting from rest, the elevator ascends, attaining its maximum speed of 1.2 m/s in 0.80 s. It travels with this constant speed for 5.0 s, undergoes a uniform *negative* acceleration for 1.5 s, and comes to rest. What does the spring scale register (a) before the elevator starts to move? (b) during the first 0.80 s? (c) while the elevator is traveling at constant speed? (d) during the negative acceleration?

67. Bob and Kathy, two construction workers on the roof of a building, are about to raise a keg of nails from the ground by means of a light rope passing over a light frictionless pulley 10.0 m above the ground. Bob weighs 900 N, Kathy 600 N, the keg 300 N, and the nails 600 N. Both workers slip off the roof, and the following unfortunate sequence of events takes place. Hanging together on the rope, Bob and Kathy strike the ground just as the keg hits the pulley. Unnerved by his fall, Bob lets go of the rope, and the keg pulls Kathy up to the roof, where she cracks her head against the pulley but gamely hangs on. However, the nails spill out of the keg when it strikes the ground, and the empty keg rises as Kathy returns to the ground. Finally, she has had enough, lets go of the rope, and remains on the ground, only to be hit by the empty keg again. Ignoring the possible mid-air collisions that merely added insult to injury, how long did it take this industrial accident to run its course? Assume that all collisions with the ground and pulley serve to start each subsequent motion from rest, so that each trip up and down begins with zero velocity.

68. A bag of cement hangs from three wires, as shown in Figure P4.68. Two of the wires make angles θ_1 and θ_2 with the horizontal. If the system is in equilibrium, (a) show that

$$T_1 = \frac{w \cos \theta_2}{\sin(\theta_1 + \theta_2)}$$

(b) Given that $w = 325$ N, $\theta_1 = 10.0°$ and $\theta_2 = 25.0°$, find the tensions T_1, T_2, and T_3 in the wires.

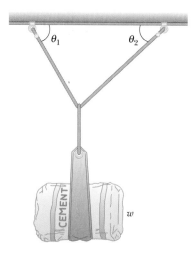

Figure P4.68

69. An inventive child named Pat wants to reach an apple in a tree without climbing the tree. Sitting in a chair connected to a rope that passes over a frictionless pulley (Fig. P4.69), Pat pulls on the loose end of the rope with such a force that the spring scale reads 250 N. Pat's true weight is 320 N, and the chair weighs 160 N. (a) Show that the acceleration of the system of Pat and the chair is *upward* and find its magnitude. (b) Find the force Pat exerts on the chair.

70. A fire helicopter carries a 620-kg bucket of water at the end of a 20.0-m long cable. Flying back from a fire at a constant speed of 40.0 m/s, the cable makes an angle of 40.0° with respect to the vertical. Determine the force of air resistance on the bucket.

Figure P4.69

Work and Energy

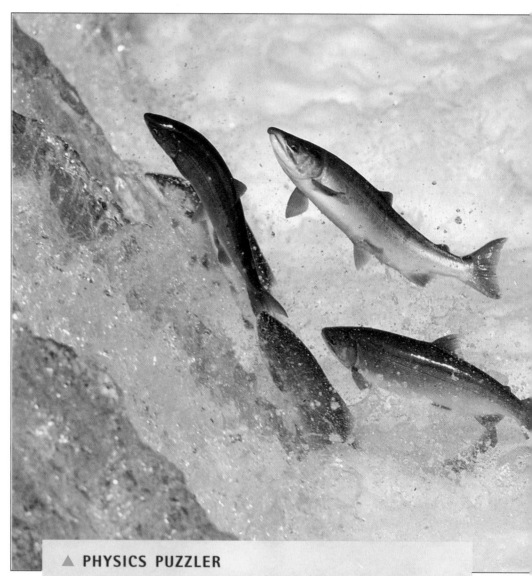

▲ **PHYSICS PUZZLER**

Chum salmon leaping up a waterfall in the McNeil River in Alaska. The salmon must reach high speeds before leaping out of the water. Why are fish ladders often built around dams? Can you estimate how fast they must swim to leap a vertical distance of 2 m? *(Daniel J. Cox/Tony Stone Images)*

The concept of energy is one of the most important in the world of science. In everyday use, the term *energy* has to do with the cost of fuel for transportation and heating, electricity for lights and appliances, and the foods we consume. However, these ideas do not really define energy. They tell us only that fuels are needed to do a job and that those fuels provide us with something we call energy.

Energy is present in the Universe in a variety of forms, including mechanical energy, chemical energy, electromagnetic energy, and nuclear energy. Although energy can be transformed from one form to another, the total amount of energy in the Universe remains the same. If one form of energy in an isolated system decreases, then, by the principle of conservation of energy, another form of energy in the system must increase. For example, if the system consists of a motor connected to a battery, chemical energy is converted to electrical energy, which in turn is converted to mechanical energy. The transformation of energy from one form into another is essential in the study of physics, chemistry, biology, geology, and astronomy.

In this chapter we are concerned only with mechanical energy. We introduce the concept of *kinetic energy,* which is defined as the energy associated with motion, and the concept of *potential energy,* the energy associated with position. We shall see that the ideas of work and energy can be used in place of Newton's laws to solve certain problems.

We begin by defining *work,* a concept that provides a link between force and energy. With this as a foundation, we can then discuss the principle of conservation of energy and apply it to problems.

Figure 5.1 If an object undergoes a displacement of **s,** the work done by the force **F** is $(F \cos \theta) s$.

5.1 WORK

Almost all the terms we have used thus far have conveyed the same meaning in physics as in everyday life. Now, however, we encounter a term whose meaning in physics is distinctly different from its meaning in our day-to-day affairs. This new term is **work.** It can be defined with the help of Figure 5.1. Here we see an object that undergoes a displacement of **s** along a straight line while acted on by a constant force, **F,** that makes an angle of θ with **s.**

> The work W done by an agent exerting a constant force is defined as the product of the component of the force along the direction of displacement and the magnitude of the displacement:

$$W \equiv (F \cos \theta) s \qquad [5.1]$$

As an example of the distinction between this definition of *work* and our everyday understanding of the word, consider holding a heavy chair at arm's length for 10 min. At the end of this time interval, your tired arms may lead you to think that you have done a considerable amount of work. According to our definition, how-

Figure 5.2 No work is done when a bucket is moved horizontally, because the applied force, **F,** is perpendicular to the displacement.

Figure 5.3 Positive work is done by this person when the box is lifted, because the applied force, **F,** is in the same direction as the displacement. When the box is lowered to the floor, the work done by the person is negative.

ever, you have done no work on the chair.[1] You have exerted a force in order to support the chair, but you have not moved it. A force does no work on an object if the object does not move. This can be seen by noting that if $s = 0$, Equation 5.1 gives $W = 0$.

Also note from Equation 5.1 that the work done by a force is zero when the force is perpendicular to the displacement. That is, if $\theta = 90°$, then $W = 0$ because $\cos 90° = 0$. For example, no work is done when a bucket of water is carried horizontally at constant velocity, because the upward force exerted to support the bucket is perpendicular to the displacement of the bucket, as shown in Figure 5.2. Likewise, the work done by the force of gravity during the horizontal displacement is also zero, for the same reason.

The sign of the work depends on the direction of **F** relative to **s.** Work is positive when the component $F \cos \theta$ is in the same direction as the displacement. For example, when you lift a box as in Figure 5.3, the work done by the force you exert on the box is positive because that force is upward, in the same direction as the displacement. Work is negative when **F** $\cos \theta$ is in the direction opposite the displacement. The negative sign comes from the fact that $\theta = 180°$ and $\cos 180° = -1$, which, from Equation 5.1, gives a negative value for W. Finally, if an

---

[1] Actually, you do burn calories while holding a chair at arm's length, because your muscles are continuously contracting and relaxing while the chair is being supported. Thus, work is being done on your body—but not on the chair.

TABLE 5.1 Units of Work in the Three Common Systems of Measurement

System	Unit of Work	Name of Combined Unit
SI	newton-meter (N · m)	joule (J)
cgs	dyne-centimeter (dyne · cm)	erg
British engineering (conventional)	foot-pound (ft · lb)	foot-pound

applied force acts along the direction of the displacement, then $\theta = 0°$. Because $\cos 0° = 1$, in this case Equation 5.1 becomes

$$W = Fs \qquad [5.2]$$

Work is a scalar quantity, and its units are force times length. Therefore, the SI unit of work is the **newton-meter (N·m).** Another name for the newton-meter is the **joule (J).** The unit of work in the cgs system is the **dyne-centimeter (dyne·cm),** which is also called the **erg,** and the unit in the conventional (British engineering) system is the **foot-pound (ft·lb).** These are summarized in Table 5.1. Note that $1 \, J = 10^7$ ergs.

Thinking Physics 1

A person lifts a cement block of mass m a vertical height h, and then walks horizontally a distance d while holding the block, as in Figure 5.4. Determine the work done by the person and by the force of gravity in this process.

Explanation Assuming that the person lifts the block with a force of magnitude equal to the weight of the block, mg, the work done by the person during the vertical displacement is mgh, because the force in this case is in the direction of the displacement. The work done by the person during the horizontal displacement of the block is zero because the applied force in this process is perpendicular to the displacement. Thus, the net work done by the person is mgh. The work done by the force of gravity during the vertical displacement of the block is $-mgh$, because this force is opposite the displacement. The work done by the force of gravity is zero during the horizontal displacement because this force is also perpendicular to the displacement. Hence, the net work done by the force of gravity is $-mgh$. The net work on the block is zero $(+mgh - mgh = 0)$.

Figure 5.4 (Thinking Physics 1) A person lifts a cement block of mass m a vertical distance h and then walks horizontally a distance d.

EXAMPLE 5.1 Mr. Clean

A man cleaning his apartment pulls the canister of a vacuum cleaner with a force of magnitude $F = 50.0 \, N$ at an angle of $30.0°$, as shown in Figure 5.5. He moves the vacuum cleaner a distance of 3.00 m. Calculate the work done by the 50.0-N force.

Figure 5.5   (Example 5.1) A vacuum cleaner being pulled at an angle of 30.0° with the horizontal.

Solution   We can use $W = (F \cos \theta)\, s$, with $F = 50.0$ N, $\theta = 30.0°$, and $s = 3.00$ m, to get

$$W = (50.0 \text{ N})(\cos 30.0°)(3.00 \text{ m}) = \boxed{130 \text{ J}}$$

Note that the normal force, **n,** the weight, *m***g,** and the upward component of the applied force (50.0 N) cos 30.0° do *no* work because they are perpendicular to the displacement.

Exercise   Find the work done on the vacuum cleaner by the man if he pulls it 3.00 m with a horizontal force of $F = 50.0$ N.

Answer   150 J

5.2   KINETIC ENERGY AND THE WORK–KINETIC ENERGY THEOREM

Solving problems with Newton's second law can be difficult if the forces involved are complex. An alternative approach to such problems is to relate the speed of an object to its displacement under the influence of some net force. If the work done by the net force on the object can be calculated for a given displacement, the change in the object's speed is easy to evaluate.

Figure 5.6 shows a particle of mass m moving to the right under the action of a constant net force, **F.** Because the force is constant, we know from Newton's second law that the particle moves with a constant acceleration, **a.** If the particle is displaced a distance of s, the work done by **F** is

$$W_{\text{net}} = Fs = (ma)\, s \qquad [5.3]$$

However, in Chapter 2 we found that the following relationship holds when an object undergoes constant acceleration:

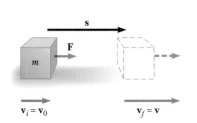

Figure 5.6   An object undergoing a displacement and a change in velocity under the action of a constant net force, **F.**

$$v^2 = v_0^2 + 2as \qquad \text{or} \qquad as = \frac{v^2 - v_0^2}{2}$$

This expression is now substituted into Equation 5.3 to give

$$W_{\text{net}} = m\left(\frac{v^2 - v_0^2}{2}\right)$$

$$W_{\text{net}} = \tfrac{1}{2}mv^2 - \tfrac{1}{2}mv_0^2 \qquad\qquad\qquad \textbf{[5.4]}$$

The quantity $mv^2/2$ has a special name in physics: **kinetic energy.** Any object of mass m and speed v is defined to have a kinetic energy, KE, of

$$KE \equiv \tfrac{1}{2}mv^2 \qquad\qquad \textbf{[5.5]} \qquad \blacktriangleleft \text{ Kinetic energy}$$

Kinetic energy is a scalar quantity and has the same units as work. For example, a 1.0-kg mass moving with a speed of 4.0 m/s has a kinetic energy of 8.0 J. We can think of kinetic energy as the energy associated with the motion of an object.

It is often convenient to write Equation 5.4 as

$$W_{\text{net}} = KE_f - KE_i \qquad\qquad \textbf{[5.6]}$$

According to this result, the work done by a *net* force acting on an object is to change the kinetic energy of the object from some initial value, KE_i, to some final value, KE_f. Equation 5.6 is an important result known as the **work-kinetic energy theorem.** Thus, we conclude that

$\blacktriangleleft$ Work-kinetic energy theorem

> **The net work done on an object by a net force acting on it is equal to the change in the kinetic energy of the object.**

From the work-kinetic energy theorem, we also find that the speed of the object increases if the net work done on it is positive, because the final kinetic energy is greater than the initial kinetic energy. The object's speed decreases if the net work is negative, because the final kinetic energy is less than the initial kinetic energy. Notice that the speed and kinetic energy of an object will change only if work is done on the object by some external force.

Consider the relationship between the work done on an object and the change in its kinetic energy as expressed by Equation 5.4. Because of this connection, we can also think of kinetic energy as the work the object can do in coming to rest. For example, suppose a hammer is on the verge of striking a nail, as in Figure 5.7. The moving hammer has kinetic energy and can do work on the nail. The work done on the nail is Fs, where F is the average force exerted on the nail by the hammer and s is the distance the nail is driven into the wall.

For convenience, Equation 5.4 was derived under the assumption that the net force acting on the object was constant. A more general derivation would show that this equation is valid under all circumstances, including that of a variable force.

Figure 5.7 The moving hammer has kinetic energy and thus can do work on the nail, driving it into the wall.

Thinking Physics 2

A car traveling at a speed v skids a distance d after its brakes lock. Estimate how far it will skid if its brakes lock when its initial speed is $2v$. What happens to the car's kinetic energy as it stops?

Explanation Let us assume that the force of kinetic friction between car and road surface is constant and the same in both cases. The net force times the displacement of the car is equal its initial kinetic energy. If the speed is doubled, as in this example, the kinetic energy of the car is quadrupled. For a given applied force (in this case, the frictional force), the distance traveled is four times as great when the initial speed is doubled, so the estimated distance it skids is $4d$. The kinetic energy of the car is changed into internal energy associated with the tires, brake pads, and road as they increase in temperature.

Thinking Physics 3

More energy is expended by walking downstairs than by walking horizontally at the same speed. Why do you think this is so?

Explanation When walking downstairs, a person has to stop the motion of her entire body at each step. This requires work to be done by the muscles. The work done by the muscles is negative, because the forces applied are opposite to the displacements representing the movement of the body. Negative work represents a transfer of energy out of the muscles (from the store of energy from the food eaten by the walker). When walking horizontally, the body can be kept in motion with an almost constant horizontal speed—only the feet and legs start and stop. Thus, more energy is transferred from the body in bringing the entire body to a stop in the trip downstairs than in walking.

EXAMPLE 5.2 Towing a Car

A 1400-kg car has a net forward force of 4500 N applied to it. The car starts from rest and travels down a horizontal highway. What are its kinetic energy and speed after it has traveled 100 m? (Ignore losses in kinetic energy because of friction and air resistance.)

Solution With the initial velocity given as zero, Equation 5.4 reduces to

$$(1) \qquad W_{\text{net}} = \tfrac{1}{2}mv^2$$

The work done by the net force on the car is

$$W_{\text{net}} = Fs = (4500 \text{ N})(100 \text{ m}) = 4.50 \times 10^5 \text{ J}$$

This work all goes into changing the kinetic energy of the car; thus the final value of the kinetic energy, from (1), is also 4.50×10^5 J.

The speed of the car can be found from (1) as follows:

$$\tfrac{1}{2}mv^2 = 4.50 \times 10^5 \text{ J}$$

$$v^2 = \frac{2(4.50 \times 10^5 \text{ J})}{1400 \text{ kg}} = 643 \text{ m}^2/\text{s}^2$$

$$v = 25.4 \text{ m/s}$$

5.3 POTENTIAL ENERGY

In the preceding section we saw that an object that has kinetic energy can do work on another object, as illustrated by the moving hammer driving a nail into the wall. Now we introduce another form of energy associated with the position of an object.

Gravitational Potential Energy

As an object falls freely in a gravitational field, the field exerts a force on it, doing positive work on it and thereby increasing its kinetic energy. Consider Figure 5.8, which shows a brick of mass m held at a height of y_i above a nail in a board lying on the ground. When the brick is released, it falls toward the ground, gaining speed and thereby gaining kinetic energy. As a result of its position in space, the brick has potential energy (it has the *potential* to do work), and this potential energy is converted to kinetic energy as the brick falls. When the brick reaches the ground, it does work on the nail, driving it into the board. The energy that an object has as a result of its position in space near the surface of the Earth is called **gravitational potential energy.**

Let us now derive an expression for the gravitational potential energy of an object at a given location. Consider a block of mass m at an initial height of y_i above the ground, as in Figure 5.9, and let us neglect air resistance. As the block falls, the only force that does work on it is the gravitational force, $m\mathbf{g}$. The work done by the gravitational force as the block undergoes a downward displacement of $s = y_i - y_f$ is

$$W_g = mgs = mgy_i - mgy_f$$

We define the quantity mgy to be the gravitational potential energy, *PE*:

$$PE \equiv mgy \qquad\qquad \text{[5.7]}$$

Thus, the gravitational potential energy associated with any object at a point in space is the product of the object's weight and its vertical coordinate. It is important to note that this relationship is valid only for objects near the Earth's surface, where g is approximately constant.

If we substitute this expression for *PE* into the expression for W_g, we have

$$W_g = PE_i - PE_f \qquad\qquad \text{[5.8]}$$

The work done on any object by the force of gravity is equal to the object's initial potential energy minus its final potential energy.

The units of gravitational potential energy are the same as those of work: joules, ergs, or foot-pounds. Potential energy, like work and kinetic energy, is a scalar quantity.

Brick

Figure 5.8 Because of its position in space, the brick can do work on the nail.

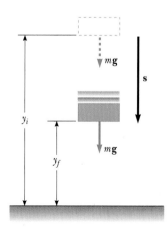

Figure 5.9 The work done by the gravitational force as the block falls from y_i to y_f is equal to $mgy_i - mgy_f$.

Figure 5.10 Any reference level could be used for measuring the gravitational potential energy of the book.

Reference Levels for Gravitational Potential Energy

In working problems involving the concept of gravitational potential energy, you must choose a location at which the gravitational potential energy is equal to zero. The location of this zero level is completely arbitrary. This is true because the quantity of importance is the *difference* in potential energy, and the difference in potential energy between two points is independent of the choice of zero level. It is often convenient to choose the surface of the Earth as the reference position for zero potential energy, but again this is not essential. Often the statement of a problem will suggest a convenient level to use. As an example, consider a book at several possible locations, as in Figure 5.10. When the book is at *A*, above the surface of a desk, a logical zero level for potential energy would be the surface of the desk. When the book is at *B*, however, the floor might be a more appropriate zero reference level. Finally, a location such as *C*, where the book is held out a window, would suggest choosing the surface of the Earth as the zero level of potential energy. The choice, however, makes no difference — any of the reference levels mentioned could be used as the zero level, regardless of whether the book is at *A*, *B*, or *C*. Example 5.3 illustrates this important point.

APPLICATION

"Ringing the Bell" at a Carnival.

Applying Physics 1

A common scene at a carnival is the Ring-the-Bell attraction, in which the player swings a hammer downward in an attempt to project a mass upward to ring a bell. The goal is to provide the mass with enough gravitational potential energy to lift it at least

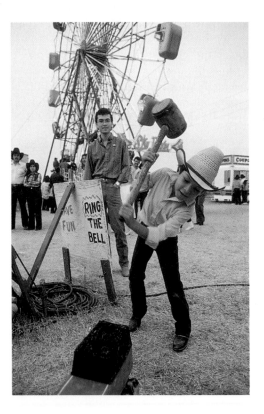

"Step right up, folks!" In this carnival event, part of the kinetic energy of the heavy hammer is transferred to a pivoted target, which in turn lifts a weight and hopefully rings the bell. *(Robert E. Daemmrich/ Tony Stone Images)*

as high as the bell. This gravitational potential energy will be converted from kinetic energy of the mass as it moves upward from its lowest position near the ground. In turn, this kinetic energy is related to how much work was done by the player in striking the device with the hammer. To maximize this work, and ultimately the height of the mass, the player must strike with as much force as possible, and move the hammer with as much distance as possible while it is in contact with the device.

EXAMPLE 5.3 Wax Your Skis

A 60.0-kg skier is at the top of a slope, as shown in Figure 5.11a. At the initial point A, the skier is 10.0 m vertically above point B.
(a) Setting the zero level for gravitational potential energy at B, find the gravitational potential energy of the skier at A and B, and then find the difference in potential energy between these two points.

(a)

(b)

Figure 5.11 (Example 5.3) *(b, Gregg Adams/Tony Stone Worldwide)*

Solution The gravitational potential energy at B is zero by choice. Hence, the potential energy at A is

$$PE_i = mgy_i = (60.0 \text{ kg})(9.80 \text{ m/s}^2)(10.0 \text{ m}) = \boxed{5880 \text{ J}}$$

Because $PE_f = 0$, the difference in potential energy is

$$PE_i - PE_f = 5880 \text{ J} - 0 \text{ J} = \boxed{5880 \text{ J}}$$

(b) Repeat this problem with the zero level at point A.

Solution   In this case, the initial potential energy is zero because of the chosen reference level. The final potential energy is

$$PE_f = mgy_f = (60.0 \text{ kg})(9.80 \text{ m/s}^2)(-10.0 \text{ m}) = \boxed{-5880 \text{ J}}$$

The distance y_f is negative because the final point is 10.0 m below the zero reference level. The difference in potential energy is now

$$PE_i = PE_f = 0 \text{ J} - (-5880 \text{ J}) = \boxed{5880 \text{ J}}$$

These calculations show that the potential energy of the skier at the top of the slope is greater than the potential energy at the bottom by 5880 J, *regardless of the zero level selected.*

Exercise   If the zero level for gravitational potential energy is selected to be midway down the slope, at a height of 5.00 m, find the initial potential energy, the final potential energy, and the difference in potential energy between points A and B.

Answer   2940 J, -2940 J, 5880 J

5.4   CONSERVATIVE AND NONCONSERVATIVE FORCES

If you slide down a very slick hill on a sled, your speed, and hence your kinetic energy, increases dramatically. However, if the hill is not very icy, your speed and kinetic energy do not increase as rapidly. What has happened to this "lost" kinetic energy? To answer this question, we shall examine the properties of two categories of forces that exist in nature — conservative and nonconservative forces.

Conservative Forces

Definition of a conservative force ▶

A force is conservative if the work it does on an object moving between two points is independent of the path the object takes between the points. In other words, the work done on an object by a conservative force depends only on the initial and final positions of the object. Furthermore, **a force is conservative if the work it does on an object moving through any closed path is zero.**

    The force of gravity is conservative. As we learned in the preceding section, the work done by the gravitational force on an object moving between any two points near the Earth's surface is

$$W_g = mgy_i - mgy_f$$

From this we see that W_g depends only on the initial and final coordinates of the object and hence is independent of path. Furthermore, note that W_g is zero when the object moves over any closed path (where $y_i = y_f$).

PHYSICS IN ACTION

On the left, these cyclists are working hard and expending energy as they pedal uphill in Marin County, California.

On the right, Twin Falls on the Island of Kauai, Hawaii. The potential energy of the water at the top of the falls is converted to kinetic energy at the bottom. In many locations, this mechanical energy is used to produce electrical energy.

Work and Energy in Sports and Nature

(David Madison, Tony Stone Images)

(Bruce Byers, FPG)

We can associate a potential energy function with any conservative force. In the preceding section, the potential energy function associated with the gravitational force was found to be

$$PE = mgy$$

Potential energy functions can be defined only for conservative forces. In general, the work, W_c, done on an object by a conservative force is equal to the initial potential energy of the object minus the final value:

$$W_c = PE_i - PE_f \qquad \qquad \textbf{[5.9]}$$

Nonconservative Forces

A force is nonconservative if it leads to a dissipation of mechanical energy. If you moved an object on a horizontal surface, returning it to the same location and same state of motion, but found it necessary to do net work on the object, then

◄ Definition of a nonconservative force

Figure 5.12 The loss in mechanical energy due to the force of friction depends on the path taken as the book is moved from *A* to *B*.

something must have dissipated the energy transferred to the object. That dissipative force is recognized as friction between object and surface. Friction is a dissipative, or nonconservative, force.

Suppose you displace a book between two points on a table. If you displace it in a straight line along the blue path shown in Figure 5.12, the change in mechanical energy due to friction is $-fd$, where d is the distance between the two points. However, if you move the book along any other path between the two points, the change in mechanical energy due to friction is greater (in absolute magnitude) than $-fd$. For example, the change in mechanical energy due to friction along the red semicircular path shown in Figure 5.12 is equal to $-f(\pi d/2)$, where d is the diameter of the circle.

5.5   CONSERVATION OF MECHANICAL ENERGY

Conservation principles play a very important role in physics, and you will encounter a number of them as you continue in this course. The first, conservation of energy, is one of the most important. Before we describe its mathematical details, let us pause briefly to examine what is meant by conservation. When we say that a physical quantity is *conserved,* we simply mean that the value of the quantity remains constant. Although the form of the quantity may change in some manner, its final value is the same as its initial value. For example, the energy in an isolated system may change from gravitational potential energy to kinetic energy or to one of a variety of other forms, but energy is never lost from the system.

In order to develop the principle of conservation of energy, let us return to the work-kinetic energy theorem, Equation 5.6, which states that the net work done on a system equals the change in the system's kinetic energy. Let us assume that the only force doing work on the system is conservative. In this case the net work on the system is equal to W_c, and from Equation 5.9 we have

$$W_{\text{net}} = W_c = PE_i - PE_f$$

We now substitute this expression into Equation 5.6 for W_{net} to find

$$PE_i - PE_f = KE_f - KE_i$$

or

$$KE_i + PE_i = KE_f + PE_f \qquad [5.10]$$

More formally, the principle of conservation of mechanical energy states that **the total mechanical energy in any isolated system of objects remains constant if the objects interact only through conservative forces.** It is important to note that Equation 5.10 is valid *provided* no energy is added to or removed from the system. Furthermore, there must be no nonconservative forces within the system. Therefore, if a conservative system (one subject only to conservative forces) has some initial combined amount of kinetic and potential energy, some of that energy may later transform from one kind into the other, but the total mechanical energy remains constant. This is equivalent to saying that, if the kinetic energy of a conservative system increases (or decreases) by some amount, the potential energy of the system must decrease (or increase) by the same amount.

As this skillful rock climber moves cautiously up the mountain, her internal (biochemical) energy is transformed into gravitational potential energy. Can you identify other forms of energy transformations if she descends the mountain with the help of supporting ropes? *(Telegraph Colour Library/FPG)*

If the force of gravity is the *only* force doing work on an object, then the total mechanical energy of the object remains constant, and the principle of conservation of mechanical energy takes the form

$$\tfrac{1}{2}mv_i^2 + mgy_i = \tfrac{1}{2}mv_f^2 + mgy_f \qquad [5.11]$$

Because there are other forms of energy besides kinetic and gravitational potential energy, we shall modify Equation 5.11 to include these new forms as we learn about them.

QUICKLAB

Find a rubber band and roughly measure its spring constant. Then estimate the maximum amount of energy you can store in it (before it breaks).

Applying Physics 2

You have graduated from college and are designing roller coasters for a living. Your boss asks you to design a roller coaster that is supposed to release a car from rest at the top of a hill of height *h*, and then roll freely down the hill and reach the peak of the next hill of height 1.1*h*. What would you tell your boss in this situation?

Explanation   You should politely tell your boss to study some physics. It is impossible to design such a roller coaster because it would violate the principle of conservation of mechanical energy. At the starting point, the roller coaster has no kinetic energy and gravitational potential energy *mgh* relative to ground level. If it were able to reach the top of the second hill, it would have gravitational potential energy 1.1*mgh*, which is greater than its initial potential energy. If this roller coaster were actually built, the car would move up the second hill to a height *h* (neglecting friction), stop short of the peak, and then start rolling backward, becoming trapped between the two peaks.

APPLICATION

Roller Coaster Design.

Thinking Physics 4

Three identical balls are thrown from the top of a building, all with the same initial speed. Ball 1 is thrown horizontally, ball 2 is thrown at some angle above the horizontal, and ball 3 is thrown at some angle below the horizontal as in Figure 5.13. Neglecting air resistance, describe their motions and compare their speeds as they reach the ground.

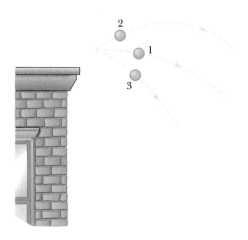

Figure 5.13   (Thinking Physics 4) Three identical balls are thrown with the same initial speed from the top of a building.

Explanation Ball 1 and ball 3 speed up after they are thrown, while ball 2 first slows down and then speeds up after reaching its peak. The paths of all three are portions of parabolas. The three take different times to reach the ground. However, all have the same impact speed because all start with the same kinetic energy and undergo the same change in gravitational potential energy. That is, $E_{total} = \frac{1}{2}mv^2 + mgh$ is the same for all three balls.

Problem-Solving Strategy

Conservation of Energy

Take the following steps in applying the principle of conservation of energy:

1. Define your system, which may consist of more than one object.
2. Select a reference position for the zero point of gravitational potential energy.
3. Determine whether or not nonconservative forces are present.
4. If mechanical energy is conserved (that is, if only conservative forces are present), you can write the total initial energy at some point as the sum of the kinetic and potential energies at that point. Then, write an expression for the total final energy, $KE_f + PE_f$, at the final point of interest. Because mechanical energy is conserved, you can equate the two total energies and solve for the unknown.

EXAMPLE 5.4   The Daring Diver

A 755-N diver drops from a board 10.0 m above the water surface, as in Figure 5.14.
(a) Use conservation of mechanical energy to find his speed 5.00 m above the water surface.

Reasoning As the diver falls toward the water, only one force acts on him: the force of gravity. (This assumes that air resistance can be neglected.) Therefore, we can be assured that only the force of gravity does any work on him and that mechanical energy is conserved. In order to find the diver's speed at the 5-m mark, we will choose the water surface as the zero level for potential energy. Also, note that the diver drops from the board; that is, he leaves with zero velocity and therefore zero kinetic energy.

Solution Conservation of mechanical energy, Equation 5.11, gives

$$\tfrac{1}{2}mv_i^2 + mgy_i = \tfrac{1}{2}mv_f^2 + mgy_f$$

$$0 + (755 \text{ N})(10.0 \text{ m}) = \tfrac{1}{2}(77.0 \text{ kg})v_f^2 + (755 \text{ N})(5.00 \text{ m})$$

$$v_f = \ 9.90 \text{ m/s}$$

(b) Find the speed of the diver just before he strikes the water.

Solution With the final position of the diver at the surface of the pool, Equation 5.11 gives

Figure 5.14 (Example 5.4) The kinetic energy and potential energy of a diver at various heights. The zero of potential energy is taken to be at the surface of the pool.

10.0 m

$KE_i = 0$
$PE_i = 7500 \text{ J}$

5.00 m

$KE = 3750 \text{ J}$
$PE = 3750 \text{ J}$

$KE_f = 7500 \text{ J}$
$PE_f = 0$

0

$$0 + (755 \text{ N})(10.0 \text{ m}) = \tfrac{1}{2}(77.0 \text{ kg})v_f^2 + 0$$

$$v_f = 14.0 \text{ m/s}$$

Exercise If the diver pushes off so that he leaves the board with an initial speed of 2.00 m/s, find his speed when he strikes the water. Use conservation of energy.

Answer 14.1 m/s

EXAMPLE 5.5   Sliding Down a Frictionless Hill

A sled and its rider together weigh 800 N. They move down a frictionless hill through a vertical distance of 10.0 m, as shown in Figure 5.15a. Use conservation of mechanical energy to find the speed of the sled–rider system at the bottom of the hill, assuming the rider pushes off with an initial speed of 5.00 m/s.

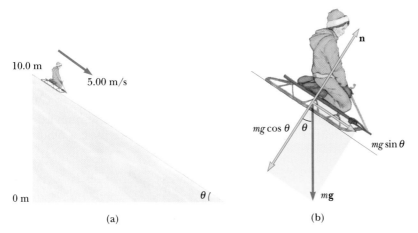

(a)   (b)

Figure 5.15 (Example 5.5) (a) A sled and rider start from the top of a frictionless hill with a speed of 5.00 m/s. (b) A free-body diagram for the system (sled plus rider).

Reasoning The forces acting on the sled and rider as they move down the hill are shown in Figure 5.15b. In the absence of a frictional force, the only forces acting are the normal force, **n,** and the gravitational force. At all points along the path, **n** is perpendicular to the direction of travel and hence does no work. Likewise, the component of weight perpendicular to the incline ($mg \cos \theta$) does no work. The only force that does any work is the component of the gravitational force, $mg \sin \theta$, along the slope of the hill. As a result, we are justified in using Equation 5.11.

Solution In this case the initial energy includes kinetic energy because of the initial speed:

$$\tfrac{1}{2}mv_i^2 + mgy_i = \tfrac{1}{2}mv_f^2 + mgy_f$$

or, after canceling m throughout the equation,

$$\tfrac{1}{2}v_i^2 + gy_i = \tfrac{1}{2}v_f^2 + gy_f$$

If we set the origin at the bottom of the incline, we see that the initial and final y coordinates are $y_i = 10.0$ m and $y_f = 0$. Thus we get

$$\tfrac{1}{2}(5.00 \text{ m/s})^2 + (9.80 \text{ m/s}^2)(10.0 \text{ m}) = \tfrac{1}{2}v_f^2 + (9.80 \text{ m/s}^2)(0)$$

$$v_f = \quad 14.9 \text{ m/s}$$

Notice that the solution of this problem does not depend on the shape of the hill. We only needed to know the rider's initial speed and the vertical distance traveled. If the hill had a varying curvature with bumps and troughs, the conservation of energy approach would yield the same result. If you try to solve the problem using Newton's laws, the varying slope of the hill would cause the acceleration to vary along the hill, making this approach very complicated. As you progress further with your study of physics, you will find many problems that are difficult or impossible to solve using Newton's laws, whereas conservation of energy provides a simple, yet powerful, alternative.

Exercise If the sled and rider start at the bottom of the incline and are given an initial speed of 5.00 m/s up the incline, how high will they rise vertically, and what will their speed be when they return to the bottom of the hill?

Answer 1.28 m; 5.00 m/s

Potential Energy Stored in a Spring

The concept of potential energy is of tremendous value in descriptions of certain types of mechanical motion. One of these, the motion of a mass attached to a stretched or compressed spring, is discussed more completely in Chapter 13, but it is instructive to examine it here.

    Consider Figure 5.16a, which shows a spring in its equilibrium position — that is, the spring is neither compressed nor stretched. If we push a block against the

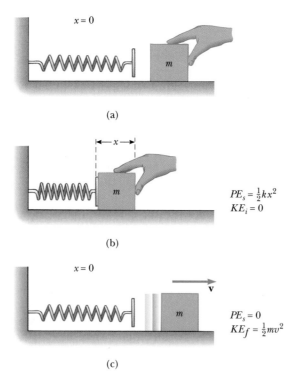

(a)

$$PE_s = \tfrac{1}{2}kx^2$$
$$KE_i = 0$$

(b)

$$PE_s = 0$$
$$KE_f = \tfrac{1}{2}mv^2$$

(c)

Figure 5.16 A block of mass m on a frictionless surface is pushed against a spring and released from rest. If x is the compression in the spring, as in (b), the potential energy stored in the spring is $\tfrac{1}{2}kx^2$. This energy is transferred to the block in the form of kinetic energy, as in (c).

spring, as in Figure 5.16b, the spring is compressed a distance of x. In order to compress the spring, we must exert on the block a force of magnitude $F = kx$, where k is a constant for a particular spring called the **spring constant.** For a flexible spring, k is a small number, whereas for a stiff spring, k is large. The equation $F = kx$ says that the more we compress the spring, the larger the force we must exert. Because the block is in equilibrium in Figure 5.16b, the spring must be exerting on it a force of $F_s = -kx$. This equation for F_s, describing the force exerted by a stretched or compressed spring, is often called **Hooke's law** after Sir Robert Hooke, who discovered the relationship between F_s and x.

To find an expression for the potential energy associated with the spring force, called the **elastic potential energy,** let us determine the work required to compress a spring from its equilibrium position to some final arbitrary position x. The force that must be applied to compress the spring varies from $F = 0$ to $F = kx$ at maximum compression. Because the force increases linearly with position (that is, $F \propto x$), the average force that must be applied is

$$\overline{F} = \frac{F_0 + F_x}{2} = \frac{0 + kx}{2} = \tfrac{1}{2}kx$$

Therefore, the work *done by the applied force* is

$$W = \overline{F}x = \tfrac{1}{2}kx^2$$

This work is stored in the compressed spring as elastic potential energy. Thus, we define the elastic potential energy, PE_s, as

$$PE_s \equiv \tfrac{1}{2}kx^2 \qquad\qquad [5.12]$$

Figure 5.16 shows how this stored elastic potential energy can be recovered. When the block is released, the spring snaps back to its original length and the stored elastic potential energy is converted to kinetic energy of the block. The elastic potential energy stored in the spring is zero when the spring is in equilibrium ($x = 0$). Note that energy is also stored in the spring when it is stretched and is given by Equation 5.12. Furthermore, the elastic potential energy is a maximum when the spring has reached its maximum compression or extension. Finally, the potential energy in the spring is always positive because it is proportional to x^2.

This new form of energy is included as another term in our equation for conservation of mechanical energy:

$$(KE + PE_g + PE_s)_i = (KE + PE_g + PE_s)_f$$

where PE_g is the gravitational potential energy.

Applying Physics 3

An old-fashioned manual phonograph, which you may have seen at an antique store, is a device that operates by use of stored elastic potential energy. A crank on the side of the record player was turned several times. This stored elastic potential energy in a spring in the motor mechanism. This energy was released through a series of gears to appear as kinetic energy of rotation of the turntable and record. Some of this energy ultimately left the phonograph by sound, allowing you to hear the information on the record. This same process is still used in nonelectric wind-up watches and wind-up toys.

APPLICATION

Wind-Up Devices.

EXAMPLE 5.6 A Block Projected on a Frictionless Surface

A 0.50-kg block rests on a horizontal, frictionless surface as in Figure 5.16. The block is pressed against a light spring having a spring constant of $k = 80.0$ N/m. The spring is compressed a distance of 2.0 cm and released. Consider the system to be the spring and block, and find the speed of the block when the block is at the $x = 0$ position.

Reasoning   Our expression for the conservation of mechanical energy is

$$(KE + PE_g + PE_s)_i = (KE + PE_g + PE_s)_f$$

The initial kinetic energy of the block is zero, and because the block remains at the same level throughout its motion, the gravitational potential energy terms on both sides of the equation are equal. Thus,

$$PE_{si} = KE_f$$

Solution   From the last expression we have

$$\tfrac{1}{2}kx_i^2 = \tfrac{1}{2}mv_f^2$$

$$\tfrac{1}{2}(80.0\ \text{N/m})(2.0 \times 10^{-2}\ \text{m})^2 = \tfrac{1}{2}(0.50\ \text{kg})v_f^2$$

$$v_f = \boxed{0.25\ \text{m/s}}$$

5.6   NONCONSERVATIVE FORCES AND THE WORK–KINETIC ENERGY THEOREM

In realistic situations, nonconservative forces such as friction are usually present. In such situations, the total mechanical energy of the system is not constant, and one cannot apply Equation 5.10. In order to account for nonconservative forces, let us return to the work-kinetic energy theorem:

$$W_{net} = \tfrac{1}{2}mv_f^2 - \tfrac{1}{2}mv_i^2$$

Let us separate the net work into two parts, that done by the nonconservative forces, W_{nc}, and that done by the conservative forces, W_c. The work-kinetic energy relationship then becomes

$$W_{nc} + W_c = \tfrac{1}{2}mv_f^2 - \tfrac{1}{2}mv_i^2 \qquad [5.13]$$

Solving for W_{nc} and substituting the expression for W_c provided by Equation 5.9, we have

$$W_{nc} = \tfrac{1}{2}mv_f^2 - \tfrac{1}{2}mv_i^2 - (PE_i - PE_f)$$

$$W_{nc} = (KE_f + PE_f) - (KE_i + PE_i) \qquad [5.14]$$

The work done by all nonconservative forces equals the change in mechanical energy of the system.

Thinking Physics 5

An automobile with kinetic energy, and potential energy carried within the gasoline, strikes a tree and comes to rest. The total energy in the system (the car) is now less than before. How did the energy leave the system, and in what form is energy left in the system?

Explanation There are a number of ways that energy transferred out of the system. The automobile did some *work* on the tree during the collision, causing the tree to deform. There was a large crash during the collision, representing transfer of energy by *sound*. If the gasoline tank leaks after the collision, the automobile is losing energy by means of *mass transfer*. There will also be some *electromagnetic radiation*, because the car has a temperature, as well as the possibility of radiation from any sparks that occur during the collision. After the car has come to rest, there is no more kinetic energy. There is some *potential energy* left in any gas remaining within the tank. There is also more *internal energy* in the automobile, because its temperature is likely to be higher after the collision than before.

Problem-Solving Strategy

Conservation of Energy Revisited

If nonconservative forces such as friction act on a system, mechanical energy is not conserved. Hence, the problem-solving strategy given immediately before Example 5.4 must be modified as follows. First, write expressions for the total initial and total final mechanical energies. In this case, the difference between the two total energies is equal to the work done by the nonconservative force(s). These steps are summarized in Equation 5.14.

EXAMPLE 5.7   A Crate Sliding Down a Ramp

A 3.00-kg crate slides down a ramp at a loading dock. The ramp is 1.00 m long and inclined at an angle of 30.0°, as shown in Figure 5.17. The crate starts from rest at the top, experiences a constant frictional force of magnitude 5.00 N, and continues to move a short distance on the flat floor. Use energy methods to determine the speed of the crate when it reaches the bottom of the ramp.

Reasoning   The crate's initial energy is all in the form of gravitational potential energy, and its final energy is all kinetic. However, we cannot say that $PE_i = KE_f$, because there is an external nonconservative force that removes mechanical energy from the crate: the force of friction. Thus, we must use Equation 5.13 to find the final speed.

Solution   Because $v_i = 0$, the initial kinetic energy is zero. If the y coordinate is measured from the bottom of the incline, then $y_i = 0.500$ m. Therefore, the total mechanical energy of the crate at the top is all potential energy, given by

$$PE_i = mgy_i = (3.00 \text{ kg})(9.80 \text{ m/s}^2)(0.500 \text{ m}) = 14.7 \text{ J}$$

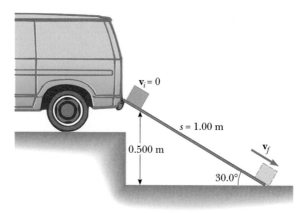

Figure 5.17 (Example 5.7) A crate slides down an incline under the influence of gravity. The potential energy of the crate decreases and its kinetic energy increases.

When the crate reaches the bottom, its potential energy is zero because its elevation is $y_f = 0$. Therefore, the total mechanical energy at the bottom is all kinetic energy:

$$KE_f = \tfrac{1}{2}mv_f^2$$

In this case, $W_{nc} = -fs$, where s is the displacement along the ramp. (Recall that the forces normal to the ramp do no work on the crate because they are perpendicular to the displacement.) With $f = 5.00$ N and $s = 1.00$ m, we have

$$W_{nc} = -fs = (-5.00 \text{ N})(1.00 \text{ m}) = -5.00 \text{ J}$$

This says that some mechanical energy is lost because of the presence of the retarding frictional force. Applying Equation 5.13 gives

$$-fs = \tfrac{1}{2}mv_f^2 - mgy_i$$

$$\tfrac{1}{2}mv_f^2 = 14.7 \text{ J} - 5.00 \text{ J} = 9.7 \text{ J}$$

$$v_f^2 = \frac{19.4 \text{ J}}{3.00 \text{ kg}} = 6.47 \text{ m}^2/\text{s}^2$$

$$v_f = \boxed{2.54 \text{ m/s}}$$

Exercise Find the speed at the bottom of the ramp if the ramp surface is frictionless.

Answer $v_f = 3.13$ m/s

Figure 5.18 (Example 5.8) If the slide is frictionless, the speed of the child at the bottom depends on the height of the slide and is independent of its shape.

EXAMPLE 5.8 Fun on the Water Slide

A child of mass 20.0 kg takes a ride on an irregularly curved water slide of height 6.00 m, as in Figure 5.18. The child starts from rest at the top. Determine her speed at the bottom, assuming no friction is present.

Reasoning The normal force does no work on the child because this force is always perpendicular to the displacement. Furthermore, because there is no friction, $W_{nc} = 0$ and we can apply the principle of conservation of mechanical energy.

Solution If we measure the y coordinate from the bottom of the slide, then $y_i = 6.00$ m, $y_f = 0$, and we get

$$KE_i + PE_i = KE_f + PE_f$$

$$0 + (20.0\text{ kg})(9.80\text{ m/s}^2)(6.00\text{ m}) = \tfrac{1}{2}(20.0\text{ kg})(v_f^2) + 0$$

$$v_f = \boxed{10.8\text{ m/s}}$$

Note that this speed is the same as if the child had fallen vertically a distance of 6.00 m! From the free-fall equation, $v_f^2 = v_i^2 - 2gy$, we get the same result:

$$v_f = \sqrt{2(9.80\text{ m/s}^2)(6.00\text{ m})} = 10.8\text{ m/s}$$

The effect of the slide is to direct the motion at the bottom horizontally rather than vertically.

Exercise If the child's speed at the bottom is 8.00 m/s rather than 10.8 m/s, how much mechanical energy is removed from the system?

Answer 536 J

This small plant, found in warm Southern waters, exhibits bioluminescence, a process in which chemical energy is converted to light. The red areas are chlorophyll, which fluoresces when excited by blue light. (*Jan Hinsch/Science Photo Library/ Photo Researchers, Inc.*)

5.7 CONSERVATION OF ENERGY IN GENERAL

We can generalize the energy conservation principle to include all forces, both conservative and nonconservative, acting on a system. In the study of thermodynamics we shall find that mechanical energy can be transformed into internal energy of the system. For example, when a block slides over a rough surface, the mechanical energy lost is transformed into internal energy stored in the block and the surface, as evidenced by measurable increases in the temperatures of both. We shall see that, on an atomic scale, this internal energy is associated with the vibration of atoms about their equilibrium positions. Such internal atomic motion has kinetic and potential energy, and so one can say that frictional forces arise fundamentally from forces that are conservative at the atomic level. Therefore, if we include this increase in internal energy in our energy expression, total energy is conserved.

The principle of conservation of energy is not solely confined to physics. In biology, energy transformations take place in myriad ways inside bacteria. Examples include the transformation of chemical energy to mechanical energy that cause flagella to move and propel an organism. Some bacteria use chemical energy to produce light. Although the mechanisms that produce these light emissions are not well understood, living creatures often rely on this light for their existence. For example, certain fish have sacs beneath their eyes filled with light-emitting bacteria. The emitted light attracts other smaller creatures that become food for the fish.

If you analyze an isolated system, you will always find that its total energy does not change, as long as you account for all forms of energy. That is, **energy can never be created or destroyed. Energy may be transformed from one form into another, but the total energy of an isolated system is always constant.** From a universal point of view, we can say that **the total energy of the Universe is**

APPLICATION

Flagellar Movement; Bioluminescence.

◁ Total energy is always conserved.

constant: If one part of the Universe gains energy in some form, another part must lose an equal amount of energy. No violation of this principle has been found.

5.8   POWER

From a practical viewpoint, it is interesting to know not only the amount of energy transferred to or from a system, but also the rate at which the transfer occurred. **Power is defined as the time rate of energy transfer.**

Let us use our familiar concept of work as an example of energy transfer. If an external force is applied to an object and if the work done by this force is W in the time interval Δt, then the average power, $\bar{P}$, during this interval is defined as the ratio of the work done to the time interval:

$$\bar{P} = \frac{W}{\Delta t} \qquad [5.15]$$

It is sometimes useful to rewrite Equation 5.15 by substituting $W = F \Delta s$ and noting that $\Delta s / \Delta t$ is the average speed of the object during the time Δt:

Average power ▶

$$\bar{P} = \frac{W}{\Delta t} = \frac{F \Delta s}{\Delta t} = F \bar{v} \qquad [5.16]$$

That is, the average power delivered either to or by an object is equal to the product of the force acting on the object during some time interval and the object's average speed during this time interval. In Equation 5.16, F is the component of force in the direction of the average velocity.

A result of the same form as Equation 5.16 is obtained for instantaneous values. That is, the instantaneous power delivered to an object is equal to the product of the force on the object at that instant and the instantaneous speed, or $P = Fv$.

The units of power in the SI system are joules per second, which are also called **watts** (W) after James Watt:

$$1 \text{ W} = 1 \text{ J/s} = 1 \text{ kg} \cdot \text{m}^2/\text{s}^3 \qquad [5.17]$$

The unit of power in the British engineering system is the horsepower (hp), in which

$$1 \text{ hp} \equiv 550 \frac{\text{ft} \cdot \text{lb}}{\text{s}} = 746 \text{ W} \qquad [5.18]$$

The watt is commonly used in electrical applications, but it can be used in other scientific areas as well. For example, European sports-car engines are now rated in kilowatts.

We now define units of energy (or work) in terms of the unit of power, the watt. One kilowatt-hour (kWh) is the energy converted or consumed in 1 h at the constant rate of 1 kW = 1000 J/s. Therefore,

$$1 \text{ kWh} = (10^3 \text{ W})(3600 \text{ s}) = (10^3 \text{ J/s})(3600 \text{ s}) = 3.60 \times 10^6 \text{ J}$$

It is important to realize that a kilowatt-hour is a unit of energy, not power. When you pay your electric bill, you are buying energy, and the amount of electricity used by an appliance is usually expressed in multiples of kilowatt-hours. For example, an electric bulb rated at 100 W would "consume" 3.6×10^5 J of energy in 1 h.

Thinking Physics 6

A lightbulb is described by some individuals as "having 60 watts." What is wrong with this phrase?

Explanation The number given represents the power of the lightbulb, which is the rate at which energy passes through it, entering as electrical energy from the power source. It is not a value that the lightbulb "possesses," however. It is only the approximate power rating of the bulb when it is connected to a normal 120-volt household supply of electricity. If the bulb were connected to a 12-volt supply, energy would enter the bulb at a rate other than 60 watts. Thus, the 60 watts is not an intrinsic property of the bulb, such as density is for a given solid material.

EXAMPLE 5.9 Power Delivered by an Elevator Motor

A 1000-kg elevator carries a maximum load of 800 kg. A constant frictional force of 4000 N retards its motion upward, as in Figure 5.19. What minimum power, in kilowatts and in horsepower, must the motor deliver to lift the fully loaded elevator at a constant speed of 3.00 m/s?

Solution The motor must supply the force, **T,** that pulls the elevator upward. From Newton's second law and from the fact that **a** = 0 because **v** is constant, we get

$$T - f - Mg = 0$$

where M is the total mass (elevator plus load), equal to 1800 kg. Therefore,

$$\begin{aligned}
T &= f + Mg \\
&= 4.00 \times 10^3 \text{ N} + (1.80 \times 10^3 \text{ kg})(9.80 \text{ m/s}^2) \\
&= 2.16 \times 10^4 \text{ N}
\end{aligned}$$

From $P = Fv$ and the fact that **T** is in the same direction as **v**, we have

$$\begin{aligned}
P &= Tv \\
&= (2.16 \times 10^4 \text{ N})(3.00 \text{ m/s}) = 6.48 \times 10^4 \text{ W} \\
&= 64.8 \text{ kW} = 86.9 \text{ hp}
\end{aligned}$$

Figure 5.19 (Example 5.9) The motor provides an upward force, **T,** on the elevator. A frictional force, **f,** and the force of gravity, $M\mathbf{g}$, act downward.

SUMMARY

The **work** done by a *constant* force, **F,** acting on an object is defined as the product of the component of the force in the direction of the object's displacement and the magnitude of the displacement. If the force makes an angle of θ with the displacement **s,** the work done by **F** is

$$W \equiv (F \cos \theta)\, s \qquad \text{[5.1]}$$

The **kinetic energy** of an object of mass m moving with speed v is defined as

$$KE \equiv \tfrac{1}{2} m v^2 \qquad \text{[5.5]}$$

The **work-kinetic energy theorem** states that the net work done on an object by external forces equals the change in kinetic energy of the object:

$$W_{\text{net}} = KE_f - KE_i \qquad \text{[5.6]}$$

The **gravitational potential energy** of an object of mass m that is elevated a distance of y above the Earth's surface is given by

$$PE \equiv mgy \qquad \text{[5.7]}$$

A force is **conservative** if the work it does on an object depends only on the initial and final positions of the object and not on the path taken between those positions. A force for which this is not true is said to be **nonconservative.**

The **principle of conservation of mechanical energy** states that if the only force acting on a system is conservative, total mechanical energy remains constant.

$$KE_i + PE_i = KE_f + PE_f \qquad \text{[5.10]}$$

The **elastic potential energy** stored in a spring is given by

$$PE_s \equiv \tfrac{1}{2} k x^2 \qquad \text{[5.12]}$$

where k is the spring constant and x is the distance the spring is compressed or extended from its unstretched (equilibrium) position.

Energy can never be created or destroyed. Energy may be transformed from one form into another, but the total energy of an isolated system is always constant. The work done by all nonconservative forces acting on a system equals the change in the total mechanical energy of the system:

$$W_{nc} = (KE_f + PE_f) - (KE_i + PE_i) \qquad \text{[5.14]}$$

Average power is the ratio of energy transfer to the time interval during which the transfer occurs. For work, this expression is

$$\overline{P} = \frac{W}{\Delta t} \qquad \text{[5.15]}$$

If a force, **F,** acts on an object moving with an average speed of $\bar{v}$, the average power delivered to the object is

$$\bar{P} = F\bar{v} \qquad\qquad [5.16]$$

MULTIPLE-CHOICE QUESTIONS

1. Bobby, of mass m, drops from a tree limb at the same time that Sally, also of mass m, begins her descent down a frictionless slide. The slide is in the shape of a quadrant of a circle. If they both start at the same height above the ground, which of the following is true about their kinetic energies as they reach the Earth?
   (a) The kinetic energy of Bobby is greater than that of Sally.
   (b) The kinetic energy of Sally is greater than that of Bobby.
   (c) Both have the same kinetic energy.
   (d) It is impossible to say from the information given.

2. A whale has a length of about 18 m and weighs about 40 000 N. Assume that its leap from the water carries him about half his height out of the water and that all the upward surge is achieved solely by its speed at the instant of leaving the water. How fast was it going as it left the water?
   (a) 9 m/s   (b) 13 m/s   (c) 18 m/s   (d) 176 m/s

(Multiple-Choice Question 2)
(Fred Felleman/Tony Stone Images)

3. The analysis of a pole vault requires one to consider the kinetic energy of the runner, elastic potential energy of the pole, and the gravitational potential energy of the vaulter. Other factors include the internal energy of the vaulter associated with his muscles, tendons, and ligaments. For this problem, however, consider a simplified version of all these factors to get an estimate of how high a vaulter can ever hope to reach. To do this assume the vaulter's center of gravity is at a height of 1.1 m above the ground when he takes off. A reasonable estimate of the maximum takeoff speed is 11.0 m/s when carrying a pole. Use conservation of energy to estimate the maximum height. (*Hint:* The answer you will get is about a meter higher than the current record.)
   (a) 6.2 m   (b) 7.3 m   (c) 11.0 m   (d) 14.6 m

(Multiple-Choice Question 3)
(Estate of Harold Edgerton, courtesy of Palm Press, Inc.)

4. A 40.0-N crate starting at rest slides down a rough 6.0-m-long ramp, inclined at 30° with the horizontal. The force of friction between crate and ramp is 6.0 N.

What will be the speed of the crate at the bottom of the incline?

(a) 1.6 m/s   (b) 3.3 m/s   (c) 4.5 m/s   (d) 6.4 m/s

5. A worker pushes a wheelbarrow with a force of 50 N over a level distance of 5.0 m. If a frictional force of 43 N acts on the wheelbarrow in a direction opposite to that of the worker, what net work is done on the wheelbarrow?

(a) 250 J   (b) 215 J   (c) 35 J   (d) 10 J

CONCEPTUAL QUESTIONS

1. Consider a tug-of-war in which two teams pulling on a rope are evenly matched, so that no motion takes place. Is work done on the rope? On the pullers? On the ground? Is work done on anything?

2. Discuss whether any work is being done by each of the following agents and, if so, whether the work is positive or negative: (a) a chicken scratching the ground, (b) a person studying, (c) a crane lifting a bucket of concrete, (d) the force of gravity on the bucket in part (c), (e) the leg muscles of a person in the act of sitting down.

3. A team of furniture movers wishes to load a truck using a ramp from the ground to the rear of the truck. One of the movers claims that less work would be required to load the truck if the length of the ramp were increased, reducing the angle with respect to the horizontal. Is his claim valid? Explain.

4. (a) Can the kinetic energy of an object be negative? (b) Can the gravitational potential energy of an object be negative? Explain.

5. Roads going up mountains are formed into switchbacks with the road weaving back and forth along the face of the slope, such that there is only a gentle rise on any portion of the roadway. Does this require any less work to be done by an automobile climbing the mountain, compared to driving on a roadway that is straight up the slope? Why are switchbacks used?

6. (a) If the speed of a particle is doubled, what happens to its kinetic energy? (b) If the net work done on a particle is zero, what can be said about its speed?

7. As a simple pendulum swings back and forth, the forces acting on the suspended mass are the force of gravity, the tension in the supporting cord, and air resistance. (a) Which of these forces, if any, does no work on the pendulum? (b) Which of these forces does negative work at all times during its motion? (c) Describe the work done by the force of gravity while the pendulum is swinging.

8. A bowling ball is suspended from the ceiling of a lecture hall by a strong cord. The ball is drawn away from its equilibrium position and released from rest at the tip of the demonstrator's nose as in Figure Q5.8. If the dem-

onstrator remains stationary, explain why she is not struck by the ball on its return swing. Would this demonstrator be safe if the ball were given a slight push from its starting position at her nose?

Figure Q5.8

9. An older model car accelerates from 0 to speed v in 10 seconds. A newer, more powerful sports car accelerates from 0 to $2v$ in the same time period. What is the ratio of powers expended by the two cars? Consider the energy coming from the engine to appear only as kinetic energy of the cars.

10. During a stress test of the cardiovascular system, a patient walks and runs on a treadmill. (a) Is the energy expended by the patient equivalent to the energy of walking and running on the ground? Explain. (b) If the treadmill is tilted upward, what effect does this have? Discuss.

11. In most situations we have encountered in this chapter, frictional forces tend to reduce the kinetic energy of an object. However, frictional forces can sometimes in-

crease an object's kinetic energy. Describe a few situations in which friction causes an increase in kinetic energy.

12. Discuss the energy transformations that occur during the pole vault event pictured in the multiflash photograph shown on page 139. Ignore rotational motion.

13. A weight is connected to a spring that is suspended vertically from the ceiling. If the weight is displaced downward from its equilibrium position and released, it will oscillate up and down. If air resistance is neglected, will the total energy of the system (weight plus spring) be conserved? How many forms of potential energy are there for this situation?

14. The driver of a car slams on her brakes to avoid colliding with a deer crossing the highway. What happens to the car's kinetic energy as it comes to rest?

15. You are reshelving books in a library. You lift a book from the floor to the top shelf. The kinetic energy of the book on the floor was zero, and the kinetic energy of the book on the top shelf is zero, so there is no change in kinetic energy. Yet you did some work in lifting the book. Is the work–kinetic energy theorem violated?

16. A ball is thrown straight up into the air. At what position is its kinetic energy a maximum? At what position is the gravitational potential energy a maximum?

17. An Earth satellite is in a circular orbit at an altitude of 500 km. Explain why the work done by the gravitational force acting on the satellite is zero. Using the work–kinetic energy theorem, what can you say about the speed of the satellite?

PROBLEMS

1, 2, 3 = straightforward, intermediate, challenging   □ = full solution available in Study Guide/Student Solutions Manual
WEB = solution posted at **http://www.saunderscollege.com/physics/college**   🔬 = biomedical application   🖥 = Interactive Physics

Review Problem

A 10-kg mass is being pulled down an incline by a force $F = 80$ N parallel to the incline as indicated in the figure. The mass starts from rest, and a frictional force of 20 N opposes this motion. When the mass has moved 4.0 m, find its speed by using (a) energy concepts and (b) Newton's second law of motion.

Section 5.1 Work

1. A weight lifter lifts a 350-N set of weights from ground level to a position over his head, a vertical distance of 2.00 m. How much work does the weight lifter do, assuming he moves the weights at constant speed?

Figure P5.1 *(Gerard Vandystadt/Photo Researchers)*

2. If a woman lifts a 20.0-kg bucket from a well and does 6.00 kJ of work, how deep is the well? Assume that the speed of the bucket remains constant as it is lifted.

3. A tugboat exerts a constant force of 5.00×10^3 N on a ship moving at constant speed through a harbor. How much work does the tugboat do on the ship if each moves a distance of 3.00 km?

4. A shopper in a supermarket pushes a cart with a force of 35 N directed at an angle of 25° downward from the horizontal. Find the work done by the shopper as he moves down a 50-m length of aisle.

5. Starting from rest, a 5.0-kg block slides 2.5 m down a rough 30.0° incline. The coefficient of kinetic friction between the block and the incline is $\mu_k = 0.436$. Determine (a) the work done by the force of gravity, (b) the work done by the friction force between block and incline, and (c) the work done by the normal force.

6. A block of mass 2.50 kg is pushed 2.20 m along a frictionless horizontal table by a constant 16.0 N force directed 25.0° below the horizontal. Determine the work done by (a) the applied force, (b) the normal force exerted by the table, (c) the force of gravity, and (d) the net force on the block.

Section 5.2   Kinetic Energy and the Work–Kinetic Energy Theorem

7. A mechanic pushes a 2.50×10^3-kg car from rest to a speed of v, doing 5000 J of work in the process. During this time, the car moves 25.0 m. Neglecting friction between car and road, find (a) v and (b) the horizontal force exerted on the car.

8. A 7.00-kg bowling ball moves at 3.00 m/s. How fast must a 2.45-g Ping-Pong ball move so that the two balls have the same kinetic energy?

9. A 70.0-kg base runner begins his slide into second base when moving at a speed of 4.0 m/s. The coefficient of friction between his clothes and Earth is 0.70. He slides so that his speed is zero just as he reaches the base. (a) How much mechanical energy is lost due to friction acting on the runner? (b) How far does he slide?

10. A 0.60-kg particle has a speed of 2.0 m/s at point A and kinetic energy of 7.5 J at point B. What is (a) its kinetic energy at A? (b) its speed at point B? (c) the total work done on the particle as it moves from A to B?

WEB 11. A person doing a chin-up weighs 700 N exclusive of her arms. During the first 25.0 cm of the lift, each arm exerts an upward force of 355 N on the torso. If the upward movement starts from rest, what is her speed at this point?

12. An outfielder throws a 0.150-kg baseball at a speed of 40.0 m/s and an initial angle of 30.0°. What is the kinetic energy of the ball at the highest point of its motion?

13. A 2.0-g bullet leaves the barrel of a gun at a speed of 300 m/s. (a) Find its kinetic energy. (b) Find the average force exerted on the bullet by the expanding gases as the bullet moves the length of the 50-cm-long barrel.

14. A 10.0-kg crate is pulled up a rough 20° incline by a 100-N force parallel to the incline. The initial speed of the crate is 1.50 m/s, the coefficient of kinetic friction is 0.40, and the crate is pulled a distance of 5.00 m. Determine how much work is done by (a) the gravitational force, and (b) the 100-N force? (c) What is the change in kinetic energy of the crate? (d) What is the speed of the crate after it is pulled 5.00 m?

15. A 2000-kg car moves down a level highway under the actions of two forces. One is a 1000-N forward force exerted on the drive wheels by the road; the other is a 950-N resistive force. Use the work-kinetic energy theorem to find the speed of the car after it has moved a distance of 20 m, assuming it starts from rest.

16. On a frozen pond, a 10-kg sled is given a kick that imparts to it an initial speed of $v_0 = 2.0$ m/s. The coefficient of kinetic friction between sled and ice is $\mu_k = 0.10$. Use the work-kinetic energy theorem to find the distance the sled moves before coming to rest.

Section 5.3   Potential Energy

Section 5.4   Conservative and Nonconservative Forces

17. A 1.00×10^3-kg roller-coaster car is initially at the top of a rise, at point A. It then moves 50.0 m at an angle of 40.0° below the horizontal to a lower point, B. (a) Choosing point B as the zero level for gravitational potential energy, find the potential energy of the car at A and B and the difference in potential energy between these points. (b) Repeat part (a), choosing point A as the zero reference level.

18. A 2.00-kg ball is attached to a ceiling by a 1.00-m-long string. The height of the room is 3 m. What is the gravitational potential energy of the ball relative to (a) the ceiling? (b) the floor? (c) a point at the same elevation as the ball?

19. A 2.0-m-long pendulum is released from rest when the support string is at an angle of 25° with the vertical. What is the speed of the bob at the bottom of the swing?

20. A softball pitcher rotates a 0.250-kg ball around a vertical circular path of radius 0.6 m before releasing it. The pitcher exerts a 30.0-N force directed parallel to the motion of the ball around the complete circular path. The speed of the ball at the top of the circle is 15.0 m/s. If the ball is released at the bottom of the circle, what is its speed on release?

21. A 40-N child is in a swing that is attached to ropes 2.0 m long. Find the gravitational potential energy of the child relative to her lowest position (a) when the ropes are horizontal, (b) when the ropes make a 30° angle with the vertical, and (c) at the bottom of the circular arc.

Section 5.5   Conservation of Mechanical Energy

22. A 50-kg pole vaulter running at 10 m/s vaults over the bar. Her speed when she is over the bar is 1.0 m/s. Neglect air resistance, as well as any energy absorbed by the pole, and determine her altitude as she crosses the bar.

WEB 23. A child and sled with a combined mass of 50.0 kg slide down a frictionless hill. If the sled starts from rest and has a speed of 3.00 m/s at the bottom, what is the height of the hill?

24. A 0.400-kg bead slides on a curved wire, starting from rest at point A in Figure P5.24. If the wire is frictionless, find the speed of the bead **(a)** at B and **(b)** at C.

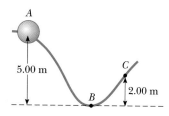

Figure P5.24

25. A 200-g particle is released from rest at point A on the inside of a smooth hemispherical bowl of radius $R = 30.0$ cm (Fig. P5.25). Calculate **(a)** its gravitational potential energy at A relative to B, **(b)** its kinetic energy at B, **(c)** its speed at B, and **(d)** its potential energy at C relative to B and its kinetic energy at C.

Figure P5.25

26. Tarzan swings on a 30.0-m-long vine initially inclined at an angle of 37.0° with the vertical. What is his speed at the bottom of the swing **(a)** if he starts from rest? **(b)** if he pushes off with a speed of 4.00 m/s?

27. Three masses, $m_1 = 5.0$ kg, $m_2 = 10.0$ kg, and $m_3 = 15.0$ kg, are attached by strings over frictionless pulleys

Figure P5.27   (Problems 27 and 38)

as indicated in Figure P5.27. The horizontal surface is frictionless and the system is released from rest. Using energy concepts, find the speed of m_3 after it moves down 4.0 m.

28. The launching mechanism of a toy gun consists of a spring of unknown spring constant, as shown in Figure P5.28a. If the spring is compressed a distance of 0.120 m and the gun fired vertically as shown, the gun can launch a 20.0-g projectile from rest to a maximum height of 20.0 m above the starting point of the projectile. Neglecting all resistive forces, determine **(a)** the spring constant and **(b)** the speed of the projectile as it moves through the equilibrium position of the spring (where $x = 0$), as shown in Figure P5.28b.

(a)   (b)

Figure P5.28

29. A projectile is launched with a speed of 40 m/s at an angle of 60° above the horizontal. Find the maximum height reached by the projectile during its flight by using conservation of energy.

30. A 0.250-kg block is placed on a vertical spring ($k = 5.00 \times 10^3$ N/m) and pushed downward, compressing the spring 0.100 m. As the block is released, it leaves the spring and continues to travel upward. What height above the point of release will the block reach if air resistance is negligible?

Section 5.6   Nonconservative Forces and the Work-Kinetic Energy Theorem

31. If the wire in Problem 24 (Fig. P5.24) is frictionless between points A and B and rough between B and C, and if the 0.400-kg bead starts from rest at A, **(a)** find its speed at B. **(b)** If the bead comes to rest at C, find the loss in mechanical energy as it goes from B to C.

32. An 80.0-N box is pulled 20.0 m up a 30° incline by an applied force of 100 N that points upward, parallel to the incline. If the coefficient of kinetic friction between box and incline is 0.220, calculate the change in the kinetic energy of the box.

WEB 33. A 70-kg diver steps off a 10-m tower and drops, from rest, straight down into the water. If he comes to rest 5.0 m beneath the surface, determine the average resistance force exerted on him by the water.

34. An airplane of mass 1.5×10^4 kg is moving at 60 m/s. The pilot then revs up the engine so that the forward thrust of the propeller becomes 7.5×10^4 N. If the force of air resistance has a magnitude of 4.0×10^4 N, find the speed of the airplane after it has traveled 500 m. Assume that the airplane is in level flight throughout this motion.

35. A 2.10×10^3-kg car starts from rest at the top of a 5.0-m-long driveway that is sloped at 20° with the horizontal. If an average friction force of 4.0×10^3 N impedes the motion, find the speed of the car at the bottom of the driveway.

36. A 25.0-kg child on a 2.00-m-long swing is released from rest when the swing supports make an angle of 30.0° with the vertical. **(a)** Neglecting friction, find the child's speed at the lowest position. **(b)** If the speed of the child at the lowest position is 2.00 m/s, what is the mechanical energy lost due to friction?

37. Starting from rest, a 10.0-kg block slides 3.00 m down a frictionless ramp (inclined at 30.0° from the floor) to the bottom. The block then slides an additional 5.00 m along the floor before coming to a stop. Determine **(a)** the speed of the block at the bottom of the ramp, **(b)** the coefficient of kinetic friction between block and floor, and **(c)** the mechanical energy lost due to friction.

38. Three masses, $m_1 = 5.0$ kg, $m_2 = 10.0$ kg, and $m_3 = 15.0$ kg, are attached by strings over frictionless pulleys, as indicated in Figure P5.27. The horizontal surface exerts a force of friction of 30 N on m_2. If the system is released from rest, use energy concepts to find the speed of m_3 after it moves 4.0 m.

39. A skier starts from rest at the top of a hill that is inclined at 10.5° with the horizontal. The hillside is 200 m long, and the coefficient of friction between snow and skis is 0.0750. At the bottom of the hill, the snow is level and the coefficient of friction is unchanged. How far does the skier move along the horizontal portion of the snow before coming to rest?

40. In a circus performance, a monkey is strapped to a sled and both are given an initial speed of 4.0 m/s up a 20° inclined track. The combined mass of monkey and sled is 20 kg, and the coefficient of kinetic friction between sled and incline is 0.20. How far up the incline do the monkey and sled move?

Section 5.8   Power

41. A skier of mass 70 kg is pulled up a slope by a motor-driven cable. **(a)** How much work is required to pull him 60 m up a 30° slope (assumed frictionless) at a constant speed of 2.0 m/s? **(b)** How many horsepower must a motor have to perform this task?

42. A 50.0-kg student climbs a 5.00-m-long rope and stops at the top. **(a)** What must her average speed be in order to match the power output of a 200-W lightbulb? **(b)** How much work does she do?

43. Water flows over a section of Niagara Falls at the rate of 1.2×10^6 kg/s and falls 50 m. How much power is generated by the falling water?

Figure P5.43 *(Ron Dorman/FPG)*

44. A 1.50×10^3-kg car accelerates uniformly from rest to 10.0 m/s in 3.00 s. Find **(a)** the work done on the car in this time interval, **(b)** the average horsepower delivered by the engine in this time interval, and **(c)** the instantaneous horsepower delivered by the engine at $t = 2.00$ s.

45. A 1.50×10^3-kg car starts from rest and accelerates uniformly to 18.0 m/s in 12.0 s. Assume that air resistance remains constant at 400 N during this time. Find **(a)** the average power developed by the engine and **(b)** the instantaneous power output of the engine at $t = 12.0$ s.

46. A 650-kg elevator starts from rest. It moves upward for 3.00 s with constant acceleration until it reaches its cruising speed, 1.75 m/s. **(a)** What is the average power of the elevator motor during this period? **(b)** How does this compare with its power during an upward cruise with constant speed?

47. While running, a person dissipates about 0.60 J of mechanical energy per step per kilogram of body mass. If a 60-kg person develops a power of 70 W during a race, how fast is the person running? Assume that a running step is 1.5 m long.

ADDITIONAL PROBLEMS

48. A car traveling at 50 km/h skids a distance of 35 m after its brakes lock. Estimate how far it will skid if its brakes lock when its initial speed is 100 km/h. What happens to the car's kinetic energy as it comes to rest?

49. A 98.0-N grocery cart is pushed 12.0 m by a shopper who exerts a constant horizontal force of 40.0 N. If all frictional forces are neglected and the cart starts from rest, what is its final speed?

50. (a) A 75-kg man steps out a window and falls (from rest) 1.0 m to a sidewalk. What is his speed just before his feet strike the pavement? (b) If the man falls with his knees and ankles locked, the only cushion for his fall is an approximately 0.50-cm give in the pads of his feet. Calculate the average force exerted on him by the ground in this situation. This average force is sufficient to cause cartilage damage in the joints or to break bones.

WEB 51. A toy gun uses a spring to project horizontally a 5.3-g soft rubber sphere. The spring constant is 8.0 N/m, the barrel of the gun is 15 cm long, and a constant frictional force of 0.032 N exists between barrel and projectile. With what speed does the projectile leave the barrel if the spring was compressed 5.0 cm for this launch?

52. Two masses are connected by a light string passing over a light, frictionless pulley as in Figure P5.52. The 5.00-kg mass is released from rest at a point 4.00 m above the floor. (a) Determine the speed of each mass when the two pass each other. (b) Determine the speed of each mass at the moment the 5.00-kg mass hits the floor. (c) How much higher does the 3.00-kg mass travel after the 5.00-kg mass hits the floor?

Figure P5.52

53. Two blocks, A and B (with mass 50 kg and 100 kg, respectively), are connected by a string, as shown in Figure P5.53. The pulley is frictionless and of negligible mass. The coefficient of kinetic friction between block A and the incline is $\mu_k = 0.25$. Determine the change in the kinetic energy of block A as it moves from C to D, a distance of 20 m up the incline if the system starts from rest.

Figure P5.53

54. The light horizontal spring in Figure 5.16 has a force constant of $k = 100$ N/m. A 2.00-kg block is pressed against one end of the spring, compressing it 0.100 m. After the block is released, it starts from rest and moves 0.250 m to the right of its starting position before coming to rest. Show that the coefficient of kinetic friction between the horizontal surface and the block is 0.102.

55. A catcher "gives" with the ball when he catches a 0.15-kg baseball moving at 25 m/s. (a) If he moves his glove a distance of 2.0 cm, what is the average force acting on his hand? (b) Repeat for the case in which his glove and hand move 10 cm.

56. A 2.0×10^3-kg car starts from rest and accelerates along a horizontal roadway to $+20$ m/s in 15 s. Assume that air resistance remains constant at -500 N during this time. Find (a) the average power developed by the engine and (b) the instantaneous power developed at $t = 15$ s.

57. A ski jumper starts from rest 50.0 m above the ground on a frictionless track, and flies off the track at an angle of 45.0° above the horizontal and at a height of 10.0 m from the ground. Neglect air resistance. (a) What is his speed when he leaves the track? (b) What is the maximum altitude he attains after leaving the track? (c) Where does he land relative to the end of the track?

58. A spring of length 0.80 m rests along a frictionless 30° incline, as in Figure P5.58a. A 2.0-kg mass, at rest against the end of the spring, compresses the spring by 0.10 m (Fig. P5.58b). (a) Determine the spring constant k. (b) The mass is pushed down, compressing the spring an *additional* 0.60 m, and then released. If the incline is

2.0 m long, determine how far beyond the rightmost edge of the incline the mass lands.

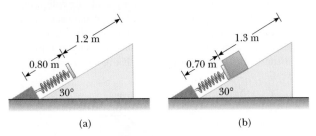

(a)   (b)

Figure P5.58

59. A 5.0-kg block is pushed 3.0 m up a vertical wall with constant speed by a constant force of magnitude F applied at an angle of $\theta = 30°$ with the horizontal, as shown in Figure P5.59. If the coefficient of kinetic friction between block and wall is 0.30, determine the work done by (a) **F**, (b) gravity, and (c) the normal force between block and wall. (d) By how much does the gravitational potential energy of the block increase?

Figure P5.59

60. A 0.300-kg piece of putty is slowly placed on the top of a massless vertical spring of constant 19.6 N/m. (a) Determine the maximum compression of the spring. (b) What is the answer to part (a) if the putty is dropped onto the spring from a height of 0.500 m from the top of the spring?

61. The masses of the javelin, discus, and shot are 0.80 kg, 2.0 kg, and 7.2 kg, respectively, and record throws in the corresponding track events are about 89 m, 69 m, and 21 m, respectively. Neglecting air resistance, (a) calculate the minimum initial kinetic energies that would produce these throws, and (b) estimate the average force exerted on each object during the throw, assuming the force acts over a distance of 2.0 m. (c) Do your results suggest that air resistance is an important factor?

62. Two identical massless springs of constant $k = 200$ N/m are fixed at opposite ends of a level track, as shown in Figure P5.62. A 5.00-kg block is pressed against the left spring, compressing it by 0.150 m. The block (initially at rest) is then released, as shown in Figure P5.62a. The entire track is frictionless *except* for the section between A and B. Given that the coefficient of kinetic friction between block and track along AB is $\mu_k = 0.0800$, and given that the length AB is 0.250 m, (a) determine the maximum compression of the spring on the right (see Fig. P5.62b). (b) Determine where the block eventually comes to rest, as measured from A (see Fig. P5.62c).

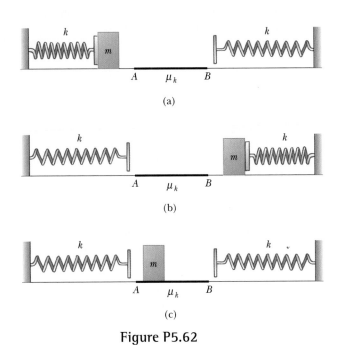

(a)

(b)

(c)

Figure P5.62

63. Jane, whose mass is 50.0 kg, needs to swing across a river filled with man-eating crocodiles in order to save Tarzan from danger. However, she must swing into a *constant* horizontal wind force, **F**, on a vine that is initially at an angle of θ with the vertical (see Fig. P5.63). $D = 50.0$ m, $F = 110$ N, $L = 40.0$ m, and $\theta = 50.0°$. (a) With what minimum speed must Jane begin her swing in order to just make it to the other side? (*Hint:* First determine the potential energy associated with the wind force. Because the wind force is constant, use an analogy with the constant gravitational force.) (b) Once the rescue is complete, Tarzan and Jane must swing back across the river. With what minimum speed must they begin their swing? Assume that Tarzan has a mass of 80.0 kg.

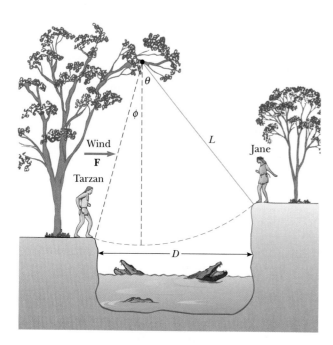

Figure P5.63

64. The ball launcher in a pinball machine has a spring that has a force constant of 1.20 N/cm (Fig. P5.64). The surface on which the ball moves is inclined 10.0° with respect to the horizontal. If the spring is initially compressed 5.00 cm, find the launching speed of a 0.100-kg ball when the plunger is released. Friction and the mass of the plunger are negligible.

Figure P5.64

65. Suppose a car is modeled as a cylinder with cross-sectional area A moving with a speed v, as in Figure P5.65. In a time Δt, a column of air of mass Δm must be

Figure P5.65

moved a distance $v \, \Delta t$ and hence must be given a kinetic energy $\frac{1}{2}(\Delta m)v^2$. Using this model, show that the power loss due to air resistance is $\frac{1}{2}\rho A v^3$ and the resistive force is $\frac{1}{2}\rho A v^2$, where ρ is the density of air.

66. An 80.0-kg skydiver jumps out of an airplane at an altitude of 1000 m and opens the parachute at an altitude of 200.0 m. (a) Assuming that the total retarding force on the diver is constant at 50.0 N with the parachute closed and constant at 3600 N with the parachute open, what is the speed of the diver when he lands on the ground? (b) Do you think the skydiver will get hurt? Explain. (c) At what height should the parachute be opened so that the final speed of the skydiver when he hits the ground is 5.00 m/s? (d) How realistic is the assumption that the total retarding force is constant? Explain.

67. A child's pogo stick (Fig. P5.67) stores energy in a spring ($k = 2.50 \times 10^4$ N/m). At position A ($x_1 = -0.100$ m), the spring compression is a maximum and the child is momentarily at rest. At position B ($x = 0$), the spring is relaxed and the child is moving upward. At position C, the child is again momentarily at rest at the top of the jump. Assuming that the combined mass of child and pogo stick is 25.0 kg, (a) calculate the total energy of the system if both potential energies are zero at $x = 0$, (b) determine x_2, (c) calculate the speed of the child at $x = 0$, (d) determine the value of x for which the kinetic energy of the system is a maximum, and (e) obtain the child's maximum upward speed.

Figure P5.67

68. A 2.00-kg block situated on a rough incline is connected to a spring of negligible mass having a spring constant of 100 N/m (Fig. P5.68). The block is released from rest when the spring is unstretched, and the pulley is frictionless. The block moves 20.0 cm down the incline before coming to rest. Find the coefficient of kinetic friction between block and incline.

Figure P5.68

69. In the dangerous "sport" of bungee jumping, a daring student jumps from a balloon with a specially designed elastic cord attached to his ankles, as shown in Figure P5.69. The unstretched length of the cord is 25.0 m, the student weighs 700 N, and the balloon is 36.0 m above the surface of a river below. Calculate the required force constant of the cord if the student is to stop safely 4.00 m above the river.

Figure P5.69   Bungee jumping. *(Gamma)*

70. An object of mass m is suspended from the top of a cart by a string of length L, as in Figure P5.70a. The cart and object are initially moving to the right at constant speed v_0. The cart comes to rest after colliding and sticking to a bumper, as in Figure P5.70b, and the suspended object swings through an angle θ. **(a)** Show that the initial speed $v_0 = \sqrt{2gL(1 - \cos\theta)}$. **(b)** If $L = 1.20$ m and $\theta = 35.0°$, find the initial speed of the cart. (*Hint:* The force exerted by the string on the object does no work on the object.)

(a)   (b)

Figure P5.70

Momentum and Collisions

Chapter Outline

6.1 Momentum and Impulse

6.2 Conservation of Momentum

6.3 Collisions

6.4 Glancing Collisions

▲ PHYSICS PUZZLER

As these charging rams crash together head-on, they exert equal and opposite forces on each other and are nearly at rest after the collision. Both rams have kinetic energy before colliding, but their kinetic energy is almost zero after colliding. What happens to the kinetic energy of the rams? *(Stephen Krasemann/ Tony Stone Images)* onsider what happens when a golf ball is struck by a club. As you well know, the ball receives a high velocity and goes a long distance because of the force exerted on it by the club. But why doesn't the club also receive a high velocity? It has the same force acting on it. Why does a good follow-through on the swing help the ball to go farther? Does the deformation of the ball on contact have any effect on its subsequent motion?

As a first step in answering such questions, we shall introduce the term *momentum*, a term that may have slightly different meaning in physics than it does in everyday life. For example, a candidate moving rapidly in the opinion polls is said to be gaining momentum, but from the standpoint of physics, is she? As another example, a very massive football player is said to have a great deal of momentum as he runs down the field. Can a much less massive player have the same momentum? What must he do to achieve this momentum? Can a common housefly have the same momentum as the charging rams in the chapter opener?

The concept of momentum will lead us to a second conservation law: conservation of momentum. This law is especially useful for treating problems that involve collisions between objects.

6.1   MOMENTUM AND IMPULSE

The linear momentum of an object of mass m moving with a velocity v is defined as the product of the mass and the velocity:

Linear momentum ▷

$$\mathbf{p} \equiv m\mathbf{v} \qquad [6.1]$$

As its definition shows, momentum is a vector quantity, with its direction matching that of the velocity. Momentum has dimensions ML/T, and its SI units are kilogram-meters per second (kg·m/s).

Often we shall find it advantageous to work with the components of momentum. For two-dimensional motion, these are

$$p_x = mv_x \qquad p_y = mv_y$$

where p_x represents the momentum of an object in the x direction and p_y its momentum in the y direction.

The definition of momentum in Equation 6.1 coincides with our everyday use of the word. When we think of a massive object moving with a high velocity, we often say that the object has a large momentum, in accordance with Equation 6.1. Likewise, a small object moving slowly is said to have a small momentum. On the other hand, a small object moving with a high velocity can have a large momentum.

When Newton first expressed the second law in mathematical form, he wrote it not as $\mathbf{F} = m\mathbf{a}$ but as

Newton's second law ▷

$$\mathbf{F} = \frac{\text{change in momentum}}{\text{time interval}} = \frac{\Delta \mathbf{p}}{\Delta t} \qquad [6.2]$$

where Δt is the time interval during which the momentum changes by $\Delta \mathbf{p}$. This equation states that **the time rate of change of momentum of an object is equal to the net force acting on the object.** To see that this is equivalent to $\mathbf{F} = m\mathbf{a}$ for an object of constant mass, consider a constant force, **F,** acting on a particle and producing a constant acceleration. We can write Equation 6.2 as

$$\mathbf{F} = \frac{\Delta \mathbf{p}}{\Delta t} = \frac{m\mathbf{v}_f - m\mathbf{v}_i}{\Delta t} = \frac{m(\mathbf{v}_f - \mathbf{v}_i)}{\Delta t} \qquad [6.3]$$

Now recall that the velocity of an object moving with constant acceleration varies with time as

$$\mathbf{v}_f = \mathbf{v}_i + \mathbf{a}t$$

If we take $\Delta t = t$ and substitute for v_f in Equation 6.3, we see that **F** reduces to the familiar equation

$$\mathbf{F} = m\mathbf{a}$$

Note from Equation 6.2 that if the net force **F** is zero, the momentum of the object does not change. In other words, the linear momentum and velocity of a particle are conserved when **F** = 0. This property of momentum is important in the analysis of collisions, a subject we shall take up in a later section.

Equation 6.2 can be written as **F** $\Delta t = \Delta \mathbf{p}$ or

$$\mathbf{F}\,\Delta t = \Delta \mathbf{p} = m\mathbf{v}_f - m\mathbf{v}_i \qquad [6.4]$$

◀ Impulse-momentum theorem

This result is often called the **impulse-momentum theorem.** The term on the left side of the equation, **F** Δt, is called the **impulse** of the force **F** for the time interval Δt. According to this result, **the impulse of the force acting on an object equals the change in momentum of that object.**

This equation tells us that if we exert a force on an object for time interval Δt, the effect of this force is to change the momentum of the object from some initial value, $m\mathbf{v}_i$, to some final value, $m\mathbf{v}_f$. For example, suppose a pitcher throws a baseball with a velocity of $\mathbf{v}_i$, and a batter hits the ball head on so as to reverse the direction of its velocity. The force, **F,** that the bat exerts on the ball can change both the direction and the magnitude of the initial velocity to a higher value, $\mathbf{v}_f$.

A word of caution. If you tried to solve a problem such as this using Newton's second law, you would encounter some difficulty in measuring the value to be used for **F,** because the force exerted on the ball is large, of short duration, and not constant. Instead, it might be represented by a curve like that in Figure 6.1a. The

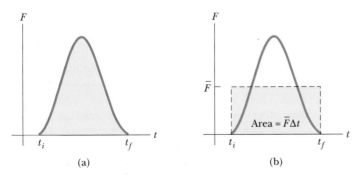

Figure 6.1   (a) A force acting on an object may vary in time. The impulse is the area under the force-time curve. (b) The average force (horizontal dashed line) gives the same impulse to the object in the time interval Δt as the real time-varying force described in (a).

Figure 6.2   An apple being pierced by a 30-caliber bullet traveling at a supersonic speed of 900 m/s. This collision was photographed with a microflash stroboscope using an exposure time of 0.33 μs. Shortly after the photograph was taken, the apple disintegrated completely. Note that the points of entry and exit of the bullet are visually explosive. *(Harold E. Edgerton, Palm Press, Inc.)*

force starts out small as the bat comes in contact with the ball, rises to a maximum value when they are firmly in contact, and then drops off as the ball leaves the bat. In such instances, it is necessary to define an **average force, $\overline{F}$,** shown as a dashed line in Figure 6.1b. This average force can be thought of as the constant force that would give the same impulse to the object in the time interval Δt as the actual time-varying force gives in this interval. The brief collision between a bullet and an apple is illustrated in Figure 6.2. **The impulse imparted by a force during the time interval Δt is equal to the area under the force-time graph from the beginning to the end of the time interval.**

APPLICATION

Cushioning Your Jump.

Think about what you do when you jump from a high position to the ground. As you strike the ground, you bend your knees. If you were to land on the ground with your legs locked, you would receive a painful shock in your legs as well as along your spine. The impulse-momentum theorem helps us understand why the landing is much less painful if you bend your knees. When you land, your momentum changes from its value just before you hit the ground to zero. By bending your knees, this change in momentum occurs over a longer time interval than with the knees locked. Thus, the force on the body is less than with the knees locked. For relatively short jumps, this force is low enough to do no damage to the bone structure.

Applying Physics 1

APPLICATION

Padded Gloves for Boxing.

In boxing matches of the nineteenth century, bare fists were used. In modern boxing, fighters wear padded gloves. How does this protect the brain of the boxer from injury? Boxers often "roll with the punch." How does this protect their health?

Explanation   The brain is immersed in a cushioning fluid inside the skull. If the head is struck suddenly by a bare fist, there is a rapid acceleration of the skull. The brain matches this acceleration only because of the large impulsive force of the skull on the brain. This large and sudden force can cause severe brain injury. Padded gloves extend the time over which the force is applied to the head. Thus, for a given impulse, the gloves provide a longer time interval than the bare fist, decreasing the average force. Because the average force is decreased, the acceleration of the skull is decreased, reducing (but not eliminating) the chance for brain injury. The same argument can be made for "rolling with the punch." If the head is held steady while being struck, the time interval over which the force is applied is relatively short and the average force is large. If the head is allowed to move in the same direction of the punch, the time interval is lengthened, and the average force reduced.

EXAMPLE 6.1 Teeing Off

A 50-g golf ball is struck with a club, as in Figure 6.3. The force on the ball varies from zero when contact is made up to some maximum value (when the ball is deformed) and then back to zero when the ball leaves the club. Thus, the force-time graph is somewhat like that in Figure 6.1. Assume that the ball leaves the club face with a velocity of +44 m/s.
(a) Estimate the impulse due to the collision.

Solution Before the club hits the ball, the ball is at rest on the tee, and as a result its initial momentum is zero. The magnitude of the momentum immediately after the collision is

$$p_f = mv_f = (50 \times 10^{-3} \text{ kg})(44 \text{ m/s}) = +2.2 \text{ kg·m/s}$$

Thus the impulse imparted to the ball, which equals its change in momentum, is

$$\Delta p = mv_f - mv_i = \boxed{+2.2 \text{ kg·m/s}}$$

(b) Estimate the duration of the collision and the average force on the ball.

Solution From Figure 6.3, it appears that a reasonable estimate of the distance the ball travels while in contact with the club is the radius of the ball, about 2.0 cm. The time it takes the club to move this distance (the contact time) is then

$$\Delta t = \frac{\Delta x}{\overline{v}_i} = \frac{2.0 \times 10^{-2} \text{ m}}{22 \text{ m/s}} = \boxed{9.1 \times 10^{-4} \text{ s}}$$

Finally, the magnitude of the average force is estimated to be

$$\overline{F} = \frac{\Delta p}{\Delta t} = \frac{2.2 \text{ kg·m/s}}{9.1 \times 10^{-4} \text{ s}} = \boxed{2.4 \times 10^3 \text{ N}}$$

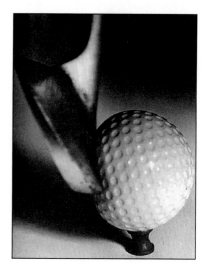

Figure 6.3 (Example 6.1) A golf ball being struck by a club. (© *Harold E. Edgerton. Courtesy of Palm Press, Inc.*)

EXAMPLE 6.2 How Good Are the Bumpers?

In a particular crash test, a 1.50×10^3 kg automobile collides with a wall, as in Figure 6.4a. The initial and final velocities of the automobile are $v_i = -15.0$ m/s and $v_f = +2.60$ m/s, respectively. If the collision lasts for 0.150 s, find the impulse due to the collision and the average force exerted on the automobile.

Solution The initial and final momenta of the automobile are

$$p_i = mv_i = (1.50 \times 10^3 \text{ kg})(-15.0 \text{ m/s}) = -2.25 \times 10^4 \text{ kg·m/s}$$

$$p_f = mv_f = (1.50 \times 10^3 \text{ kg})(2.60 \text{ m/s}) = +0.390 \times 10^4 \text{ kg·m/s}$$

Hence, the impulse, which equals the change in momentum, is

$$\overline{F}\Delta t = \Delta p = p_f - p_i$$
$$= 0.390 \times 10^4 \text{ kg·m/s} - (-2.25 \times 10^4 \text{ kg·m/s})$$

$$\overline{F}\Delta t = \boxed{+2.64 \times 10^4 \text{ kg·m/s}}$$

Before

−15.0 m/s

After

2.60 m/s

(a)

(b)

Figure 6.4   (Example 6.2) (a) This car's momentum changes as a result of its collision with the wall. (b) In a crash test (an inelastic collision), much of the car's initial kinetic energy is transformed into the energy it took to damage the vehicle. *(b, Courtesy of General Motors)*

The average force exerted on the automobile is

$$\bar{F} = \frac{\Delta p}{\Delta t} = \frac{2.64 \times 10^4 \ \text{kg} \cdot \text{m/s}}{0.150 \ \text{s}} = +1.76 \times 10^5 \ \text{N}$$

6.2   CONSERVATION OF MOMENTUM

An important feature of all isolated collisions is that momentum is conserved. In this section we shall see how the laws of motion lead us to this important conservation law.

A collision may be the result of physical contact between two objects, as illustrated in Figure 6.5a. This is a common observation when two macroscopic objects, such as two billiard balls or a baseball and a bat, strike each other. But "contact" on a submicroscopic scale is ill defined and hence meaningless, and so the notion of *collision* must be generalized. More accurately, forces between two bodies arise from the electrostatic interaction of the electrons in the surface atoms of the bodies. To understand the distinction between macroscopic and microscopic collisions, consider the collision between two positive charges as in Figure 6.5b. Because the two particles are positively charged, they repel each other and do not touch.

Figure 6.6 shows two isolated particles before and after they collide. By *isolated* we mean that no external forces, such as the gravitational force or friction, are present. Before the collision, the velocities of the two particles are $\mathbf{v}_{1i}$ and $\mathbf{v}_{2i}$; after the collision, the velocities are $\mathbf{v}_{1f}$ and $\mathbf{v}_{2f}$. The impulse-momentum theorem applied to m_1 becomes

$$\bar{\mathbf{F}}_1 \, \Delta t = m_1 \mathbf{v}_{1f} - m_1 \mathbf{v}_{1i}$$

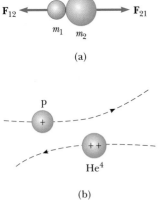

Figure 6.5   (a) A collision between two objects resulting from direct contact. (b) A collision between two charged objects (in this case, a proton and a helium nucleus).

Before collision

(a)

After collision

(b)

During collision

Figure 6.6 Before and after a head-on collision between two objects. The momentum of each object changes as a result of the collision, but the total momentum of the system remains constant.

Figure 6.7 When two objects collide, the force $\mathbf{F}_1$ exerted on m_1 is equal in magnitude and opposite in direction to the force $\mathbf{F}_2$ exerted on m_2.

The force from a nitrogen-propelled, hand-controlled device allows an astronaut to move about freely in space without restrictive tethers. *(Courtesy of NASA)*

Likewise, for m_2 we have

$$\overline{\mathbf{F}}_2 \, \Delta t = m_2\mathbf{v}_{2f} - m_2\mathbf{v}_{2i}$$

where $\overline{\mathbf{F}}_1$ is the force on m_1 due to m_2 during the collision and $\overline{\mathbf{F}}_2$ is the force on m_2 due to m_1 during the collision (Fig. 6.7).

We are using average values for $\overline{\mathbf{F}}_1$ and $\overline{\mathbf{F}}_2$ even though the actual forces may vary in time in a complicated way, as is the case in Figure 6.8. Newton's third law states that at all times these two forces are equal in magnitude and opposite in direction ($\mathbf{F}_1 = -\mathbf{F}_2$). In addition, the two forces act for the same time interval. Thus,

$$\overline{\mathbf{F}}_1 \, \Delta t = -\overline{\mathbf{F}}_2 \, \Delta t$$

or

$$m_1\mathbf{v}_{1f} - m_1\mathbf{v}_{1i} = -(m_2\mathbf{v}_{2f} - m_2\mathbf{v}_{2i})$$

from which we find

$$m_1\mathbf{v}_{1i} + m_2\mathbf{v}_{2i} = m_1\mathbf{v}_{1f} + m_2\mathbf{v}_{2f} \qquad \text{[6.5]}$$

This result is known as **conservation of momentum.**

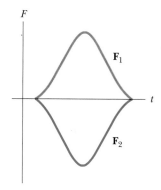

Figure 6.8 Force as a function of time for the two colliding particles in Figure 6.7. Note that $\mathbf{F}_1 = -\mathbf{F}_2$.

The principle of conservation of momentum states that, when no external forces act on a system consisting of two objects that collide with each other, the total momentum of the system before the collision is equal to the total momentum of the system after the collision.

◀ Conservation of momentum

Note that momentum is conserved for a *system* of objects. In the example used to derive Equation 6.5, the system was taken to be two colliding objects. More

A squid propels itself by expelling water at a high velocity. *(Mike Severns/Tony Stone Images)*

APPLICATION

Squid Propulsion.

generally, a system includes all the objects that are interacting with one another. In addition, we have assumed that the only forces acting during the collision are internal forces, meaning the forces that arise between the interacting objects of the system. For example, in the collision of two objects shown in Figure 6.7, the internal forces are $\mathbf{F}_1$ and $\mathbf{F}_2$. If a third object outside the system (consisting of m_1 and m_2) were to exert a force on either m_1 and m_2 (or both objects) during the collision, momentum would not be conserved for the system. In all the example problems that we shall consider, the system will be assumed to be isolated.

Our derivation has assumed that only two objects interact, but the result remains valid regardless of the number involved. In its most general form, conservation of momentum can be stated as follows: **The total momentum of an isolated system of objects is conserved regardless of the nature of the forces between the objects.**

To understand what this statement means, imagine that you initially stand at rest and then jump upward, leaving the ground with velocity $\mathbf{v}$. Obviously, your momentum is not conserved, because before the jump it was zero and it became $m\mathbf{v}$ as you began to rise. However, the total momentum of the system is conserved if the system selected includes all objects that exert forces on one another. You must include the Earth as part of this system because you exert a downward force on the Earth when you jump. The Earth in turn exerts on you an upward force of the same magnitude, as required by Newton's third law. Momentum is conserved for the system consisting of you and the Earth. Thus, as you move upward with some momentum $m\mathbf{v}$, the Earth moves in the opposite direction with momentum of the same magnitude. The recoil velocity of the Earth due to this event is imperceptibly small, of course, because the Earth is so massive, but its momentum is not zero.

Conservation of momentum is in operation in the movement of squids through the water. To accelerate, a squid fills areas in its body walls with water and then squirts the water backward through a tube beneath its head. Because the squid is increasing the momentum of the water by exerting a force on it in the backward direction, the momentum of the squid increases in the forward direction. This conserves the momentum of the system of squid and ejected water. The same principle is used for rocket engines on spacecraft and jet engines on aircraft.

Applying Physics 2

A baseball is projected into the air at an angle to the ground. As it moves through its trajectory, its velocity and, therefore, its momentum constantly changes. Is this a violation of conservation of momentum?

Explanation   The principle of conservation of momentum states that the momentum of a particle or a system of particles is conserved *in the absence of external forces*. A ball projected through the air is subject to the external force of gravity, so we would not expect its momentum to be conserved. The gravitational force is in the vertical direction, so it is actually only the vertical component of the momentum that changes due to this force. In the horizontal direction, there is no force (ignoring air resistance), so the horizontal component of the ball's momentum is conserved. We used the same idea earlier in stating that the horizontal component of the *velocity* of a projectile remains constant.

If we consider the baseball and the Earth as a system of particles, then the gravitational force is an internal force to this system. The momentum of the ball-Earth system is conserved—as the ball rises and falls, the Earth sinks (due to the downward force from the thrower's feet) and rises (as the ball pulls it back upward, although imperceptibly!), so that the total momentum of the system remains unchanged.

EXAMPLE 6.3   The Recoiling Pitching Machine

A baseball player attempts to use a pitching machine to help him improve his batting average. He places the 50-kg machine on a frozen pond, as in Figure 6.9. The machine fires a 0.15-kg baseball horizontally with a speed of 36 m/s. What is the recoil velocity of the machine?

Figure 6.9   (Example 6.3) When the baseball is fired to the right, the pitching machine recoils to the left.

Solution   We take the system to consist of the baseball and the pitching machine. Because of the force of gravity and the normal force, the system is not really isolated. However, both forces are directed perpendicularly to the motion of the system. Therefore, momentum is conserved in the x direction, because there are no external forces in this direction (assuming the surface is frictionless).

Because the baseball and pitching machine are at rest before firing, the total momentum of the system is zero. Therefore, the total momentum after firing must also be zero; that is,

$$m_1\mathbf{v}_1 + m_2\mathbf{v}_2 = 0$$

With $m_1 = 0.15$ kg, $\mathbf{v}_1 = +36$ m/s, and $m_2 = 50$ kg, we find the recoil velocity of the pitching machine to be

$$\mathbf{v}_2 = -\frac{m_1}{m_2}\mathbf{v}_1 = -\left(\frac{0.15 \text{ kg}}{50 \text{ kg}}\right)(36 \text{ m/s}) = \boxed{-0.11 \text{ m/s}}$$

The negative sign for $\mathbf{v}_2$ indicates that the pitching machine is moving to the left after firing, in the direction *opposite* the motion of the baseball.

6.3   COLLISIONS

We have seen that, for any type of collision, the total momentum of the system just before collision equals the total momentum just after collision. We can say that the total momentum is always conserved for any type of collision. However, the total kinetic energy is generally not conserved in a collision, because some of the kinetic

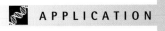

APPLICATION

Glaucoma Testing.

energy is converted to thermal energy and internal elastic potential energy when the objects deform.

We define an inelastic collision as a collision in which momentum is conserved but kinetic energy is not. The collision of a rubber ball with a hard surface is inelastic, because some of the kinetic energy is lost when the ball is deformed during contact with the surface. When two objects collide and stick together, the collision is called **perfectly inelastic.** For example, if two pieces of putty collide, they stick together and move with some common velocity after the collision. If a meteorite collides head on with the Earth, it becomes buried in the Earth and the collision is considered perfectly inelastic. Not all of the initial kinetic energy is necessarily lost in a perfectly inelastic collision.

An elastic collision is defined as one in which both momentum and kinetic energy are conserved. Billiard ball collisions and the collisions of air molecules with the walls of a container at ordinary temperatures are highly elastic. Macroscopic collisions such as those between billiard balls are only approximately elastic, because some permanent deformation, and hence some loss of kinetic energy, takes place. Perfectly elastic collisions do occur, however, between atomic and subatomic particles. Elastic and perfectly inelastic collisions are *limiting* cases; most actual collisions fall into a category between them.

A practical example of an inelastic collision is used to detect glaucoma, a disease in which the pressure inside the eye builds up and leads to blindness by damaging the cells of the retina. In this application, medical professionals use a device called a tonometer to measure the pressure inside the eye. This device releases a puff of air against the outer surface of the eye and measures the speed of the air after reflection from the eye. At normal pressure, the eye is slightly spongy, and the pulse is reflected at low speed. As the pressure inside the eye increases, the outer surface becomes more rigid, and the speed of the reflected pulse increases. Thus, the speed of the reflected puff of air is used to measure the internal pressure of the eye.

We can summarize the types of collisions as follows:

Elastic collision ▶

Inelastic collision ▶

- An elastic collision is one in which both momentum and kinetic energy are conserved.
- An inelastic collision is one in which momentum is conserved but kinetic energy is not.
- A perfectly inelastic collision is an inelastic collision in which the two objects stick together after the collision, so that their final velocities are the same and the momentum of the system is conserved.

In the remainder of this section, we shall treat perfectly inelastic and elastic collisions in one dimension.

Perfectly Inelastic Collisions

Consider two objects of masses m_1 and m_2 moving with initial velocities v_{1i} and v_{2i} along a straight line, as in Figure 6.10. We shall assume that the objects collide head on so that they move along the same line of motion after the collision. If the two objects stick together and move with some common velocity, v_f, after the collision, then only the momentum of the system is conserved, and we can say that the total momenta before and after the collision are equal:

$$m_1 v_{1i} + m_2 v_{2i} = (m_1 + m_2) v_f \qquad \text{[6.6]}$$

It is important to note that v_{1i}, v_{2i}, and v_f represent the x components of the vectors $\mathbf{v}_{1i}$, $\mathbf{v}_{2i}$, and $\mathbf{v}_f$, so one must be careful with signs. For example, in Figure 6.10, v_{1i} would have a positive value (m_1 moving to the right), whereas v_{2i} would have a negative value (m_2 moving to the left).

In a typical inelastic collision problem, only one quantity in the preceding equation is unknown, and so conservation of momentum is sufficient to tell us what we need to know.

Before collision

After collision

Figure 6.10 (a) Before and (b) after a perfectly inelastic head-on collision between two objects.

EXAMPLE 6.4 The Cadillac Versus the Compact Car

An 1800-kg luxury sedan stopped at a traffic light is struck from the rear by a compact car with a mass of 900 kg. The two cars become entangled as a result of the collision. If the compact car were moving at a velocity of $+20.0$ m/s before the collision, what is the velocity of the entangled mass after the collision?

Solution The momentum before the collision is that of the compact car alone, because the large car is initially at rest. Thus, for the momentum before the collision we have

$$p_i = m_1 v_i = (900 \text{ kg})(20.0 \text{ m/s}) = +1.80 \times 10^4 \text{ kg} \cdot \text{m/s}$$

After the collision, the mass that moves is the sum of the masses of the large car and the compact car, and the momentum of the combination is

$$p_f = (m_1 + m_2) v_f = (2700 \text{ kg})(v_f)$$

Equating the momentum before the collision to the momentum after the collision and solving for v_f, the velocity of the wreckage, we get

$$v_f = \frac{p_i}{m_1 + m_2} = \frac{1.80 \times 10^4 \text{ kg} \cdot \text{m/s}}{2700 \text{ kg}} = +6.67 \text{ m/s}$$

EXAMPLE 6.5 Here's Mud in Your Eye

Two balls of mud collide head on in a perfectly inelastic collision, as in Figure 6.10. Suppose $m_1 = 0.500$ kg, $m_2 = 0.250$ kg, $v_{1i} = +4.00$ m/s, and $v_{2i} = -3.00$ m/s. (a) Find the velocity of the composite ball of mud after the collision.

Solution Writing Equation 6.6 for conservation of momentum with the positive direction for velocity to the right, we can find the velocity of the combined mass after the collision:

$$m_1 v_{1i} + m_2 v_{2i} = (m_1 + m_2) v_f$$

$$(0.500 \text{ kg})(4.00 \text{ m/s}) + (0.250 \text{ kg})(-3.00 \text{ m/s}) = (0.750 \text{ kg})(v_f)$$

$$v_f = +1.67 \text{ m/s}$$

(b) How much kinetic energy is lost in the collision?

Solution   The kinetic energy of the system before the collision is

$$KE_i = KE_1 + KE_2 = \tfrac{1}{2}m_1 v_{1i}^2 + \tfrac{1}{2}m_2 v_{2i}^2$$
$$= \tfrac{1}{2}(0.500 \text{ kg})(4.00 \text{ m/s})^2 + \tfrac{1}{2}(0.250 \text{ kg})(-3.00 \text{ m/s})^2$$
$$= 5.13 \text{ J}$$

The kinetic energy of the system after the collision is

$$KE_f = \tfrac{1}{2}(m_1 + m_2)v_f^2 = \tfrac{1}{2}(0.750 \text{ kg})(1.67 \text{ m/s})^2 = 1.05 \text{ J}$$

Hence, the loss in kinetic energy is

$$KE_i - KE_f = \boxed{4.08 \text{ J}}$$

Most of this lost energy is converted to thermal energy and internal elastic potential energy as the two objects collide and deform.

EXAMPLE 6.6   The Ballistic Pendulum

The ballistic pendulum (Fig. 6.11) is a device used to measure the velocity of a fast-moving projectile such as a bullet. The bullet is fired into a large block of wood suspended from some light wires. The bullet is stopped by the block, and the entire system swings through

(a)

(b)

Figure 6.11   (Example 6.6) (a) Diagram of a ballistic pendulum. Note that $\mathbf{v}_f$ is the velocity of the system just after the perfectly inelastic collision. (b) Multiflash photograph of a laboratory ballistic pendulum. *(Courtesy of Central Scientific Company)* the vertical distance h. It is possible to obtain the initial velocity of the bullet by measuring h and the two masses. As an example of the technique, assume that the mass of the bullet, m_1, is 5.00 g, the mass of the pendulum, m_2, is 1.00 kg, and h is 5.00 cm. Find the initial velocity of the bullet, v_{1i}.

Solution The collision between the bullet and the block (the system) is perfectly inelastic. Writing the conservation of momentum for the collision in the form of Equation 6.6, we have

$$m_1 v_{1i} = (m_1 + m_2) v_f$$
$$(1) \qquad (5.00 \times 10^{-3} \text{ kg})(v_{1i}) = (1.0050 \text{ kg})(v_f)$$

There are two unknowns in this equation, v_{1i} and v_f; the latter is the velocity of the block plus embedded bullet *immediately after the collision*. We must look for additional information if we are to complete the problem. Kinetic energy is not conserved during an inelastic collision. However, mechanical energy is conserved after the collision, and so the kinetic energy of the system at the bottom is transformed into the potential energy of the bullet plus block at height h:

$$\tfrac{1}{2}(m_1 + m_2) v_f^2 = (m_1 + m_2) gh$$
$$\tfrac{1}{2}(1.0050 \text{ kg})(v_f^2) = (1.0050 \text{ kg})(9.80 \text{ m/s}^2)(5.00 \times 10^{-2} \text{ m})$$

giving

$$v_f = 0.990 \text{ m/s}$$

With v_f now known, (1) yields v_{1i}:

$$v_{1i} = \frac{(1.0050 \text{ kg})(0.990 \text{ m/s})}{5.00 \times 10^{-3} \text{ kg}} = \boxed{199 \text{ m/s}}$$

Exercise Explain why it would be incorrect to equate the initial kinetic energy of the incoming bullet to the final gravitational potential energy of the bullet-block combination.

Elastic Collisions

Now consider two objects that undergo an elastic head-on collision (Fig. 6.12). In this situation, *both momentum and kinetic energy are conserved;* we can write these conditions as

$$m_1 v_{1i} + m_2 v_{2i} = m_1 v_{1f} + m_2 v_{2f} \qquad \text{[6.7]}$$

$$\tfrac{1}{2} m_1 v_{1i}^2 + \tfrac{1}{2} m_2 v_{2i}^2 = \tfrac{1}{2} m_1 v_{1f}^2 + \tfrac{1}{2} m_2 v_{2f}^2 \qquad \text{[6.8]}$$

where v is positive if an object moves to the right and negative if it moves to the left.

In a typical problem involving elastic collisions, there are two unknown quantities, and Equations 6.7 and 6.8 can be solved simultaneously to find them. An alternative approach, employing a little mathematical manipulation of Equation 6.8, often simplifies this process. To see this, let's cancel the factor $\tfrac{1}{2}$ in Equation 6.8 and rewrite it as

$$m_1(v_{1i}^2 - v_{1f}^2) = m_2(v_{2f}^2 - v_{2i}^2)$$

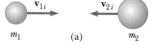

Before collision

m_1 (a) m_2

After collision

(b)

Figure 6.12 (a) Before and (b) after an elastic head-on collision between two hard spheres.

As a result of the collision between the bowling ball and pin, part of the ball's momentum is transferred to the pin. As a consequence, the pin acquires momentum and kinetic energy, and the ball loses momentum and kinetic energy. However, the total momentum of the system (ball and pin) remains constant. *(Ben Rose/ The Image Bank)*

Here we have moved the terms containing m_1 to one side of the equation and those containing m_2 to the other. Next, let us factor both sides of the equation:

$$m_1(v_{1i} - v_{1f})(v_{1i} + v_{1f}) = m_2(v_{2f} - v_{2i})(v_{2f} + v_{2i}) \qquad [6.9]$$

We now separate the terms containing m_1 and m_2 in the equation for the conservation of momentum (Eq. 6.7) to get

$$m_1(v_{1i} - v_{1f}) = m_2(v_{2f} - v_{2i}) \qquad [6.10]$$

To obtain our final result, we divide Equation 6.9 by Equation 6.10 and get

$$v_{1i} + v_{1f} = v_{2f} + v_{2i}$$

$$v_{1i} - v_{2i} = -(v_{1f} - v_{2f}) \qquad [6.11]$$

This equation, in combination with the condition for conservation of momentum, will be used to solve problems dealing with perfectly elastic, head-on collisions. According to Equation 6.11, the relative velocity of the two objects before the collision, $v_{1i} - v_{2i}$, equals the negative of the relative velocity of the two objects after the collision, $-(v_{1f} - v_{2f})$.

Applying Physics 3

A great deal of expense is involved in repairing automobiles after collisions. It seems that much of this money could be saved if cars were built from hard rubber, so that they made approximately elastic collisions and simply bounced off each other without denting. Why isn't this done?

Explanation   Making cars out of hard rubber would save on automobile repair bills, but the results for the drivers and passengers would be disastrous. Imagine two identical cars traveling toward each other at identical speeds collide head on. In the approximation that the collision is elastic, the cars bounce off each other and leave the collision with their velocities reversed in direction. Thus, the inhabitants of the cars would suffer a large change in momentum in a short period of time. As a result, the force on the inhabitants would become very large, causing significant bodily injury. For real cars in this situation, the cars would come to rest during the collision, as the front sections of the cars crumpled. Thus, the change in momentum is only half what it was in the elastic case, and the change occurs over the longer time period during which the cars crumpled. The resulting smaller force on the inhabitants will result in fewer injuries.

PROBLEM-SOLVING STRATEGY

Conservation of Momentum

The following procedure is recommended for problems involving collisions between two objects:

1. Set up a coordinate system and define your velocities with respect to that system. That is, objects moving in the direction selected as the positive di-

rection of the *x* axis are considered to have a positive velocity, and those moving in the negative *x* direction, a negative velocity. It is convenient to make the *x* axis coincide with one of the initial velocities.

2. In your sketch of the coordinate system, draw all velocity vectors with labels and include all the given information.

3. Write expressions for the momenta of each object before and after the collision. (In two-dimensional collision problems, write expressions for the *x* and *y* components of momentum before and after the collision.) Remember to include the appropriate signs for the velocity vectors.

4. Now write expressions for the *total* momentum *before* and *after* the collision and equate the two. (For two-dimensional collisions, this expression should be written for the momentum in both the *x* and *y* directions.) Remember, it is the momentum of the *system* (the two colliding objects) that is conserved, *not* the momenta of the individual objects.

5. If the collision is *inelastic*, kinetic energy is not conserved. Proceed to solve the momentum equations for the unknown quantities.

6. If the collision is *elastic*, kinetic energy is conserved, so you can equate the total kinetic energies before and after the collision. This gives an additional relationship between the velocities.

Q U I C K L A B

Hold a Ping-Pong ball or tennis ball on top of a basketball. Drop them both at the same time so that the basketball hits the floor, bounces up, and hits the smaller falling ball. What happens and why?

EXAMPLE 6.7 Let's Play Pool

Two billiard balls move toward one another, as in Figure 6.12. The balls have identical masses, and assume that the collision between them is elastic. If the initial velocities of the balls are $+30$ cm/s and -20 cm/s, what is the velocity of each ball after the collision?

Solution   We turn first to Equation 6.7. The equal masses cancel on each side, and after substituting the appropriate values for the initial velocities, we have

$$30 \text{ cm/s} + (-20 \text{ cm/s}) = v_{1f} + v_{2f}$$

(1) $$10 \text{ cm/s} = v_{1f} + v_{2f}$$

Because kinetic energy is also conserved, we can apply Equation 6.11, which gives

$$30 \text{ cm/s} - (-20 \text{ cm/s}) = v_{2f} - v_{1f}$$

(2) $$50 \text{ cm/s} = v_{2f} - v_{1f}$$

Solving (1) and (2) simultaneously, we find

$$v_{1f} = \boxed{-20 \text{ cm/s}} \qquad v_{2f} = \boxed{+30 \text{ cm/s}}$$

That is, the balls *exchange velocities!* This is always the case when two objects of equal mass collide elastically head on.

Exercise   Find the final velocity of the two balls if the ball with initial velocity -20 cm/s has a mass equal to half that of the ball with initial velocity $+30$ cm/s.

Answer $v_{1f} = -3.0$ cm/s; $v_{2f} = +47$ cm/s

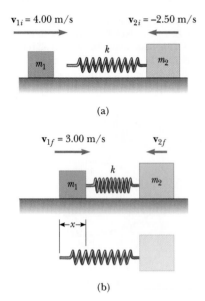

$\mathbf{v}_{1i} = 4.00$ m/s $\mathbf{v}_{2i} = -2.50$ m/s

k

m_1 m_2

(a)

$\mathbf{v}_{1f} = 3.00$ m/s $\mathbf{v}_{2f}$

k

m_1 m_2

x

(b)

Figure 6.13   (Example 6.8)

EXAMPLE 6.8   A Two-Body Collision with Spring

A block of mass $m_1 = 1.60$ kg, moving to the right with a speed of 4.00 m/s on a friction-less, horizontal track, collides with a spring attached to a second block, of mass $m_2 = 2.10$ kg, that is moving to the left with a speed of 2.50 m/s (Fig. 6.13a). The spring constant is 600 N/m. For the instant when m_1 is moving to the right with a speed of 3.00 m/s, determine (a) the velocity of m_2 and (b) the distance, x, that the spring is compressed.

Solution   (a) First, note that the initial velocity of m_2 is $v_{2i} = -2.50$ m/s because it is moving to the left. Because the total momentum of the system is constant, we have

$$m_1 v_{1i} + m_2 v_{2i} = m_1 v_{1f} + m_2 v_{2f}$$

$$(1.60 \text{ kg})(4.00 \text{ m/s}) + (2.10 \text{ kg})(-2.50 \text{ m/s}) = (1.60 \text{ kg})(3.00 \text{ m/s}) + (2.10 \text{ kg})v_{2f}$$

$$v_{2f} = \boxed{-1.74 \text{ m/s}}$$

This result shows that m_2 is still moving to the left at that instant.
(b) To determine the compression, x, in the spring in Figure 6.13b, we can make use of conservation of energy because no friction forces are acting on the system. Thus, we have

$$\tfrac{1}{2}m_1 v_{1i}^2 + \tfrac{1}{2}m_2 v_{2i}^2 = \tfrac{1}{2}m_1 v_{1f}^2 + \tfrac{1}{2}m_2 v_{2f}^2 + \tfrac{1}{2}kx^2$$

Substitution of the given values and the result of part (a) into this expression gives

$$x = \boxed{0.173 \text{ m}}$$

Exercise   Find the velocity of m_1 and the compression in the spring at the instant m_2 is at rest.

Answer   0.719 m/s; 0.251 m

EXAMPLE 6.9   Executive Stress Reliever

An ingenious device that illustrates conservation of momentum and conservation of kinetic energy is shown in Figure 6.14a. It consists of five identical hard balls supported by strings of equal lengths. When one ball is pulled out and released, after the almost elastic collision one ball moves out on the opposite side as in Figure 6.14b. If two balls are pulled out and released, two balls swing out on the opposite side, and so forth—but how do the balls know? For example, is it possible that on occasion when one ball is released, two will swing out on the opposite side traveling with half the speed of the incoming ball, as in Figure 6.14c? Assume the collision is elastic and use conservation of momentum and conservation of kinetic energy to show that the collision shown in Figure 6.14c can never occur.

Solution   First, let us apply conservation of momentum to the collision. The momentum of the system before the collision is mv, where m is the mass of the one ball released and v is its speed just before the collision. After the collision, we have two balls each of mass m, moving with speed $v/2$. Thus, the total momentum of the system after the collision is $m(v/2) + m(v/2) = mv$. Thus, momentum is conserved. However, let us now consider the kinetic energy of the system before and after the collision. The kinetic energy just before the collision is $KE_{\text{before}} = \tfrac{1}{2}mv^2$ and the kinetic energy just after the collision is

(a)

Figure 6.14 (Example 6.9) An executive stress reliever.

$KE_{after} = \frac{1}{2}m(v/2)^2 + \frac{1}{2}m(v/2)^2 = \frac{1}{4}mv^2$. Thus, kinetic energy is *not* conserved in this collision. The only way to have both momentum and kinetic energy conserved is for one ball to move out when one ball is released, or two balls to move out when two are released, and so on. Try other combinations for yourself to verify this statement.

6.4 GLANCING COLLISIONS

The collisions we have considered until now have been head-on collisions, in which the incident mass strikes a second mass head on and both rebound along a straight-line path that coincides with the line of motion of the incident mass. Anyone who has ever played billiards knows that such collisions are the exception rather than the rule. A more common type of collision is a *glancing collision,* in which the colliding masses rebound at some angle relative to the line of motion of the incident mass. Figure 6.15a shows a blue ball that travels with an initial speed of v_{1i} and strikes a red ball obliquely (off center). After the collision, the blue ball caroms off

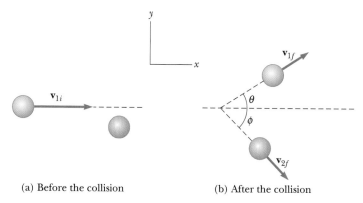

(a) Before the collision 　　　　　 (b) After the collision

Figure 6.15 (a) Before and (b) after a glancing collision between two balls.

at an angle of θ relative to its incident line of motion, and the red ball rebounds at an angle of ϕ.

As we emphasized earlier, **momentum is conserved in all collisions when no external forces are acting,** and glancing collisions are no exception. Because momentum is a vector quantity, the conservation of momentum principle must be written $\mathbf{p}_i = \mathbf{p}_f$. That is, the total initial momentum of the system (the two balls) must equal the total final momentum of the system. For a collision in two dimensions, such as that in Figure 6.16, this implies that the total momentum is conserved along the x direction *and* along the y direction. We can state this in equation form as

$$\Sigma p_{ix} = \Sigma p_{fx} \quad \text{and} \quad \Sigma p_{iy} = \Sigma p_{fy} \qquad [6.12]$$

The following example illustrates how to use the principle of conservation of momentum to treat glancing collisions.

EXAMPLE 6.10   Collision at an Intersection

At an intersection, a 1500-kg car traveling east at 25 m/s collides with a 2500-kg van traveling north at 20 m/s, as shown in Figure 6.16. Find the direction and magnitude of the velocity of the wreckage immediately after the collision, assuming that the vehicles undergo a perfectly inelastic collision (that is, they stick together).

Solution   Let us choose the positive x direction to be east and the positive y direction to be north, as in Figure 6.16. Before the collision, the only object having momentum in the x direction is the car. Thus, the total initial momentum of the system (car plus van) in the x direction is

$$\Sigma p_{ix} = (1500 \text{ kg})(25 \text{ m/s}) = 37\,500 \text{ kg·m/s}$$

Now let us assume that the wreckage moves at an angle of θ and a speed of v after the collision, as in Figure 6.16. The total momentum in the x direction after the collision is

$$\Sigma p_{fx} = (4000 \text{ kg})(v \cos \theta)$$

Because momentum is conserved in the x direction, we have

$$\Sigma p_{ix} = \Sigma p_{fx}$$
$$(1) \qquad 37\,500 \text{ kg·m/s} = (4000 \text{ kg})(v \cos \theta)$$

Similarly, the total initial momentum of the system in the y direction is that of the van, which equals $(2500 \text{ kg})(20 \text{ m/s})$. Applying conservation of momentum to the y direction, we have

$$\Sigma p_{iy} = \Sigma p_{fy}$$
$$(2500 \text{ kg})(20 \text{ m/s}) = (4000 \text{ kg})(v \sin \theta)$$
$$(2) \qquad 50\,000 \text{ kg·m/s} = (4000 \text{ kg})(v \sin \theta)$$

If we divide (2) by (1), we get

$$\tan \theta = \frac{50\,000}{37\,500} = 1.33$$
$$\theta = 53°$$

When this angle is substituted into (2)—or, alternatively, into (1)—the value of v is

$$v = \frac{50\,000 \text{ kg·m/s}}{(4000 \text{ kg})(\sin 53°)} = 16 \text{ m/s}$$

Figure 6.16   (Example 6.10) A top view of a perfectly inelastic collision between a car and a van.

PHYSICS IN ACTION

A Glancing Collision and Life on Mars

About 15 million years ago, a meteorite of unknown origin came hurtling into the atmosphere of Mars at a glancing angle. On impact with the Martian surface a lot of debris was tossed into space and out of the gravitational grasp of Mars. The energy of the collision was sufficiently great to melt a portion of one particular rock about the size of a potato that was a part of that debris. This melting created bubbles in the rock that captured traces of the Martian atmosphere. The makeup of this atmosphere is well-known because of samples taken by the Viking landers. This particular ejected rock eventually became a meteorite for Earth and landed on its surface. In 1984, this inconsequential appearing rock was collected by a team searching for meteorites in Antarctica and given the name ALH84001, a designation signifying that this was the first meteorite (001) collected from the Allen Hills Region of Antarctica (ALH). Analysis of the gas from these bubbles was later to provide convincing proof of the Martian origin of the rock. Between 1994 and 1996, David McKay and a team of scientists from the University of Georgia, Johnson Space Center, McGill University, and Stanford University subjected the meteorite to a wide array of tests in an effort to find evidence of life. In 1996, they announced that they had found four pieces of evidence to support the conclusion that tiny, cylindrical shaped organisms once lived in the interstices of the rock. Specifically, they found (1) traces of carbonate mineral. This mineral forms only in the presence of water. Thus, one of the presumed necessities of life, water, had at one time percolated through the microscopic channels of the rock. (2) Spectroscopic analysis of the materials in the rock showed the presence of a type of organic molecule called polycyclic aromatic hydrocarbons. This is a substance often formed when microbes die. (3) Minute specimens of magnetite, gregite, and pyrrhotite were revealed by a powerful electron microscope. All of these materials can be produced by various microscopic lifeforms. Finally, (4) this same microscope revealed cylindrical shapes that appear to be the fossilized remains of primitive lifeforms. The fact that all of these life signs increase with depth inside the rock indicate that they did not arise from contamination while the meteorite was on Earth.

The evidence is not conclusive to all scientists, but regardless of the opinion held, a glancing collision eons ago has certainly brought new interest in that age-old question, "Are we alone?"

(NASA)

SUMMARY

The **linear momentum** of an object of mass m moving with a velocity of $\mathbf{v}$ is defined to be

$$\mathbf{p} \equiv m\mathbf{v} \qquad [6.1]$$

The **impulse** of a force, $\mathbf{F}$, acting on an object is equal to the product of the force and the time interval during which the force acts:

$$\text{Impulse} = \mathbf{F}\,\Delta t$$

The **impulse-momentum theorem** states that the impulse of a force on an object is equal to the change in momentum of the object:

$$\mathbf{F}\,\Delta t = \Delta \mathbf{p} = m\mathbf{v}_f - m\mathbf{v}_i \qquad [6.4]$$

Conservation of momentum applied to a system of two interacting objects states that, when no external forces act on the system, the total momentum of the system before the collision is equal to the total momentum of the system after the collision.

$$m_1\mathbf{v}_{1i} + m_2\mathbf{v}_{2i} = m_1\mathbf{v}_{1f} + m_2\mathbf{v}_{2f} \qquad [6.5]$$

An **inelastic collision** is one in which momentum is conserved but kinetic energy is not. A **perfectly inelastic collision** is one in which the colliding objects stick together after the collision. An **elastic collision** is one in which both momentum and kinetic energy are conserved.

In glancing collisions, conservation of momentum can be applied along two perpendicular directions — that is, along an x axis and a y axis.

MULTIPLE-CHOICE QUESTIONS

1. An object of mass m moves to the right with a speed v. It collides head-on in a perfectly inelastic collision with an object of twice the mass but half the speed moving in the opposite direction. What is the speed of the combined mass after the collision?
   (a) 0   (b) $v/2$ (c) v (d) $2v$

2. A baseball pitching machine of mass M (including the mass m of a baseball) rests on the pitcher's mound. What is the speed of recoil of the machine when it fires the baseball at a speed v?
   (a) 0   (b) $mv/(M - m)$
   (c) $mv/(M + m)$ (d) mv/M
   (e) $2mv/(M - m)$

3. A 0.10-kg object moving initially with a velocity of 0.20 m/s eastward makes an elastic head-on collision with a 0.15-kg object initially at rest. What is the final velocity of the 0.10-kg object after the collision?

   (a) 0.16 m/s eastward
   (b) 0.16 m/s westward
   (c) 0.04 m/s eastward
   (d) 0.04 m/s westward

4. When one catches a baseball, it hurts the hand less if one gives with the ball. The reason this helps the hand is that
   (a) this makes the kinetic energy change less.
   (b) this makes the momentum change less.
   (c) this makes the time interval for stopping greater.
   (d) this increases the impulse of the collision.

5. A 10.0-g bullet is fired into a 200-g block of wood at rest on a horizontal surface. After impact, the block slides 8.00 m before coming to rest. If the coefficient of friction is $\mu_k = 0.400$, find the speed of the bullet before impact.
   (a) 106 m/s   (b) 166 m/s   (c) 226 m/s   (d) 286 m/s

CONCEPTUAL QUESTIONS

1. In perfectly inelastic collisions between two objects, there are some instances in which all of the original kinetic energy is transformed to other forms than kinetic. Give an example of such a situation.

2. If two objects collide and one is initially at rest, is it possible for both to be at rest after the collision? Is it possible for one to be at rest after the collision? Explain.

3. A ball of clay of mass m is thrown with a speed v against a brick wall. The clay sticks to the wall and stops. Is the principle of conservation of momentum violated in this example?

4. A skater is standing still on a frictionless ice rink. Her friend throws a Frisbee straight at her. In which of the following cases is the largest momentum transferred to the skater? **(a)** The skater catches the Frisbee and holds onto it. **(b)** The skater catches the Frisbee momentarily but then drops it vertically downward. **(c)** The skater catches the Frisbee, holds it momentarily, and throws it back to her friend.

5. You are standing perfectly still and then you take a step forward. Before the step your momentum was zero, but afterward you have some momentum. Is the conservation of momentum violated in this case?

6. If two particles have equal kinetic energies, are their momenta necessarily equal? Explain.

7. A boy stands at one end of a floating raft that is stationary relative to the shore. He then walks to the opposite end of the raft, away from the shore. Does the raft move? Explain.

8. If two automobiles collide, they usually do not stick together. Does this mean the collision is perfectly elastic? Explain why a head-on collision is likely to be more dangerous than other types of collisions.

9. An open box slides across a frictionless, icy surface of a frozen lake. What happens to the speed of the box as water from a rain shower collects in it, assuming that the rain falls vertically downward into the box? Explain.

10. Consider a perfectly inelastic collision between a car and a large truck. Which vehicle loses more kinetic energy as a result of the collision?

11. Your physical education teacher, who knows something about physics, throws you a tennis ball at a certain velocity, and you catch it. You are now given the following choice: The teacher can throw you a medicine ball, which is much more massive than the tennis ball, with the same velocity as the tennis ball, the same momentum, or the same kinetic energy. Which choice would you make in order to make the easiest catch, and why?

12. You are watching a movie about a superhero and notice that the superhero hovers in the air and throws a piano at some bad guys while remaining stationary in the air. What's wrong with this scenario?

13. In golf, novice players are often advised to be sure to "follow through" with their swing. Why does this make the ball travel a longer distance? If a shot is taken near the green, little follow-through is required. Why?

14. An airbag is inflated when a collision occurs, which protects the passenger (the dummy, in this case) from serious injury (Fig. Q6.14). Why does the airbag soften the blow? Discuss the physics involved in this dramatic photograph.

Figure Q6.14 *(Courtesy of Saab)*

15. A sharpshooter fires a rifle while standing with the butt of the gun against his shoulder. If the forward momentum of a bullet is the same as the backward momentum of the gun, why isn't it as dangerous to be hit by the gun as by the bullet?

16. A large bedsheet is held vertically by two students. A third student, who happens to be the star pitcher on the baseball team, throws a raw egg at the sheet. Explain why the egg does not break when it hits the sheet, regardless of its initial speed. (If you try this, make sure the pitcher hits the sheet near its center, and do not allow the egg to fall on the floor after being caught by the sheet.)

PROBLEMS

1, 2, 3 = straightforward, intermediate, challenging   □ = full solution available in Study Guide/Student Solutions Manual
WEB = solution posted at **http://www.saunderscollege.com/physics/college**   = biomedical application   = Interactive Physics

Section 6.1   Momentum and Impulse

1. Show that the kinetic energy of a particle of mass m is related to the magnitude of the momentum p of that particle by $KE = p^2/2m$. *Note:* This expression is invalid for very high speed particles.

2. Calculate the magnitude of the linear momentum for the following cases: **(a)** a proton with mass 1.67×10^{-27} kg, moving with a speed of 5.00×10^6 m/s; **(b)** a 15.0-g bullet moving with a speed of 300 m/s; **(c)** a 75.0-kg sprinter running with a speed of 10.0 m/s; **(d)** the Earth (mass $= 5.98 \times 10^{24}$ kg) moving with an orbital speed equal to 2.98×10^4 m/s.

3. A 0.10-kg ball is thrown straight up into the air with an initial speed of 15 m/s. Find the momentum of the ball **(a)** at its maximum height and **(b)** halfway to its maximum height.

4. A pitcher claims he can throw a 0.145-kg baseball with as much momentum as a 3.00-g bullet moving with a speed of 1.50×10^3 m/s. **(a)** What must the baseball's speed be if the pitcher's claim is valid? **(b)** Which has greater kinetic energy, the ball or the bullet?

5. A 1500-kg car moving with a speed of 15 m/s collides with a utility pole and is brought to rest in 0.30 s. Find the magnitude of the average force exerted on the car during the collision.

6. A car is stopped for a traffic signal. When the light turns green, the car accelerates, increasing its speed from 0 to 5.20 m/s in 0.832 s. What are the magnitudes of the linear impulse and the average force experienced by a 70.0-kg passenger in the car during this time?

7. A 0.500-kg football is thrown with a speed of 15.0 m/s. A stationary receiver catches the ball and brings it to rest in 0.020 s. **(a)** What is the impulse delivered to the ball? **(b)** What is the average force exerted on the receiver?

8. The force shown in the force-time diagram in Figure P6.8 acts on a 1.5-kg mass. Find **(a)** the impulse of the

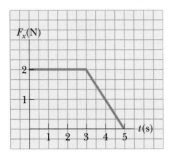

Figure P6.8

force, **(b)** the final velocity of the mass if it is initially at rest, and **(c)** the final velocity of the mass if it is initially moving along the x axis with a velocity of -2.0 m/s.

9. The force, $\mathbf{F}_x$, acting on a 2.00-kg particle varies in time as shown in Figure P6.9. Find **(a)** the impulse of the force, **(b)** the final velocity of the particle if it is initially at rest, and **(c)** the final velocity of the particle if it is initially moving along the x axis with a velocity of -2.00 m/s.

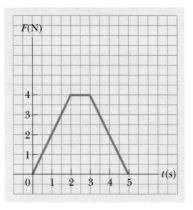

Figure P6.9   (Problems 9 and 38)

10. The forces shown in the force-time diagram in Figure P6.10 act on a 1.5-kg mass. Find **(a)** the impulse for the interval $t = 0$ to $t = 3.0$ s and **(b)** the impulse for the interval $t = 0$ to $t = 5.0$ s. **(c)** If the forces act on a 1.5-kg particle that is initially at rest, find the particle's speed at $t = 3.0$ s and at $t = 5.0$ s.

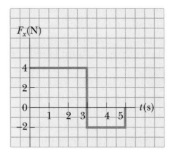

Figure P6.10

11. The front of a 1400-kg car is designed to absorb the shock of a collision by having a "crumple zone" in which the front 1.20 m of the car collapses in absorbing the shock of a collision. If a car traveling 25.0 m/s stops uni-

formly in 1.20 m, **(a)** how long does the collision last, **(b)** what is the magnitude of the average force on the car, and **(c)** what is the acceleration of the car in g's?

12. A 0.50-kg object is at rest at the origin of a coordinate system. A 3.0-N force in the $+x$ direction acts on the object for 1.50 s. **(a)** What is the velocity at the end of this interval? **(b)** At the end of this interval, a constant force of 4.0 N is applied in the $-x$ direction for 3.0 s. What is the velocity at the end of the 3.0 s?

WEB 13. A 0.15-kg baseball is thrown with a speed of 20 m/s. It is hit straight back at the pitcher with a final speed of 22 m/s. **(a)** What is the impulse delivered to the ball? **(b)** Find the average force exerted by the bat on the ball if the two are in contact for 2.0×10^{-3} s.

14. A 3.0-kg steel ball strikes a massive wall at 10 m/s at an angle of 60° with the plane of the wall. It bounces off with the same speed and angle (Fig. P6.14). If the ball is in contact with the wall for 0.20 s, what is the average force exerted on the ball by the wall?

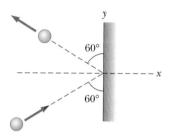

Figure P6.14

Section 6.2 Conservation of Momentum

15. A 730-N man stands in the middle of a frozen pond of radius 5.0 m. He is unable to get to the other side because of a lack of friction between his shoes and the ice. To overcome this difficulty, he throws his 1.2-kg physics textbook horizontally toward the north shore, at a speed of 5.0 m/s. How long does it take him to reach the south shore?

16. High-speed stroboscopic photographs show that the head of a 200-g golf club is traveling at 55 m/s just before it strikes a 46-g golf ball at rest on a tee. After the collision, the club head travels (in the same direction) at 40 m/s. Find the speed of the golf ball just after impact.

17. A rifle with a weight of 30 N fires a 5.0-g bullet with a speed of 300 m/s. **(a)** Find the recoil speed of the rifle. **(b)** If a 700-N man holds the rifle firmly against his shoulder, find the recoil speed of man and rifle.

18. A 45.0-kg girl is standing on a 150-kg plank. The plank, originally at rest, is free to slide on a frozen lake, which is a flat, frictionless surface. The girl begins to walk along the plank at a constant velocity of 1.50 m/s relative to the plank. **(a)** What is her velocity relative to the ice surface? **(b)** What is the velocity of the plank relative to the ice surface?

19. A 65.0-kg person throws a 0.0450-kg snowball forward with a ground speed of 30.0 m/s. A second person, with a mass of 60.0 kg, catches the snowball. Both people are on skates. The first person is initially moving forward with a speed of 2.50 m/s, and the second person is initially at rest. What are the velocities of the two people after the snowball is exchanged? Disregard the friction between the skates and the ice.

20. A 7.00-kg bowling ball is dropped from rest at an initial height of 3.00 m. **(a)** What is the speed of the Earth coming up to meet the ball just before the ball hits the ground? Use 5.98×10^{24} kg as the mass of the Earth. **(b)** Use your answer to part (a) to justify ignoring the motion of the Earth when dealing with the motions of terrestrial objects.

Section 6.3 Collisions

Section 6.4 Glancing Collisions

21. A 7.00-kg bowling ball collides head-on with a 2.00-kg bowling pin that was originally at rest. The pin flies forward with a speed of 3.00 m/s. If the ball continues forward with a speed of 1.80 m/s, what was the initial speed of the ball? Ignore rotation of the ball.

22. A 3.00-kg sphere makes a perfectly inelastic collision with a second sphere that is initially at rest. The composite system moves with a speed equal to one third the original speed of the 3.00-kg sphere. What is the mass of the second sphere?

23. A railroad car of mass 2.00×10^4 kg moving at 3.00 m/s collides and couples with two coupled railroad cars, each of the same mass as the single car and moving in the same direction at 1.20 m/s. **(a)** What is the speed of the three coupled cars after the collision? **(b)** How much kinetic energy is lost in the collision?

24. A 7.0-g bullet is fired into a 1.5-kg ballistic pendulum. The bullet emerges from the block with a speed of 200 m/s, and the block rises to a maximum height of 12 cm. Find the initial speed of the bullet.

WEB 25. A 0.030-kg bullet is fired vertically at 200 m/s into a 0.15-kg baseball that is initially at rest. How high does the combination rise after the collision, assuming the bullet embeds in the ball?

26. An 8.00-g bullet is fired into a 250-g block that is initially at rest at the edge of a table of 1.00-m height (Fig.

P6.26). The bullet remains in the block, and after the impact the block lands 2.00 m from the bottom of the table. Determine the initial speed of the bullet.

8.00 g

250 g

1.00 m

2.00 m

Figure P6.26

27. A 12.0-g bullet is fired horizontally into a 100-g wooden block that is initially at rest on a frictionless horizontal surface and connected to a spring of constant 150 N/m. If the bullet-block system compresses the spring by a maximum of 80.0 cm, what was the velocity of the bullet at impact with the block?

28. (a) Three carts of masses 4.0 kg, 10 kg, and 3.0 kg move on a frictionless horizontal track with speeds of 5.0 m/s, 3.0 m/s, and 4.0 m/s, as shown in Figure P6.28. The carts stick together after colliding. Find the final velocity of the three carts. (b) Does your answer require that all carts collide and stick together at the same time?

5.0 m/s   3.0 m/s   −4.0 m/s

4.0 kg   10 kg   3.0 kg

Figure P6.28

29. A 5.00-g object moving to the right at 20.0 cm/s makes an elastic head-on collision with a 10.0-g object that is initially at rest. Find (a) the velocity of each object after the collision and (b) the fraction of the initial kinetic energy transferred to the 10.0-g object.

30. A 10.0-g object moving to the right at 20.0 cm/s makes an elastic head-on collision with a 15.0-g object moving in the opposite direction at 30.0 cm/s. Find the velocity of each object after the collision.

31. A 25.0-g object moving to the right at 20.0 cm/s overtakes and collides elastically with a 10.0-g object moving in the same direction at 15.0 cm/s. Find the velocity of each object after the collision.

32. A billiard ball rolling across a table at 1.50 m/s makes a head-on elastic collision with an identical ball. Find the speed of each ball after the collision (a) when the second ball is initially at rest, (b) when the second ball is moving toward the first at a speed of 1.00 m/s, and (c) when the second ball is moving away from the first at a speed of 1.00 m/s.

33. A 90-kg fullback moving east with a speed of 5.0 m/s is tackled by a 95-kg opponent running north at 3.0 m/s. If the collision is perfectly inelastic, calculate (a) the velocity of the players just after the tackle and (b) the kinetic energy lost as a result of the collision. Can you account for the missing energy?

34. An 8.00-kg mass moving east at 15.0 m/s on a frictionless horizontal surface collides with a 10.0-kg mass that is initially at rest. After the collision, the 8.00-kg mass moves south at 4.00 m/s. (a) What is the velocity of the 10.0-kg mass after the collision? (b) What percentage of the initial kinetic energy is lost in the collision?

35. A 2000-kg car moving east at 10.0 m/s collides with a 3000-kg car moving north. The cars stick together and move as a unit after the collision, at an angle of 40.0° north of east and at a speed of 5.22 m/s. Find the velocity of the 3000-kg car before the collision.

36. A 0.30-kg puck, initially at rest on a frictionless horizontal surface, is struck by a 0.20-kg puck that is initially moving along the x axis with a velocity of 2.0 m/s. After the collision, the 0.20-kg puck has a speed of 1.0 m/s at an angle of $\theta = 53°$ to the positive x axis. (a) Determine the velocity of the 0.30-kg puck after the collision. (b) Find the fraction of kinetic energy lost in the collision.

WEB 37. A billiard ball moving at 5.00 m/s strikes a stationary ball of the same mass. After the collision, the first ball moves at 4.33 m/s at an angle of 30° with respect to the original line of motion. (a) Find the velocity (magnitude and direction) of the second ball after collision. (b) Was this an inelastic collision or an elastic collision?

ADDITIONAL PROBLEMS

38. Consult the force-time graph in Figure P6.9. Find the average force exerted on the particle for the time interval $t_i = 0$ to $t_f = 3.0$ s.

39. A 0.400-kg soccer ball approaches a player horizontally with a speed of 15.0 m/s. The player illegally strikes the ball with her hand and causes it to move in the opposite direction with a speed of 22.0 m/s. What impulse was delivered to the ball by the player?

40. An 80.0-kg astronaut is working on the engines of his spaceship, which is drifting through space with a con-

stant velocity. The astronaut, wishing to get a better view of the Universe, pushes against the ship and later finds himself 30.0 m behind the ship and moving so slowly that he can be considered to be at rest. Without a thruster, the only way to return to the ship is to throw his 0.500-kg wrench directly away from the ship. If he throws the wrench with a speed of 20.0 m/s, how long does it take him to reach the ship?

41. A 2.0-g particle moving at 8.0 m/s makes an elastic head-on collision with a resting 1.0-g object. (a) Find the speed of each after the collision. (b) If the stationary particle has a mass of 10 g, find the speed of each particle after the collision. (c) Find the final kinetic energy of the incident 2.0-g particle in the situations described in (a) and (b). In which case does the incident particle lose more kinetic energy?

42. Consider the ballistic pendulum described in Example 6.6 and shown in Figure 6.11. If the mass of the bullet is 8.00 g and the mass of the pendulum is 2.00 kg, find the ratio of the total kinetic energy after the collision to the kinetic energy before the collision. What accounts for the missing energy?

43. A 0.400-kg bead slides on a curved frictionless wire, starting from rest at point A in Figure P6.43. At point B the bead collides elastically with a 0.600-kg bead at rest. Find the distance the bead moves up the wire.

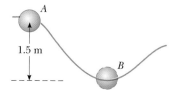

Figure P6.43

44. A standing 80-kg man steps off a 3.0-m high diving platform and begins to fall from rest. Two (2.0) seconds after reaching the water, the man comes to rest. What average force did the water exert on him?

WEB 45. A 12.0-g bullet is fired horizontally into a 100-g wooden block initially at rest on a horizontal surface. After impact, the block slides 7.5 m before coming to rest. If the coefficient of kinetic friction between block and surface is 0.650, what was the speed of the bullet immediately before impact?

46. Two blocks of masses $m_1 = 2.00$ kg and $m_2 = 4.00$ kg are each released from rest at a height of 5.00 m on a frictionless track, as shown in Figure P6.46, and undergo an elastic head-on collision. (a) Determine the velocity of each block just before the collision. (b) Determine the velocity of each block immediately after the collision.

(c) Determine the maximum heights to which m_1 and m_2 rise after collision.

Figure P6.46

47. A 0.500-kg block is released from rest at the top of a frictionless track 2.50 m above the top of a table. It then collides elastically with a 1.00-kg mass that is initially at rest on the table, as shown in Figure P6.47. (a) Determine the speeds of the two masses just after the collision. (b) How high up the track does the 0.500-kg mass travel back after the collision? (c) How far away from the bottom of the table does the 1.00-kg mass land, given that the table is 2.00 m high? (d) How far away from the bottom of the table does the 0.500-kg mass eventually land?

$m_1 = 0.500$ kg
$m_2 = 1.00$ kg
$h_1 = 2.50$ m
$h_2 = 2.00$ m

Figure P6.47

48. An 8.0-g bullet is fired into a 2.5-kg ballistic pendulum and becomes embedded in it. If the pendulum rises a vertical distance of 6.0 cm, calculate the initial speed of the bullet.

49. A small block of mass $m_1 = 0.500$ kg is released from rest at the top of a curved wedge of mass $m_2 = 3.00$ kg, which sits on a frictionless horizontal surface, as in Figure P6.49a. When the block leaves the wedge, its velocity

is measured to be 4.00 m/s to the right, as in Figure P6.49b. **(a)** What is the velocity of the wedge after the block reaches the horizontal surface? **(b)** What is the height, h, of the wedge?

(a)   (b)

Figure P6.49

50. Two carts of equal mass, $m = 0.250$ kg, are placed on a frictionless track that has a light spring of force constant $k = 50.0$ N/m attached to one end of it, as in Figure P6.50. The red cart is given an initial velocity of $\mathbf{v}_0 = 3.00$ m/s to the right, and the blue cart is initially at rest. If the carts collide elastically, find **(a)** the velocity of the carts just after the first collision and **(b)** the maximum compression in the spring.

Figure P6.50

51. A cannon, initially resting on a frictionless surface, of mass $m_1 = 800$ kg (when unloaded) is loaded with a "shot" of mass $m_2 = 10.0$ kg. The cannon is aimed at mass $m_3 = 7990$ kg, which is connected to a massless spring of spring constant $k = 4500$ N/m, as in Figure P6.51a. The cannon is then fired, and the shot inelastically collides with mass m_3 and sticks in it, as shown in Figure P6.51b. The combined system then compresses the spring a maximum distance of $d = 0.5$ m, as in Figure P6.51c. **(a)** Determine the speed of m_2 just before it collides with m_3. (You may assume that m_2 travels in a straight line.) **(b)** Determine the recoil speed of the cannon. **(c)** The cannon recoils toward the right, and when it passes point A there is friction (with $\mu_k = 0.600$) between the cannon and the ground. How far to the right of point A does the cannon slide before coming to rest?

Figure P6.51

52. Water falls at the rate of 250 g/s from a height of 60 m into a 750-g bucket on a scale (without splashing). If the bucket is originally empty, what does the scale read after 3.0 s?

53. The bird perched on the swing in Figure P6.53 has a mass of 52.0 g, and the base of the swing has a mass of 153 g. Assume that the swing and bird are originally at rest and that the bird then takes off horizontally at 2.00 m/s. If the base can swing freely (i.e., without friction) around the pivot, how high will the base of the swing rise above its original level?

8.00 cm

Figure P6.53

54. A 2000-kg car traveling at 10.0 m/s collides with a 3000-kg car that is initially at rest at a stoplight. The cars stick together and move 2.00 m before friction causes them to stop. Determine the coefficient of kinetic friction between the cars and the road, assuming that the negative acceleration is constant and all wheels on both cars lock at the time of impact.

55. As shown in Figure P6.55, a bullet of mass m and speed v passes completely through a pendulum bob of mass M. The bullet emerges with a speed of $v/2$. The pendulum bob is suspended by a stiff rod of length ℓ and negligible mass. What is the minimum value of v such that the pendulum bob will barely swing through a complete vertical circle?

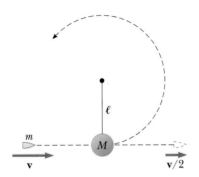

Figure P6.55

56. Two particles of masses m and $3m$ are moving toward each other along the x axis with the same initial speeds v_0. Mass m is traveling to the left, and mass $3m$ is traveling to the right. They undergo an elastic glancing collision such that mass m is moving downward after the collision at right angles from its initial direction. (a) Find the final speeds of the two masses. (b) What is the angle θ at which the mass $3m$ is scattered?

57. How fast can you set the Earth moving? In particular, when you jump straight up as high as you can, you give the Earth a maximum recoil speed of what order of magnitude? Visualize the Earth as a perfectly solid object. In your solution state the physical quantities you take as data and the values you measure or estimate for them.

58. A neutron in a reactor makes an elastic collision head on with a carbon atom that is initially at rest. (The mass of the carbon nucleus is about 12 times that of the neutron.) (a) What fraction of the neutron's kinetic energy is transferred to the carbon nucleus? (b) If the neutron's initial kinetic energy is 1.6×10^{-13} J, find its final kinetic energy and the kinetic energy of the carbon nucleus after the collision.

59. A cue ball traveling at 4.0 m/s makes a glancing, elastic collision with a target ball of equal mass that is initially at rest. The cue ball is deflected so that it makes an angle of $30°$ with its original direction of travel. Find (a) the angle between the velocity vectors of the two balls after the collision and (b) the speed of each ball after the collision.

60. A block of mass m lying on a rough horizontal surface is given an initial velocity of $\mathbf{v}_0$. After traveling a distance d, it makes a head-on elastic collision with a block of mass $2m$. How far does the second block move before coming to rest? (Assume the coefficient of friction μ is the same for both blocks.)

Circular Motion and the Law of Gravity

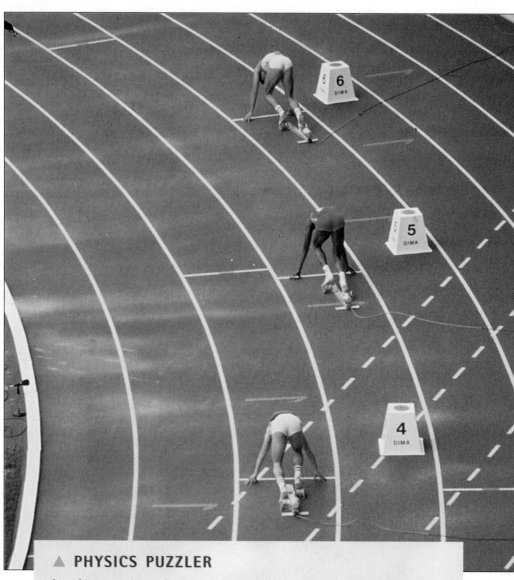

▲ **PHYSICS PUZZLER**

In a short race at a track event, such as a 200 m or 400 m sprint, the runners begin from staggered positions on the track. Why don't all the runners begin from the same line? *(© Gerard Vandystadt/Photo Researchers, Inc.)*

I n this chapter we shall investigate circular motion, a specific type of two-dimensional motion. We shall encounter such terms as *angular speed, angular acceleration,* and *centripetal acceleration.* The results derived will help you understand the motions of a diverse range of objects in our environment, from a car moving around a circular race track to clusters of galaxies orbiting a common center.

We shall also introduce Newton's universal law of gravitation, one of the fundamental laws in nature, and show how this law, together with Newton's laws of motion, enables us to understand a variety of familiar phenomena.

Circular motion and the universal law of gravitation are related historically in that Newton discovered the law of gravity as a result of attempting to explain the circular motion of the Moon about the Earth and the motions of the planets about the Sun. Thus, it is appropriate to consider these two important physical topics together. A more general treatment of gravitational potential energy and the concept of escape velocity are provided in an optional section.

7.1 ANGULAR SPEED AND ANGULAR ACCELERATION

We began our study of linear motion by defining the terms *displacement, velocity,* and *acceleration.* We will take the same basic approach now as we turn to a study of rotational motion. Consider Figure 7.1a, a top view of a rotating compact disc. The axis of rotation is at the center of the disc, at *O.* A point *P* on the disc is at the distance *r* from the origin and moves about *O* in a circle of radius *r.* In fact, every point on the disc undergoes circular motion about *O.* To analyze such motion, it is convenient to set up a *fixed* reference line, as shown in Figure 7.1a. Let us assume that at time *t* = 0, the point *P* is on the reference line, as in Figure 7.1a, and a line is drawn on the disc from the origin out to *P.* After an interval of Δ*t* has elapsed, *P* has advanced to a new position (Fig. 7.1b). In this time interval, the line *OP* has moved through the angle θ with respect to the reference line. Likewise, *P* has moved a distance of *s* measured along the circumference of the circle; *s* is called an *arc length.*

In situations we have encountered thus far, angles have been measured in degrees. However, in scientific work angles are often measured in *radians* (rad) rather than degrees, for the effect of simplifying certain equations. In fact, almost all of

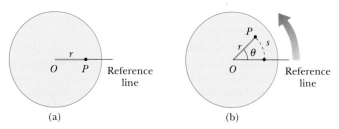

Figure 7.1 (a) The point *P* on a rotating compact disc at *t* = 0. (b) As the disc rotates, the point *P* moves through an arc length of *s.*

the equations derived in this chapter and the next require that angles be measured in radians. With reference to Figure 7.1b, when the arc length s is equal to the radius r, the angle θ swept out by r is equal to one radian. In general, any angle θ, measured in radians, is defined by the relation

$$\theta \equiv \frac{s}{r} \qquad [7.1]$$

The radian is a pure number with no dimensions. This can be seen from Equation 7.1, because θ is the ratio of an arc length (a distance) to the radius of the circle (also a distance).

To convert degrees to radians, note that when point P in Figure 7.1 moves through an angle of 360° (one revolution), the arc length s is equal to the circumference of the circle, $2\pi r$. From Equation 7.1 we see that the corresponding angle in radians is $2\pi r/r = 2\pi$ rad. Hence,

$$1 \text{ rad} \equiv \frac{360°}{2\pi} \approx 57.3°$$

From this definition, it follows that any angle in degrees can be converted to an angle in radians with the expression

$$\theta(\text{rad}) = \frac{\pi}{180°}\,\theta(\text{deg})$$

For example, 60° equals $\pi/3$ rad and 45° equals $\pi/4$ rad.

Returning to the compact disc, we see from Figure 7.2 that, as the disc rotates and a point on it moves from P to Q in a time of Δt, the angle through which the disc rotates is $\Delta\theta = \theta_2 - \theta_1$. We define $\Delta\theta$ as the **angular displacement. The average angular speed,** $\overline{\omega}$ (ω is the Greek letter omega), **of a rotating rigid object is the ratio of the angular displacement,** $\Delta\theta$, **to the time interval** Δt:

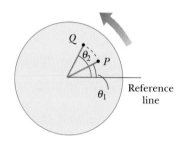

Figure 7.2   As a point on the compact disc moves from P to Q, the disc rotates through the angle $\Delta\theta = \theta_2 - \theta_1$.

Average angular speed ▶

$$\overline{\omega} \equiv \frac{\theta_2 - \theta_1}{t_2 - t_1} = \frac{\Delta\theta}{\Delta t} \qquad [7.2]$$

By analogy with linear speed, **the instantaneous angular speed,** ω, **is defined as the limit of the average speed,** $\Delta\theta/\Delta t$, **as the time interval** Δt **approaches zero:**

Instantaneous angular speed ▶

$$\omega \equiv \lim_{\Delta t \to 0} \frac{\Delta\theta}{\Delta t} \qquad [7.3]$$

Angular speed has the units radians per second (rad/s). We shall take ω to be positive when θ is increasing (counterclockwise motion) and negative when θ is decreasing (clockwise motion). When the angular speed is constant, the instantaneous angular speed is equal to the average angular speed.

EXAMPLE 7.1 Whirlybirds

The rotor on a helicopter turns at an angular speed of 320 revolutions per minute (in this book, we shall sometimes use the abbreviation rpm, but in most cases we shall use rev/min). Express this in radians per second.

Solution We shall use the conversion factors 1 rev = 2π rad and 60 s = 1 min, to give

$$320 \frac{\text{rev}}{\text{min}} = 320 \frac{\text{rev}}{\text{min}} \left(\frac{2\pi \text{ rad}}{\text{rev}} \right) \left(\frac{1 \text{ min}}{60 \text{ s}} \right) = 10.7\pi \frac{\text{rad}}{\text{s}}$$

Figure 7.3 shows a bicycle turned upside down so that a repair person can work on the rear wheel. The bicycle pedals are turned so that at time t_1 the wheel has angular speed ω_1 (Fig. 7.3a), and at a later time, t_2, it has angular speed ω_2 (Fig. 7.3b). **The average angular acceleration, $\overline{\alpha}$** (α is the Greek letter alpha), **of an object is defined as the ratio of the change in the angular speed to the time, Δt, it takes the object to undergo the change:**

$$\overline{\alpha} \equiv \frac{\text{change in angular speed}}{\text{time interval}} = \frac{\omega_2 - \omega_1}{t_2 - t_1} = \frac{\Delta\omega}{\Delta t} \qquad \text{[7.4]}$$

◀ Average angular acceleration

The instantaneous angular acceleration is defined as the limit of the ratio $\Delta\omega/\Delta t$ as Δt approaches zero. Angular acceleration has the units radians per second per second (rad/s²). **When a rigid object rotates about a fixed axis, as does the bicycle wheel, every portion of the object has the same angular speed**

◀ Instantaneous angular acceleration

t_1

ω_1

(a)

t_2

ω_2

(b)

Figure 7.3 An accelerating bicycle wheel rotates with (a) angular speed ω_1 at time t_1 and (b) angular speed ω_2 at time t_2.

and the same angular acceleration. This, in fact, is precisely what makes these variables so useful for describing rotational motion.

The following argument should convince you that ω and α are the same for every point on the wheel. If a point on the rim of the wheel had a greater angular speed than a point nearer the center, the shape of the wheel would be changing. The wheel remains circular (symmetrically distributed about the axle) only if all points have the same angular speed and the same angular acceleration.

7.2   ROTATIONAL MOTION UNDER CONSTANT ANGULAR ACCELERATION

Let us pause for a moment to consider some similarities between the equations we have found thus far for rotational motion and those we found for linear motion in earlier chapters. For example, compare the defining equation for average angular speed,

$$\bar{\omega} \equiv \frac{\theta_f - \theta_i}{t_f - t_i} = \frac{\Delta\theta}{\Delta t}$$

with the defining equation for average linear speed,

$$\bar{v} \equiv \frac{x_f - x_i}{t_f - t_i} = \frac{\Delta x}{\Delta t}$$

The equations are similar in the sense that θ replaces x and ω replaces v. Take careful note of such similarities as you study rotational motion, because virtually every linear quantity we have encountered thus far has a corresponding "twin" in rotational motion. Once you are adept at recognizing such analogies, you will find it unnecessary to memorize many of the equations in this chapter. In addition, the techniques for solving rotational motion problems are quite similar to those you have already learned for linear motion. For example, problems concerned with objects that rotate with constant angular acceleration can be solved in much the same manner as those dealing with linear motion under constant acceleration. If you understand problems involving objects that move with constant linear acceleration, these rotational motion problems should be little more than a review for you.

An additional analogy between linear motion and rotational motion is revealed when we compare the defining equation for average angular acceleration,

$$\bar{\alpha} \equiv \frac{\omega_f - \omega_i}{t_f - t_i} = \frac{\Delta\omega}{\Delta t}$$

with the defining equation for average linear acceleration,

$$\bar{a} \equiv \frac{v_f - v_i}{t_f - t_i} = \frac{\Delta v}{\Delta t}$$

In light of the analogies between variables in linear motion and those in rotational motion, it should not surprise you that the equations of rotational motion involve the variables θ, ω, and α. In Chapter 2, Section 2.6, we developed a set of kinematic equations for linear motion under constant acceleration. The same procedure can be used to derive a similar set of equations for rotational motion under constant angular acceleration. The resulting equations of rotational kinematics, along with the corresponding equations for linear motion under constant acceleration, are as follows:

Rotational Motion About a Fixed Axis with α Constant (Variables: θ and ω)	Linear Motion with a Constant (Variables: x and v)	
$\omega = \omega_0 + \alpha t$	$v = v_0 + at$	[7.5]
$\theta = \omega_0 t + \frac{1}{2}\alpha t^2$	$x = v_0 t + \frac{1}{2}at^2$	[7.6]
$\omega^2 = \omega_0{}^2 + 2\alpha\theta$	$v^2 = v_0{}^2 + 2ax$	[7.7]

Again, note the one-to-one correspondence between the rotational equations involving the angular variables θ, ω, and α and the equations of linear motion involving the variables x, v, and a.

EXAMPLE 7.2 The Rotating Wheel

The bicycle wheel of Figure 7.3 rotates with a constant angular acceleration of 3.5 rad/s^2. If the initial angular speed of the wheel is 2.0 rad/s at $t_0 = 0$, (a) through what angle does the wheel rotate in 2.0 s?

Solution Because we are given $\omega_0 = 2.0$ rad/s and $\alpha = 3.5$ rad/s^2, we use

$$\theta = \omega_0 t + \tfrac{1}{2}\alpha t^2$$

$$\theta = (2.0 \text{ rad/s})(2.0 \text{ s}) + \tfrac{1}{2}(3.5 \text{ rad/s}^2)(2.0 \text{ s})^2$$

$$= 11 \text{ rad} = \quad 630°$$

(b) What is the angular speed at $t = 2.0$ s?

Solution Making use of Equation 7.5, we find that

$$\omega = \omega_0 + \alpha t = 2.0 \text{ rad/s} + (3.5 \text{ rad/s}^2)(2.0 \text{ s}) = \quad 9.0 \text{ rad/s}$$

Exercise Find the angular speed of the wheel at $t = 2.0$ s by making use of Equation 7.7 and the results of part (a).

7.3 RELATIONS BETWEEN ANGULAR AND LINEAR QUANTITIES

In this section we shall derive some useful relations between the angular speed and acceleration of a rotating object and the linear speed and acceleration of an arbitrary point in the object. Consider the arbitrarily shaped object in Figure 7.4, rotating about the z axis through the point O. Assume that the object rotates through

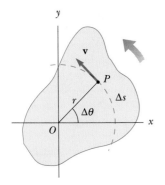

Figure 7.4   Rotation of an object about an axis through O that is perpendicular to the plane of the figure (the z axis). Note that a point, P, on the object rotates in a circle of radius r centered at O.

the angle $\Delta\theta$, and hence point P moves through the arc length Δs in the interval Δt. We know from the defining equation for radian measure that

$$\Delta\theta = \frac{\Delta s}{r}$$

Let us now divide both sides of this equation by Δt, the interval during which the rotation occurred:

$$\frac{\Delta\theta}{\Delta t} = \frac{1}{r}\frac{\Delta s}{\Delta t}$$

If Δt is very small, then the angle $\Delta\theta$ through which the object rotates is small and the ratio $\Delta\theta/\Delta t$ is the instantaneous angular speed, ω. Also, Δs is very small when Δt is very small, and the ratio $\Delta s/\Delta t$ equals the instantaneous linear speed, v, for small values of Δt. Hence, the preceding equation is equivalent to

$$\omega = \frac{v}{r}$$

Figure 7.4 allows us to interpret this equation. The distance Δs is traversed along an arc of the circular path followed by the point P as it moves during the time Δt. Thus, v must be the linear speed of a point lying along this arc, where the direction of v is *tangent to the circular path*. As a result, we often refer to this linear speed as the *tangential speed* of a particle moving in a circular path, and write

Tangential speed ▷

$$v_t = r\omega \qquad [7.8]$$

That is, **the tangential speed of a point on a rotating object equals the distance of that point from the axis of rotation multiplied by the angular speed.** Equation 7.8 shows that the linear speed of a point on the rotating object increases with movement outward from the center of rotation toward the rim, as one would intuitively expect. **Although every point on the rotating object has the same angular speed, not every point has the same linear (tangential) speed.**

Exercise caution when using Equation 7.8. It has been derived using the defining equation for radian measure and hence is valid only when ω is measured in radians per unit time. Other measures of angular speed, such as degrees per second and revolutions per second, are not to be used in Equation 7.8.

To find a second equation relating linear and angular quantities, imagine that an object rotating about a fixed axis (Fig. 7.4) changes its angular speed by $\Delta\omega$ in the interval Δt. At the end of this interval, the speed of a point on the object, such as P, has changed by the amount Δv_t. From Equation 7.8 we have

$$\Delta v_t = r\,\Delta\omega$$

Dividing by Δt gives

$$\frac{\Delta v_t}{\Delta t} = r\frac{\Delta\omega}{\Delta t}$$

If the time interval Δt is very small, then the ratio $\Delta v_t/\Delta t$ is the tangential acceleration of that point and $\Delta\omega/\Delta t$ is the angular acceleration. Therefore, we see that

Tangential acceleration ▷

$$a_t = r\alpha \qquad [7.9]$$

QUICKLAB

Experiment with a bicycle wheel, like the one in Figure 7.3. Wrap tape around the middle of one of the spokes so that you can easily see it and draw a chalk mark on a point on the rim at the end of the spoke. Make the tire rotate by gently turning the cranks by hand. Compare the linear speed of the tape with that of the chalk mark. Why do these two points move at different speeds? What happens to the speed of the tape as you slide it closer to the center of the wheel?

That is, **the tangential acceleration of a point on a rotating object equals the distance of that point from the axis of rotation multiplied by the angular acceleration.** Again, radian measure must be used for the angular acceleration term in this equation.

There is one more equation that relates linear quantities to angular quantities, but we shall defer its derivation to the next section.

Thinking Physics 1

The launch area for the European Space Agency is not in Europe—it is in South America. Why?

Explanation Placing a satellite in Earth orbit requires providing a large tangential speed to the satellite. This is the task of the rocket propulsion system. Anything that reduces the requirements on the rocket system is a welcome contribution. The surface of the Earth is already traveling toward the east at a high speed, due to the rotation of the Earth. Thus, if rockets are launched toward the east, the rotation of the Earth provides some initial tangential speed, reducing the requirements on the rocket. If rockets were launched from Europe, which is at a relatively large-angle latitude, the contribution of the Earth's rotation is relatively small, because the distance between Europe and the rotation axis of the Earth is relatively small. The ideal place for launching is at the Equator, which is as far as one can be from the rotation axis of the Earth and still be on the surface of the Earth. This results in the largest possible tangential speed due to the Earth's rotation. The European Space Agency exploits this advantage by launching from French Guiana, which is only a few degrees north of the equator.

Applying Physics 1

When a tall smokestack falls over, it often breaks somewhere along its length before it hits the ground, as in Figure 7.5. Why does this happen?

Explanation As the smokestack rotates around its base, each higher portion of the smokestack falls with an increasing tangential acceleration. The tangential acceleration of a given point on the smokestack is proportional to the distance of that portion from the base. As the acceleration increases, eventually higher portions of the smokestack will experience an acceleration greater than that which could result from gravity alone. This can only happen if these portions are being pulled downward by a force in addition to the gravitational force. The force that causes this to occur is the shear force from lower portions of the smokestack. Eventually, the shear force that provides this acceleration is larger than the smokestack can withstand, and it breaks.

Figure 7.5 (Applying Physics 1) A falling smokestack.

EXAMPLE 7.3 Computer Disks

A floppy disk in a computer rotates from rest up to an angular speed of 31.4 rad/s in a time of 0.892 s.
(a) What is the angular acceleration of the disk, assuming the angular acceleration is uniform?

Solution If we use $\omega = \omega_0 + \alpha t$ and the fact that $\omega_0 = 0$ at $t = 0$, we get

$$\alpha = \frac{\omega}{t} = \frac{31.4 \text{ rad/s}}{0.892 \text{ s}} = \boxed{35.2 \text{ rad/s}^2}$$

(b) How many rotations does the disk make while coming up to speed?

Solution Equation 7.6 enables us to find the angular displacement during the 0.892-s time interval. Taking $\omega_0 = 0$ gives

$$\theta = \omega_0 t + \tfrac{1}{2}\alpha t^2 = \tfrac{1}{2}(35.2 \text{ rad/s}^2)(0.892 \text{ s})^2 = 14.0 \text{ rad}$$

Because 2π rad $= 1$ rev, this angular displacement corresponds to $\boxed{2.23 \text{ rev.}}$

(c) If the radius of the disk is 4.45 cm, find the final linear speed of a microbe riding on the rim of the disk.

Solution The relation $v_t = r\omega$ and the given values lead to

$$v_t = r\omega = (0.0445 \text{ m})(31.4 \text{ rad/s}) = \boxed{1.40 \text{ m/s}}$$

(d) What is the magnitude of the tangential acceleration of the microbe at this time?

Solution We use $a_t = r\alpha$, which gives

$$a_t = r\alpha = (0.0445 \text{ m})(35.2 \text{ rad/s}^2) = \boxed{1.57 \text{ m/s}^2}$$

Exercise What is the angular speed and angular displacement of the disk 0.300 s after it begins to rotate?

Answer 10.6 rad/s; 1.58 rad

APPLICATION

Phonograph Records and
Compact Discs.

Before compact discs became the medium of choice for recorded music, phonograph records were popular. There are similarities between the rotational motion of phonograph records and compact discs, and there are some interesting differences. A phonograph record, for instance, rotates at a constant angular speed. Popular angular speeds were 33 ⅓ rev/min for long-playing albums (hence the nickname "LP"), 45 rev/min for "singles," and 78 rev/min used in very early recordings. At the outer edge of the record, the pickup needle (stylus) was moving over the vinyl material at a faster tangential speed than when the needle was close to the center of the record. As a result, the sound information was compressed into a smaller length of track near the center of the record than near the outer edge.

Compact discs, on the other hand, are designed such that the laser pickup moves over the information in the tracks at a constant tangential speed. Because the pickup moves radially as it follows the tracks of information, this requires that the angular speed of the compact disc must vary according to the radial position of the laser. Because the tangential speed is fixed, the information density (per length of track) anywhere on the disc is the same. Example 7.4 demonstrates numerical calculations for both compact discs and phonograph records.

EXAMPLE 7.4 Length of a Compact Disc Track

(a) A compact disc is designed such that as the read head moves out from the center of the disc, the angular speed of the disc changes so that the linear speed at the position of the head will always be at a constant value of about 1.3 m/s. Find the angular speed of the disc when the read head is at a distance of 5.0 cm from the center of the disc and at 8.0 cm.

Solution From $v_t = r\omega$, we have at $r = 5.0$ cm,

$$\omega = \frac{v_t}{r} = \frac{1.3 \text{ m/s}}{5.0 \times 10^{-2} \text{ m}} = \boxed{26.0 \text{ rad/s}}$$

and at 8.0 cm,

$$\omega = \frac{v_t}{r} = \frac{1.3 \text{ m/s}}{8.0 \times 10^{-2} \text{ m}} = \boxed{16.3 \text{ rad/s}}$$

(b) On the other hand, an old-fashioned record player rotated at a constant angular speed while the linear speed of the detector, a stylus, changed. Find the linear speed of a 45.0 rpm record at points 5.0 and 8.0 cm from the center.

Solution Because 45.0 rev/min = 4.7 rad/s, the linear speed at 5.0 cm is

$$v_t = r\omega = (5.0 \times 10^{-2} \text{ m})(4.7 \text{ rad/s}) = \boxed{0.24 \text{ m/s}}$$

At 8.0 cm, we have

$$v_t = r\omega = (8.0 \times 10^{-2} \text{ m})(4.7 \text{ rad/s}) = \boxed{0.38 \text{ m/s}}$$

(c) In both the CD and record, information is recorded in a continuous spiral track as one moves inward from the edge of the disc. Use the information above to calculate the total length of the track for a CD designed to play for 1 hour.

Solution The read head moves across the CD with a constant linear speed of 1.3 m/s. Thus, the total length of the track that it moves over in 3600 seconds is

$$x = v_t\, t = (1.3 \text{ m/s})(3600 \text{ s}) = \boxed{4680 \text{ m}}$$

7.4   CENTRIPETAL ACCELERATION

Figure 7.6a shows a car moving in a circular path with *constant linear speed v*. Students are often surprised to find that **even though the car moves at a constant speed, it still has an acceleration.** To see why this occurs, consider the defining equation for average acceleration:

$$\mathbf{a} = \frac{\mathbf{v}_f - \mathbf{v}_i}{t_f - t_i} \tag{7.10}$$

Note that the average acceleration depends on the *change in the velocity vector*. Because velocity is a vector, there are two ways in which an acceleration can be produced: by a change in the *magnitude* of the velocity and by a change in the *direction*

(a)

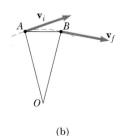

(b)

Figure 7.6   (a) Circular motion of a car moving with constant speed. (b) As the car moves along the circular path from A to B, the direction of its velocity vector changes, so the car undergoes a centripetal acceleration.

(a)

(b)

Figure 7.7   (a) As the particle moves from A to B, the direction of its velocity vector changes from $\mathbf{v}_i$ to $\mathbf{v}_f$. (b) The construction for determining the direction of the change in velocity, $\Delta\mathbf{v}$, which is toward the center of the circle.

of the velocity. It is the latter situation that occurs for the car moving in a circular path with constant speed (see Fig. 7.6b). We shall show that the acceleration vector in this case is perpendicular to the path and always points toward the center of the circle. An acceleration of this nature is called a **centripetal** (center-seeking) **acceleration.** Its magnitude is given by

$$a_c = \frac{v^2}{r} \qquad [7.11]$$

To derive Equation 7.11, consider Figure 7.7a. Here an object is seen first at point A with velocity $\mathbf{v}_i$ at time t_i and then at point B with velocity $\mathbf{v}_f$ at a later time, t_f. Let us assume that here $\mathbf{v}_i$ and $\mathbf{v}_f$ differ only in direction; their magnitudes are the same (that is, $v_i = v_f = v$). To calculate the acceleration, we begin with Equation 7.10, which indicates that we must vectorially subtract $\mathbf{v}_i$ from $\mathbf{v}_f$:

$$\mathbf{a} = \frac{\mathbf{v}_f - \mathbf{v}_i}{t_f - t_i} = \frac{\Delta\mathbf{v}}{\Delta t} \qquad [7.12]$$

where $\Delta\mathbf{v} = \mathbf{v}_f - \mathbf{v}_i$ is the change in velocity. This can be accomplished graphically, as shown by the vector triangle in Figure 7.7b. Note that when Δt is very small, Δs and $\Delta\theta$ will also be very small. In this case, $\mathbf{v}_f$ will almost parallel $\mathbf{v}_i$, and the vector $\Delta\mathbf{v}$ will be approximately perpendicular to them, pointing toward the center of the circle. In the limiting case where Δt becomes vanishingly small, $\Delta\mathbf{v}$ will point exactly toward the center of the circle. Furthermore, in this limiting case the acceleration will also be directed toward the center of the circle because it is in the direction $\Delta\mathbf{v}$.

Now consider the triangle in Figure 7.7a, which has sides Δs and r. This triangle and the one formed by the vectors in Figure 7.7b are similar. This enables us to write a relationship between the lengths of the sides:

$$\frac{\Delta v}{v} = \frac{\Delta s}{r}$$

or

$$\Delta v = \frac{v}{r} \Delta s$$

This can be substituted into Equation 7.12 for Δv to give

$$a = \frac{v}{r} \frac{\Delta s}{\Delta t} \qquad [7.13]$$

In this situation, Δs is a small distance measured along the arc of the circle (a tangential distance), so $v = \Delta s/\Delta t$, where v is the tangential speed. Therefore, Equation 7.13 reduces to Equation 7.11:

$$a_c = \frac{v^2}{r}$$

Because the tangential speed is related to the angular speed through the relation $v_t = r\omega$ (Eq. 7.8), an alternative form of Equation 7.11 is

$$a_c = \frac{r^2\omega^2}{r} = r\omega^2 \qquad [7.14]$$

Thus, we conclude that

> In circular motion the centripetal acceleration is directed inward toward the center of the circle and has a magnitude given by either v^2/r or $r\omega^2$.

◀ Centripetal acceleration

You should show that the dimensions of a_c are L/T^2, as required.

In order to clear up any misconceptions that might exist concerning centripetal and tangential acceleration, let us consider a car moving around a circular race track. Because the car is moving in a circular path, it always has a centripetal component of acceleration because its direction of travel, and hence the direction of its velocity, is continuously changing. If the car's speed is increasing or decreasing, it also has a tangential component of acceleration. To summarize, the tangential component of acceleration is due to changing speed; the centripetal component of acceleration is due to changing direction.

When both components of acceleration exist simultaneously, the tangential acceleration is tangent to the circular path and the centripetal acceleration points toward the center of the circular path. Because these components of acceleration are perpendicular to each other, we can find the magnitude of the **total acceleration** using the Pythagorean theorem:

$$a = \sqrt{a_t^2 + a_c^2}$$ [7.15]   ◀ Total acceleration

EXAMPLE 7.5 Let's Go for a Spin

A test car moves at a constant speed of 10 m/s around a circular road of radius 50 m. Find the car's (a) centripetal acceleration and (b) angular speed.

Solution (a) From Equation 7.11, the magnitude of the centripetal acceleration of the car is found to be

$$a_c = \frac{v^2}{r} = \frac{(10 \text{ m/s})^2}{50 \text{ m}} = \boxed{2.0 \text{ m/s}^2}$$

By definition, the direction of $\mathbf{a}_c$ is toward the center of curvature of the road.
(b) The angular speed of the car can be found using the expression $v_t = r\omega$, which gives

$$\omega = \frac{v_t}{r} = \frac{10 \text{ m/s}}{50 \text{ m}} = \boxed{0.20 \text{ rad/s}}$$

Note that we can also find the centripetal acceleration by using this value of ω, in an alternative method to part (a). From Equation 7.14 we have

$$a_c = r\omega^2 = (50 \text{ m})(0.20 \text{ rad/s})^2 = 2.0 \text{ m/s}^2$$

As expected, the two methods for finding the centripetal acceleration give the same answer.

Exercise Find the car's tangential acceleration and total acceleration.

Answer The tangential acceleration is zero because the speed of the car remains constant. Because $a_t = 0$, the total acceleration in this example equals the centripetal acceleration, 2.0 m/s², found in part (a).

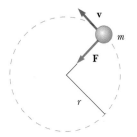

Figure 7.8   A ball attached to a string of length, r, rotating in a circular path at constant speed.

7.5   FORCES CAUSING CENTRIPETAL ACCELERATION

Consider a ball of mass m tied to a string of length r and being whirled in a horizontal circular path, as in Figure 7.8. Let us assume that the ball moves with constant speed. Because the velocity vector, **v,** changes its direction continuously during the motion, the ball experiences a centripetal acceleration directed toward the center of motion, as described in Section 7.4, with magnitude

$$a_c = \frac{v_t^2}{r}$$

The inertia of the ball tends to maintain motion in a straight-line path; however, the string overcomes this by exerting a force on the ball that makes it instead follow a circular path. This force (equal to the force of tension) is directed along the length of the string toward the center of the circle in Figure 7.8. Thus, the equation for Newton's second law along the radial direction is

$$F = ma_c = m\frac{v_t^2}{r} \qquad [7.16]$$

Because it acts at right angles to the motion, a force that produces a centripetal acceleration causes a change in the direction of the velocity. Beyond this, such a force is no different from any of the other forces we have studied. For example, friction between tires and road provides the force that enables a race car to travel in a circular path on a flat road, and the gravitational force exerted on the Moon by the Earth provides the force necessary to keep the Moon in its orbit. (We shall discuss the gravitational force in more detail in Section 7.7.)

Regardless of the example used, if the force that produces a centripetal acceleration vanishes, the object does not continue to move in its circular path; instead, it moves along a straight-line path tangent to the circle. But as the race car in the previous paragraph travels faster and faster on the curved flat road, the centripetal acceleration will increase, as will the friction force needed to cause this acceleration. At some speed, the friction force needed to keep the car on a circular path will be greater than the maximum possible static friction force allowed by the static coefficient of friction. At this point, the car will skid off the road.

In order to prevent this from happening, roadways are often banked instead of being built with flat curves. By banking the roadway toward the inside of the circular path, there is a component of the normal force of the road on the car that is directed toward the center of the path. This component of the normal force adds to the centrally directed friction force. Thus, the car can negotiate a banked curve at a higher speed than it can an unbanked curve. The banking of roadways is very evident at high speed race tracks, in bicycle tracks in velodromes, and in bobsled tracks.

APPLICATION

Banked Roadways.

Thinking Physics 2

Suppose you are the only rider on a Ferris wheel. The drive mechanism breaks after the ride starts, so that you continue to rotate freely, without friction. As you pass through the lowest point (let us call it the 6 o'clock point), you will feel the heaviest— that is, the seat will exert the largest upward force on you. As you pass over the highest point (the 12 o'clock point), you will feel the lightest. At the 3 o'clock and 9 o'clock points, will you feel as if you have the "correct" weight? If not, where *do* you feel as if your weight is "correct"?

Explanation If the wheel were rotating at a constant speed, the answer to the question would be *yes*. However, because you are the only rider, the wheel is *unbalanced* and does not rotate with a constant speed. As you descend from the top point, your speed increases. After you pass through the lowest point, your speed decreases as you move back to the top. Thus, you will experience tangential acceleration as well as centripetal acceleration. At the 3 o'clock position, as you are moving downward, you have a downward tangential acceleration. Thus, just as in a downward accelerating elevator, you will feel lighter. At the 9 o'clock position, as you are moving upward, you are accelerating downward tangentially. Again, you will feel lighter at this point. In order to counteract the *downward* component of tangential acceleration near both positions, you need some *upward* component from the centripetal acceleration. This will be present if you are at positions *below* 3 o'clock or 9 o'clock, so that the centripetal acceleration has an upward component. The exact positions will depend on your velocity and the details of the Ferris wheel.

As these speed skaters round a curve, the force of friction exerted on their skates provides the centripetal acceleration. (*Bill Bachman/Photo Researchers, Inc.*)

Problem-Solving Strategy

Forces that Cause Centripetal Acceleration

Use the following steps when dealing with centripetal accelerations and forces that produce them:

1. Draw a free-body diagram of the object(s) under consideration, showing and labeling all forces that act on it.
2. Choose a coordinate system that has one axis perpendicular to the circular path followed by the object and one axis tangent to the circular path.
3. Find the net force toward the center of the circular path. This is the force that causes the centripetal acceleration.
4. From here on the steps are virtually identical to those encountered when solving Newton's second law problems with $\Sigma \mathbf{F} = m\mathbf{a}$. Also, note that the magnitude of the centripetal acceleration can always be written $a_c = v_t^2/r$.

(a)

 EXAMPLE 7.6 Buckle Up for Safety

A car travels at a constant speed of 30.0 mi/h (13.4 m/s) on a level circular turn of radius 50.0 m, as shown in the bird's-eye view in Figure 7.9a. What is the minimum coefficient of static friction between the tires and roadway in order that the car makes the circular turn without sliding?

Reasoning The force in the radial direction acting on the car is the force of static friction directed toward the center of the circular path, as in Figure 7.9a. Thus, the objective in this example will be to find an expression for the maximum frictional force $f = \mu_s n$ and use this in Equation 7.16.

(b)

Figure 7.9 (Example 7.6) (a) Top view of a car on a curved path. (b) A free-body diagram for the car, showing an end view.

Solution   Thus, Equation 7.16 becomes

$$(1) \qquad f = m\frac{v_t^2}{r}$$

All the forces acting on the car are shown in Figure 7.9b. Because equilibrium is in the vertical direction, the normal force upward is balanced by the force of gravity downward, so that

$$n = mg$$

From this expression we can find the maximum force of static friction, as follows:

$$(2) \qquad f = \mu_s n = \mu_s mg$$

By setting the right-hand sides of (1) and (2) equal to each other, we find that

$$\mu_s \cancel{m} g = \cancel{m}\frac{v_t^2}{r}$$

or

$$(3) \qquad \mu_s = \frac{v_t^2}{rg}$$

Therefore, the minimum coefficient of static friction that is required for the car to make the turn without sliding outward is

$$\mu_s = \frac{v_t^2}{rg} = \frac{(13.4 \text{ m/s})^2}{(50.0 \text{ m})(9.80 \text{ m/s}^2)} = \boxed{0.366}$$

The value of μ_s for rubber on dry concrete is very close to 1; thus, the car can negotiate the curve with ease. If the road were wet or icy, however, the value for μ_s could be 0.2 or lower. Under such conditions, the radial force provided by static friction would not be great enough to enable the car to follow the circular path, and it would slide off the roadway.

EXAMPLE 7.7   Having Fun with a Yo-Yo

A child swings a yo-yo of weight mg in a horizontal circle so that the cord makes an angle of 30.0° with the vertical, as in Figure 7.10a. Find the centripetal acceleration of the yo-yo.

Reasoning   Two forces act on the yo-yo: the force of gravity and the force of tension in the cord, **T**, as in Figure 7.10b. The first condition for equilibrium applied in the y direction, $\Sigma F_y = 0$, will enable us to find T. The component of **T** toward the center of the circular path will be the force that produces the centripetal acceleration, and we can finally find a_c from $F = ma_c$.

Solution   Applying ΣF_y gives

$$T\cos 30.0° - mg = 0$$

$$T = \frac{mg}{\cos 30.0°}$$

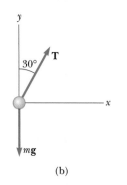

(a)

(b)

Figure 7.10   (Example 7.7) (a) A yo-yo swinging in a horizontal circle so that the cord makes a constant angle of 30° with the vertical. (b) A free-body diagram for the yo-yo.

The *horizontal component* of **T** causes the centripetal acceleration.

$$F = T \sin 30.0° = \frac{mg \sin 30.0°}{\cos 30.0°} = mg \tan 30.0°$$

The centripetal acceleration can now be found by applying Newton's second law along the horizontal direction:

$$F = ma_c$$

$$a_c = \frac{F}{m} = \frac{mg \tan 30.0°}{m} = g \tan 30.0° = (9.80 \text{ m/s}^2)(\tan 30.0°)$$

$$= \boxed{5.66 \text{ m/s}^2}$$

7.6 DESCRIBING MOTION OF A ROTATING SYSTEM

We have seen that an object moving in a circle of radius r with constant speed v_t has a centripetal acceleration whose magnitude is v_t^2/r and whose direction is toward the center of rotation. The force necessary to maintain this centripetal acceleration must also act toward the center of rotation. In the case of a ball rotating at the end of a string, the force exerted on the ball by the string (equal in magnitude to the tension in the string) is the force causing the centripetal acceleration; for a satellite in a circular orbit around the Earth, the force of gravity plays that role, and for a car rounding a curve on a level road, it is the force of friction between the tires and the pavement; and so forth.

In order to better understand the motion of a rotating system, consider a car approaching a curved exit ramp at high speed as in Figure 7.11. As the car takes the sharp left turn onto the ramp, a person sitting in the passenger seat slides to the right across the seat and hits the door. At that point, the force of the door keeps her from being ejected from the car. What causes the passenger to move toward the door? A popular but *erroneous* explanation is that some mysterious force pushes her outward. (This is sometimes erroneously called the "centrifugal" force.)

The phenomenon is correctly explained as follows. Before the car enters the ramp, the passenger is moving in a straight-line path. As the car enters the ramp and travels a curved path, the passenger, because of inertia, tends to move along the original straight-line path. This is in accordance with Newton's first law: The natural tendency of an object in motion is to continue moving in a straight line. However, if a sufficiently large force toward the center of curvature acts on the passenger, she moves in a curved path along with the car. The origin of this force is the force of friction between the passenger and the car seat. If this frictional force is not great enough, the passenger slides across the seat as the car turns under her. Because of the passenger's inertia, she continues to move in a straight-line path. Eventually the passenger encounters the door, which provides a large enough force to enable her to follow the same curved path as the car. The passenger slides toward the door not because of some mysterious outward force but because there is no force great enough to enable her to travel along the circular path followed by the car.

Figure 7.11 A car approaching a curved exit ramp.

As a second example, consider what happens when you run clothes through the rinse cycle of a washing machine. In the last phase of this cycle, the drum spins rapidly to remove water from the clothes. Why is the water thrown off? An *erroneous* explanation is that the rotating system creates some mysterious outward force on each drop of water, and this force causes the water to be hurled to the outer drum of the machine. The correct explanation is as follows. When the clothes are at rest in the machine, water is held to them by molecular forces between the water and the fabric. During the spin cycle, the clothes rotate and the molecular forces are not great enough to provide the necessary radial force to keep the water molecules moving in a circular path along with the clothes. Hence, the drops of water, because of their inertia, move in straight-line paths until they encounter the sides of the spinning drum.

In summary, in describing motion in an accelerating frame, one must be very careful to distinguish real forces from fictitious ones. An observer in a car rounding a curve is in an accelerating frame and invents a fictitious outward force to explain why he or she is thrown outward. A stationary observer outside the car, however, considers only real forces on the passenger. To the observer, the mysterious outward force *does not exist!* The only real external force on the passenger is the radial force due to friction or the contact force of the door.

EXAMPLE 7.8 Riding the Rails

Figure 7.12a shows a roller coaster moving around an almost circular loop of radius R. What speed must the car have at the bottom of the track so that it will just make it over the top of the loop?

(a)   (b)

Figure 7.12 (Example 7.8) (a) A roller coaster traveling around a nearly circular track. (b) Passengers on a roller coaster at Knott's Berry Farm experience the thrill of a loop-the-loop. *(Superstock)*

Solution   If the car is to just make it over the top and be able to complete the loop, the force of the track on the car must become zero at the top of the loop. Therefore, the only force exerted on the car at the top of the loop is the force of gravity, mg, and it is this force that provides the centripetal acceleration of the car. Thus, we have

$$mg = m\frac{v_{\text{top}}^2}{R}$$

where v_{top} is the minimum speed the car must have at the top to stay in contact with the tracks. Solving this equation for v_{top} gives

$$v_{\text{top}} = \sqrt{gR}$$

At the top of the loop, the car has both kinetic energy and potential energy. If we select the zero level for potential energy to be at the bottom of the loop, the mechanical energy of the car at the top is

$$E_{\text{top}} = \tfrac{1}{2}mv_{\text{top}}^2 + mg2R = \tfrac{1}{2}mgR + 2mgR = 2.5mgR$$

The mechanical energy of the car at the bottom of the loop is solely kinetic energy:

$$E_{\text{bot}} = \tfrac{1}{2}mv_{\text{bot}}^2$$

Because mechanical energy is conserved, the energy at the bottom must equal the energy at the top:

$$\tfrac{1}{2}mv_{\text{bot}}^2 = 2.5mgR$$

Solving for v_{bot} gives

$$v_{\text{bot}} = \sqrt{5gR}$$

For example, if the radius of the loop is 10.0 m, the velocity at the bottom must be at least

$$v_{\text{bot}} = \sqrt{5(9.8 \text{ m/s}^2)(10.0 \text{ m})} = 22.1 \text{ m/s}$$

Such a high speed can make the trip very uncomfortable for the passengers. The required speed can be reduced by changing the shape of the loop from a circular shape to one that closes down slightly at the top as shown in Figure 7.12b.

7.7   NEWTON'S UNIVERSAL LAW OF GRAVITATION

Prior to 1686 a great mass of data had been collected on the motions of the Moon and the planets, but a clear understanding of the forces that cause these celestial bodies to move the way they do was not available. In that year, Isaac Newton provided the key that unlocked the secrets of the heavens. He knew, from the first law, that a net force had to be acting on the Moon. If it were not, the Moon would move in a straight-line path rather than in its almost circular orbit around the Earth. Newton reasoned that this force arises as a result of an attractive field force between Moon and Earth, which we call the force of gravity. He also concluded that there could be nothing special about the Earth-Moon system or the Sun and its planets that would cause gravitational forces to act on them alone. In other words, he saw that the same force of attraction that causes the Moon to follow its path also causes an apple to fall to Earth from a tree. He wrote, "I deduced that the forces which keep the planets in their orbs must be reciprocally as the squares of their distances from the centers about which they revolve; and thereby compared the force requisite to keep the Moon in her orb with force of gravity at the surface of the Earth; and found them answer pretty nearly."

In 1687 Newton published his work on the universal law of gravitation, which states that

> **Every particle in the Universe attracts every other particle with a force that is directly proportional to the product of their masses and inversely proportional to the square of the distance between them.**

If the particles have masses m_1 and m_2 and are separated by the distance r, the magnitude of the gravitational force between them is

Universal law of gravitation ▶

$$F = G \frac{m_1 m_2}{r^2} \qquad [7.17]$$

where G is a universal constant called the **constant of universal gravitation,** which has been measured experimentally. Its value in SI units is

$$G = 6.673 \times 10^{-11} \frac{\text{N} \cdot \text{m}^2}{\text{kg}^2} \qquad [7.18]$$

This force law is an example of an **inverse-square law** in that it varies as the inverse square of the separation. The force acts so that the objects are always attracted to one another. From Newton's third law, we also know that the force on m_2 due to m_1, designated $\mathbf{F}_{21}$ in Figure 7.13, is equal in magnitude to the force on m_1 due to m_2, $\mathbf{F}_{12}$, but in the opposite direction. That is, these forces form an action-reaction pair.

Several features of the universal law of gravitation deserve some attention:

> 1. The gravitational force is a field force that always exists between two particles regardless of the medium that separates them.
> 2. The force varies as the inverse square of the distance between the particles and therefore decreases rapidly with increasing separation.
> 3. The force is proportional to the product of the particles' masses.

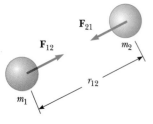

Figure 7.13 The gravitational force between two particles is attractive. Note that, according to Newton's third law, $\mathbf{F}_{12} = -\mathbf{F}_{21}$.

Another important fact is that **the gravitational force exerted by a spherical mass on a particle outside the sphere is the same as if the entire mass of the sphere were concentrated at its center.** For example, the force on a particle of mass m at the Earth's surface has the magnitude

$$F = G \frac{M_E m}{R_E^2}$$

where M_E is the Earth's mass and R_E is its radius. This force is directed toward the center of the Earth.

Thinking Physics 3

A student drops a ball and considers why the ball falls to the ground as opposed to the situation in which the ball stays stationary and the Earth moves up to meet it. The student comes up with the following explanation: "The Earth is much more massive than the ball, so the Earth pulls much harder on the ball than the ball pulls on the Earth. Thus, the ball falls while the Earth remains stationary." What do you think about this explanation?

Explanation   According to Newton's universal law of gravity, the force between the ball and the Earth depends on the product of their masses, so both forces, that of the ball on the Earth, and that of the Earth on the ball, are equal in magnitude. This follows also, of course, from Newton's third law. The ball has large motion compared to the Earth because according to Newton's second law, the force gives a much greater acceleration to the small mass of the ball.

Measurement of the Gravitational Constant

The gravitational constant was first measured in an important experiment by Henry Cavendish in 1798. His apparatus consisted of two small spheres, each of mass m, fixed to the ends of a light horizontal rod suspended by a thin metal wire, as in Figure 7.14. Two large spheres, each of mass M, were placed near the smaller spheres. The attractive force between the smaller and larger spheres caused the rod to rotate in a horizontal plane and the wire to twist. The angle through which the suspended rod rotated was measured with a light beam reflected from a mirror attached to the vertical suspension. (Such a moving spot of light is an effective technique for amplifying the motion.) The experiment was carefully repeated with

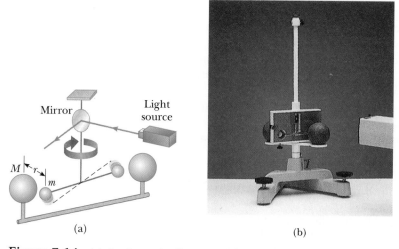

(a)   (b)

Figure 7.14   (a) A schematic diagram of the Cavendish apparatus for measuring G. The smaller spheres of mass m are attracted to the large spheres of mass M, and the rod rotates through a small angle. A light beam reflected from a mirror on the rotating apparatus measures the angle of rotation. (b) A student Cavendish apparatus. *(Courtesy of PASCO Scientific)*

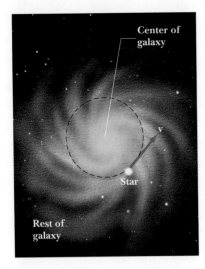

Figure 7.15   The star revolving at a speed v about the center of the galaxy is affected only by the mass inside its orbit (shaded region).

different masses at various separations. In addition to providing a value for G, the results showed that the force is attractive, proportional to the product mM, and inversely proportional to the square of the distance r.

Dark Matter

It is of interest to consider some facts from astronomy that can be understood based on the results of this section. In recent years astronomers have been interested in so-called missing mass in the Universe. To see why missing mass has been proposed, consider Fig. 7.15 which shows a star revolving at a speed v in a circular path of radius r about the center of a galaxy. We know that the force holding the star in its orbit is $G\dfrac{mM}{r^2}$ where m is the mass of the star and M is the total mass inside the circular orbit, light blue in the figure. From Newton's second law, we have

$$G\frac{mM}{r^2} = m\frac{v^2}{r}$$

or

$$v = \sqrt{\frac{GM}{r}}$$

From this equation we see that the greater the mass inside the orbit of the star, the faster it must move in order to remain at a given distance r from the center. However, when astronomers look at distant galaxies, they see stars moving at a speed v *larger* than expected compared to the amount of mass inside the orbit! The only explanation for this is that there is some form of mass present inside the orbit that cannot be seen by visible light. This material, known as **dark matter,** has been the source of intense experimental investigation in recent years. The interest in dark matter has come about in an effort to explain the future of the Universe — namely, will it continue to expand forever or will its present expansion slow, come to a halt, and begin to collapse? The determining factor will be the amount of mass present in the Universe. If the amount of mass is too low, the expansion of the Universe will continue forever, and the ultimate fate of the universe will be one in which all the stars die, cool, and move outward ad infinitum. On the other hand, if the amount of mass is large enough, the Universe will collapse on itself, perhaps to explode once again in another Big Bang to create yet another Universe. At present, even including dark matter, only about 10 percent of the amount of mass needed to produce a collapse has been found, but the search continues.

EXAMPLE 7.9   Billiards, Anyone?

Three 0.300-kg billiard balls are placed on a table at the corners of a right triangle, as shown from overhead in Figure 7.16. Find the net gravitational force on the ball designated as m_1 due to the forces exerted on it by the other two balls.

Solution   To find the net gravitational force on m_1, we first calculate the force exerted on m_1 due to m_2. Then we find the force on m_1 due to m_3. Finally, we add these two forces *vectorially* to obtain the net force on m_1.

The force exerted on m_1 due to m_2, denoted by $\mathbf{F}_1$ in Figure 7.16, is upward. Its magnitude is calculated using Equation 7.17:

$$F_1 = G\frac{m_1 m_2}{r^2} = (6.67 \times 10^{-11}\ \text{N}\cdot\text{m}^2/\text{kg}^2)\frac{(0.300\ \text{kg})(0.300\ \text{kg})}{(0.400\ \text{m})^2}$$

$$= 3.75 \times 10^{-11}\ \text{N}$$

This result shows that gravitational forces between the common objects that surround us have extremely small magnitudes.

Now let us calculate the gravitational force exerted on m_1 due to m_3, denoted by $\mathbf{F}_2$ in Figure 7.16. It is directed to the right, and its magnitude is

$$F_2 = G\frac{m_1 m_3}{r^2} = (6.67 \times 10^{-11}\ \text{N}\cdot\text{m}^2/\text{kg}^2)\frac{(0.300\ \text{kg})(0.300\ \text{kg})}{(0.300\ \text{m})^2}$$

$$= 6.67 \times 10^{-11}\ \text{N}$$

The net gravitational force exerted on m_1 is found by adding $\mathbf{F}_1$ and $\mathbf{F}_2$ as *vectors*. The magnitude of this net force is given by

$$F = \sqrt{F_1^2 + F_2^2} = \sqrt{(3.75)^2 + (6.67)^2} \times 10^{-11}\ \text{N} = \boxed{7.65 \times 10^{-11}\ \text{N}}$$

Exercise   Find the direction of the resultant force on m_1.

Answer   The vector $\mathbf{F}$ makes an angle of 29.3° with the line joining m_1 and m_3.

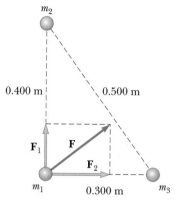

Figure 7.16   (Example 7.9)

EXAMPLE 7.10   The Mass of the Earth

Use the gravitational force law to find an approximate value for the mass of the Earth.

Solution   Consider a baseball of mass m_b falling toward the Earth at a location where the free-fall acceleration is g. We know that the magnitude of the gravitational force exerted on the baseball by the Earth is the same as the weight of the ball and that this is given by $w = m_b g$. Because the magnitude of the gravitational force is given by Equation 7.17, we find that

$$m_b g = G\frac{M_E m_b}{R_E^2}$$

We can divide each side of this equation by m_b and solve for the mass of the Earth, M_E:

$$M_E = \frac{gR_E^2}{G}$$

The falling baseball is close enough to the Earth so that the distance of separation between the center of the ball and the center of the Earth can be taken as the radius of the Earth, 6.38×10^6 m. Thus, the mass of the Earth is

$$M_E = \frac{(9.80\ \text{m/s}^2)(6.38 \times 10^6\ \text{m})^2}{6.67 \times 10^{-11}\ \text{N}\cdot\text{m}^2/\text{kg}^2} = \boxed{5.98 \times 10^{24}\ \text{kg}}$$

TABLE 7.1
Free-Fall Acceleration, g, at
Various Altitudes

Altitude (km) [a]	g (m/s^2)
1000	7.33
2000	5.68
3000	4.53
4000	3.70
5000	3.08
6000	2.60
7000	2.23
8000	1.93
9000	1.69
10 000	1.49
50 000	0.13

[a] All values are distances above the Earth's surface.

EXAMPLE 7.11   Gravity and Altitude

Derive an expression that shows how the acceleration due to gravity varies with distance from the center of the Earth at an exterior point.

Solution   The falling baseball of Example 7.10 can be used here also. Now, however, assume that the ball is located at some arbitrary distance r from the Earth's center. The first equation in Example 7.10, with r replacing R_E and m_b removed from both sides, becomes

$$g = G\frac{M_E}{r^2}$$

This indicates that the free-fall acceleration at an exterior point decreases as the inverse square of the distance from the center of the Earth. Our assumption of Chapter 2 that objects fall with constant acceleration is obviously incorrect in light of the present example. For short falls, however, this change in g is so small that neglecting the variation does not introduce a significant error in the result.

Because the weight of an object is mg, we see that a change in the value of g produces a change in the weight of the object. For example, if you weigh 800 N at the surface of the Earth, you will weigh only 200 N at a height above the Earth equal to the radius of the Earth. Also, we see that if the distance of an object from the Earth becomes infinitely large, the weight approaches zero. Values of g at various altitudes are listed in Table 7.1.

Exercise   If an object weighs 270 N at the Earth's surface, what will it weigh at an altitude equal to twice the radius of the Earth?

Answer   30.0 N

Optional Section

7.8   GRAVITATIONAL POTENTIAL ENERGY REVISITED

In Chapter 5 we introduced the concept of gravitational potential energy and found that the potential energy of an object could be calculated using the equation $PE = mgh$, where h is the height of the object above or below some reference level. This equation is actually valid only when the object is near the Earth's surface. For objects high above the Earth's surface, such as a satellite, an alternative expression must be used to compute the gravitational potential energy. The general expression for the gravitational potential energy for an object of mass m at a distance r from the center of the Earth can be shown (using integral calculus) to be

General form of gravitational ▶
potential energy

$$PE = -G\frac{M_E m}{r} \tag{7.19}$$

where M_E is the mass of the Earth.

This equation assumes that the zero level for potential energy is at an infinite distance from the center of the Earth. This point is necessary because the gravitational force goes to zero when r is set equal to infinity.

Thinking Physics 4

Why is the Sun hot?

Explanation   The Sun was formed when a cloud of gas and dust coalesced, due to gravitational attraction, into a massive astronomical object. Before this occurred, the particles were widely scattered, representing a large amount of gravitational potential energy. As the particles came together to form the Sun, the gravitational potential energy decreased. According to the principle of conservation of energy, this potential energy must be transformed to another form. The form to which it transformed was internal energy, representing an increase in temperature. If enough particles come together, the temperature can rise to a point at which nuclear fusion occurs, and the object becomes a star. If there are not enough particles, the temperature rises, but not to a point at which fusion occurs—the object is a planet, if it is in orbit around a star. Jupiter is an example of a large planet that might have been a star if more particles were available to be collected.

EXAMPLE 7.12   Does the Potential Energy Reduce to *mgh*?

(a) Find expressions for the gravitational potential energy of an object at the surface of the Earth and for the same object when at a height h above the surface of the Earth (Fig. 7.17). (b) From the answers to part (a), show that the difference in potential energy between these two points reduces to the familiar expression $PE = mgh$ when h is small compared to the Earth's radius.

Solution   Equation 7.19 gives the potential energy at the surface of the Earth as

$$PE_1 = -G\frac{M_E m}{R_E}$$

We can also use Equation 7.19 to find the potential energy of the object at the height h above the surface of the Earth. Taking $r = R_E + h$, we find

$$PE_2 = -G\frac{M_E m}{(R_E + h)}$$

(b) The difference in potential energy between these two points is found as follows:

$$PE_2 - PE_1 = -G\frac{M_E m}{(R_E + h)} - \left(-G\frac{M_E m}{R_E}\right)$$

$$= -GM_E m\left[\frac{1}{(R_E + h)} - \frac{1}{R_E}\right]$$

After finding a common denominator and applying some algebra to the equation above, we find

$$PE_2 - PE_1 = \frac{GM_E mh}{R_E(R_E + h)}$$

When the height h is very small compared to R_E, the denominator in the expression above is approximately equal to $R_E{}^2$. Thus, we have

Figure 7.17   (Example 7.12)

Figure 7.18 An object of mass m projected upward from the Earth's surface with an initial speed v_i reaches a maximum altitude h (where $M_E \gg m$).

$$PE_2 - PE_1 \cong \frac{GM_E}{R_E^2} \, mh$$

Now note that the free-fall acceleration at the surface of the Earth is given by $g = GM_E/R_E^2$ (see Example 7.10). Thus,

$$PE_2 - PE_1 \cong mgh$$

Escape Speed

If an object is projected upward with a large enough speed it can escape the gravitational pull of the Earth and go soaring off into space. This particular speed is called the escape speed of an object from the Earth. We shall now proceed to derive an expression that will enable us to calculate the escape speed.

Suppose an object of mass m is projected vertically upward from the Earth's surface with an initial speed v_i, as in Figure 7.18. The initial mechanical energy (kinetic plus potential energy) is given by

$$KE_i + PE_i = \frac{1}{2} \, mv_i^2 - G \frac{M_E m}{R_E}$$

We neglect air resistance and assume that the initial speed is just large enough to allow the object to reach infinity with a speed of zero. We call this value of v_i the escape speed v_e. When the object is at an infinite distance from the Earth, its kinetic energy is zero, because $v_f = 0$, and the gravitational potential energy is also zero because our zero level of potential energy was selected at $r = \infty$. Hence, the total mechanical energy is zero, and the law of conservation of energy gives

$$\frac{1}{2} \, mv_e^2 - G \frac{M_E m}{R_E} = 0$$

and

Escape speed ▶

$$v_e = \sqrt{\frac{2GM_E}{R_E}} \qquad\qquad [7.20]$$

From this equation, one finds that the escape speed for Earth is about 11.2 km/s, which corresponds to about 25 000 mi/h (see Example 7.13). Note that this expression for v_e is independent of the mass of the object projected from the Earth. For example, a spacecraft has the same escape speed as a molecule. A list of escape speeds for the planets and the Moon is given in Table 7.2. The data presented in this table help in understanding why some planets have atmospheres and others do not. For example, on the very hot planet Mercury, gas molecules often reach speeds that are greater than the escape speed of 4.3 km/s. Consequently, any gases that might have been present on the surface of the planet at its formation have long since wandered off into space. Likewise, in the atmosphere of our own Earth, hydrogen and helium molecules through collisions often gain speeds greater than 11.2 km/s, the escape speed on Earth. Therefore, these gases are not retained in the Earth's atmosphere. On the other hand, heavier gases such as oxygen and nitrogen in the Earth's atmosphere have average speeds of less than 11.2 km/s and do not escape.

TABLE 7.2

Escape Speeds for the Planets and the Moon

Planet	v_e(km/s)
Mercury	4.3
Venus	10.3
Earth	11.2
Moon	2.3
Mars	5.0
Jupiter	60.0
Saturn	36.0
Uranus	22.0
Neptune	24.0
Pluto	1.1

EXAMPLE 7.13 Escape Speed of a Rocket

Calculate the escape speed from the Earth for a 5000-kg spacecraft.

Solution Using Equation 7.20 with $M_E = 5.98 \times 10^{24}$ kg and $R_E = 6.38 \times 10^6$ m gives

$$v_e = \sqrt{\frac{2GM_E}{R_E}} = \sqrt{\frac{2(6.68 \times 10^{-11})(5.98 \times 10^{24})}{6.37 \times 10^6}} = \boxed{1.12 \times 10^4 \text{ m/s}}$$

This corresponds to about 25 000 mi/h or about 7 mi/s. Note that the mass of the spacecraft was not required for this calculation.

7.9 KEPLER'S LAWS

The movements of the planets, stars, and other celestial bodies have been observed for thousands of years. In early history, scientists regarded the Earth as the center of the Universe. This *geocentric model* was developed extensively by the Greek astronomer Claudius Ptolemy in the second century A.D. and was accepted for the next 1400 years. In 1543 the Polish astronomer Nicolaus Copernicus (1473–1543) showed that the Earth and the other planets revolve in circular orbits about the Sun (the *heliocentric hypothesis*).

The Danish astronomer Tycho Brahe (pronounced Brah or BRAH-huh; 1546–1601) made accurate astronomical measurements over a period of 20 years and provided the data for the currently accepted model of the Solar System. Brahe's precise observations of the planets and 777 stars were carried out with nothing more elaborate than a large sextant and compass; the telescope had not yet been invented.

The German astronomer Johannes Kepler, who was Brahe's assistant, acquired Brahe's astronomical data and spent about 16 years trying to deduce a mathematical model for the motions of the planets. After many laborious calculations, he found that Brahe's precise data on the motion of Mars about the Sun provided the answer. Kepler's analysis first showed that the concept of circular orbits about the Sun had to be abandoned. He eventually discovered that the orbit of Mars could be accurately described by an ellipse with the Sun at one focus. He then generalized this analysis to include the motions of all planets. The complete analysis is summarized in three statements known as **Kepler's laws:**

1. **All planets move in elliptical orbits with the Sun at one of the focal points.**
2. **A line drawn from the Sun to any planet sweeps out equal areas in equal time intervals.**
3. **The square of the orbital period of any planet is proportional to the cube of the average distance from the planet to the Sun.**

Newton later demonstrated that these laws are consequences of a simple force that exists between any two masses. Newton's universal law of gravity, together with

Johannes Kepler, German astronomer (1571–1630)

Kepler is best known for developing the laws of planetary motion based on the careful observations of Tycho Brahe. Throughout his life, Kepler was sidetracked by mystic ideas dating back to the ancient Greeks. For example, he believed in the notion of the "music of the spheres" proposed by Pythagoras, in which each planet in its motion sounds out an exact musical note. After spending several years trying to work out a "regular-solid theory" of the planets, he concluded that the Copernican view of circular planetary orbits had to be abandoned for the view that the planetary orbits are ellipses with the Sun always at one of the foci. (*Art Resource*)

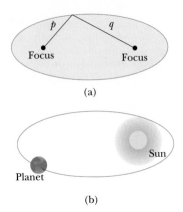

(a)

(b)

Figure 7.19 (a) The sum $p + q$ is the same for every point on the ellipse. (b) In the Solar System, the Sun is at one focus of the elliptical orbit of each planet and the other focus is empty.

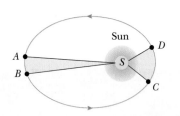

Figure 7.20 The two areas swept out by the planet in its elliptical orbit about the Sun are equal if the time interval between points A and B is equal to the time interval between points C and D.

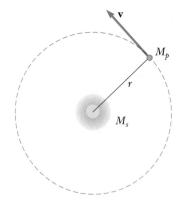

Figure 7.21 A planet of mass M_p moving in a circular orbit about the Sun. The orbits of all planets except Mercury and Pluto are nearly circular.

his laws of motion, provides the basis for a full mathematical solution to the motions of planets and satellites. More important, Newton's universal law of gravity correctly describes the gravitational attractive force between *any* two masses.

Kepler's First Law

We shall not attempt to derive Kepler's first law. It can be shown that the first law arises as a natural consequence of the inverse-square nature of Newton's law of gravitation. That is, any object bound to another by a force that varies as $1/r^2$ will move in an elliptical orbit. As shown in Figure 7.19a, an ellipse is a curve drawn so that the sum of the distances from any point on the curve to two internal points called focal points or foci (singular, focus) is always the same. For the Sun-planet configuration (Fig. 7.19b), the Sun is at one focus and the other focus is empty. Because the orbit is an ellipse, the distance from the Sun to the planet continuously changes.

Kepler's Second Law

Kepler's second law states that a line drawn from the Sun to any planet sweeps out equal areas in equal time intervals. Consider a planet in an elliptical orbit about the Sun, as in Figure 7.20. Imagine that at some instant of time we draw a line from the Sun to a planet, tracing out the line AS where S represents the position of the Sun. Exactly 30 days later, we repeat the process and draw the line BS. The area swept out is shown in color in the diagram. When the planet is at point C, we repeat the process, drawing the line CS from the planet to the Sun. We now wait 30 days, exactly the same amount of time we waited before, and draw the line DS. The area swept out in this interval is also shown in color in the figure. The two colored areas are equal.

Kepler's Third Law

The derivation of Kepler's third law is simple enough to carry out here. Consider a planet of mass M_p moving about the Sun, which has a mass of M_S, in a circular orbit (Fig. 7.21). (The assumption of a circular rather than an elliptical orbit will not introduce serious error into our approach, because the orbits of all planets except Mercury and Pluto are very nearly circular.) Because the gravitational force on the planet is the force needed to keep it moving in a circle,

$$\frac{GM_S M_p}{r^2} = \frac{M_p v^2}{r}$$

The speed, v, of the planet in its orbit is equal to the circumference of the orbit divided by the time required for one revolution, T, called the **period** of the planet. That is, $v = 2\pi r/T$, and the preceding expression becomes

$$\frac{GM_S}{r^2} = \frac{(2\pi r/T)^2}{r}$$

Kepler's third law ▶ $$T^2 = \left(\frac{4\pi^2}{GM_S}\right) r^3 = K_S r^3 \qquad [7.21]$$

PHYSICS IN ACTION

Views of the Planets

A**View of Saturn from Voyager 2.**   In the photograph on the left, Saturn's rings are bright and its northern hemisphere defined by bright features as NASA's Voyager 2 approaches the planet, which it encountered on August 25, 1981. Three images, taken through ultraviolet, violet, and green filters on July 12, 1981, were combined to make this photograph. Voyager 2 was 43 million km (27 million miles) from Saturn when it took this photograph.

A View of Pluto and Charon.   In the middle photograph, the Hubble Space Telescope's Faint Object Camera has obtained the clearest image ever of Pluto and its moon, Charon. Pluto is the bright object at the center of the frame; Charon is the fainter object in the lower left. Charon's orbit around Pluto is a circle seen nearly edge-on from Earth.

A Comet Colliding with Jupiter.   On the right, a computer has put together separate views of Jupiter and of fragments of Comet Shoemaker-Levy 9, both taken with the Hubble Space Telescope about two months before Jupiter and the comet collided in July 1994. Their relative sizes and distance were altered. The black spot on Jupiter is the shadow of its moon Io.

(All photos courtesy of NASA)

where K_S is a constant given by

$$K_S = \frac{4\pi^2}{GM_S} = 2.97 \times 10^{-19} \text{ s}^2/\text{m}^3$$

Equation 7.21 is Kepler's third law. It is also valid for elliptical orbits if r is replaced by a length equal to the semimajor axis of the ellipse. Note that K_S is independent of the mass of the planet. Therefore, Equation 7.21 is valid for any

TABLE **7.3** Useful Planetary Data

Body	Mass (kg)	Mean Radius (m)	Period (s)	Distance from Sun (m)	$\dfrac{T^2}{r^3}\left(\dfrac{s^2}{m^3}\right)$
Mercury	3.18×10^{23}	2.43×10^6	7.60×10^6	5.79×10^{10}	2.97×10^{-19}
Venus	4.88×10^{24}	6.06×10^6	1.94×10^7	1.08×10^{11}	2.99×10^{-19}
Earth	5.98×10^{24}	6.37×10^6	3.156×10^7	1.496×10^{11}	2.97×10^{-19}
Mars	6.42×10^{23}	3.37×10^6	5.94×10^7	2.28×10^{11}	2.98×10^{-19}
Jupiter	1.90×10^{27}	6.99×10^7	3.74×10^8	7.78×10^{11}	2.97×10^{-19}
Saturn	5.68×10^{26}	5.85×10^7	9.35×10^8	1.43×10^{12}	2.99×10^{-19}
Uranus	8.68×10^{25}	2.33×10^7	2.64×10^9	2.87×10^{12}	2.95×10^{-19}
Neptune	1.03×10^{26}	2.21×10^7	5.22×10^9	4.50×10^{12}	2.99×10^{-19}
Pluto	$\approx 1.4 \times 10^{22}$	$\approx 1.5 \times 10^6$	7.82×10^9	5.91×10^{12}	2.96×10^{-19}
Moon	7.36×10^{22}	1.74×10^6	—	—	—
Sun	1.991×10^{30}	6.96×10^8	—	—	—

planet. If we consider the orbit of a satellite, such as the Moon, about the Earth, then the constant has a different value, with the mass of the Sun replaced by the mass of the Earth. In this case, K_E equals $4\pi^2/GM_E$.

Kepler's third law gives us a method for measuring the mass of the Sun or of any celestial object around which at least one other object moves in orbit. Notice that the constant K_S in Equation 7.20 includes the mass of the Sun. The value of this constant can be found by substituting the values of the period and orbital radius and solving for K_S. The mass of the Sun is then

$$M_S = \frac{4\pi^2}{GK_S}$$

This same process can be used to calculate the mass of the Earth (by considering the period and orbital radius of the Moon) and the mass of other planets in the Solar System that possess moons.

A collection of useful planetary data is presented in Table 7.3, where the last column verifies that T^2/r^3 is a constant.

Thinking Physics 5

Some science fiction stories fantasize about a sister planet to the Earth, which is in the same orbit as the Earth but 180° ahead of us. Thus, we could never see this planet from Earth because it will always be behind the Sun, hidden from our point of view. How would you plan a visit to this proposed planet if you start from an Earth orbit?

Explanation The quickest way to arrive at the planet is to fire your rocket so that you slow down in your orbit around the Sun. This will transfer you to an elliptical orbit around the Sun with a smaller major axis than that of the Earth's orbit. Thus, you will complete this orbit in a time less than one year. If you design your orbit so that the period is 0.5 year, then when you return to your original point, the sister planet will also be there.

A photomontage of the Earth and the surface of the Moon. The prominent colors are the blues of oceans and the white swirls of clouds. *(© Index Stock Photography, Inc.)*

A calculation using Kepler's third law shows that your spacecraft will dip inside the orbit of Venus, so you must be sure not to collide with Venus and that you have the appropriate thermal protection to deal with increased radiation from the Sun.

Another possibility is to fire your rockets so that you speed up in your orbit around the Sun, transferring to an elliptical orbit with a period of 1.5 years. This avoids the thermal problems of going near the Sun but takes three times as long.

Applying Physics 2

NASA employs a method called either "slingshotting" or "gravitational boosting" to add velocity to a spaceship as it passes a planet. To understand how this works, consider a spacecraft moving with a speed of 10.0 km/s as shown in Figure 7.22 approaching the planet Jupiter, which moves in its orbit at an average speed of about 13.1 km/s. Find the speed of the spacecraft after the "collision" with Jupiter assuming that it leaves traveling in a direction directly opposite to its incoming velocity.

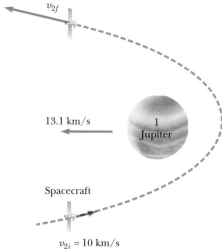

v_{2f}

13.1 km/s

1
Jupiter

Spacecraft

$v_{2i} = 10$ km/s

Figure 7.22 (Applying Physics 2)

Explanation This is a collision in the pure sense of the word; the forces are not contact forces but a gravitational interaction. As the gravitational force is conservative, the collision is perfectly elastic. Because Jupiter is so massive, its velocity will be unaffected, so before and after the collision it will have a velocity of -13.1 km/s. To find the velocity of the spacecraft after the collision, we call Jupiter body 1 and the ship body 2 and apply Equation 6.11:

$$v_{1i} - v_{2i} = -(v_{1f} - v_{2f})$$

$$-13.1 \text{ km/s} - 10.0 \text{ km/s} = -(-13.1 \text{ km/s} - v_{2f})$$

$$v_{2f} = -36.2 \text{ km/s}$$

Thus, the spacecraft leaves the vicinity of Jupiter traveling more than three times faster than before its encounter with the planet.

Example 7.14   Geosynchronous Orbit

Satellite dishes do not have to change directions in order to stay focused on a signal from a satellite. This means that the satellite always has to be found at the same location with respect to the surface of the Earth. In order for this to occur, the satellite must be at a height such that its revolution period is the same as that of the Earth, 24 h. At what height must a satellite be to achieve this?

Reasoning and Solution   The force that produces the centripetal acceleration of the satellite is the gravitational force, so

$$(1) \qquad G\frac{M_E m}{r^2} = \frac{mv^2}{r}$$

where M_E is the mass of the Earth and r is the distance of the satellite from the center of the Earth.

Also, we find the velocity of the satellite to be

$$(2) \qquad v = \frac{d}{T} = \frac{2\pi r}{T}$$

where T is the orbital period of the satellite. Solving (1) and (2) simultaneously for r yields

$$r = \left(\frac{T^2}{4\pi^2} GM_E\right)^{1/3}$$

Exercise   Taking $T = 24$ h $= 86{,}400$ s and $M_E = 5.98 \times 10^{24}$ kg, show that the height above the Earth is approximately 22 000 miles. (Be careful; will $r = 22\,000$ miles?)

EXAMPLE 7.15   An Earth Satellite

A satellite of mass m moves in a circular orbit about the Earth with a constant speed of v and a height of $h = 1000$ km above the Earth's surface, as in Figure 7.23. (For clarity, this figure is not drawn to scale.) Find the orbital speed of the satellite. The radius of the Earth is 6.38×10^6 m, and its mass is 5.98×10^{24} kg.

Reasoning   The only force acting on the satellite is the gravitational force directed toward the center of the circular orbit. This can be equated to mv^2/r to find v. (The distance r is the distance from the center of the Earth to the satellite.)

Solution   The only external force on the satellite is that of the gravitational attraction exerted by the Earth. This force is directed toward the center of the satellite's circular path and is the force that produces the centripetal acceleration of the satellite. Because the force of gravity is $GM_E m/r^2$ we find that

$$F = G\frac{M_E m}{r^2} = m\frac{v^2}{r}$$

$$v^2 = \frac{GM_E}{r}$$

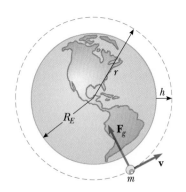

Figure 7.23 (Example 7.15) A satellite of mass m moving in a circular orbit of radius r and with constant speed v around the Earth (not drawn to scale). The force causing the centripetal acceleration is provided by the gravitational force acting on the satellite.

In this expression, the distance r is the Earth's radius plus the height of the satellite. That is, $r = R_E + h = 7.38 \times 10^6$ m, and so

$$v^2 = \frac{(6.67 \times 10^{-11}\ \text{N} \cdot \text{m}^2/\text{kg}^2)(5.98 \times 10^{24}\ \text{kg})}{7.38 \times 10^6\ \text{m}} = 5.40 \times 10^7\ \text{m}^2/\text{s}^2$$

Therefore,

$$v = \boxed{7.35 \times 10^3\ \text{m/s} \approx 16\ 400\ \text{mi/h}}$$

Note that v *is independent of the mass of the satellite!*

Exercise Calculate the period of revolution (the time required for one revolution about the Earth), T, of the satellite.

Answer 105 min

7.10 THE VECTOR NATURE OF ANGULAR QUANTITIES   *Optional Section*

In this chapter we considered only the magnitudes of the angular velocity and angular acceleration. However, these are in fact vector quantities, having both magnitude and direction. Let us first describe how to find the direction of the angular velocity, $\boldsymbol{\omega}$.

When a rigid object such as a disk rotates about a fixed axis, the angular velocity, $\boldsymbol{\omega}$, points along this axis. A convenient right-hand rule, illustrated in Figure 7.24a, can be used to determine the direction of $\boldsymbol{\omega}$. Grasp the axis of rotation with your right hand so that your fingers wrap in the direction of rotation. Your extended thumb points in the direction of $\boldsymbol{\omega}$. Figure 7.24b illustrates that $\boldsymbol{\omega}$ is also in the direction of advance of a rotating right-handed screw.

Let us apply this rule to a rotating disk when viewed along the axis of rotation as in Figure 7.25. When the disk rotates clockwise, as in Figure 7.25a, applying the right-hand rule shows that the direction of $\boldsymbol{\omega}$ is into the page. When the disk rotates counterclockwise, as in Figure 7.25b, the direction of $\boldsymbol{\omega}$ is out of the page.

Finally, the direction of the angular acceleration vector, $\boldsymbol{\alpha}$, is in the direction of $\boldsymbol{\omega}$ if the angular speed (the magnitude of $\boldsymbol{\omega}$) is increasing in time and antiparallel to $\boldsymbol{\omega}$ if the angular speed is decreasing in time.

(a)

(b)

Figure 7.25 An end view of a disk rotating about an axis perpendicular to the page. (a) When the disk rotates clockwise, $\boldsymbol{\omega}$ is into the page. (b) When the disk rotates counterclockwise, $\boldsymbol{\omega}$ is out of the page.

Figure 7.24 (a) The right-hand rule for determining the direction of the angular velocity vector. (b) The direction of $\boldsymbol{\omega}$ is in the direction of advance of a right-handed screw.

SUMMARY

The **average angular speed,** $\bar{\omega}$, of a rigid object is defined as the ratio of the angular displacement, $\Delta\theta$, to the time interval, Δt:

$$\bar{\omega} \equiv \frac{\theta_2 - \theta_1}{t_2 - t_1} = \frac{\Delta\theta}{\Delta t} \qquad [7.2]$$

where $\bar{\omega}$ is in radians per second (rad/s).

The **average angular acceleration,** $\bar{\alpha}$, of a rotating object is defined as the ratio of the change in angular speed, $\Delta\omega$, to the time interval, Δt.

$$\bar{\alpha} \equiv \frac{\omega_2 - \omega_1}{t_2 - t_1} = \frac{\Delta\omega}{\Delta t} \qquad [7.4]$$

where $\bar{\alpha}$ is in radians per second per second (rad/s^2).

If an object undergoes rotational motion about a fixed axis under constant angular acceleration α, one can describe its motion with the following set of equations:

$$\omega = \omega_0 + \alpha t \qquad [7.5]$$

$$\theta = \omega_0 t + \tfrac{1}{2}\alpha t^2 \qquad [7.6]$$

$$\omega^2 = \omega_0^2 + 2\alpha\theta \qquad [7.7]$$

When an object rotates about a fixed axis, the angular speed and angular acceleration are related to the tangential speed and tangential acceleration through the relationships

$$v_t = r\omega \qquad [7.8]$$

$$a_t = r\alpha \qquad [7.9]$$

Any object moving in a circular path has an acceleration directed toward the center of the circular path, called a **centripetal acceleration.** Its magnitude is given by

$$a_c = \frac{v_t^2}{r} = r\omega^2 \qquad [7.11, 7.14]$$

Any object moving in a circular path must have a net force exerted on it that is directed toward the center of the circular path. From Newton's second law, this force and the centripetal acceleration are related by

$$F = m\frac{v_t^2}{r} \qquad [7.16]$$

Some examples of forces that cause centripetal acceleration are the force of gravity (as in the motion of a satellite) and the force of tension in a string.

Newton's law of universal gravitation states that every particle in the Universe attracts every other particle with a force that is directly proportional to the product of their masses and inversely proportional to the square of the distance, r, between them:

$$F = G\frac{m_1 m_2}{r^2} \qquad [7.17]$$

where $G = 6.673 \times 10^{-11}$ N·m^2/kg^2 is the **constant of universal gravitation.**

Kepler's laws of planetary motion state that

1. All planets move in elliptical orbits with the Sun at one of the focal points.

2. A line drawn from the Sun to any planet sweeps out equal areas in equal time intervals.

3. The square of the orbital period of a planet is proportional to the cube of the mean distance from the planet to the Sun:

$$T^2 = \left(\frac{4\pi^2}{GM_S} \right) r^3 \qquad\qquad \text{[7.21]}$$

MULTIPLE-CHOICE QUESTIONS

A race track is constructed such that two semicircles of radius 80 m at *A* and 40 m at *B* are joined by two stretches of straight track, *C* and *D*. In a particular trial run, a driver traveled at a constant speed of 50 m/s for one complete lap. Answer the next three questions concerning this one lap.

1. The ratio of the tangential acceleration at *A* to that at *B* is
   (a) 1/2 (b) 1/4 (c) 2 (d) 4 (e) Undefined. The tangential acceleration is zero at both points.

2. The ratio of the centripetal acceleration at *A* to that at *B* is
   (a) 1/2 (b) 1/4 (c) 2 (d) 4 (e) Undefined. The centripetal acceleration is zero at both points.

3. The angular speed is greatest at
   (a) *A*   (b) *B*   (c) It is equal at both *A* and *B*.

4. Superman circles the Earth at a radius of $2R$, where *R* is the radius of the Earth. He then moves to a radius of $4R$. The gravitational force on him at this second orbit as compared to the first orbit is
   (a) the same.
   (b) twice as great.
   (c) four times as great.
   (d) half as great.
   (e) one fourth as great.

5. A 0.400-kg object is swung in a circular path and in a vertical plane on a 0.500-m length string. If a constant angular speed of 8.00 rad/s is maintained, what is the tension in the string when the object is at the top of the circle?
   (a) 8.88 N   (b) 10.5 N   (c) 12.8 N   (d) 19.6 N

CONCEPTUAL QUESTIONS

1. An object executes circular motion with a constant speed whenever a net force of constant magnitude acts perpendicular to the velocity. What happens to the speed if the force is not perpendicular to the velocity?

2. Explain why the Earth is not spherical in shape and bulges at the equator.

3. When a wheel of radius *R* rotates about a fixed axis do all points on the wheel have the same angular speed? Do they all have the same linear speed? If the angular speed is constant and equal to ω, describe the linear speeds and linear accelerations of the points located at $r = 0$, $r = R/2$, and $r = R$, where the points are measured from the center of the wheel.

4. If a car's wheels are replaced with wheels of greater diameter, will the reading of the speedometer change? Explain.

5. (1) At night, you are farther away from the Sun than during the day. (2) The force from the Sun on you is downward into the Earth at night and upward into the sky during the day. Based on these two facts, if you had a sensitive enough bathroom scale, would you appear to weigh more at night than during the day?

6. Correct the following statement: "The racing car rounds the turn at a constant velocity of 90 miles per hour."

7. A satellite in orbit is not truly traveling through a vacuum—it is moving through very thin air. Does the resulting air friction cause the satellite to slow down?

8. Why does an astronaut in a space capsule orbiting the Earth experience a feeling of weightlessness?

9. How would you explain the fact that Saturn and Jupiter have periods much greater than one year?

10. Because of the Earth's rotation about its axis, you would weigh slightly less at the Equator than at the poles. Why?

11. It has been suggested that rotating cylinders about 10 miles long and 5.0 miles in diameter be placed in space for colonies. The purpose of their rotations is to simulate gravity for the inhabitants. Explain the concept behind this proposal.

12. Describe the path of a moving body in the event that the acceleration is constant in magnitude at all times and (a) perpendicular to the velocity; (b) parallel to the velocity.

13. An object moves in a circular path with constant speed *v*. (a) Is the object's velocity constant? (b) Is its acceleration constant? Explain.

14. Centrifuges are often used in dairies to separate the cream from the milk. Which remains on the bottom?

15. Is it possible for a car to move in a circular path in such a way that it has a tangential acceleration but no centripetal acceleration?

16. Use Kepler's second law to convince yourself that the Earth must orbit faster during January, when it is closest to the Sun, than during July, when it is farthest from the Sun.

17. Why does a pilot sometimes black out when pulling out of a steep dive?

18. Consider the yo-yo in Figure 7.10. Is it possible to whirl it fast enough so that the cord makes an angle of 90° with the vertical? Explain.

19. A pail of water can be whirled in a vertical path such that no water is spilled. Why does the water remain in the pail, even when the pail is above your head?

PROBLEMS

1, 2, 3 = straightforward, intermediate, challenging   ☐ = full solution available in Study Guide/Student Solutions Manual
WEB = solution posted at **http://www.saunderscollege.com/physics/college**   ◩ = biomedical application   ◪ = Interactive Physics

Review Problem

The Mars *Pathfinder* probe is designed to drop the instrument package from a height of 20 m above the surface, the package speed being brought to zero by a combination parachute–rocket system at that height. To cushion the landing, giant airbags surround the package. The mass of Mars is 0.1074 times that of Earth and the radius of Mars is 0.5282 that of Earth. Find (a) the acceleration due to gravity at the surface of Mars and (b) how long it takes for the instrument package to fall the last 20 m.

Section 7.1   Angular Speed and Angular Acceleration

1. The tires on a new compact car have a diameter of 2.0 ft and are warranted for 60 000 miles. (a) Determine the angle (in radians) through which one of these tires will rotate during the warranty period. (b) How many revolutions of the tire are equivalent to your answer in (a)?

2. A wheel has a radius of 4.1 m. How far (path length) does a point on the circumference travel if the wheel is rotated through angles of 30°, 30 rad, and 30 rev, respectively?

3. A rotating body has a constant angular speed of 33 rev/min. (a) What is its angular speed in rad/s? (b) Through what angle, in radians, does it rotate in 1.5 s?

4. Find the angular speed of the Earth about the Sun in radians per second and degrees per day.

5. A potter's wheel moves from rest to an angular speed of 0.20 rev/s in 30 s. Find its angular acceleration in radians per second per second.

Section 7.2   Rotational Motion Under Constant Angular Acceleration

Section 7.3   Relations Between Angular and Linear Quantities

6. A tire placed on a balancing machine in a service station starts from rest and turns through 4.7 revolutions in 1.2 s before reaching its final angular speed. Calculate its angular acceleration.

7. A machine part rotates at an angular speed of 0.60 rad/s; its speed is then increased to 2.2 rad/s at an angular acceleration of 0.70 rad/s². Find the angle through which the part rotates before reaching this final speed.

8. A dentist's drill starts from rest. After 3.20 s of constant angular acceleration it turns at a rate of 2.51×10^4 rev/min. (a) Find the drill's angular acceleration. (b) Determine the angle (in radians) through which the drill rotates during this period.

9. An electric motor rotating a workshop grinding wheel at a rate of 100 rev/min is switched off. Assume constant negative angular acceleration of magnitude 2.00 rad/s². (a) How long does it take for the grinding wheel to stop? (b) Through how many radians has the wheel turned during the interval found in (a)?

10. The tub of a washer goes into its spin-dry cycle, starting from rest and reaching an angular speed of 5.0 rev/s in 8.0 s. At this point the person doing the laundry opens the lid, and a safety switch turns off the washer. The tub slows to rest in 12.0 s. Through how many revolutions does the tub turn during this 20.0-s interval? Assume constant angular acceleration while it is starting and stopping.

11. A rotating wheel requires 3.00 s to rotate 37.0 revolutions. Its angular velocity at the end of the 3.00-s interval is 98.0 rad/s. What is the constant angular acceleration of the wheel?

12. A mass attached to a 50.0-cm-long string starts from rest and is rotated 40 times in 1.00 min before reaching a final angular speed. (a) Determine the angular acceleration of the mass, assuming that it is constant. (b) What is the angular speed of the mass after 1.00 min?

13. A coin with a diameter of 2.40 cm is dropped onto a horizontal surface. The coin starts out with an initial angular speed of 18.0 rad/s and rolls in a straight line without slipping. If the rotation slows with an angular

acceleration of magnitude 1.90 rad/s², how far does the coin roll before coming to rest?

14. A car is traveling at 17.0 m/s on a straight horizontal highway. The wheels of the car have radii of 48.0 cm. If the car speeds up with an acceleration of 2.00 m/s² for 5.00 s, find the number of revolutions of the wheels during this period.

15. The turntable of a record player rotates initially at 33 rev/min and takes 20 s to come to rest. **(a)** What is the angular acceleration of the turntable, assuming it is uniform? **(b)** How many rotations does the turntable make before coming to rest? **(c)** If the radius of the turntable is 0.14 m, what is the initial linear speed of a bug riding on the rim?

Section 7.4 Centripetal Acceleration

16. Find the centripetal accelerations of **(a)** a point on the Equator of the Earth and **(b)** a point at the North Pole of the Earth.

17. It has been suggested that rotating cylinders about 10 mi long and 5.0 mi in diameter be placed in space and used as colonies. What angular speed must such a cylinder have so that the centripetal acceleration at its surface equals Earth's gravity?

18. A tire of 2.0 ft diameter is placed on a balancing machine where it is spun so that its tread is moving at a constant speed of 60 mph. A small stone is stuck in the tread of the tire. What is the acceleration of the stone as the tire is being balanced?

19. **(a)** What is the tangential acceleration of a bug on the rim of a 10.0-in. diameter disk if the disk moves from rest to an angular speed of 78 revolutions per minute in 3.0 s? **(b)** When the disk is at its final speed, what is the tangential velocity of the bug? **(c)** One second after the bug starts from rest, what are its tangential acceleration, radial acceleration, and total acceleration?

20. A race car starts from rest on a circular track of radius 400 m. The car's speed increases at the constant rate of 0.500 m/s². At the point at which the magnitudes of the centripetal and tangential accelerations are equal, determine **(a)** the speed of the race car, **(b)** the distance traveled, and **(c)** the elapsed time.

Section 7.5 Forces Causing Centripetal Acceleration

21. A sample of blood is placed in a centrifuge of radius 15.0 cm. The mass of a red corpuscle is 3.0×10^{-16} kg, and the magnitude of the force required to make it settle out of the plasma is 4.0×10^{-11} N. At how many revolutions per second should the centrifuge be operated?

22. A 2000-kg car rounds a circular turn of radius 20 m. If the road is flat and the coefficient of friction between tires and road is 0.70, how fast can the car go without skidding?

23. A 50.0-kg child stands at the rim of a merry-go-round of radius 2.00 m, rotating with an angular speed of 3.00 rad/s. **(a)** What is the child's centripetal acceleration? **(b)** What is the minimum force between her feet and the floor of the merry-go-round that is required to keep her in the circular path? **(c)** What minimum coefficient of static friction is required? Is the answer you found reasonable? In other words, is she likely to be able to stay on the merry-go-round?

24. An airplane is flying in a horizontal circle at a speed of 100 m/s. The 80.0-kg pilot does not want her radial acceleration to exceed 7g. **(a)** What is the minimum radius of the circular path? **(b)** At this radius, what is the force causing the centripetal acceleration of the pilot?

25. An engineer wishes to design a curved exit ramp for a toll road in such a way that a car will not have to rely on friction to round the curve without skidding. He does so by banking the road in such a way that the necessary force causing the centripetal acceleration will be supplied by the component of the normal force toward the center of the circular path. **(a)** Show that for a given speed of v and a radius of r, the curve must be banked at the angle θ such that $\tan \theta = v^2/rg$. **(b)** Find the angle at which the curve should be banked if a typical car rounds it at a 50.0-m radius and a speed of 13.4 m/s.

26. An air puck of mass 0.25 kg is tied to a string and allowed to revolve in a circle of radius 1.0 m on a frictionless horizontal table. The other end of the string passes through a hole in the center of the table, and a mass of 1.0 kg is tied to it (Fig. P7.26). The suspended mass remains in equilibrium while the puck on the tabletop revolves. **(a)** What is the tension in the string? **(b)** What is the force causing the centripetal acceleration of the puck? **(c)** What is the speed of the puck?

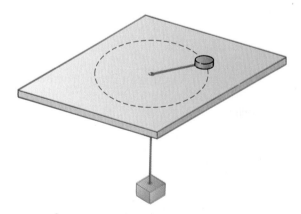

Figure P7.26

27. Tarzan (m = 85 kg) tries to cross a river by swinging from a 10-m-long vine. His speed at the bottom of the swing (as he just clears the water) is 8.0 m/s. Tarzan doesn't know that the vine has a breaking strength of 1000 N. Does he make it safely across the river? Justify your answer.

28. A 40.0-kg child takes a ride on a Ferris wheel that rotates four times each minute and has a diameter of 18.0 m. (a) What is the centripetal acceleration of the child? (b) What force (magnitude and direction) does the seat exert on the child at the lowest point of the ride? (c) What force does the seat exert on the child at the highest point of the ride? (d) What force does the seat exert on the child when the child is halfway between the top and bottom?

29. A roller-coaster car has a mass of 500 kg when fully loaded with passengers (Fig. P7.29). (a) If the vehicle has a speed of 20.0 m/s at point A, what is the force of the track on the vehicle at this point? (b) What is the maximum speed the vehicle can have at B in order for gravity to hold it on the track?

Figure P7.29

30. A pail of water is rotated in a vertical circle of radius 1.00 m (the approximate length of a person's arm). What must be the minimum speed of the pail at the top of the circle if no water is to spill out?

Section 7.7   Newton's Universal Law of Gravitation

Section 7.8   Gravitational Potential Energy Revisited (Optional)

Section 7.9   Kepler's Laws

31. The average distance separating the Earth and the Moon is 384 000 km. Use the data in Table 7.3 to find the net gravitational force the Earth and the Moon exerts on a 3.00×10^4-kg spaceship located halfway between them.

32. During a solar eclipse, the Moon, Earth, and Sun all lie on the same line, with the Moon between the Earth and the Sun. (a) What force is exerted on the Moon by the Sun? (b) What force is exerted on the Moon by the Earth? (c) What force is exerted on the Earth by the Sun? (See Table 7.3 and Problem 31 above.)

33. A coordinate system (in meters) is constructed on the surface of a pool table, and three masses are placed on the coordinate system as follows: a 2.0-kg mass at the origin, a 3.0-kg mass at (0, 2.0), and a 4.0-kg mass at (4.0, 0). Find the resultant gravitational force exerted on the mass at the origin by the other two masses.

34. Use the data of Table 7.3 to find the point between the Earth and the Sun at which an object can be placed so that the net gravitational force exerted on it by these two objects is zero.

35. Given that the Moon's period about the Earth is 27.32 days and the distance from the Earth to the Moon is 3.84×10^8 m, estimate the mass of the Earth. Assume the orbit is circular. Why do you suppose your estimate is high?

36. A satellite moves in a circular orbit around the Earth at a speed of 5000 m/s. Determine (a) the satellite's altitude above the surface of the Earth and (b) the period of the satellite's orbit.

WEB 37. A 600-kg satellite is in a circular orbit about the Earth at a height above the Earth equal to the Earth's mean radius. Find (a) the satellite's orbital speed, (b) the period of its revolution, and (c) the gravitational force acting on it.

38. A satellite is in a circular orbit just above the surface of the Moon. (See Table 7.3.) What are the satellite's (a) acceleration and (b) speed? (c) What is the period of the satellite orbit?

39. A satellite has a mass of 100 kg and is in an orbit at 2.00×10^6 m above the surface of the Earth. (a) What is the potential energy of the satellite at this location? (b) What is the magnitude of the gravitational force on the satellite?

40. Use data from Table 7.3 to determine the escape speed from (a) the Moon, (b) Mercury, and (c) Jupiter.

41. A satellite of mass 200 kg is launched from a site on the Earth's Equator into an orbit at 200 km above the surface of Earth. (a) Assuming a circular orbit, what is the orbital period of this satellite? (b) What is the satellite's speed in orbit? (c) What is the minimum energy necessary to place this satellite in orbit, assuming no air friction?

ADDITIONAL PROBLEMS

42. A rotating bicycle wheel has an angular speed of 3.00 rad/s at some instant of time. It is then given an angular acceleration of 1.50 rad/s². A chalk line drawn on the wheel is horizontal at $t = 0$. (a) What angle does this line make with its original direction at $t = 2.00$ s? (b) What is the angular speed of the wheel at $t = 2.00$ s?

43. Three masses are aligned along the x axis of a rectangular coordinate system so that a 2.0-kg mass is at the origin, a 3.0-kg mass is at (2.0, 0) m, and a 4.0-kg mass is at (4.0, 0) m. Find (a) the gravitational force exerted on

the 4.0-kg mass by the other two masses and **(b)** the magnitude and direction of the gravitational force exerted on the 3.0-kg mass by the other two.

44. An athlete swings a 5.00-kg ball horizontally on the end of a rope. The ball moves in a circle of radius 0.800 m at an angular speed of 0.500 rev/s. What are **(a)** the tangential speed of the ball and **(b)** its centripetal acceleration? **(c)** If the maximum tension the rope can withstand before breaking is 100 N, what is the maximum tangential speed the ball can have?

45. Io, a small moon of the giant planet Jupiter, has an orbital period of 1.77 days and an orbital radius equal to 4.22×10^5 km. From these data, determine the mass of Jupiter.

46. A high-speed sander has a disk 6.00 cm in radius that rotates about its axis at a constant rate of 1200 rev/min. Determine **(a)** the angular speed of the disk in radians per second, **(b)** the linear speed of a point 2.00 cm from the disk's center, **(c)** the centripetal acceleration of a point on the rim, and **(d)** the total distance traveled by a point on the rim in 2.00 s.

47. The Solar Maximum Mission Satellite was placed in a circular orbit about 150 mi above the Earth. Determine **(a)** the orbital speed of the satellite and **(b)** the time required for one complete revolution.

48. Geosynchronous satellites have an angular velocity that matches the rotation of the Earth and follow circular orbits in the equatorial plane of the Earth. (Almost all communications satellites are geosynchronous and appear to be stationary above a point on the Equator.) **(a)** What must be the radius of the orbit of a geosynchronous satellite? **(b)** How high (in miles) above the surface of the Earth are geosynchronous satellites located?

WEB 49. A car moves at speed v across a bridge made in the shape of a circular arc of radius r. **(a)** Find an expression for the normal force acting on the car when it is at the top of the arc. **(b)** At what minimum speed will the normal force become zero (causing occupants of the car to seem weightless) if $r = 30.0$ m?

50. A 0.400-kg pendulum bob passes through the lowest part of its path at a speed of 3.00 m/s. **(a)** What is the tension in the pendulum cable at this point if the pendulum is 80.0 cm long? **(b)** When the pendulum reaches its highest point, what angle does the cable make with the vertical? **(c)** What is the tension in the pendulum cable when the pendulum reaches its highest point?

51. **(a)** Find the acceleration of gravity at the surface of a neutron star of mass 1.5 solar masses and having a radius of 10.0 km. **(b)** Find the weight of a 0.120-kg baseball on this star. **(c)** Assume the equation $PE = mgh$ applies and calculate the energy that a 70.0-kg person would expend climbing a 1.00-cm-tall mountain on this star.

52. Because of the Earth's rotation about its axis, a point on the Equator experiences a centripetal acceleration of 0.0340 m/s² while a point at the poles experiences no centripetal acceleration. **(a)** Show that at the Equator the gravitational force on an object (the true weight) must exceed the object's apparent weight. **(b)** What are the apparent weights at the Equator and at the poles of a 75.0-kg person? (Assume the Earth is a uniform sphere, and take $g = 9.800$ m/s².)

53. In a popular amusement park ride, a rotating cylinder of radius 3.00 m is set in rotation at an angular speed of 5.00 rad/s, as in Figure P7.53. The floor then drops away, leaving the riders suspended against the wall in a vertical position. What minimum coefficient of friction between a rider's clothing and the wall is needed to keep the rider from slipping? (*Hint:* Recall that the magnitude of the maximum force of static friction is equal to μn, where n is the normal force—in this case, the force causing the centripetal acceleration.)

Figure P7.53

54. A stunt man whose mass is 70 kg swings from the end of a 4.0-m-long rope along the arc of a vertical circle. Assuming he starts from rest when the rope is horizontal, find the tensions in the rope that are required to make him follow his circular path, **(a)** at the beginning of his motion, **(b)** at a height of 1.5 m above the bottom of the circular arc, and **(c)** at the bottom of his arc.

55. A 0.50-kg ball that is tied to the end of a 1.5-m light cord is revolved in a horizontal plane with the cord making a 30° angle with the vertical (see Fig. 7.10).

(a) Determine the ball's speed. (b) If the ball is revolved so that its speed is 4.0 m/s, what angle does the cord make with the vertical? (c) If the cord can withstand a maximum tension of 9.8 N, what is the highest speed at which the ball can move?

56. A skier starts at rest at the top of a large hemispherical hill (Fig. P7.56). Neglecting friction, show that the skier will leave the hill and become airborne at a distance of $h = R/3$ below the top of the hill. (*Hint:* At this point, the normal force goes to zero.)

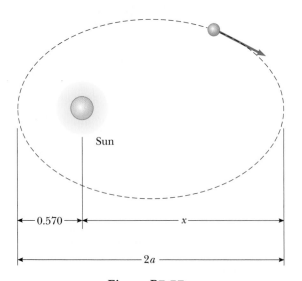

Figure P7.56

57. Halley's comet approaches the Sun to within 0.570 A.U., and its orbital period is 75.6 years. (A.U. is the abbreviation for astronomical unit, where 1 A.U. = 1.50 × 10^{11} m is the mean Earth–Sun distance.) How far from the Sun will Halley's comet travel before it starts its return journey? (See Fig. P.7.57.)

Figure P7.57

58. Two planets X and Y travel counterclockwise in circular orbits about a star, as in Figure P7.58. The radii of their orbits is in the ratio 3:1. At some time, they are aligned, as in Figure P7.58a, making a straight line with the star. Five years later, planet X has rotated through 90°, as in Figure P7.58b. Where is planet Y at this time?

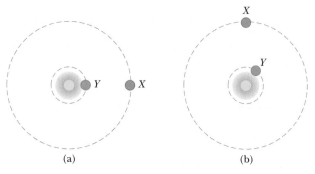

Figure P7.58

59. A spacecraft in the shape of a long cylinder has a length of 100 m and its mass with occupants is 1000 kg. It has strayed too close to a 1.0-m radius black hole having a mass 100 times that of the Sun (Fig. P7.59). If the nose of the spacecraft points toward the center of the black hole, and if distance between the nose of the spacecraft and the black hole's center is 10 km, (a) determine the total force on the spacecraft. (b) What is the difference in the force per kilogram of mass felt by the occupants in the nose of the ship and those in the rear of the ship farthest from the black hole?

Figure P7.59

60. In Robert Heinlein's *The Moon Is a Harsh Mistress,* the colonial inhabitants of the moon threaten to launch rocks down onto the Earth if they are not given independence (or at least representation). Assuming that a rail gun could launch a rock of mass m at twice the lunar escape speed, calculate the speed of the rock as it enters the Earth's atmosphere.

61. Show that the escape speed from the surface of a planet of uniform density is directly proportional to the radius of the planet.

62. A massless spring of constant $k = 78.4$ N/m is fixed on the left side of a level track. A block of mass $m = 0.50$ kg is pressed against the spring and compresses it a distance of d, as in Figure P7.62. The block (initially at rest) is then released and travels toward a circular loop-the-loop of radius $R = 1.5$ m. The entire track and the loop-the-loop are frictionless except for the section of track between points A and B. Given that the coefficient of kinetic friction between the block and the track along AB is $\mu_k = 0.30$, and that the length of AB is 2.5 m, determine the minimum compression, d, of the spring that enables the block to just make it through the loop-the-loop at point C. (*Hint:* The force of the track on the block will be zero if the block barely makes it through the loop-the-loop.)

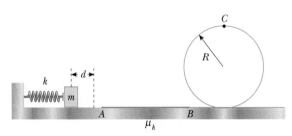

Figure P7.62

63. A small block of mass $m = 0.50$ kg is fired with an initial speed of $v_0 = 4.0$ m/s along a horizontal section of frictionless track, as shown in the top portion of Figure P7.63. The mass then moves along the frictionless semicircular, *vertical* tracks of radius $R = 1.5$ m. **(a)** Determine the force of the track on the block at points A and B. **(b)** The bottom of the track consists of a section ($L = 0.40$ m) with friction. Determine the coefficient of kinetic friction between the block and that portion of the bottom track if the mass just makes it to point C on the first trip. (*Hint:* If the block just makes it to point C, the force of contact exerted on the block by the track at that point should be zero.)

Figure P7.63

64. A frictionless roller coaster is given an initial velocity of v_0 at height h, as in Figure P7.64. The radius of curvature of the track at point A is R. **(a)** Find the maximum value of v_0 so that the roller coaster stays on the track at A solely because of gravity. **(b)** Using the value of v_0 calculated in (a), determine the value of h' that is necessary if the roller coaster is to just make it to point B.

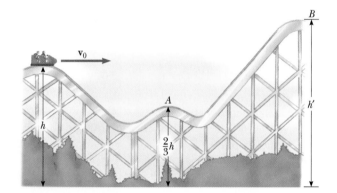

Figure P7.64

65. A car rounds a banked curve where the radius of curvature of the road is R, the banking angle is θ, and the coefficient of static friction is μ. **(a)** Determine the range of speeds the car can have without slipping up or down the road. **(b)** What is the range of speeds possible if $R = 100$ m, $\theta = 10°$, and $\mu = 0.10$ (slippery conditions)?

Rotational Equilibrium and Rotational Dynamics

Chapter Outline

▲ PHYSICS PUZZLER

This student is holding the axle of a spinning bicycle wheel while seated on a pivoted stool. The student and stool are initially at rest while the wheel spins in a horizontal plane. When the wheel is inverted about its center by 180°, the student and stool begin to rotate. If the wheel is inverted back to its original orientation, the student and stool stop rotating. Can you explain this set of events? *(Courtesy of Central Scientific Company)*

This chapter completes our study of equilibrium begun in Chapter 4. An understanding of equilibrium problems is important in a variety of fields. For example, students of architecture or industrial technology benefit from an understanding of the forces that act on buildings or large machines, and biology students should understand the forces at work in muscles and bones. Newton's first law, as discussed in Chapter 4, is the basis for much of our work in this chapter. In addition, in order to fully understand objects in equilibrium, we must consider torque. This concept will also play a key role in our discussion of rotational motion.

In this chapter we will also complete our study of rotational motion. We shall build on the definitions of angular speed and angular acceleration encountered in Chapter 7 by examining the relationship between these concepts and the forces that produce rotational motion. Specifically, we shall find the rotational analog of Newton's second law and define a term that needs to be added to our equation for conservation of mechanical energy: rotational kinetic energy. One of the central purposes of this chapter is to develop the concept of angular momentum, a quantity that plays a key role in rotational motion. Finally, just as we found that linear momentum is conserved, we shall also find that the angular momentum of any isolated system is always conserved.

8.1 TORQUE

Consider Figure 8.1, an overhead view of a door hinged at point O. From this viewpoint, the door is free to rotate about a line perpendicular to the page and passing through O. When the force $\mathbf{F}$ is applied at the outer edge, as shown, the door can easily be caused to rotate counterclockwise; that is, the rotational effect of the force is quite large. On the other hand, the same force applied at a point nearer the hinges produces a smaller rotational effect on the door.

The tendency of a force to rotate a body about some axis is measured by a quantity called the torque, τ. The torque due to a force of $\mathbf{F}$ has the magnitude

$$\tau = Fd \qquad [8.1]$$

In this equation, τ (the Greek letter tau) is the torque, and the distance d is the **lever arm** (or moment arm) of the force $\mathbf{F}$. **The lever arm is the perpendicular distance from the axis of rotation to a line drawn along the direction of the force.**

As an example, consider the wrench pivoted about the axis O in Figure 8.2a. In this case, the applied force $\mathbf{F}$ acts at an angle of ϕ with the horizontal. If you examine the definition of lever arm just given, you will see that in this case the lever arm is the distance d shown in the figure and not L, the length of the wrench. That is, d is the perpendicular distance from the axis of rotation to the line along which the applied force acts. In this case, d is related to L by the expression $d = L \sin \phi$. Thus, the net torque produced by the force $\mathbf{F}$ is given by $\tau = FL \sin \phi$. Actually, computing torque in situations such as this is usually best accomplished by resolving the force into components, as shown in Figure 8.2b. The torque about the axis of rotation produced by the component $F \cos \phi$ is zero because the lever arm of this component is zero. That is, the distance from the pivot to the line along which the

Figure 8.1 A bird's-eye view of a door hinged at O, with a force applied perpendicularly to the door.

Figure 8.2 (a) A force, **F**, acting at an angle of ϕ with the horizontal produces a torque of magnitude Fd about the pivot O. (b) The component $F \sin \phi$ tends to rotate the system about O. The component $F \cos \phi$ produces no torque about O.

(a)

(b)

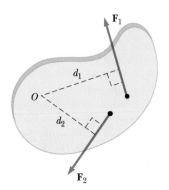

Figure 8.3 The force $\mathbf{F}_1$ tends to rotate the body counterclockwise about O, and $\mathbf{F}_2$ tends to rotate the body clockwise.

force acts is zero, because the line along which the force acts passes through the axis of rotation. The component $F \sin \phi$ has a lever arm of L, and the torque produced by this component is, as before, $\tau = FL \sin \phi$.

If two or more forces are acting on an object, as in Figure 8.3, then each has a tendency to produce a rotation about the pivot O. For example, $\mathbf{F}_2$ tends to rotate the object clockwise and $\mathbf{F}_1$ tends to rotate the object counterclockwise. We shall use the convention that the sign of the torque resulting from a force is positive if its turning tendency is counterclockwise and negative if its turning tendency is clockwise. In Figure 8.3, then, the torque resulting from $\mathbf{F}_1$, which has a moment arm of d_1, is positive and equal to $F_1 d_1$; the torque associated with $\mathbf{F}_2$ is negative and equal to $-F_2 d_2$. The *net torque* acting on the object about O is found by summing the torques:

$$\Sigma \tau = \tau_1 + \tau_2 = F_1 d_1 - F_2 d_2$$

Notice that the units of torque are units of force times length, such as the newton-meter (N·m) or the pound-foot (lb·ft).

EXAMPLE 8.1 The Spinning Crate

Figure 8.4 is a top view of a packing crate being pushed by two equal and opposite forces acting as shown. Find the net torque exerted on the crate if its width is 1.0 m. Assume an axis of rotation through the center of the crate.

Solution   The torque produced by $\mathbf{F}_1$ is

$$\tau_1 = F_1 d_1 = -(500 \text{ N})(0.50 \text{ m}) = -250 \text{ N·m}$$

and the torque produced by $\mathbf{F}_2$ is

$$\tau_2 = F_2 d_2 = -(500 \text{ N})(0.50 \text{ m}) = -250 \text{ N·m}$$

Because each force produces clockwise rotation, the torques are both negative. Thus, the net torque is -500 N·m.

Figure 8.4 (Example 8.1) A top view of a packing crate being pushed with equal but opposite forces.

EXAMPLE 8.2 The Swinging Door

Find the torque produced by the 300-N force applied at an angle of 60° to the door of Figure 8.5a.

Reasoning In Figure 8.5b, the 300-N force has been replaced by its horizontal and vertical components:

$$F_x = (300 \text{ N})\cos 60° = 150 \text{ N}$$

$$F_y = (300 \text{ N})\sin 60° = 260 \text{ N}$$

A convenient and obvious location for the axis of rotation is the hinge of the door.

Solution In this case, the 150-N force produces zero torque about the axis of rotation, because the line along which the force acts passes through the axis of rotation and hence the lever arm is zero. The 260-N force has a lever arm of 2.0 m and thus produces a torque of

$$\tau = (260 \text{ N})(2.0 \text{ m}) = \boxed{520 \text{ N·m}}$$

(a)

(b)

Figure 8.5 (Example 8.2) (a) A top view of a door being pulled by a 300-N force. (b) The components of the 300-N force.

8.2 TORQUE AND THE SECOND CONDITION FOR EQUILIBRIUM

In Chapter 4 we examined some situations, under the heading of the first condition of equilibrium, in which an object has no net external force acting on it; that is, $\Sigma F_x = 0$ and $\Sigma F_y = 0$. We found there that such an object either remains at rest if it is at rest or moves with a constant velocity if it is in motion. However, the first condition is not sufficient to ensure that an object is in complete equilibrium. This can be understood by considering the situation illustrated in Figure 8.4. In this situation, the two applied forces acting on the crate are equal in magnitude and opposite in direction. The first condition for equilibrium is satisfied because the two external forces balance each other, and yet the crate can still move — it will rotate clockwise because the forces do not act through a common point.

This case illustrates that, to fully understand the effect of a force or group of forces on an object, one must know not only the magnitude and direction of the force(s) but the point of application. That is, the net torque acting on an object must be considered. An object in rotational equilibrium has no angular acceleration.

The second condition for equilibrium asserts that if an object is in rotational equilibrium, the net torque acting on it about any axis must be zero. That is,

$$\Sigma \tau = 0 \qquad \text{[8.2]}$$

You can see that a body in static equilibrium must satisfy two conditions:

1. The resultant external force must be zero. $\Sigma \mathbf{F} = 0$

2. The resultant external torque must be zero. $\Sigma \tau = 0$

◀ The two conditions for static equilibrium

The Chinese acrobats in this difficult formation represent a balanced mechanical system. The external forces acting on the system, as shown by the blue vectors, are the weights of the acrobats, $\mathbf{w}_1$ and $\mathbf{w}_2$, and the upward force of the support, $\mathbf{n}$, on the lower acrobat. The vector sum of these external forces must be zero in such a balanced system. The net external torque acting on such a balanced system must also be zero. (*J. P. Lafont, Sygma*)

The first condition is a statement of translational equilibrium; the second is a statement of rotational equilibrium. We shall discuss torque and its relation to rotational motion in more detail later in the chapter.

Position of the Axis of Rotation

In the cases we have been describing so far, the axes of rotation for calculating torques have been selected without explanation. Often the nature of a problem suggests a convenient location for the axis, but just as often no single location stands out as being preferable. You might ask, "What axis should I choose in calculating the net torque?" The answer is as follows: **If the object is in equilibrium, it does not matter where you put the axis of rotation for calculating the net torque; the location of the axis is completely arbitrary.**

Figure 8.6 (Applying Physics 1)

Applying Physics 1

When an automobile driver steps on the accelerator, the nose of the car moves upward. When the driver brakes, the nose moves downward. Why do these effects occur?

Explanation When the driver steps on the accelerator, there is an increased force on the tires from the roadway. This force is parallel to the roadway and directed toward the front of the automobile, as suggested in Figure 8.6a. This force provides a torque that tends to cause the car to rotate in the counterclockwise direction in the diagram. The result of this rotation is a "nosing up" of the car. When the driver steps on the brake, there is an increased force on the tires from the roadway, directed toward the rear of the automobile, as suggested in Figure 8.6b. This force results in a torque that causes a clockwise rotation and the subsequent "nosing down" of the automobile.

8.3 THE CENTER OF GRAVITY

One of the forces that must be considered in dealing with a rigid object is the force of gravity acting on the object. To compute the torque due to the weight force, all of the weight can be thought of as concentrated at a single point called either the center of gravity or the center of mass.

Consider an object of arbitrary shape lying in the xy plane, as in Figure 8.7. The object is divided into a large number of very small particles of weight m_1g, m_2g, m_3g, . . . having coordinates (x_1, y_1), (x_2, y_2), (x_3, y_3), ... Each particle contributes a torque about the origin that is equal to its weight multiplied by its lever arm. For example, the torque due to the weight m_1g is m_1gx_1, and so forth.

We wish to locate the one position of the single force of magnitude w—the total weight of the object—whose effect on the rotation of the object is the same as that of the individual particles. This point is called the **center of gravity** of the object. Equating the torque exerted by w at the center of gravity to the sum of the torques acting on the individual particles gives

$$(m_1g + m_2g + m_3g + . . .)x_{cg} = m_1gx_1 + m_2gx_2 + m_3gx_3 + . . .$$

If we assume that g is uniform over the object (which will be the case in all the situations we will examine), then the g terms in the preceding equation cancel and we get

$$x_{cg} = \frac{m_1x_1 + m_2x_2 + m_3x_3 + . . .}{m_1 + m_2 + m_3 + . . .} = \frac{\Sigma m_ix_i}{\Sigma m_i} \qquad [8.3]$$

Similarly, the y coordinate of the center of gravity of the system can be found from

$$y_{cg} = \frac{\Sigma m_iy_i}{\Sigma m_i} \qquad [8.4]$$

The center of gravity of a homogeneous, symmetric body must lie on the axis of symmetry. For example, the center of gravity of a homogeneous rod must lie midway between the ends of the rod. The center of gravity of a homogeneous sphere or a homogeneous cube must lie at the geometric center of the object. One can determine the center of gravity of an irregularly shaped object, such as a wrench, experimentally by suspending the wrench from two different points (Fig. 8.8). The wrench is first hung from point A, and a vertical line, AB (which can be established with a plumb bob), is drawn when the wrench is in equilibrium. The wrench is then hung from point C, and a second vertical line, CD, is drawn. The

Figure 8.7 The center of gravity of an object is the point where all the weight of the object can be considered to be concentrated.

Figure 8.8 An experimental technique for determining the center of gravity of a wrench. The wrench is hung freely from two different pivots, A and C. The intersection of the two vertical lines, AB and CD, locates the center of gravity.

This one-bottle wine holder is an interesting example of a balanced mechanical system, which seems to defy gravity. The system (wine holder + bottle) is balanced when its center of gravity is directly over the lowest support point so that the net torque acting on the system is zero. *(Courtesy of Charles Winters)*

center of gravity coincides with the intersection of these two lines. In fact, if the wrench is hung freely from any point, the vertical line through that point must pass through the center of gravity.

In several examples in Section 8.4, we shall be concerned with homogeneous, symmetric objects whose centers of gravity coincide with their geometric centers. A rigid object in a uniform gravitational field can be balanced by a single force equal in magnitude to the weight of the object, as long as the force is directed upward through the object's center of gravity.

Thinking Physics 1

Why can't you put the back of your heels firmly against a wall and then bend over without falling?

Explanation In order for you to remain in equilibrium, your center of gravity must always be over your point of support, the feet. When you bend over with your heels against the wall, your center of gravity shifts forward in front of your feet. This produces a torque on your body rotating you forward.

EXAMPLE 8.3 Where Is the Center of Gravity?

Three particles are located in a coordinate system as shown in Figure 8.9. Find the center of gravity.

Reasoning The y coordinate of the center of gravity is zero because all the particles are on the x axis. To find the x coordinate of the center of gravity, we use Equation 8.3:

$$x_{cg} = \frac{\sum m_i x_i}{\sum m_i}$$

Solution For the numerator, we find

$$\sum m_i x_i = m_1 x_1 + m_2 x_2 + m_3 x_3$$
$$= (5.00 \text{ kg})(-0.500 \text{ m}) + (2.00 \text{ kg})(0 \text{ m}) + (4.00 \text{ kg})(1.00 \text{ m})$$
$$= 1.50 \text{ kg} \cdot \text{m}$$

The denominator is $\sum m_i = 11.0$ kg; therefore,

$$x_{cg} = \frac{1.50 \text{ kg} \cdot \text{m}}{11.0 \text{ kg}} = \boxed{0.136 \text{ m}}$$

Exercise If a fourth particle of mass 2.00 kg is placed at $x = 0$, $y = 0.250$ m, find the x and y coordinates of the center of gravity for this system of four particles.

Answer $x_{cg} = 0.115$ m; $y_{cg} = 0.0380$ m

Figure 8.9 (Example 8.3) Locating the center of gravity for a system of three particles.

8.4 EXAMPLES OF OBJECTS IN EQUILIBRIUM

In Chapter 4 we discussed some techniques for solving problems concerned with objects in equilibrium. Recall that when the objects were treated as geometric points, it was sufficient to simply apply the condition that the net force on the object must be zero. In this chapter we have shown that for objects of finite dimensions, a second condition for equilibrium must be satisfied—namely that the net torque on the object must also be zero. The following general procedure is recommended for solving problems that involve objects in equilibrium.

Problem-Solving Strategy

Objects in Equilibrium

1. Draw a simple, neat diagram of the system.
2. Isolate the object that is being analyzed. Draw a free-body diagram showing all external forces that are acting on this object. For systems that contain more than one object, draw a *separate* diagram for each object. Do not include forces that the object exerts on its surroundings.
3. Establish convenient coordinate axes for each body and find the components of the forces along these axes. Now apply the first condition of equilibrium (the net force on the object in the *x* and *y* direction must be zero) for each object under consideration.
4. Choose a convenient origin for calculating the net torque on the object. Now apply the second condition of equilibrium (the net torque on the object about any origin must be zero). Remember that the choice of the origin for the torque equation is arbitrary: therefore, choose an origin that will simplify your calculation as much as possible. Note that a force that acts along a line passing through the point chosen as the axis of rotation gives zero contribution to the torque.
5. The first and second conditions for equilibrium will give a set of simultaneous equations with several unknowns. All that is left to complete your solution is to solve for the unknowns in terms of the known quantities.

EXAMPLE 8.4 The Seesaw

A uniform 40.0-N board supports two children weighing 500 N and 350 N (Fig. 8.10a). The support (often called the *fulcrum*) is under the center of gravity of the board, and the 500-N child is 1.50 m from the center.
(a) Determine the upward force, **n,** exerted on the board by the support.

Reasoning First note in Figure 8.10b that, in addition to **n,** the external forces acting on the board are the forces of gravity acting on the children and the board, all of which act downward. We can assume that the board's center of gravity is at its geometric center because we were told that the board is uniform. Because the system is in equilibrium, the upward force **n** must balance all the downward forces.

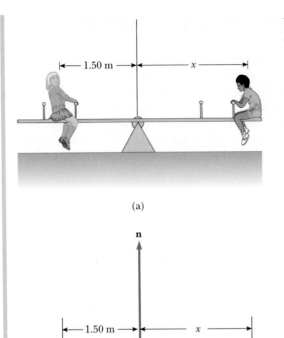

Figure 8.10   (Example 8.4) Two children balanced on a seesaw.

(a)

(b)

Solution   From $\Sigma F_y = 0$, we have

$$n - 500 \text{ N} - 350 \text{ N} - 40.0 \text{ N} = 0 \quad \text{or} \quad n = \boxed{890 \text{ N}}$$

Although the equation $\Sigma F_x = 0$ also applies to this situation, it is unnecessary to consider it because no forces are acting horizontally on the board.

(b) Determine where the 350-N child should sit to balance the system.

Reasoning   To find this position, we must invoke the second condition for equilibrium. We take the center of gravity of the board as the axis for the torque equation. This choice simplifies the problem, because the torques produced by both **n** and the 40.0-N weight are zero about this axis (because the lever arm of each is zero).

Solution   We apply $\Sigma \tau = 0$ to find

$$(500 \text{ N})(1.50 \text{ m}) - (350 \text{ N})(x) = 0$$

$$x = \boxed{2.14 \text{ m}}$$

(c) Repeat part (b), using another axis for the torque computations.

Reasoning   We stated that when an object is in equilibrium, the choice for the axis about which to compute torques is completely arbitrary. To illustrate this point, let us choose an axis perpendicular to the page and passing through the location of the 500-N child.

Solution In this case $\Sigma\tau = 0$ yields

$$n(1.50 \text{ m}) - (40.0 \text{ N})(1.50 \text{ m}) - (350 \text{ N})(1.50 + x) = 0$$

From part (a) we know that $n = 890$ N. Thus, we can solve for x to find $x = 2.14$ m, in agreement with the result of part (b).

EXAMPLE 8.5 A Weighted Forearm

A 50.0-N weight is held in a person's hand with the forearm horizontal, as in Figure 8.11a. The biceps muscle is attached 0.0300 m from the joint, and the weight is 0.350 m from the joint. Find the upward force exerted on the forearm (the ulna) by the biceps and the downward force on the forearm (the humerus) acting at the joint. Neglect the weight of the forearm.

Solution The forces acting on the forearm are equivalent to those acting on a bar of length 0.350 m, as shown in Figure 8.11b, where **F** is the upward force of the biceps and **R** is the downward force at the joint. From the first condition for equilibrium, we have

$$(1) \qquad \Sigma F_y = F - R - 50.0 \text{ N} = 0$$

From the second condition for equilibrium, we know that the sum of the torques about any point must be zero. With the joint O as the axis, we have

$$F(0.0300 \text{ m}) - (50.0 \text{ N})(0.350 \text{ m}) = 0$$

$$F = \boxed{583 \text{ N}}$$

This value for F can be substituted into (1) to give $R = 533 \text{ N}$. The two values correspond to $F = 131$ lb and $R = 119$ lb. Clearly, the forces at joints and in muscles can be extremely large.

EXAMPLE 8.6 Walking a Horizontal Beam

A uniform, horizontal 300-N beam, 5.00 m long, is attached to a wall by a pin connection that allows the beam to rotate. Its far end is supported by a cable that makes an angle of 53.0° with the horizontal (Fig. 8.12a). If a 600-N person stands 1.50 m from the wall, find the tension in the cable and the force exerted on the beam by the wall.

Reasoning First we must identify all the external forces acting on the beam and sketch them on a free-body diagram. This is shown in Figure 8.12b. The forces on the beam consist of the downward force of gravity acting on the beam, which has a magnitude of 300 N; the downward force exerted on the beam by the man, which is equal in magnitude to his weight, 600 N; the tension force, **T,** in the cable; and the force of the wall on the beam, **R.** We now resolve the forces **T** and **R** into their horizontal and vertical components, as shown in Figure 8.12c. Note that the x component of the tension force ($T\cos 53.0°$) is to the left, whereas the y component ($T\sin 53.0°$) is upward. The horizontal and vertical components of **R** are denoted by R_x and R_y, respectively. The first condition

Figure 8.11 (Example 8.5) (a) A weight held with the forearm horizontal. (b) The mechanical model for the system.

(a)

(c)

Figure 8.12 (Example 8.6) (a) A uniform beam attached to a wall and supported by a cable. (b) A free-body diagram for the beam. (c) The component form of the free-body diagram.

for equilibrium can now be applied in the x and y direction to give us two equations in terms of our unknowns R_x, R_y, and T. The necessary third equation can be found from the second condition of equilibrium.

Solution From the first condition for equilibrium, we find

$$(1) \qquad F_x = R_x - T\cos 53.0° = 0$$

$$(2) \qquad F_y = R_y + T\sin 53.0° - 600\ \text{N} - 300\ \text{N} = 0$$

The unknowns are R_x, R_y, and T. Because there are three unknowns and only two equations, we cannot find a solution from just the first condition of equilibrium.

Now let us use the second condition of equilibrium. The axis that passes through the pivot at the wall is a convenient one to choose for the torque equation because the forces R_x, R_y, and $T\cos 53.0°$ all have lever arms of zero and hence have zero torque about this pivot. Recalling our sign convention for the torque about an axis and noting that the lever arms of the 600-N, 300-N, and $T\sin 53.0°$ forces are 1.50 m, 2.50 m, and 5.00 m, respectively, we get

$$(3) \qquad \Sigma\tau_0 = (T\sin 53.0°)(5.00\ \text{m}) - (300\ \text{N})(2.50\ \text{m}) - (600\ \text{N})(1.50\ \text{m}) = 0$$

$$T = \boxed{413\ \text{N}}$$

Thus, the torque equation using this axis gives us one of the unknowns immediately! This value for T is then substituted into (1) and (2) to give

$$R_x = \boxed{249\ \text{N}} \qquad R_y = \boxed{570\ \text{N}}$$

If we selected some other axis for the torque equation, the solution would be the same. For example, if the axis were to pass through the center of gravity of the beam, the torque equation would involve both T and R_y; together with (1) and (2), however, it could still be solved for the unknowns. Try it!

Exercise Repeat this problem, but with the direction of R_x opposite that shown in Figure 8.12c. What answers do you get for T, R_x, and R_y?

Answer $T = 413$ N, $R_x = -249$ N, and $R_y = 570$ N. The negative sign for R_x means its direction was chosen incorrectly. The direction of R_x must be to the right, as shown in Figure 8.12c.

EXAMPLE 8.7   Don't Climb the Ladder

A uniform 10-m-long, 50-N ladder rests against a smooth vertical wall, as in Figure 8.13a. If the ladder is just on the verge of slipping when the angle it makes with the ground is 50°, find the coefficient of static friction between the ladder and ground.

Reasoning Figure 8.13b is the free-body diagram for the ladder, showing all external forces acting on it. At the base of the ladder, the Earth exerts an upward normal force, **n,** and a force of static friction, **f,** acts to the right. The wall exerts the force **P** to the left. Note that **P** is horizontal because the wall is smooth. (If the wall were rough, an upward frictional force would be exerted on the ladder.) The first condition for equilibrium can now be applied in the x and y directions to give us two equations in terms of our unknowns f, P, and n. The necessary third equation can be found from the second condition of equilibrium.

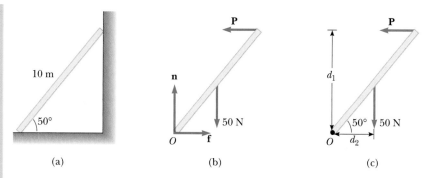

Figure 8.13 (Example 8.7) (a) A ladder leaning against a frictionless wall. (b) A free-body diagram for the ladder. (c) Lever arms for the force of gravity and **P**.

Solution From the first condition for equilibrium applied to the ladder, we have

$$(1) \qquad \Sigma F_x = f - P = 0$$

$$(2) \qquad \Sigma F_y = n - 50 \text{ N} = 0$$

From (2) we see that $n = 50$ N. Furthermore, when the ladder is on the verge of slipping, the force of static friction must be maximum and given by the relation $f_{s,\max} = \mu_s n = \mu_s(50 \text{ N})$. Thus, (1) reduces to

$$(3) \qquad \mu_s(50 \text{ N}) = P$$

Let us now apply the second condition of equilibrium and take the torques about the axis O at the bottom of the ladder, as in Figure 8.13c. The force **P** and the force of gravity acting on the ladder are the only forces that contribute to the torque about this axis, and their lever arms are shown in Figure 8.13c. Note that because the length of the ladder is 10 m, the lever arm for **P** is $d_1 = 10$ m sin 50°. Likewise, the lever arm for the 50-N force of gravity is $d_2 = 5.0$ m cos 50°, where the force of gravity acts through the center because the ladder is uniform. Thus, we find that

$$\Sigma \tau_0 = P(10 \text{ m sin } 50°) - (50 \text{ N})(5.0 \text{ m cos } 50°) = 0$$

$$P = 21 \text{ N}$$

Now that P is known, we can substitute its value into (3) to find μ_s:

$$\mu_s = \frac{21 \text{ N}}{50 \text{ N}} = \boxed{0.42}$$

8.5 RELATIONSHIP BETWEEN TORQUE AND ANGULAR ACCELERATION

Earlier in this chapter we considered the situation in which both the net force and the net torque acting on an object are zero. Such objects are said to be in equilibrium. We shall now examine the behavior of an object when the net torque acting on it is not zero. As you shall see, when a rigid object is acted on by a net torque, it undergoes an angular acceleration. Furthermore, the angular acceleration is directly proportional to the net torque. The end result of our investigation will be an expression that is analogous to $\mathbf{F} = m\mathbf{a}$ in translational motion.

Figure 8.14 A mass, m, attached to a light rod of length r moves in a circular path on a frictionless horizontal surface while the tangential force $\mathbf{F}_t$ acts on it.

Let us begin by considering the system shown in Figure 8.14, which consists of a mass m connected to a very light rod of length r. The rod is pivoted at the point O, and its movement is confined to rotation on a frictionless *horizontal* table. Now let us assume that a force, $\mathbf{F}_t$, perpendicular to the rod and hence tangent to the circular orbit, is acting on m. Because there is no force to oppose this tangential force, the mass undergoes a tangential acceleration according to Newton's second law:

$$F_t = ma_t$$

Multiplying the left and right sides of this equation by r gives

$$F_t r = mra_t$$

In Chapter 7 we found that the tangential acceleration and angular acceleration for a particle rotating in a circular path are related by the expression

$$a_t = r\alpha$$

so now we find that

$$F_t r = mr^2 \alpha \qquad [8.5]$$

The left side of Equation 8.5, which should be familiar to you, is the torque acting on the mass about its axis of rotation. That is, the torque is equal in magnitude to the force on m multiplied by the perpendicular distance from the pivot to the line along which the force acts, or $\tau = F_t r$. Hence, we can write Equation 8.5 as

$$\tau = mr^2 \alpha \qquad [8.6]$$

Equation 8.6 shows that the torque on the system is proportional to angular acceleration, where the constant of proportionality, mr^2, is called the **moment of inertia** of the mass m. (Because the rod is very light, its moment of inertia can be neglected.)

Torque on a Rotating Object

Now consider a solid disk rotating about its axis as in Figure 8.15a. The disk consists of many particles at various distances from the axis of rotation, as in Figure 8.15b. The torque on each one of these particles is given by Equation 8.6. The *total* torque on the disk is given by the sum of the individual torques on all the particles:

$$\Sigma \tau = (\Sigma mr^2)\alpha \qquad [8.7]$$

Note that, because the disk is rigid, all particles have the *same* angular acceleration, so α is not involved in the sum. If the masses and distances of the particles are labeled with subscripts as in Figure 8.15b, then

$$\Sigma mr^2 = m_1 r_1^2 + m_2 r_2^2 + m_3 r_3^2 + \ldots$$

This quantity is called the **moment of inertia** of the whole body and is given the symbol I:

Moment of inertia ▶

$$I \equiv \Sigma mr^2 \qquad [8.8]$$

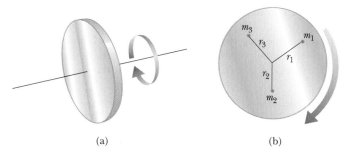

(a)   (b)

Figure 8.15 (a) A solid disk rotating about its axis. (b) The disk consists of many particles, all with the same angular acceleration.

The moment of inertia has the SI units $kg \cdot m^2$. Using this result in Equation 8.7, we see that the total torque on a rigid body rotating about a fixed axis is given by

$$\Sigma \tau = I\alpha \qquad [8.9]$$

◀ Relationship between net torque and angular acceleration

The angular acceleration of an object is proportional to the net torque acting on it. The proportionality constant, I, between the net torque and angular acceleration is the moment of inertia.

It is important to note that the equation $\Sigma \tau = I\alpha$ (Eq. 8.9) is the rotational counterpart to Newton's second law, $\Sigma F = ma$. Thus, the correspondence between rotational motion and linear motion continues. Recall from Chapter 7 that the linear variables x, v, and a are replaced in rotational motion by the variables θ, ω, and α. Likewise, we now see that **the force and mass in linear motion correspond to torque and moment of inertia in rotational motion.** In this chapter we shall develop other equations for the rotational kinetic energy and the angular momentum of a body rotating about a fixed axis. Based on the analogies already presented, you should be able to predict the form of these equations.

Applying Physics 2

The gear system on a bicycle provides an easily visible example of the relationship between torque and angular acceleration. Consider first a five-speed gear system in which the drive-chain can be adjusted to wrap around any of five gears attached to the back wheel (Figure 8.16). The gears are concentric with the wheel hub and are of different radii. When the cyclist begins from rest, the chain is attached to the largest gear. Because it has the largest radius, this gear provides the largest torque to the drive wheel. Large torque is required for starting, because the bicycle must change its motion from rest to some speed. As the bicycle rolls faster, the tangential speed of the chain increases and becomes too fast for the cyclist to maintain by pushing the pedals. The chain is then moved to the gear with the next smallest radius, and the chain now has a smaller tangential speed, which the cyclist can maintain by pushing the pedals. This gear does not provide as much torque as the first, but the cyclist does not need as much—he or she only needs to accelerate from some speed to a somewhat higher speed. This process continues as the bicycle moves faster and faster, and the cyclist shifts through all five gears. The fifth gear supplies the lowest torque, but now the main function of the supplied torque is to counter the frictional torque from the rolling tires, which will tend to slow the bicycle. The small radius of this gear allows the

APPLICATION

Bicycle Gears.

Figure 8.16 (Applying Physics 2) The drive wheel and gears of a bicycle. *(Semple Design & Photography)*

cyclist to keep up with the chain's movement by pushing the pedals, despite the rapid movement of the bicycle.

A ten-speed bicycle has the same gear structure on the drive wheel, but has two gears on the sprocket connected to the pedals. By combining different positions of the chain on the rear gears and the sprocket gears, ten different torques are available.

More on the Moment of Inertia

Because the moment of inertia of a body—as defined by $I = \Sigma mr^2$ (Eq. 8.8)—will be used throughout the remainder of this chapter, it will be useful to examine it in more detail before discussing other aspects of rotational motion. As seen earlier, a small object (or particle) has a moment of inertia equal to mr^2 about some axis. Now consider a somewhat more complicated system, the baton being twirled by a majorette in Figure 8.17. Let us assume that the baton can be modeled as a very light rod of length 2ℓ with a heavy mass at each end. (The rod of a real baton has significant mass relative to its ends.) Because we are neglecting the mass of the rod, the moment of inertia of the baton about an axis through its center and perpendicular to its length is given by Equation 8.8:

$$I = \Sigma mr^2$$

Because in this system there are two equal masses equidistant from the axis of rotation, we see that $r = \ell$ for each mass, and the sum is

$$I = \Sigma mr^2 = m\ell^2 + m\ell^2 = 2m\ell^2$$

We pointed out earlier that I is the rotational counterpart of m. However, there are some important distinctions between the two. For example, mass is an intrinsic property of an object that does not change, whereas **the moment of inertia of a**

Figure 8.17   A baton of length 2ℓ and mass $2m$ (the mass of the connecting rod is neglected). The moment of inertia about the axis through the baton's center and perpendicular to its length is $2m\ell^2$.

system depends on the axis of rotation and on the manner in which the mass is distributed. Examples 8.8 and 8.9 illustrate this point.

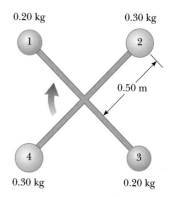

EXAMPLE 8.8 The Baton Twirler

In an effort to be the star of the half-time show, a majorette twirls a highly unusual baton made up of four masses fastened to the ends of light rods (Fig. 8.18). Each rod is 1.0 m long. Find the moment of inertia of the system about an axis perpendicular to the page and passing through the point where the rods cross.

Solution Applying Equation 8.8, we get

$$I = \Sigma mr^2 = m_1 r_1{}^2 + m_2 r_2{}^2 + m_3 r_3{}^2 + m_4 r_4{}^2$$

$$= (0.20 \text{ kg})(0.50 \text{ m})^2 + (0.30 \text{ kg})(0.50 \text{ m})^2 + (0.20 \text{ kg})(0.50 \text{ m})^2$$

$$+ (0.30 \text{ kg})(0.50 \text{ m})^2$$

$$= \boxed{0.25 \text{ kg} \cdot \text{m}^2}$$

Figure 8.18 (Example 8.8) Four masses connected to light rods rotating in the plane of the page.

EXAMPLE 8.9 The Baton Twirler: Second Act

Not satisfied with the crowd reaction from the baton twirling of Example 8.8, the majorette tries spinning her strange baton about the axis 00′, as shown in Figure 8.19. Calculate the moment of inertia about this axis.

Solution Again applying $I = \Sigma mr^2$, we have

$$I = (0.20 \text{ kg})(0)^2 + (0.30 \text{ kg})(0.50 \text{ m})^2 + (0.20 \text{ kg})(0)^2 + (0.30 \text{ kg})(0.50 \text{ m})^2$$

$$= \boxed{0.15 \text{ kg} \cdot \text{m}^2}$$

Calculation of Moments of Inertia for Extended Objects

The method used for calculating moments of inertia in Examples 8.8 and 8.9 is simple enough when you have only a few small masses rotating about an axis. The situation becomes much more complex when the object is an extended mass, such as a sphere, a cylinder, or a cone. One type of extended object that is amenable to a simple solution is a hoop rotating about an axis perpendicular to its plane and passing through its center, as shown in Figure 8.20. A bicycle tire, for example, would fit in this category.

To evaluate the moment of inertia of the hoop, we can still use the equation $I = \Sigma mr^2$ (Eq. 8.8) and imagine that the hoop is divided into a number of small segments having masses $m_1, m_2, m_3, \ldots$ as in Figure 8.20. This approach is just an extension of the baton problem described in the preceding examples, except that now we have a large number of small masses in rotation instead of only four.

We can express the sum for I as

$$I = \Sigma mr^2 = m_1 r_1{}^2 + m_2 r_2{}^2 + m_3 r_3{}^2 + \ldots$$

Figure 8.19 (Example 8.9) A double baton rotating about the axis *OO′*.

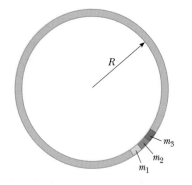

Figure 8.20 A uniform hoop can be divided into a large number of small segments that are equidistant from the center of the hoop.

All of the segments around the hoop are at the *same distance, R,* from the axis of rotation; thus, we can drop the subscripts on the distances and factor out the common factor R^2:

$$I = (m_1 + m_2 + m_3 + \ldots)R^2$$

We know, however, that the sum of the masses of all the segments must equal the total mass of the hoop, M:

$$M = m_1 + m_2 + m_3 + \ldots$$

and so we can express I as

$$I = MR^2 \qquad\qquad [8.10]$$

This expression can be used for the moment of inertia of any ring-shaped object rotating about an axis through its center and perpendicular to its plane. Note that the result is strictly valid only if the thickness of the ring is small relative to its inner radius.

The hoop we selected as an example is unique in that we were able to find an expression for its moment of inertia by using only simple algebra. Unfortunately, most extended objects are more difficult to work with, and the methods of integral calculus are required. Such methods are beyond the scope of this text. The moments of inertia for some common shapes are given without proof in Table 8.1.

TABLE 8.1   Moments of Inertia for Various Rigid Bodies of Uniform Composition

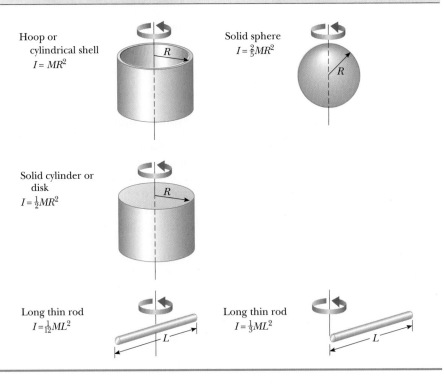

Hoop or cylindrical shell
$I = MR^2$

Solid sphere
$I = \frac{2}{5}MR^2$

Solid cylinder or disk
$I = \frac{1}{2}MR^2$

Long thin rod
$I = \frac{1}{12}ML^2$

Long thin rod
$I = \frac{1}{3}ML^2$

When the need arises, you can use this table to determine the moment of inertia of a body having any one of the listed shapes.

EXAMPLE 8.10 Warming Up

A baseball player loosening up his arm before a game tosses a 0.150-kg baseball using only the rotation of his forearm to accelerate the ball (Fig. 8.21). The ball starts at rest and is released with a speed of 30.0 m/s in 0.300 s.
(a) Find the constant angular acceleration of the arm and ball.

Solution During its acceleration, the ball moves through an arc of a circle with a radius of 0.350 m. We can determine the angular acceleration using $\omega = \omega_0 + \alpha t$. Because $\omega_0 = 0$, however, $\omega = \alpha t$, or

$$\alpha = \frac{\omega}{t}$$

We also know that $v = r\omega$, and so we get

$$\alpha = \frac{\omega}{t} = \frac{v}{rt} = \frac{30.0 \text{ m/s}}{(0.350 \text{ m})(0.300 \text{ s})} = \boxed{286 \text{ rad/s}^2}$$

(b) Find the torque exerted on the ball to give it this angular acceleration.

Solution The moment of inertia of the ball about an axis that passes through the elbow, perpendicularly to the arm, is

$$I = mr^2 = (0.150 \text{ kg})(0.350 \text{ m})^2 = 1.84 \times 10^{-2} \text{ kg} \cdot \text{m}^2$$

Thus, the required torque is

$$\tau = I\alpha = (1.84 \times 10^{-2} \text{ kg} \cdot \text{m}^2)(286 \text{ rad/s}^2) = \boxed{5.26 \text{ N} \cdot \text{m}}$$

Figure 8.21 (Example 8.10) A ball being tossed by a pitcher. The forearm is being used to accelerate the ball.

35.0 cm

EXAMPLE 8.11 The Falling Bucket

A solid, frictionless cylindrical pulley of mass $M = 3.00$ kg and radius $R = 0.400$ m is used to draw water from a well (Fig. 8.22a). A bucket of mass $m = 2.00$ kg is attached to a cord that is wrapped around the cylinder. If the bucket starts from rest at the top of the well and falls for 3.00 s before hitting the water, how far does it fall?

Reasoning Figure 8.22b shows the two forces on the bucket as it falls: **T** is the tension in the cord, and $m\mathbf{g}$ is the force of gravity acting on the bucket. We shall choose downward as the positive direction and write Newton's second law for the bucket as

$$mg - T = ma$$

When the given quantities are substituted into this equation, we have

$$(1) \qquad (2.00 \text{ kg})(9.80 \text{ m/s}^2) - T = (2.00 \text{ kg})a$$

With one equation and two unknowns, we must develop an additional equation to complete the problem. To obtain this second equation, let us consider the cylinder's rotational motion. Equation 8.9 applied to the cylinder gives the necessary expression:

$$\tau = I\alpha = \tfrac{1}{2}MR^2\alpha$$

(a)

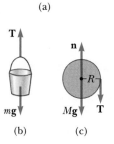

(b)   (c)

Figure 8.22 (Example 8.11) (a) A water bucket attached to a rope passing over a frictionless pulley. (b) A free-body diagram for the bucket. (c) The tension produces a torque on the cylinder about its axis of rotation.

Figure 8.22c shows that the only force producing a torque on the cylinder as it rotates about an axis through its center is **T,** the force due to the tension in the cord. Actually, two other forces act on the cylinder—the force of gravity and the upward force of the axle; but we do not have to consider them here because the lever arm of each about the axis of rotation is zero. Thus, we have

$$T(0.400 \text{ m}) = \tfrac{1}{2}(3.00 \text{ kg})(0.400 \text{ m})^2(\alpha)$$

(2) $T = (0.600 \text{ kg} \cdot \text{m})\alpha$

At this point, it is important to recognize that the downward acceleration of the bucket is equal to the tangential acceleration of a point on the rim of the cylinder. Therefore, the angular acceleration of the cylinder and the linear acceleration of the bucket are related by $a_t = r\alpha$. When this relation is used in (2), we get

(3) $T = (1.50 \text{ kg})a_t$

Solution Equations (1) and (3) can now be solved simultaneously to find a_t and T. This procedure gives

$$a_t = 5.60 \text{ m/s}^2 \qquad T = 8.40 \text{ N}$$

Finally, we turn to the equations for motion with constant linear acceleration to find the distance, d, that the bucket falls in 3.00 s. Because $v_0 = 0$, we get

$$d = v_0 t + \tfrac{1}{2}at^2 = \tfrac{1}{2}(5.60 \text{ m/s}^2)(3.00 \text{ s})^2 = \boxed{25.2 \text{ m}}$$

8.6 ROTATIONAL KINETIC ENERGY

In Chapter 5 we defined the kinetic energy of a particle moving through space with a speed v as the quantity $\frac{1}{2}mv^2$. Analogously, **a body rotating about some axis with an angular speed ω has rotational kinetic energy given by $\frac{1}{2}I\omega^2$.** To prove this, consider a rigid plane body rotating about some axis perpendicular to its plane, as in Figure 8.23. The body consists of many small particles, each of mass m. All these particles rotate in circular paths about the axis. If r is the distance of one of the particles from the axis of rotation, the speed of this particle is $v = r\omega$. Because the *total* kinetic energy of the body is the sum of all the kinetic energies associated with all the particles making up the body, we have

$$KE_r = \Sigma(\tfrac{1}{2}mv^2) = \Sigma(\tfrac{1}{2}mr^2\omega^2) = \tfrac{1}{2}(\Sigma mr^2)\omega^2$$

$$KE_r = \tfrac{1}{2}I\omega^2 \qquad\qquad \text{[8.11]}$$

where $I = \Sigma mr^2$ is the moment of inertia of the body. Note that the ω^2 term is factored out because it is the same for every particle.

We found the energy concept to be extremely useful for describing the linear motion of a system. It can be equally useful for simplifying the analysis of rotational motion. Consider a system that is described by three types of energy: **gravitational potential energy, PE_g, translational kinetic energy, KE_t,** and **rotational kinetic energy, KE_r.** We must include all these forms of energy in our equation for conservation of mechanical energy.

$$(KE_t + KE_r + PE_g)_i = (KE_t + KE_r + PE_g)_f \qquad \text{[8.12]}$$

where i and f refer to initial and final values, respectively. This relation is *only* true if we ignore dissipative forces such as friction.

Figure 8.23 A rigid plane body rotating about the z axis with angular speed ω. The kinetic energy of a particle of mass m is $\frac{1}{2}mv^2$. The total kinetic energy of the body is $\frac{1}{2}I\omega^2$.

◀ Conservation of mechanical energy

Thinking Physics 2

Two children are playing on a hill, rolling automobile tires down the slope. One child claims that the tire will roll faster if one of them curls up inside the tire and rides down inside of it. The other child claims that this will cause the tire to roll more slowly. Which child is correct?

Explanation The first child is correct. The original gravitational potential energy of the object turns into kinetic energy of translation plus kinetic energy of rotation plus work against friction. Dividing the mass out of the work-energy theorem shows that the final energy is measured by the expression

$$v^2\left(1 + \frac{I}{mR^2}\right)$$

where I is the moment of inertia of the object of mass m, and R is the radius on which it rolls. To make its center-of-mass speed v large, the children want to channel more energy into translation and less into rotation. That is, they want to make the fraction I/mR^2 small. Filling in the hole in the tire has this effect. It decreases the average distance from the axis out to bits of matter.

Q U I C K L A B

Compare the motion of an empty soup can and a filled soup can down the same incline such as a tilted table. If they are released from rest at the same height on the incline, which one reaches the bottom first? Repeat your observations with different kinds of soup (tomato, chicken noodle, etc.) and a can of beans. Compare their motions, and try to explain your observations. Finally, compare the motion of a filled soup can with a tennis ball, and explain your results to a friend.

EXAMPLE 8.12 A Ball Rolling Down an Incline

A ball of mass M and radius R starts from rest at a height of 2.00 m and rolls down a 30.0° slope, as shown in Figure 8.24. What is the linear speed of the ball when it leaves the incline? Assume that the ball rolls without slipping.

Figure 8.24 (Example 8.12) A ball starts from rest at the top of an incline and rolls to the bottom without slipping.

Reasoning The initial energy of the ball is gravitational potential energy, and when the ball reaches the bottom of the ramp, this potential energy has been converted to translational and rotational kinetic energy. The conservation of mechanical energy equation becomes

$$(PE_g)_i = (KE_t + KE_r)_f$$

$$Mgh = \tfrac{1}{2}Mv^2 + \tfrac{1}{2}(\tfrac{2}{5}MR^2)\omega^2$$

where h is the distance through which the ball's center of gravity falls and where we have used $I = \tfrac{2}{5}MR^2$ from Table 8.1 as the moment of inertia of the ball. If the ball rolls without slipping, a point on its surface must have the same instantaneous speed as its center of gravity has relative to the incline.[1] Thus, we can relate the ball's linear speed to its rotational speed:

$$v = R\omega$$

This expression can be used to eliminate ω from the equation for conservation of mechanical energy:

$$Mgh = \tfrac{1}{2}Mv^2 + \tfrac{1}{5}Mv^2$$

Solution Solving for v gives

$$v = \sqrt{\frac{10gh}{7}} = \sqrt{\frac{10(9.80 \text{ m/s}^2)(2.00 \text{ m})}{7}} = 5.29 \text{ m/s}$$

Exercise Repeat this example for a solid cylinder of the same mass and radius as the ball and released from the same height. In a race between the two objects on the incline, which one would win?

Answer $v = \sqrt{4gh/3} = 5.11$ m/s; the ball would win.

8.7   ANGULAR MOMENTUM

In Figure 8.25, an object of mass m is positioned a distance of r away from a center of rotation. Under the action of a constant torque on the object, its angular speed will increase from the value ω_0 to the value ω in a time of Δt. Thus, we can write

$$\tau = I\alpha = I\left(\frac{\omega - \omega_0}{\Delta t}\right) = \frac{I\omega - I\omega_0}{\Delta t}$$

Figure 8.25 An object of mass m rotating in a circular path under the action of a constant torque.

If we define the product

Definition of angular momentum ▶

$$L \equiv I\omega \tag{8.13}$$

as the angular momentum of the object, then we can write

$$\tau = \frac{\text{change in angular momentum}}{\text{time interval}} = \frac{\Delta L}{\Delta t} \tag{8.14}$$

---

[1] Note that a point on the surface of the ball travels a distance of $2\pi R$ (relative to the center of the ball) during one rotation. If no slippage occurs, this is also the distance traveled by the center of gravity in the same time interval. Thus, the center of gravity has the same speed relative to the incline as a point on the surface of the ball has relative to the center.

Equation 8.14 is the rotational analog of Newton's second law, $F = \Delta p/\Delta t$, and states that **the torque acting on an object is equal to the time rate of change of the object's angular momentum.**

When the net external torque ($\Sigma\tau$) acting on the system is zero, we see from Equation 8.14 that $\Delta L/\Delta T = 0$. In this case, the rate of change of the system's angular momentum is zero. Therefore, **the product $I\omega$ remains constant in time.** That is, $L_i = L_f$, or

$$I_i\omega_i = I_f\omega_f \quad \text{if} \quad \Sigma\tau = 0 \qquad [8.15]$$

> The angular momentum of a system is conserved when the net external torque acting on the system is zero. That is, when $\Sigma\tau = 0$, the initial angular momentum equals the final angular momentum.

◄ Conservation of angular momentum

In Equation 8.15 we have a third conservation law to add to our list: **conservation of angular momentum.** We can now state that **the energy, linear momentum, and angular momentum of an isolated system all remain constant.**

There are many examples of conservation of angular momentum, some of which should be familiar to you. You may have observed a figure skater spinning in the finale of her act. The skater's angular speed increases when she pulls her hands and feet close to the trunk of her body, as in Figure 8.26. That is, $\omega_2 > \omega_1$.

APPLICATION

Figure Skating.

(a)   (b)

Figure 8.26 (a) The angular speed of this skater increases when she pulls her arms in close to her body, demonstrating that angular momentum is conserved. (b) Photograph of Nancy Kerrigan in her final spinning routine. *(Reuters/Corbis-Bettmann)*

As the acrobat somersaults through the air, his angular momentum remains constant. Note that the center of mass of the acrobat follows a parabolic trajectory. *(Globus Bros. Studios/The Stock Market)*

Neglecting friction between skater and ice, we see that there are no external torques on the skater. The moment of inertia of her body decreases as her hands and feet are brought in. The resulting change in angular speed is accounted for as follows. Because angular momentum must be conserved, the product $I\omega$ has to remain constant, and a decrease of the moment of inertia of the skater must be compensated by a corresponding increase in the angular speed.

Similarly, when a diver or acrobat wishes to make several somersaults, he pulls his hands and feet close to the trunk of his body in order to rotate at a greater angular speed. In this case, the external force due to gravity acts through the center of gravity and hence exerts no torque about the axis of rotation, and the angular momentum about the center of gravity is conserved. For example, when a diver wishes to double his angular speed, he must reduce his moment of inertia to half its initial value.

APPLICATION

Aerial Somersaults.

Derek Swinson, professor of physics at the University of New Mexico, demonstrating the "gyro-ski" technique. The skier initiates a turn by lifting the axle of the rotating bicycle wheel. The direction of the turn depends on whether the left or right hand is used to lift the axle from the horizontal. Ignoring friction and gravity, the angular momentum of the system (the skier and the bicycle wheel) is conserved. *(Courtesy of Derek Swinson)*

Applying Physics 3

When a star exhausts its fuel, it tends to collapse to its center due to the mutual gravitational attraction of the mass of its material. For some massive stars, the gravitational force can be so strong as to squeeze protons and electrons together to form neutrons. The result is a neutron star—a star composed entirely of neutrons. On gravitational collapse, the star's radius becomes very small, on the order of a few kilometers. In general, a star rotates before it exhausts its fuel. As its radius becomes very small, its moment of inertia decreases even more drastically from its precollapse value. (Remember that moment of inertia is proportional to the square of the radius.) Because the star's angular momentum is conserved, its angular speed increases to an enormous value. And because stars possess magnetic fields (see Chapters 19–21), a rapidly rotating neutron star sends out an electromagnetic beacon that pulses at a very high frequency—hence the name *pulsar*. For example, the pulsar at the center of the Crab Nebula (which resulted from a supernova observed in 1054 A.D.) has a period of 33 ms, corresponding to a rotational frequency of 30 s^{-1}. Imagine a star rotating 30 times every second!

APPLICATION

Pulsars.

Thinking Physics 3

A common classroom demonstration is to spin on a rotating stool with heavy weights in your outstretched hands, and bring the weights close to your body (Fig. 8.27). The result is an increase in the angular velocity, in agreement with the conservation of angular momentum. As discussed previously, ice skaters use this effect to produce high-speed spins when they bring their arms close to the rotation axis. Let us imagine we

(a)   (b)

Figure 8.27 (Thinking Physics 3) (a) This student is given an initial angular speed while holding two masses as shown. (b) When the masses are pulled in close to the body, the angular speed of the system increases.

perform such a demonstration, and moving the weights inward results in halving the moment of inertia, and, therefore, doubling the angular velocity. If we consider the rotational kinetic energy, we see that the energy is *doubled* in this situation. Thus, angular momentum is conserved, but kinetic energy is not. Where does this extra energy come from?

Explanation   As you spin with the weights in your hands, there is a tension force in your arms that causes the weights to move in a circular path. As you bring the weights in, this force from your arms does work on the weights, because it is moving them through a radial displacement. What's more, as the angular velocity increases in response to the weights moving inward, you must apply more and more force to keep the weights moving in the circular path of smaller radius. Thus, the work that you do in bringing the weights in increases as they are brought in. This work that you do with your arms appears as the increase in rotational kinetic energy of the system. You can perform this work due to the energy stored in your body from previous meals, so you are transforming potential energy stored in your body into rotational kinetic energy.

Problem-Solving Strategy

Rotational Motion

Keep the following facts and procedures in mind when solving rotational motion problems.

1. Very few new techniques must be learned in order to solve rotational motion problems. For example, problems involving the equation $\Sigma\tau = I\alpha$ are very similar to those encountered in Newton's second law problems, $\Sigma F = ma$. Note the correspondences between linear and rotational quantities in that F is replaced by τ, m by I, and a by α.
2. Other analogs between rotational quantities and linear quantities include the replacement of x by θ and v by ω. These are helpful as memory devices for such rotational motion quantities as rotational kinetic energy $KE_r = \frac{1}{2}I\omega^2$, and angular momentum, $L = I\omega$.
3. With the analogs mentioned in the previous step, techniques for conservation of energy are the same as those examined in Chapter 5 except for the fact that the object's rotational kinetic energy must be included in the expression for the conservation of energy.
4. Likewise, the techniques for solving problems in conservation of angular momentum are essentially the same as those for solving problems in conservation of linear momentum, except that total initial angular momentum is equated to total final angular momentum as $I_i\omega_i = I_f\omega_f$.

EXAMPLE 8.13   The Spinning Stool

A student sits on a pivoted stool while holding a pair of masses (see Fig. 8.27). The stool is free to rotate about a vertical axis with negligible friction. The moment of inertia of student, masses, and stool is 2.25 kg·m². The student is set in rotation with an initial angular speed of 5.00 rad/s, with masses outstretched. As he rotates, he pulls the masses inward so that the new moment of inertia of the system (student, masses, and stool) becomes 1.80 kg·m². What is the new angular speed of the system?

Reasoning We shall apply the principle of conservation of angular momentum to find the new angular speed. The initial angular momentum, $I_i\omega_i$, will be equated to the final angular momentum, $I_f\omega_f$.

Solution The initial angular momentum of the system is

$$L_i = I_i\omega_i = (2.25 \text{ kg}\cdot\text{m}^2)(5.00 \text{ rad/s}) = 11.3 \text{ kg}\cdot\text{m}^2/\text{s}$$

When the masses are pulled in, they are closer to the axis of rotation, and as a result the moment of inertia of the system is reduced. The new angular momentum is

$$L_f = I_f\omega_f = (1.80 \text{ kg}\cdot\text{m}^2)\omega_f$$

Because the net external torque on the system is zero, angular momentum is conserved. Thus, we find that

$$(11.3 \text{ kg}\cdot\text{m}^2/\text{s}) = (1.80 \text{ kg}\cdot\text{m}^2)\omega_f$$

$$\omega_f = 6.28 \text{ rad/s}$$

EXAMPLE 8.14 The Merry-Go-Round

A student stands at the edge of a circular platform that rotates in a horizontal plane about a frictionless vertical axle (Fig. 8.28). The platform has a mass of $M = 100$ kg and a radius of $R = 2.00$ m. The student, whose mass is $m = 60.0$ kg, walks slowly from the rim of the disk toward the center. If the angular speed of the system is 2.00 rad/s when the student is at the rim, calculate the angular speed when the student reaches a point 0.500 m from the center.

Figure 8.28 (Example 8.14) As this student walks toward the center of the rotating platform, the angular speed of the system increases because the angular momentum of the system (student + platform) must remain constant.

Reasoning We shall use the principle of conservation of angular momentum. The initial angular momentum of the system is the sum of the angular momentum of the platform plus that of the student when he is at the rim of the merry-go-round. The final angular momentum is the sum of the angular momentum of the platform plus that of the student when he is 0.500 m from the center.

Solution   The moment of inertia of the platform, I_p, is

$$I_p = \tfrac{1}{2}MR^2 = \tfrac{1}{2}(100 \text{ kg})(2.00 \text{ m})^2 = 200 \text{ kg} \cdot \text{m}^2$$

Treating the student as a point mass, his initial moment of inertia is

$$I_s = mR^2 = (60.0 \text{ kg})(2.00 \text{ m})^2 = 240 \text{ kg} \cdot \text{m}^2$$

Thus, the initial angular momentum of the platform plus student is

$$L_i = (I_p + I_s)(\omega_i) = (200 \text{ kg} \cdot \text{m}^2 + 240 \text{ kg} \cdot \text{m}^2)(2.00 \text{ rad/s}) = 880 \text{ kg} \cdot \text{m}^2/\text{s}$$

When the student has walked to the position 0.500 m from the center, his moment of inertia is

$$I_s' = mr_f^2 = (60.0 \text{ kg})(0.500 \text{ m})^2 = 15.0 \text{ kg} \cdot \text{m}^2$$

No change occurs in the moment of inertia of the platform. Because there are no external torques on the *system* (student plus platform) about the axis of rotation, we can apply the law of conservation of angular momentum:

$$L_i = L_f$$

$$880 \text{ kg} \cdot \text{m}^2/\text{s} = 200\omega_f + 15.0\omega_f = 215\omega_f$$

$$\omega_f = \boxed{4.09 \text{ rad/s}}$$

Exercise   Calculate the change in kinetic energy of the system (student plus platform) for this situation. What accounts for this change in energy?

Answer $KE_f - KE_i = 918$ J. The student must perform positive work in order to walk toward the center of the platform.

SUMMARY

The tendency of a force to rotate an object about some axis is measured by a quantity called **torque, τ. The magnitude of the torque is given by**

$$\tau = Fd \qquad \text{[8.1]}$$

In this equation, d is the **lever arm** — the perpendicular distance from the axis of rotation to a line drawn along the direction of the force. The sign of the torque is negative if the turning tendency of the corresponding force is clockwise and positive if the turning tendency is counterclockwise.

An object is in **equilibrium** when the following conditions are satisfied: (1) the net external force must be zero, and (2) the net external torque must be zero about any origin. That is,

$$\Sigma\mathbf{F} = 0 \qquad \Sigma\tau = 0$$

The first equation is called the **first condition for equilibrium.** When this equation holds, an object is either at rest or moving with a constant velocity. The second equation is called the **second condition for equilibrium.** When it holds, an object is said to be in rotational equilibrium.

The **moment of inertia** of a group of particles is

$$I \equiv \Sigma mr^2 \qquad \text{[8.8]}$$

If a rigid body free to rotate about a fixed axis has a net external torque acting on it, the body undergoes an angular acceleration, α, where

$$\Sigma\tau = I\alpha \qquad\qquad [8.9]$$

If a rigid object rotates about a fixed axis with angular speed ω, its **rotational kinetic energy** is

$$KE_r \equiv \tfrac{1}{2}I\omega^2 \qquad\qquad [8.11]$$

where I is the moment of inertia about the axis of rotation.

The **angular momentum** of a rotating object is

$$L \equiv I\omega \qquad\qquad [8.13]$$

If the net external torque acting on a system is zero, the total angular momentum of the system is constant. Applying the law of conservation of angular momentum to an object whose moment of inertia changes with time gives

$$I_i\omega_i = I_f\,\omega_f \qquad\qquad [8.15]$$

MULTIPLE-CHOICE QUESTIONS

1. A constant net torque is applied to an object. One of the following will definitely *not* be a constant. It is the object's
   (a) angular acceleration
   (b) angular velocity
   (c) moment of inertia
   (d) center of mass

2. A rod 7.0 m long is pivoted at a point 2.0 m from the left end. A downward force of 50 N acts at the left end, and a downward force of 200 N acts at the right end. At what distance from the pivot can a third force of 300 N acting upward be placed to produce rotational equilibrium? Neglect the weight of the rod.
   (a) 1.0 m (b) 2.0 m (c) 3.0 m (d) 4.0 m
   (e) none of the above answers is correct

3. A horizontal disk with moment of inertia I_1 rotates with angular velocity ω_0 about a vertical frictionless axle. A second horizontal disk, with moment of inertia I_2 and

   initially not rotating, drops onto the first. Because the surfaces are rough, the two eventually reach the same angular velocity ω. The ratio ω/ω_0 is
   (a) I_1/I_2 (b) I_2/I_1 (c) $I_1/(I_1 + I_2)$ (d) $I_2/(I_1 + I_2)$

4. An 80.0 kg man is one fourth of the way up a 10.0 m ladder that is resting against a smooth, frictionless wall. If the ladder has a mass of 20.0 kg and makes an angle of 60.0° with the ground, find the force of friction of the ground on the foot of the ladder.
   (a) 784 N (b) 196 N (c) 50 N (d) 170 N

5. What must be the angular velocity of a solid cylinder rolling on the ground at the bottom of a hill so that it will be able to roll to the top of the hill if the hill is 10.0 m long and 3.00 m high? The mass of the cylinder is 2.00 kg and its radius is 0.400 m.
   (a) 15.7 rad/s (b) 27.1 rad/s (c) 19.2 rad/s
   (d) 28.6 rad/s

CONCEPTUAL QUESTIONS

1. Both torque and work are products of force and distance. How are they different? Do they have the same units?

2. Is it possible to calculate the torque acting on a rigid body without specifying the origin? Is the torque independent of the location of the origin?

3. Suppose a pencil is balanced on its end on a perfectly

   frictionless table. If it falls over, what is the path followed by the center of mass of the pencil?

4. When a high diver wants to turn a flip in midair, she will often draw her legs up against her chest. Why does this make her rotate faster? What should she do when she wants to come out of her flip?

5. In some motorcycle races, the riders drive over small hills, and the motorcycle becomes airborne for a short time. If the motorcycle racer keeps the throttle open while leaving the hill and going into the air, the motorcycle tends to nose upward. Why does this happen?

6. Why does holding a long pole help a tightrope walker stay balanced?

7. Suppose you are designing a car for a coasting race — the cars in these races have no engines; they simply coast downhill. Do you want large wheels or small wheels? Do you want solid, disk-like wheels or hoop-like wheels? Should the wheels be heavy or light?

8. Three homogeneous rigid bodies — a solid sphere, a solid cylinder, and a hollow cylinder — are placed at the top of an incline (Fig. Q8.8). If all are released from rest at the same elevation and roll without slipping, which reaches the bottom first? Which reaches last? You should try this at home and note that the result is *independent* of the masses and radii.

Figure Q8.8   Which object wins the race?

9. Stars originate as large bodies of slowly rotating gas. Because of gravity, these clumps of gas slowly decrease in size. What happens to the angular speed of a star as it shrinks? Explain.

10. Two solid spheres are rolled down a hill — a large, massive sphere and a small sphere with low mass. Which one reaches the bottom of the hill first? Next, we roll a large, low-density sphere, and a small, high-density sphere, and both spheres have the same mass. Which one wins the race?

11. If global warming occurs over the next century, it is likely that the polar ice caps of the Earth will melt and the water will be distributed closer to the Equator. How would this change the moment of inertia of the Earth? Would the length of the day (one revolution) increase or decrease?

12. If angular momentum is conserved for a propeller-driven airplane, it seems that if the propeller is turning clockwise, the airplane should be turning counterclockwise. Why doesn't this occur?

13. In a tape recorder, the tape is pulled past the read-and-write heads at a constant speed by the drive mechanism. Consider the reel from which the tape is pulled. As the tape is pulled off it, the radius of the roll of remaining tape decreases. How does the torque on the reel change with time? How does the angular velocity of the reel change with time? If the tape mechanism is suddenly turned on so that the tape is quickly pulled with a large force, is the tape more likely to break when pulled from a nearly full reel or a nearly empty reel?

14. (a) Give an example in which the net force acting on an object is zero and yet the net torque is nonzero. (b) Give an example in which the net torque acting on an object is zero and yet the net force is nonzero.

15. A mouse is initially at rest on a horizontal turntable mounted on a frictionless vertical axle. If the mouse begins to walk clockwise around the perimeter, what happens to the turntable? Explain.

16. A cat usually lands on its feet regardless of the position from which it is dropped (though they can be hurt or even killed when they fail to land on their feet when dropped from excessive heights — so don't try it!). A slow-motion film of a cat falling (Fig. Q8.16) shows that the upper half of its body twists in one direction while the lower half twists in the opposite direction. Why does this type of rotation occur?

Figure Q8.16   A falling, twisting cat. (© *Gerard Lacz/NHPA*)

17. A ladder rests inclined against a wall. Would you feel safer climbing up the ladder if you were told that the floor is frictionless but the wall is rough or that the wall is frictionless but the floor is rough? Justify your answer.

PROBLEMS

1, 2, 3 = straightforward, intermediate, challenging   ☐ = full solution available in Study Guide/Student Solutions Manual
WEB = solution posted at **http://www.saunderscollege.com/physics/college**   🔬 = biomedical application   🖥 = Interactive Physics

Review Problem

The mass of the Earth, M_E, is 81 times the mass of the Moon, M_M, the radius of the Earth, R_E, is 6400 km, and the distance, d_c, between the centers of the Earth and the Moon is 3.84×10^5 km. **(a)** Find the location of the point at which the gravitational forces exerted on an object by the Earth and the Moon will cancel each other. **(b)** Where is the location of the center of gravity (more properly called *center of mass* in this case) of the Earth-Moon system? Is this the same as your answer to **(a)**?

Section 8.1   Torque

1. If the torque required to loosen a nut that is holding a flat tire in place on a car has a magnitude of 40.0 N·m, what *minimum* force must be exerted by the mechanic at the end of a 30.0-cm lug wrench to accomplish the task?

🔬 2. A steel band exerts a horizontal force of 80.0 N on a tooth at point *B* in Figure P8.2. What is the torque on the root of the tooth about point *A*?

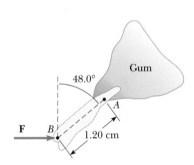

Figure P8.2

☐3. A simple pendulum consists of a 3.0-kg point mass hanging at the end of a 2.0-m-long light string that is connected to a pivot point. Calculate the magnitude of the torque (due to the force of gravity) about this pivot point when the string makes a 5.0° angle with the vertical.

4. A fishing pole is 2.00 m long and inclined to the horizontal at an angle of 20.0° (Fig. P8.4). What is the torque exerted by the fish about an axis perpendicular to the page and passing through the hand of the person holding the pole?

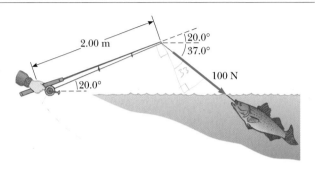

Figure P8.4

5. Calculate the net torque (magnitude and direction) on the beam in Figure P8.5 about **(a)** an axis through *O*, perpendicular to the page, and **(b)** an axis through *C*, perpendicular to the page.

Figure P8.5

Section 8.2   Torque and the Second Condition for Equilibrium

Section 8.3   The Center of Gravity

Section 8.4   Examples of Objects in Equilibrium

6. Four objects are situated along the *y* axis as follows: a 2.00-kg object is at +3.00 m, a 3.00-kg object is at +2.50 m, a 2.50-kg object is at the origin, and a 4.00-kg object is at −0.500 m. Where is the center of gravity of this system?

☐7. A water molecule consists of an oxygen atom with two hydrogen atoms bound to it, as shown in Figure P8.7. The bonds are 0.100 nm in length and the angle between the two bonds is 106°. Use the *x*-*y* axis shown and determine the location of the center of gravity of the molecule. Consider the mass of an oxygen atom to be 16 times the mass of a hydrogen atom.

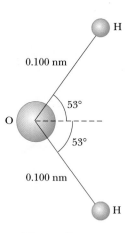

Figure P8.7

8. Consider the following mass distribution where the x-y coordinates are given in meters: 5 kg at $(0, 0)$ m, 3 kg at $(0, 4)$ m, and 4 kg at $(3, 0)$ m. Where should a fourth mass of 8 kg be placed so the center of gravity of the four-mass arrangement will be at $(0, 0)$ m?

9. Find the x- and y-coordinates of the center of gravity of a 4.00 ft by 8.00 ft uniform sheet of plywood with the upper right quadrant removed as shown in Figure P8.9.

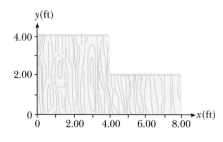

Figure P8.9

10. A meter stick is found to balance at the 49.7 cm mark when placed on a fulcrum. When a 50.0-g mass is attached at the 10.0-cm mark, the fulcrum must be moved to the 39.2-cm mark for balance. What is the mass of the meter stick?

11. A cook holds a 2.00-kg carton of milk at arm's length (Fig. P8.11). What force $\mathbf{F}_B$ must be exerted by the biceps muscle? (Ignore the weight of the forearm.)

12. A window washer is standing on a scaffold supported by a vertical rope at each end. The scaffold weighs 200 N and is 3.00 m long. What is the tension in each rope when the 700-N worker stands 1.00 m from one end?

Figure P8.11

13. The chewing muscle, the masseter, is one of the strongest in the human body. It is attached to the mandible (lower jawbone), as shown in Figure P8.13a. The jawbone is pivoted about a socket just in front of the auditory canal. The forces acting on the jawbone are equivalent to those acting on the curved bar in Figure P8.13b: **C** is the force exerted against the jawbone by the food being chewed, **T** is the tension in the masseter, and **R** is the force exerted on the mandible by the socket. Find **T** and **R** if you bite down on a piece of steak with a force of 50.0 N.

(a)

(b)

Figure P8.13

14. A hungry 700-N bear walks out on a beam in an attempt to retrieve some goodies hanging at the end (Fig. P8.14). The beam is uniform, weighs 200 N, and is 6.00 m long; the goodies weigh 80.0 N. (a) Draw a free-

Figure P8.14

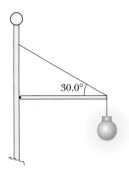

Figure P8.16

body diagram for the beam. **(b)** When the bear is at $x = 1.00$ m, find the tension in the wire and the components of the reaction force at the hinge. **(c)** If the wire can withstand a maximum tension of 900 N, what is the maximum distance the bear can walk before the wire breaks?

15. A uniform semicircular sign 1.00 m in diameter and of weight W is supported by two wires, as shown in Figure P8.15. What is the tension in each of the wires supporting the sign?

Figure P8.15

16. A 20.0-kg floodlight in a park is supported at the end of a horizontal beam of negligible mass that is hinged to a pole, as shown in Figure P8.16. A cable at an angle of 30.0° with the beam helps to support the light. Find **(a)** the tension in the cable and **(b)** the horizontal and vertical forces exerted on the beam by the pole.

17. A 500-N uniform rectangular sign 4.00 m wide and 3.00 m high is suspended from a horizontal, 6.00-m-long, uniform, 100-N rod, as indicated in Figure P8.17. The left end the rod is supported by a hinge and the right end is supported by a thin cable making a 30.0° angle with the vertical. **(a)** Find the tension, T, in the cable. **(b)** Find the horizontal and vertical components of force exerted on the left end of the rod by the hinge.

Figure P8.17

18. The arm in Figure P8.18 weighs 41.5 N. The force of gravity acting on the arm acts through point A. Determine the magnitudes of the tension force $\mathbf{F}_t$ in the deltoid muscle and the force $\mathbf{F}_s$ of the shoulder on the humerus (upper-arm bone) to hold the arm in the position shown.

Figure P8.18

19. A uniform plank of length 2.00 m and mass 30.0 kg is supported by three ropes, as indicated by the blue vectors in Figure P8.19. Find the tension in each rope when a 700-N person is 0.500 m from the left end.

Figure P8.19

20. A 15.0-m, 500-N uniform ladder rests against a frictionless wall, making an angle of 60.0° with the horizontal. (a) Find the horizontal and vertical forces exerted on the base of the ladder by the Earth when an 800-N fire fighter is 4.00 m from the bottom. (b) If the ladder is just on the verge of slipping when the fire fighter is 9.00 m up, what is the coefficient of static friction between ladder and ground?

WEB 21. An 8.0-m, 200-N uniform ladder rests against a smooth wall. The coefficient of static friction between the ladder and the ground is 0.60, and the ladder makes a 50.0° angle with the ground. How far up the ladder can an 800-N person climb before the ladder begins to slip?

22. A 1200-N uniform boom is supported by a cable perpendicular to the boom, as in Figure P8.22. The boom is hinged at the bottom, and a 2000-N weight hangs from its top. Find the tension in the supporting cable and the components of the reaction force exerted on the boom by the hinge.

Figure P8.22

23. The large quadriceps muscle in the upper leg terminates at its lower end in a tendon attached to the upper end of the tibia (Fig. P8.23a). The forces on the lower leg when the leg is extended are modeled as in Figure P8.23b, where **T** is the tension in the tendon, **C** is the force of gravity acting on the lower leg, and **F** is the force of gravity acting on the foot. Find **T** when the tendon is at an angle of 25.0° with the tibia, assuming that $C = 30.0$ N, $F = 12.5$ N, and the leg is extended at an angle of 40.0° with the vertical ($\theta = 40.0°$). Assume that the center of gravity of the lower leg is at its center, and that the tendon attaches to the lower leg at a point one fifth of the way down the leg.

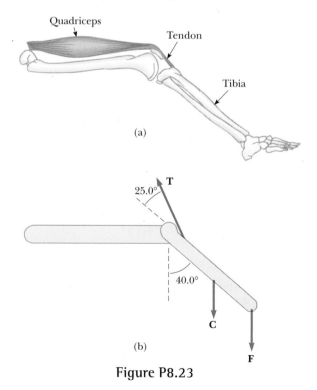

Figure P8.23

24. One end of a uniform 4.0-m long rod of weight w is supported by a cable. The other end rests against the wall, where it is held by friction (see Fig. P8.24). The coefficient of static friction between the wall and the rod is $\mu_s = 0.50$. Determine the minimum distance, x, from point A at which an additional weight w (same as the weight of the rod) can be hung without causing the rod to slip at point A.

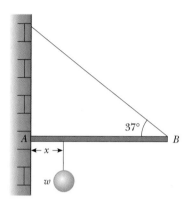

Figure P8.24

Section 8.5 Relationship Between Torque and Angular Acceleration

25. Four masses are held in position at the corners of a rectangle by light rods, as shown in Figure P8.25. Find the moment of inertia of the system about (a) the x axis, (b) the y axis, and (c) an axis through O and perpendicular to the page.

26. If the system shown in Figure P8.25 is set in rotation about each of the axes mentioned in Problem 25, find the torque that will produce an angular acceleration of 1.50 rad/s^2 in each case.

27. A potter's wheel having a radius of 0.50 m and a moment of inertia of 12.5 kg·m^2 is rotating freely at 50 rev/min. The potter can stop the wheel in 6.0 s by pressing a wet rag against the rim and exerting a radially inward force of 70 N. Find the effective coefficient of kinetic friction between the wheel and the wet rag.

28. A cylindrical fishing reel has a moment of inertia of $I = 6.8 \times 10^{-4}$ kg·m^2 and a radius of 4.0 cm. A friction clutch in the reel exerts a restraining torque of 1.3 N·m if a fish pulls on the line. The fisherman gets a bite, and the reel begins to spin with an angular acceleration of 66 rad/s^2. (a) What is the force of the fish on the line? (b) How much line unwinds in 0.50 s?

29. A 150-kg merry-go-round in the shape of a uniform, solid, horizontal disk of radius 1.50 m is set in motion by wrapping a rope about the rim of the disk and pulling on the rope. What constant force would have to be exerted on the rope to bring the merry-go-round from rest to an angular speed of 0.500 rev/s in 2.00 s?

30. A cable passes over a pulley. Because of friction, the tension in the cable is not the same on opposite sides of the pulley. The force on one side is 120 N, and the force on the other side is 100 N. Assuming that the pulley is a uniform disk of mass 2.1 kg and radius 0.81 m, determine its angular acceleration.

WEB **31.** A cylindrical 5.00-kg pulley with a radius of 0.600 m is used to lower a 3.00-kg bucket into a well (Fig. P8.31). The bucket starts from rest and falls for 4.00 s. (a) What is the linear acceleration of the falling bucket? (b) How far does it drop? (c) What is the angular acceleration of the cylinder?

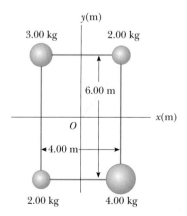

Figure P8.25 (Problems 25 and 26.)

5.00 kg

0.600 m

3.00 kg

Figure P8.31

32. An airliner lands with a speed of 50.0 m/s. Each wheel of the plane has a radius of 1.25 m and a moment of inertia of 110 kg·m². At touchdown the wheels begin to spin under the action of friction. Each wheel supports a weight of 1.40×10^4 N, and the wheels attain the angular speed of rolling without slipping in 0.480 s. What is the coefficient of kinetic friction between the wheels and the runway? Assume that the speed of the plane is constant.

33. A light string is wrapped around a solid cylindrical spool of radius 0.500 m and mass 0.500 kg. A 5.00-kg mass is hung from the string, causing the spool to rotate and the string to unwind (Fig. P8.33). Assume that the system starts from rest and no slippage takes place between the string and the spool. By direct application of Newton's second law, determine the angular speed of the spool after the mass has dropped 4.00 m.

Figure P8.33   (Problems 33 and 41.)

34. A string is wrapped around a uniform cylinder of mass m and radius r (Fig. P8.34). One end of the string is attached to the ceiling, and the cylinder is allowed to fall from rest. **(a)** Write Newton's second law for the cylinder. **(b)** Find the net torque about the center of the cylinder and equate this to $I\alpha$. **(c)** Use the equations found in (a) and (b) along with $a = r\alpha$ to show that the linear acceleration of the cylinder is given by $a = (2/3)g$.

Figure P8.34

Section 8.6   Rotational Kinetic Energy

35. A 10.0-kg cylinder rolls without slipping on a rough surface. At the instant its center of mass has a speed of 10.0 m/s, determine **(a)** the translational kinetic energy of its center of mass, **(b)** the rotational kinetic energy about its center of mass, and **(c)** its total kinetic energy.

36. The net work done in accelerating a propeller from rest to an angular speed of 200 rad/s is 3000 J. What is the moment of inertia of the propeller?

37. A horizontal 800-N merry-go-round of radius 1.50 m is started from rest by a constant horizontal force of 50.0 N applied tangentially to the merry-go-round. Find the kinetic energy of the merry-go-round after 3.00 s. (Assume it is a solid cylinder.)

38. A car is designed to get its energy from a rotating flywheel with a radius of 2.00 m and a mass of 500 kg. Before a trip, the flywheel is attached to an electric motor, which brings the flywheel's rotational speed up to 5000 rev/min. **(a)** Find the kinetic energy stored in the flywheel. **(b)** If the flywheel is to supply energy to the car as would a 10.0-hp motor, find the length of time the car could run before the flywheel would have to be brought back up to speed.

39. The top in Figure P8.39 has a moment of inertia of 4.00×10^{-4} kg·m² and is initially at rest. It is free to rotate about a stationary axis, AA'. A string, wrapped around a peg along the axis of the top, is pulled in such a manner as to maintain a constant tension of 5.57 N in the string. If the string does not slip while wound around the peg, what is the angular speed of the top after 80.0 cm of string has been pulled off the peg? (*Hint:* Consider the work done.)

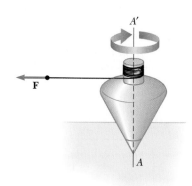

Figure P8.39

40. A 240-N sphere 0.20 m in radius rolls, without slipping, 6.0 m down a ramp that is inclined at 37° with the horizontal. What is the angular speed of the sphere at the bottom of the hill if it starts from rest?

WEB **41.** Use conservation of energy to determine the angular speed of the spool shown in Figure P8.33 after the mass *m* has fallen 4.00 m, starting from rest. The light string attached to this mass is wrapped around the spool and does not slip as it unwinds. Assume that the spool is a solid cylinder of radius 0.500 m and mass 0.500 kg, and that *m* = 5.00 kg.

Section 8.7 Angular Momentum

42. (a) Calculate the angular momentum of the Earth that arises from its spinning motion on its axis and (b) the angular momentum of the Earth that arises from its orbital motion about the Sun.

43. A light rigid rod 1.00 m in length rotates about an axis perpendicular to its length and through its center as shown in Figure P8.43. Two particles of masses 4.00 kg and 3.00 kg are connected to the ends of the rod. What is the angular momentum of the system if the speed of each particle is 5.00 m/s?

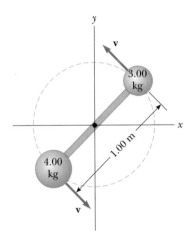

Figure P8.43

44. Halley's comet moves about the Sun in an elliptical orbit, with its closest approach to the Sun being 0.59 A.U. and its greatest distance being 35 A.U. (1 A.U. = Earth–Sun distance). If the comet's speed at closest approach is 54 km/s, what is its speed when it is farthest from the Sun? You may neglect any change in the comet's mass and assume that its angular momentum about the Sun is conserved.

45. The system of point masses shown in Figure P8.45 is rotating at an angular speed of 2.0 rev/s. The masses are connected by light, flexible spokes that can be lengthened or shortened. What is the new angular speed if the spokes are shortened to 0.50 m? (An effect similar to that illustrated in this problem occurred in the early stages of the formation of our Galaxy. As the massive cloud of dust and gas that was the source of the stars

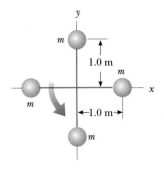

Figure P8.45

and planets contracted, an initially small rotation increased with time.)

46. A solid, horizontal cylinder of mass 10.0 kg and radius 1.00 m rotates with an angular speed of 7.00 rad/s about a fixed vertical axis through its center. A 0.250-kg piece of putty is dropped vertically onto the cylinder at a point 0.900 m from the center of rotation, and sticks to the cylinder. Determine the final angular speed of the system.

47. A student sits on a rotating stool holding two 3.0-kg masses. When his arms are extended horizontally, the masses are 1.0 m from the axis of rotation, and he rotates with an angular speed of 0.75 rad/s. The moment of inertia of the student plus stool is 3.0 kg·m² and is assumed to be constant. The student then pulls the masses horizontally to 0.30 m from the rotation axis. (a) Find the new angular speed of the student. (b) Find the kinetic energy of the student before and after the masses are pulled in.

48. The puck in Figure P8.48 has a mass of 0.120 kg. Its original distance from the center of rotation is 40.0 cm, and the puck is moving with a speed of 80.0 cm/s. The string is pulled downward 15.0 cm through the hole in the frictionless table. Determine the work done on the

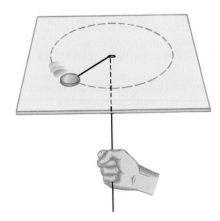

Figure P8.48

puck. (*Hint:* Consider the change of kinetic energy of the puck.)

49. A merry-go-round rotates at the rate of 0.20 rev/s with an 80-kg man standing at a point 2.0 m from the axis of rotation. **(a)** What is the new angular speed when the man walks to a point 1.0 m from the center? Assume that the merry-go-round is a solid 25-kg cylinder of radius 2.0 m **(b)** Calculate the change in kinetic energy due to this movement. How do you account for this change in kinetic energy?

ADDITIONAL PROBLEMS

50. A cylinder with moment of inertia I_1 rotates with angular velocity ω_0 about a frictionless vertical axle. A second cylinder, with moment of inertia I_2, initially not rotating, drops onto the first cylinder (Fig. P8.50). Because the surfaces are rough, the two eventually reach the same angular velocity, ω. **(a)** Calculate ω. **(b)** Show that kinetic energy is lost in this situation, and calculate the ratio of the final to the initial kinetic energy.

Figure P8.50

51. A 0.100-kg meter stick is supported at its 40.0-cm mark by a string attached to the ceiling. A 0.700-kg mass hangs vertically from the 5.00-cm mark. A mass, m, is attached somewhere on the meter stick to keep it horizontal and in rotational and translational equilibrium. If the tension in the string attached to the ceiling is 19.6 N, determine **(a)** the value of m and **(b)** its point of attachment on the stick.

52. Show that the kinetic energy of an object rotating about a fixed axis with angular momentum $L = I\omega$ can be written as $KE = L^2/2I$.

53. A person bends over and lifts a 200-N weight, as in Figure P8.53a, with his back horizontal. The muscle that attaches two thirds of the way up the spine maintains the position of the back; the angle between the spine and this muscle is 12°. Using the mechanical model in Figure P8.53b and taking the weight of the upper body to be 350 N, find the tension in the back muscle and the compressional force in the spine.

Figure P8.53

54. A 12.0-kg mass is attached to a cord that is wrapped around a wheel of radius $r = 10.0$ cm (Fig. P8.54). The acceleration of the mass down the frictionless incline is measured to be 2.00 m/s². Assuming the axle of the wheel to be frictionless, determine **(a)** the tension in the rope, **(b)** the moment of inertia of the wheel, and **(c)** the angular speed of the wheel 2.00 s after it begins rotating, starting from rest.

Figure P8.54

55. A uniform ladder of length L and weight W is leaning against a vertical wall. The coefficient of static friction between the ladder and the floor is the same as that between the ladder and the wall. If this coefficient of static friction is $\mu_s = 0.500$, determine the smallest angle the ladder can make with the floor without slipping.

56. A uniform 10.0-N picture frame is supported as shown in Figure P8.56. Find the tension in the cords and the magnitude of the horizontal force at P that are required to hold the frame in the position shown.

Figure P8.56

Figure P8.59

57. A solid 2.0-kg ball of radius 0.50 m starts at a height of 3.0 m above the surface of the Earth and *rolls* down a 20° slope. A solid disk and a ring start at the same time and the same height. The ring and disk each have the same mass and radius as the ball. Which of the three wins the race to the bottom if all roll without slipping?

58. A 40.0-kg child stands at one end of a 70.0-kg boat that is 4.00 m long (Fig. P8.58). The boat is initially 3.00 m from the pier. The child notices a turtle on a rock beyond the far end of the boat and proceeds to walk to that end to catch the turtle. **(a)** Neglecting friction between the boat and water, describe the motion of the system (child + boat). **(b)** Where will the child be relative to the pier when he reaches the far end of the boat? **(c)** Will he catch the turtle? (Assume that he can reach out 1.00 m from the end of the boat.)

Figure P8.58

59. The pulley in Figure P8.59 has a moment of inertia of 5.0 kg·m² and a radius of 0.50 m. The cord supporting the masses m_1 and m_2 does not slip, and the axle is frictionless. **(a)** Find the acceleration of each mass when $m_1 = 2.0$ kg and $m_2 = 5.0$ kg. **(b)** Find the tension in the cable supporting m_1 and the tension in the cable supporting m_2. (Note: They are different.)

60. A 4.00-kg mass is connected by a light cord to a 3.00-kg mass on a smooth surface (Fig. P8.60). The pulley rotates about a frictionless axle and has a moment of inertia of 0.500 kg·m² and a radius of 0.300 m. Assuming that the cord does not slip on the pulley, find **(a)** the acceleration of the two masses and **(b)** the tensions T_1 and T_2.

Figure P8.60

61. A model airplane whose mass is 0.750 kg is tethered by a wire so that it flies in a circle 30.0 m in radius. The airplane engine provides a net thrust of 0.800 N perpendicular to the tethering wire. **(a)** Find the torque the net thrust produces about the center of the circle. **(b)** Find the angular acceleration of the airplane when it is in level flight. **(c)** Find the linear acceleration of the airplane tangent to its flight path.

62. A puck of mass 80.0 g and radius 4.00 cm slides along an air table at a speed of 1.50 m/s, as shown in Figure P8.62a. It makes a glancing collision with a second puck

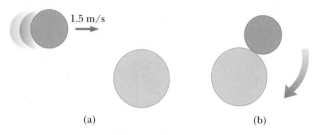

(a)   (b)

Figure P8.62

of radius 6.00 cm and mass 120 g (initially at rest) such that their rims just touch. The pucks stick together and spin after the collision (Fig. P8.62b). (a) What is the angular momentum of the system relative to the center of mass? (b) What is the angular velocity about the center of mass?

WEB 63. Two astronauts (Fig. P8.63), each having a mass of 75.0 kg, are connected by a 10.0-m rope of negligible mass. They are isolated in space, orbiting their center of mass at speeds of 5.00 m/s. Calculate (a) the magnitude of the angular momentum of the system by treating the astronauts as particles and (b) the rotational energy of the system. By pulling on the rope, the astronauts shorten the distance between them to 5.00 m. (c) What is the new angular momentum of the system? (d) What are their new speeds? (e) What is the new rotational energy of the system? (f) How much work is done by the astronauts in shortening the rope?

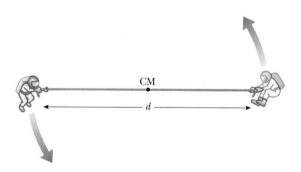

Figure P8.63   (Problems 63 and 64.)

64. Two astronauts (Fig. P8.63), each having a mass M, are connected by a rope of length d having negligible mass. They are isolated in space, orbiting their center of mass at speeds v. Calculate (a) the magnitude of the angular momentum of the system by treating the astronauts as particles and (b) the rotational energy of the system. By pulling on the rope, the astronauts shorten the distance between them to d/2. (c) What is the new angular momentum of the system? (d) What are their new speeds?

(e) What is the new rotational energy of the system?
(f) How much work is done by the astronauts in shortening the rope?

65. Two window washers Bob and Joe, are on a 3.00-m-long, 345-N scaffold supported by two cables attached to its ends. Bob, who weighs 750 N, stands 1.00 m from the left end, as shown in Figure P8.65. Two meters from the left end is the 500-N washing equipment. Joe is 0.500 m from the right end and weighs 1000 N. Given that the scaffold is in rotational and translational equilibrium, what are the forces on each cable?

Figure P8.65

66. When the motor in Figure P8.66 raises the 1000-kg mass, it produces a tension of 1.14×10^4 N in the cable on the right side of the pulley. The pulley has a mo-

1000 kg

Motor

Figure P8.66

ment of inertia of 79.8 kg·m² and a radius of 0.762 m. The cable rides over the pulley without slipping. Determine the acceleration of the 1000-kg mass. (*Hint:* Draw free-body diagrams of the mass and the pulley. Do not assume that the tension in the cable is the same on both sides of the pulley.)

67. A 60-kg woman stands at the rim of a horizontal turntable having a moment of inertia of 500 kg·m² and a radius of 2.0 m. The system is initially at rest, and the turntable is free to rotate about a frictionless vertical axle through its center. The woman then starts walking clockwise (looking downward) around the rim at a constant speed of 1.5 m/s relative to the Earth. (a) In what direction and with what angular speed does the turntable rotate? (b) How much work does the woman do to set the system in motion?

68. We have all complained that there aren't enough hours in a day. In an attempt to change that, suppose that all the people in the world lined up at the Equator and all started running east at 2.5 m/s relative to the surface of the Earth. By how much would the length of a day increase? (Assume that there are 5.5×10^9 people in the world with an average mass of 70 kg each and that the Earth is a solid homogeneous sphere. In addition, you may use the result $1/(1 - x) \approx 1 + x$ for x small.)

69. In a circus performance, a large 5.0-kg hoop of radius 3.0 m rolls without slipping. If the hoop is given an angular speed of 3.0 rad/s while rolling on the horizontal and allowed to roll up a ramp inclined at 20° with the horizontal, how far (measured along the incline) does the hoop roll?

70. A uniform, solid cylinder of mass M and radius R rotates on a frictionless horizontal axle (Fig. P8.70). Two equal masses hang from light cords wrapped around the cylinder. If the system is released from rest, find (a) the tension in each cord and (b) the acceleration of each mass after the masses have descended a distance of h.

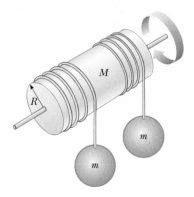

Figure P8.70

71. Figure P8.71 shows a vertical force applied tangentially to a uniform cylinder of weight w. The coefficient of static friction between the cylinder and all surfaces is 0.500. Find, in terms of w, the maximum force **F** that can be applied without causing the cylinder to rotate. (*Hint:* When the cylinder is on the verge of slipping, both friction forces are at their maximum values. Why?)

Figure P8.71

72. A common physics demonstration (Figure P8.72) consists of a ball resting at the end of a board of length ℓ that is elevated at an angle θ with the horizontal. A light cup is attached to the board at r_c so that it will catch the ball when the support stick is suddenly removed. (a) Show that the ball will lag behind the falling board when $\theta < 35.3°$, and (b) the ball will fall into the cup when the board is supported at this limiting angle and the cup is placed at

$$r_c = \frac{2\ell}{3 \cos \theta}$$

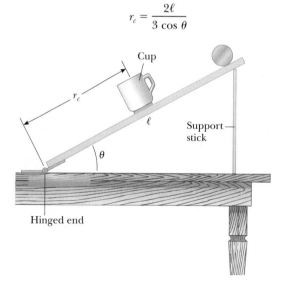

Figure P8.72

Solids and Fluids

▲ PHYSICS PUZZLER

A scuba diver plays with an octopus in Poor Knights Island, New Zealand. As the diver descends to greater depths, the water pressure increases above atmospheric pressure. Scuba divers have been able to swim at depths of more than 300 m. Can you estimate the pressure exerted on the diver at such great depths? How is the body able to withstand such pressure? Why is it necessary to breath in air from pressurized tanks? *(Darryl Torckler/Tony Stone Images)*

In this chapter we consider some properties of solids and fluids (both liquids and gases). We spend some time looking at properties that are peculiar to solids, but much of our emphasis is on the properties of fluids. We take this approach because an understanding of the behavior of fluids is fundamentally important to students in the life sciences. We open the fluids part of the chapter with a study of fluids at rest and finish with a discussion of the properties of fluids in motion. Additional topics discussed in this chapter include surface tension, viscosity, and transport phenomena.

9.1 STATES OF MATTER

Matter is normally classified as being in one of three states—solid, liquid, or gaseous. Often this classification system is extended to include a fourth state, referred to as a *plasma*. When matter is heated to high temperatures, many of the electrons surrounding each atom are freed from the nucleus. The resulting substance is a collection of free, electrically charged particles—negatively charged electrons and positively charged ions. Such a highly ionized substance containing equal amounts of positive and negative charges is a **plasma.** Plasmas exist inside stars, for example. If we were to take a grand tour of our Universe, we would find that there is far more matter in the plasma state than in the more familiar solid, liquid, and gaseous states because there are far more stars around than any other form of celestial matter. In this chapter, however, we ignore plasmas and concentrate on the more familiar solid, liquid, and gaseous forms that make up the environment on our planet.

Everyday experience tells us that a solid has definite volume and shape. A brick, for example, maintains its familiar shape and size day in and day out. We also know that a liquid has a definite volume but no definite shape. For instance, when you fill the tank on a lawn mower, the gasoline changes its shape from that of the original container to that of the tank on the mower. If there is a gallon of gasoline before you pour, however, there still is a gallon after. Finally, a gas has neither definite volume nor definite shape.

All matter consists of some distribution of atoms or molecules. The atoms in a solid are held, by forces that are mainly electrical, at specific positions with respect to one another and vibrate about these equilibrium positions. At low temperatures, however, the vibrating motion is slight and the atoms can be considered to be essentially fixed. As energy is added to the material, the amplitude of the vibrations increases. A vibrating atom can be viewed as being bound in its equilibrium position by springs attached to neighboring atoms. A collection of such atoms and imaginary springs is shown in Figure 9.1. If a solid is compressed by external forces, we can picture the forces as compressing these tiny internal springs. When the external forces are removed, the solid tends to return to its original shape and size. Consequently, a solid is said to have *elasticity*.

Solids can be classified as being either crystalline or amorphous. A **crystalline solid** is one in which the atoms have an ordered structure. For example, in the sodium chloride crystal (common table salt), sodium and chlorine atoms occupy alternate corners of a cube, as in Figure 9.2a. In an **amorphous solid,** such as glass, the atoms are arranged randomly, as in Figure 9.2b.

For any given substance, the liquid state exists at a higher temperature than the solid state. The intermolecular forces in a liquid are not strong enough to keep

Crystals of natural quartz (SiO_2), one of the most common minerals on Earth. Quartz crystals are used to make special lenses and prisms and in certain electronic applications. *(Charles Winters)*

Figure 9.1 A model of a portion of a solid. The atoms (spheres) are imagined as being attached to each other by springs, which represent the elastic nature of the interatomic forces. A solid consists of trillions of segments like this with springs connecting all of them.

Figure 9.2 (a) The NaCl structure, with the Na⁺ (red) and Cl⁻
(blue) ions at alternate corners of a
cube. (b) In an amorphous solid,
the atoms are arranged randomly.
(c) Erratic motion of a molecule in
a liquid.

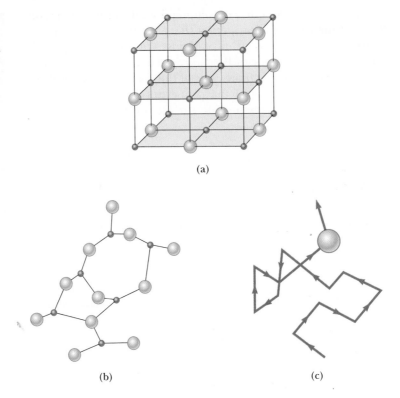

(a)

(b)   (c)

the molecules in fixed positions, and they wander through the liquid in a random
fashion (Fig. 9.2c). Solids and liquids have the following property in common:
When an attempt is made to compress them, strong repulsive atomic forces act
internally to resist compression.

In the gaseous state, the molecules are in constant random motion and exert
only weak forces on each other. The average separation distances between the molecules of a gas are quite large compared with the size of the molecules. Occasionally
the molecules collide with each other, but most of the time they move as nearly
free, noninteracting particles. We shall say more about gases in subsequent
chapters.

9.2 THE DEFORMATION OF SOLIDS

We usually think of a solid as an object having definite shape and volume. In our
study of mechanics, to keep things simple, we assumed that objects remain unde-
formed when external forces act on them. In reality, all objects are deformable.
That is, it is possible to change the shape or size of an object (or both) through
the application of external forces. When the forces are removed, the object tends
to return to its original shape and size, which means that the deformation exhibits
an elastic behavior.

The elastic properties of solids are discussed in terms of stress and strain. **Stress** is related to the force causing a deformation; **strain** is a measure of the degree of deformation. It is found that, for sufficiently small stresses, stress is proportional to strain, and the constant of proportionality depends on the material being deformed and on the nature of the deformation. We call this proportionality constant the **elastic modulus:**

$$\text{Elastic modulus} \equiv \frac{\text{stress}}{\text{strain}} \qquad \text{[9.1]}$$

Figure 9.3 A long bar clamped at one end is stretched by the amount ΔL under the action of a force, **F.**

Young's Modulus: Elasticity in Length

Consider a long bar of cross-sectional area A and length L_0, clamped at one end (Fig. 9.3). When an external force, **F,** is applied along the bar, perpendicularly to the cross-section, internal forces in the bar resist the distortion ("stretching") that **F** tends to produce, but the bar nevertheless attains an equilibrium in which (1) its length is greater than L_0 and (2) the external force is balanced by internal forces. In such a situation the bar is said to be stressed. We define the **tensile stress** as the ratio of the magnitude of the external force, F, to the cross-sectional area, A. The SI units of stress are newtons per square meter (N/m^2). One N/m^2 is also given a special name, the **pascal** (Pa):

◀ The pascal

$$1 \text{ Pa} \equiv 1 \text{ N/m}^2 \qquad \text{[9.2]}$$

The **tensile strain** in this case is defined as the ratio of the change in length, ΔL, to the original length, L_0, and is therefore a dimensionless quantity. Thus, we can use Equation 9.1 to define **Young's modulus, Y:**

$$Y \equiv \frac{\text{tensile stress}}{\text{tensile strain}} = \frac{F/A}{\Delta L/L_0} = \frac{FL_0}{A\,\Delta L} \qquad \text{[9.3]}$$

This quantity is typically used to characterize a rod or wire stressed under *either tension or compression*. Note that because the strain is a dimensionless quantity, Y is in pascals. Typical values are given in Table 9.1. Experiments show that (1) the change in length for a fixed external force is proportional to the original length

TABLE 9.1 Typical Values for Elastic Modulus

Substance	Young's Modulus (Pa)	Shear Modulus (Pa)	Bulk Modulus (Pa)
Aluminum	7.0×10^{10}	2.5×10^{10}	7.0×10^{10}
Brass	9.1×10^{10}	3.5×10^{10}	6.1×10^{10}
Copper	11×10^{10}	4.2×10^{10}	14×10^{10}
Steel	20×10^{10}	8.4×10^{10}	16×10^{10}
Tungsten	35×10^{10}	14×10^{10}	20×10^{10}
Glass	$6.5\text{--}7.8 \times 10^{10}$	$2.6\text{--}3.2 \times 10^{10}$	$5.0\text{--}5.5 \times 10^{10}$
Quartz	5.6×10^{10}	2.6×10^{10}	2.7×10^{10}
Water	—	—	0.21×10^{10}
Mercury	—	—	2.8×10^{10}

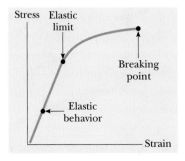

Figure 9.4 A stress-strain curve for a solid.

Figure 9.5 (a) A shear deformation in which a rectangular block is distorted by a force applied tangent to one of its faces. (b) A book under shear stress.

Bulk modulus ▶

and (2) the force necessary to produce a given strain is proportional to the cross-sectional area.

It is possible to exceed the *elastic limit* of a substance by applying a sufficiently great stress (Fig. 9.4). At the *elastic limit,* the stress-strain curve departs from a straight line. A material subjected to a stress beyond this level ordinarily does not return to its original length when the external force is removed. As the stress is increased further, the material ultimately breaks.

Shear Modulus: Elasticity of Shape

Another type of deformation occurs when a body is subjected to a force, **F,** tangential to one of its faces while the opposite face is held fixed (Fig. 9.5a). If the object is originally a rectangular block, such a tangential force results in a shape whose cross section is a parallelogram. In this situation, the stress is called a shear stress. A book pushed sideways, as in Figure 9.5b, is under a shear stress. There is no change in volume with this deformation. We define the **shear stress** as F/A, the ratio of the magnitude of the tangential force to the area, A, of the face being sheared. The **shear strain** is the ratio $\Delta x/h$, where Δx is the horizontal distance the sheared face moves and h is the height of the object. In terms of these quantities, the **shear modulus,** S, is defined as

$$S \equiv \frac{\text{shear stress}}{\text{shear strain}} = \frac{F/A}{\Delta x/h} \qquad [9.4]$$

Shear moduli for some representative materials are listed in Table 9.1. Note that the units of shear modulus are force per unit area (Pa).

Bulk Modulus: Volume Elasticity

Bulk modulus characterizes the response of a substance to uniform squeezing. Suppose that the external forces acting on an object are at right angles to all of its faces (Fig. 9.6) and distributed uniformly over all the faces. As we shall see later, this occurs when an object is immersed in a fluid. A body subject to this type of deformation undergoes a change in volume but no change in shape. The **volume stress,** ΔP, is defined as the ratio of the magnitude of the change in the normal force, ΔF, to the area, A. (When dealing with fluids, we shall refer to the quantity F/A as the **pressure.**) The volume strain is equal to the change in volume, ΔV, divided by the original volume, V. Thus, from Equation 9.1 we can characterize a volume compression in terms of the **bulk modulus,** B, defined as

$$B \equiv \frac{\text{volume stress}}{\text{volume strain}} = -\frac{\Delta F/A}{\Delta V/V} = -\frac{\Delta P}{\Delta V/V} \qquad [9.5]$$

Note that a negative sign is included in this defining equation so that B is always positive. An increase in pressure (positive ΔP) causes a decrease in volume (negative ΔV) and vice versa.

Table 9.1 lists bulk modulus values for some materials. If you look up such values in a different source, you will often find that the reciprocal of the bulk modulus, called the **compressibility** of the material, is listed. Note from Table 9.1 that both solids and liquids have bulk moduli. However, there is no shear modulus and no Young's modulus for liquids because a liquid will not sustain a shearing stress or a tensile stress (it flows instead).

EXAMPLE 9.1   Built to Last

A vertical steel beam in a building supports a load of 6.0×10^4 N. If the length of the beam is 4.0 m and its cross-sectional area is 8.0×10^{-3} m^2, find the distance it is compressed along its length.

Solution   Because the beam is under compression, we can use Equation 9.3. Taking Young's modulus for steel, $Y = 20 \times 10^{10}$ Pa, from Table 9.1, we have

$$Y = \frac{FL_0}{A\,\Delta L}$$

or

$$\Delta L = \frac{FL_0}{YA} = \frac{(6.0 \times 10^4\ \text{N})(4.0\ \text{m})}{(20 \times 10^{10}\ \text{Pa})(8.0 \times 10^{-3}\ \text{m}^2)} = \boxed{1.5 \times 10^{-4}\ \text{m}}$$

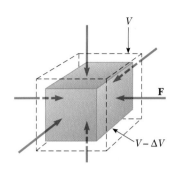

Figure 9.6   When a solid is under uniform pressure, it undergoes a change in volume but no change in shape. This cube is compressed on all sides by forces normal to its six faces.

EXAMPLE 9.2   Squeezing a Lead Sphere

A solid lead sphere of volume 0.50 m^3 is dropped in the ocean to a depth where the pressure increases by 2.0×10^7 Pa. Lead has a bulk modulus of 7.7×10^9 Pa. What is the change in volume of the sphere?

Solution   From the definition of bulk modulus (Eq. 9.5), we have

$$B = -\frac{\Delta P}{\Delta V/V}$$

$$\Delta V = -\frac{V\,\Delta P}{B}$$

In this case, when the sphere is at the surface, where its volume is 0.50 m^3, the pressure exerted on it is atmospheric pressure. The increase in pressure, ΔP, when the sphere is submerged is 2.0×10^7 Pa. Therefore, the change in its volume when it is submerged is

$$\Delta V = -\frac{(0.50\ \text{m}^3)(2.0 \times 10^7\ \text{Pa})}{7.7 \times 10^9\ \text{Pa}} = \boxed{-1.3 \times 10^{-3}\ \text{m}^3}$$

The negative sign indicates a *decrease* in volume.

9.3   DENSITY AND PRESSURE

The **density** of a substance of uniform composition is defined as its mass per unit volume.

Figure 9.7 The force exerted by a fluid on a submerged object at any point is perpendicular to the surface of the object. The force exerted by the fluid on the walls of the container is perpendicular to the walls at all points and increases with depth.

TABLE 9.2 Density of Some Common Substances

Substance	$\rho(kg/m^3)^a$	Substance	$\rho(kg/m^3)^a$
Ice	0.917×10^3	Water	1.00×10^3
Aluminum	2.70×10^3	Glycerin	1.26×10^3
Iron	7.86×10^3	Ethyl alcohol	0.806×10^3
Copper	8.92×10^3	Benzene	0.879×10^3
Silver	10.5×10^3	Mercury	13.6×10^3
Lead	11.3×10^3	Air	1.29
Gold	19.3×10^3	Oxygen	1.43
Platinum	21.4×10^3	Hydrogen	8.99×10^{-2}
Uranium	18.7×10^3	Helium	1.79×10^{-1}

 ^{a} All values are at standard atmospheric pressure and temperature (STP), defined as 0°C (273 K) and 1 atm (1.01×10^5 Pa). To convert to grams per cubic centimeter, multiply by 10^{-3}.

In symbolic form, a substance of mass M and volume V has a density, ρ (Greek rho), given by

 Density ▶

$$\rho \equiv \frac{M}{V} \qquad [9.6]$$

The units of density are kilograms per cubic meter in the SI system and grams per cubic centimeter in the cgs system. Table 9.2 lists the densities of some substances. The densities of most liquids and solids vary slightly with changes in temperature and pressure; the densities of gases vary greatly with such changes. Note that under normal conditions the densities of solids and liquids are about 1000 times greater than the densities of gases. This difference implies that the average spacing between molecules in a gas under these conditions is about ten times greater than that in a solid or liquid.

The **specific gravity** of a substance is the ratio of its density to the density of water at 4°C, which is 1.0×10^3 kg/m^3. By definition, specific gravity is a dimensionless quantity. For example, if the specific gravity of a substance is 3.0, its density is $3.0(1.0 \times 10^3$ kg/m$^3) = 3.0 \times 10^3$ kg/m^3.

We have seen that fluids do not sustain shearing stresses, and thus the only stress that can exist on an object submerged in a fluid is one that tends to compress the object. The force exerted by the fluid on the object is always perpendicular to the surfaces of the object, as shown in Figure 9.7.

The pressure at a specific point in a fluid can be measured with the device pictured in Figure 9.8—an evacuated cylinder enclosing a light piston connected to a spring. As the device is submerged in a fluid, the fluid presses down on the top of the piston and compresses the spring until the inward force of the fluid is balanced by the outward force of the spring. The fluid pressure can be measured directly if the spring is calibrated in advance. This is accomplished by applying a known force to the spring to compress it a given distance.

If F is the magnitude of the force exerted by the fluid on the piston and A is the area of the piston, then **the average pressure, P, of the fluid at the level to which the device has been submerged is defined as the ratio of force to area:**

Figure 9.8 A simple device for measuring pressure in a fluid.

$$P \equiv \frac{F}{A}$$ [9.7]

Because pressure is defined as force per unit area, it has units of pascals (newtons per square meter).

As you can see from this definition, one needs to know the magnitude of the force exerted on a surface *and* the area over which that force is applied. For example, a 700-N man can stand on a vinyl-covered floor in regular street shoes without damaging the surface. If he wears golf shoes with numerous metal cleats protruding from each sole, he does considerable damage to the floor. In both cases the net force applied to the floor is 700 N. However, when the man wears ordinary shoes, the area of his contact with the floor is considerably larger than when he wears golf shoes. (In the latter case, the only area in contact with the floor is the sum of the small cross-sectional areas of the metal cleats.) Hence, the *pressure* on the floor is much smaller when he wears ordinary shoes.

Snowshoes use this principle. The weight of the body—the force—is distributed over the very large areas of the snowshoes so that the pressure at any given point is relatively low and the person does not penetrate very deeply into the snow.

Snowshoes prevent the person from sinking into the soft snow because the person's weight is spread over a larger area, which reduces the pressure on the snow's surface. *(Earl Young/FPG)*

Thinking Physics 1

The daring physics professor, after a long lecture, stretches out for a nap on a bed of nails, as in Figure 9.9. How is this possible?

Figure 9.9   (Thinking Physics 1) Does anyone have a pillow?

Explanation   If you try to support your entire weight on a single nail, the pressure on your body is your weight divided by the very small area of the nail. This pressure is sufficiently large to penetrate the skin. However, if you distribute your weight over several hundred nails, as the professor is doing, the pressure is considerably reduced because the area that supports your weight is the total area of all nails in contact with your body. (Note that lying on a bed of nails is much more comfortable than sitting on the bed. Why? Extend the logic to show that it would be even more uncomfortable to stand on a bed of nails without shoes.)

QUICKLAB

Place a tack between your thumb and index finger, as in the figure. Now squeeze the tack and note the sensation. The sharp end of the tack causes pain, and the blunt end does not. According to Newton's third law, the force exerted on the thumb is equal and opposite the force exerted on the index finger. However, the pressure at the sharp end of the tack is much greater than the pressure at the blunt end. (Remember that pressure is a measure of force per unit area.)

Tack

EXAMPLE 9.3   The Water Bed

A water bed is 2.00 m on a side and 30.0 cm deep.
(a) Find its weight.

Solution   Because the density of water is 1000 kg/m³ (Table 9.2) and the bed's volume is $(2.00 \times 2.00 \times 0.300)$ m³ = 1.20 m³, the mass of the bed is

$$M = \rho V = (1000 \text{ kg/m}^3)(1.20 \text{ m}^3) = 1.20 \times 10^3 \text{ kg}$$

and its weight is

$$w = Mg = (1.20 \times 10^3 \text{ kg})(9.80 \text{ m/s}^2) = \boxed{1.18 \times 10^4 \text{ N}}$$

This is equivalent to approximately 2640 lb. In order to support such a heavy load, you would be well advised to keep your water bed in the basement or on a sturdy, well-supported floor.
(b) Find the pressure that the water bed exerts on the floor. Assume that the entire lower surface of the bed makes contact with the floor.

Solution   The weight of the water bed is 1.18×10^4 N. Its cross-sectional area is 4.00 m². The pressure exerted on the floor is therefore

$$P = \frac{1.18 \times 10^4 \text{ N}}{4.00 \text{ m}^2} = \boxed{2.95 \times 10^3 \text{ Pa}}$$

Exercise   Calculate the pressure exerted by the bed on the floor if the bed rests on its side.

Answer   Because the area of the bed's side is 0.600 m², the pressure is 1.96×10^4 Pa.

9.4   VARIATION OF PRESSURE WITH DEPTH

If a fluid is at rest in a container, *all* portions of the fluid must be in static equilibrium. Furthermore, **all points at the same depth must be at the same pressure.** If this were not the case, a given portion of the fluid would not be in equilibrium. For example, consider the small block of fluid shown in Figure 9.10a. If the pressure were greater on the left side of the block than on the right, $\mathbf{F}_1$ would be greater than $\mathbf{F}_2$, and the block would accelerate and thus would not be in equilibrium.

Now let us examine the portion of the fluid contained within the volume indicated by the darker region in Figure 9.10b. This region has a cross-sectional area A and extends to a depth h below the surface of the water. Three external forces act on this volume of fluid: the force of gravity, Mg; the upward force, PA, exerted by the fluid below it; and a downward force, P_0A, exerted by the atmosphere, where P_0 is atmospheric pressure. Because this volume of fluid is in equilibrium, these forces must add to zero, and so we get

$$PA - Mg - P_0A = 0 \tag{9.8}$$

From the relation $M = \rho V = \rho Ah$, the weight of the fluid in the volume is

$$w = Mg = \rho gAh \tag{9.9}$$

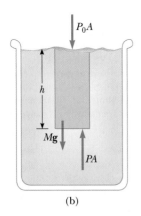

Figure 9.10   (a) If the block of fluid is to be in equilibrium, the force $\mathbf{F}_1$ must balance the force $\mathbf{F}_2$. (b) The net force on the volume of water within the darker region must be zero.

(a)

(b)

Figure 9.11 (a) Diagram of a hydraulic press. Because the pressure is the same at the left and right sides, a small force, $\mathbf{F}_1$, at the left produces a much larger force, $\mathbf{F}_2$, at the right. (b) A bus under repair is supported by a hydraulic lift in a garage. *(Superstock)*

When Equation 9.9 is substituted into Equation 9.8, we get

$$P = P_0 + \rho g h \qquad [9.10]$$

where normal atmospheric pressure at sea level is $P_0 = 1.01 \times 10^5$ Pa (equivalent to 14.7 lb/in.²). According to Equation 9.10, **the pressure, P, at a depth of h below the surface of a liquid open to the atmosphere is greater than atmospheric pressure by the amount $\rho g h$.** Moreover, the pressure is not affected by the shape of the vessel.

Pressure in a fluid depends only on depth. Any increase in pressure at the surface is transmitted to every point in the fluid. This was first recognized by the French scientist Blaise Pascal (1623–1662) and is called **Pascal's principle:**

> **Pressure applied to an enclosed fluid is transmitted undiminished to every point of the fluid and to the walls of the containing vessel.**

An important application of Pascal's principle is the hydraulic press (Fig. 9.11a). A downward force, $\mathbf{F}_1$, is applied to a small piston of area A_1. The pressure is transmitted through a fluid to a larger piston of area A_2. Because the pressure is the same on both sides, we see that $P = F_1/A_1 = F_2/A_2$. Therefore, the magnitude of the force $\mathbf{F}_2$ is larger than the magnitude of $\mathbf{F}_1$ by the factor A_2/A_1. That is why a large load, such as a car, can be supported on the large piston by a much smaller force on the smaller piston. Hydraulic brakes, car lifts, hydraulic jacks, forklifts, and other machines make use of this principle.

This photograph illustrates that the pressure in a liquid is the same at all points having the same elevation. For example, the pressure is the same at points *A*, *B*, *C* and *D*. Note that the shape of the vessel does not affect the pressure. *(Courtesy of Central Scientific Company)*

APPLICATION

Hydraulic Lifts.

Applying Physics 1

A corollary to the statement that pressure in a fluid increases with depth is that water always seeks its own level. This means that regardless of the shape of the vessel, when filled with water, the surface of the water is perfectly flat and at the same height at all points. The ancient Egyptians used this fact to level the pyramids. Devise a scheme showing how this could be done.

Explanation   There are many ways this could have been done, but Figure 9.12 shows the scheme used by the Egyptians. The builders cut grooves in the base of the pyramid as in (a) and partially filled the grooves with water. The height of the water was marked as in (b) and the rock was chiseled down to this line (c). Finally, the groove was filled with crushed rock and gravel, as in (d).

(a)   (b)   (c)   (d)

Figure 9.12   (Applying Physics 1)

9.5   PRESSURE MEASUREMENTS

(a)

(b)

Figure 9.13   Two devices for measuring pressure: (a) an open-tube manometer and (b) a mercury barometer.

Another simple device for measuring pressure is the open-tube manometer (Fig. 9.13a). One end of a U-shaped tube containing a liquid is open to the atmosphere, and the other end is connected to a system of unknown pressure P. The pressure at point B equals $P_0 + \rho g h$, where ρ is the density of the fluid. The pressure at B, however, equals the pressure at A, which is also the unknown pressure P. Therefore, we conclude that

$$P = P_0 + \rho g h$$

The pressure P is called the *absolute pressure*, and $P - P_0$ is called the *gauge pressure*. Thus, if P in the system is greater than atmospheric pressure, h is positive. If P is less than atmospheric pressure (a partial vacuum), h is negative.

A third instrument that is used to measure pressure is the *barometer* (Fig. 9.13b), invented by Evangelista Torricelli (1608–1647). A long tube that is closed at one end is filled with mercury and then inverted into a dish of mercury. The closed end of the tube is nearly a vacuum, and so its pressure can be taken as zero. It follows that $P_0 = \rho g h$, where ρ is the density of the mercury and h is the height of the mercury column.

One atmosphere of pressure is defined to be the pressure equivalent of a column of mercury that is exactly 0.76 m in height at 0°C with $g = 9.806\ 65\ \text{m/s}^2$. At this temperature mercury has a density of $13.595 \times 10^3\ \text{kg/m}^3$; therefore,

$$P_0 = \rho g h = (13.595 \times 10^3\ \text{kg/m}^3)(9.806\ 65\ \text{m/s}^2)(0.7600\ \text{m})$$
$$= 1.013 \times 10^5\ \text{Pa}$$

It is interesting to note that the force of the atmosphere on our bodies (assuming a body area of 2000 in.²) is extremely large, on the order of 30 000 lb! A natural question to raise is, "How can we exist under such great forces attempting to collapse our bodies?" The answer is that our body cavities and tissues are permeated with fluids and gases that are pushing outward with this same atmospheric pressure.

Consequently, our bodies are in equilibrium under the force of the atmosphere pushing in and an equal internal force pushing out.

Blood Pressure Measurements

A specialized manometer (called a sphygmomanometer) is often used to measure blood pressure. In this application, a rubber bulb forces air into a cuff wrapped tightly around the upper arm and simultaneously into a manometer, as in Figure 9.14. The pressure in the cuff is increased until the flow of blood through the brachial artery in the arm is stopped. A valve on the bulb is then opened and the measurer listens with a stethoscope to the artery at a point just below the cuff. When the pressure in the cuff and brachial artery is just below the maximum value produced by the heart (the systolic pressure) the artery opens momentarily on each beat of the heart. At this point, the velocity of the blood is high and turbulent, and the flow is noisy. The manometer is calibrated to read the pressure in millimeters of mercury, and the value obtained is about 120 mm for a normal heart. Values of 140 mm or above are considered to be high and medication to lower blood pressure is often prescribed. As the pressure in the cuff is lowered further, intermittent sounds are still heard until the pressure falls just below the minimum heart pressure (the diastolic pressure). At this point, continuous sounds are heard. In the normal heart, this transition occurs at about 80 mm of mercury, and values above 90 are considered to need medical intervention. Blood pressure readings are usually expressed as the ratio systolic/diastolic or 120/80 for a healthy heart.

APPLICATION

Measuring Blood Pressure.

Figure 9.14 A sphygmomanometer can be used to measure blood pressure.

Applying Physics 2

Blood pressure is normally measured with the cuff of the sphygmomanometer around the arm. Suppose that the blood pressure were measured with the cuff around the calf of the leg of a standing person. Would the reading of the blood pressure be the same here as it was for the arm?

Explanation The blood pressure measured at the calf would be larger than that measured at the arm. If we imagine the vascular system of the body to be a vessel containing a liquid (the blood), the pressure in the liquid will increase with depth. The blood at the calf is deeper in the liquid than that at the arm and is at a higher pressure.

Applying Physics 3

In a ball-point pen, ink moves down a tube to the tip where it is spread on a sheet of paper by a rolling stainless steel ball. Near the top of the ink cartridge, there is a small hole open to the atmosphere. If you seal this hole, you will find that the pen no longer works. Use your knowledge of how a barometer works to explain this behavior.

Explanation If the hole is sealed, or if it were not present, the pressure of the air above the ink would decrease as some ink is used. Consequently, atmospheric pressure exerted against the ink at the bottom of the cartridge would prevent some of the ink from flowing out. The hole allows the pressure above the ink to remain at atmospheric pressure. Why does a ball-point pen seem to run out of ink when you write on a vertical surface?

APPLICATION

Ball-Point Pens.

10.3 m

Figure 9.15 Even the man of steel is defeated if the straw is too long.

Applying Physics 4

Figure 9.15 shows Superman attempting to drink through a very long straw. Unbelievably, he finds that he is unable to lift the fluid to his lips if the straw is longer than about 10.3 m. Explain why this is the case, and then extend this reasoning to determine what would happen if Superman tried to drink through a straw on the Moon where there is no atmosphere.

Explanation   Without a doubt, Superman is able to remove all the air from above the surface of the fluid. At this point, the pressure at the top of the straw (P_0) is zero. $P = P_0 + \rho g h$ becomes, with $P_0 = 0$, $P = P_a$ and with ρ being the density of the fluid (assumed to be water), $P_a = \rho g h$ or

$$h = \frac{P_a}{\rho g} = \frac{1.01 \times 10^5 \text{ Pa}}{(1.00 \times 10^3 \text{ kg/m}^3)(9.8 \text{ m/s}^2)} = 10.3 \text{ m}$$

By the same reasoning, if P_a goes to zero as it would on the Moon, Superman would not be able to lift any height of fluid into the straw.

EXAMPLE 9.4   The Car Lift

In a car lift, compressed air exerts a force on a piston with a radius of 5.00 cm. This pressure is transmitted to a second piston, of radius 15.0 cm. What force must the compressed air exert in order to lift a car weighing 1.33×10^4 N? What air pressure produces this force? Neglect the weights of the pistons.

Solution   Because the pressure exerted by the compressed air is transmitted undiminished throughout the fluid, we have

$$F_1 = \left(\frac{A_1}{A_2}\right) F_2 = \frac{\pi (5.00 \times 10^{-2} \text{ m})^2}{\pi (15.0 \times 10^{-2} \text{ m})^2} (1.33 \times 10^4 \text{ N}) = \boxed{1.48 \times 10^3 \text{ N}}$$

The air pressure that produces this force is

$$P = \frac{F_1}{A_1} = \frac{1.48 \times 10^3 \text{ N}}{\pi (5.00 \times 10^{-2} \text{ m})^2} = \boxed{1.88 \times 10^5 \text{ Pa}}$$

This pressure is approximately twice atmospheric pressure. (Note that pressure here means gauge pressure.)

EXAMPLE 9.5   Pressure in the Ocean

Calculate the absolute pressure at an ocean depth of 1000 m. Assume that the density of water is 1.0×10^3 kg/m^3 and that $P_0 = 1.01 \times 10^5$ Pa.

Solution

$$P = P_0 + \rho g h$$
$$= 1.01 \times 10^5 \text{ Pa} + (1.0 \times 10^3 \text{ kg/m}^3)(9.80 \text{ m/s}^2)(1.00 \times 10^3 \text{ m})$$
$$P \approx \boxed{9.9 \times 10^6 \text{ Pa}}$$

This is approximately 100 times greater than atmospheric pressure! Obviously, the design and construction of vessels that can withstand such enormous pressures are not trivial matters.

Exercise Calculate the total force exerted on the outside of a 30-cm-diameter circular submarine window at this depth.

Answer 7.0×10^5 N

9.6 BUOYANT FORCES AND ARCHIMEDES'S PRINCIPLE

A fundamental principle affecting objects submerged in fluids was discovered by the Greek mathematician Archimedes. **Archimedes's principle** can be stated as follows:

> **Any body completely or partially submerged in a fluid is buoyed up by a force whose magnitude is equal to the weight of the fluid displaced by the body.**

Archimedes, Greek mathematician, physicist, and engineer (287–212 B.C.)

Archimedes was perhaps the greatest scientist of antiquity. He is well known for discovering the nature of the buoyant force and was a gifted inventor. According to legend, Archimedes was asked by King Hieron to determine whether the king's crown was made of pure gold or merely a gold alloy. The task was to be performed without damaging the crown. Archimedes allegedly arrived at a solution while taking a bath, noting a partial loss of weight after submerging his arms and legs in the water. As the story goes, he was so excited about his great discovery that he ran naked through the streets of Syracuse shouting, "Eureka!" which is Greek for "I have found it."

Everyone has experienced Archimedes's principle. For example, recall that it is relatively easy to lift someone if you are both standing in a swimming pool, whereas lifting that same individual on dry land would be a difficult task. Water provides partial support to any object placed in it. We say that an object placed in a fluid is buoyed up by the fluid, and we call this upward force the **buoyant force.** According to Archimedes's principle, **the magnitude of the buoyant force always equals the weight of the fluid displaced by the object.** The buoyant force acts vertically upward through what was the center of gravity of the fluid before the fluid was displaced.

Archimedes's principle can be verified in the following manner. Suppose we focus our attention on the cube of water that is colored red in the container of Figure 9.16. This cube of water is in equilibrium under the action of the forces on it. One of those forces is the force of gravity. What cancels that downward force? Apparently, the water beneath the cube is buoying it up and holding it in equilibrium. Thus, the buoyant force, **B,** on the cube of water must be exactly equal in magnitude to the weight of the water inside the cube: $B = w$.

Now imagine that the cube of water is replaced by a cube of steel of the same dimensions. What is the buoyant force on the steel? The water surrounding a cube behaves in the same way whether the cube is made of water or steel; therefore, **the buoyant force acting on the steel is the same as the buoyant force acting on a cube of water of the same dimensions.** This result applies for a totally submerged object of any shape, size, or density.

Let us show explicitly that the buoyant force is equal in magnitude to the weight of the displaced fluid. The pressure at the bottom of the cube in Figure 9.16 is greater than the pressure at the top by the amount $\rho_f g h$, where ρ_f is the density of the fluid and h is the height of the cube. Because the pressure difference, ΔP, is equal to the buoyant force per unit area, or, $\Delta P = B/A$, we see that $B = (\Delta P)(A) = (\rho_f g h)(A) = \rho_f g V$, where V is the volume of the cube. Because the mass of the fluid in the cube is $M = \rho_f V$, we see that

$$B = \rho_f V g = Mg = w_f \qquad [9.11]$$

where w_f is the weight of the displaced fluid.

Figure 9.16 The external forces on a cube of water are the force of gravity, **w,** and the buoyant force, **B.** Under equilibrium conditions, $B = w$.

Hot-air balloons. Because hot air is less dense than cold air, there is a net upward buoyant force on the balloons. *(Richard Megna, Fundamental Photographs, NYC)*

The weight of the submerged object is $w_0 = mg = \rho_0 Vg$, where ρ_0 is the density of the object. Because $w_f = \rho_f Vg$ and $w_0 = \rho_0 Vg$, we see that if the density of the object is greater than the density of the fluid, the unsupported object sinks. If the density of the object is less than that of the fluid, the unsupported submerged object accelerates upward and ultimately floats. When a *floating* object is in equilibrium, only part of it is submerged. In this case, the magnitude of the buoyant force equals the weight of the object.

It is instructive to compare the forces on a totally submerged object with those on a floating object.

Case I: A Totally Submerged Object. When an object is *totally* submerged in a fluid of density ρ_f, the upward buoyant force has a magnitude of $B = \rho_f V_0 g$, where V_0 is the volume of the object. If the object has density ρ_0, the downward force is equal to $mg = \rho_0 V_0 g$, and the net force on it is $B - w = (\rho_f - \rho_0)V_0 g$. Hence, if the density of the object is less than the density of the fluid, as in Figure 9.17a, the net force is positive (upward) and the unsupported object accelerates upward. If the density of the object is greater than the density of the fluid, as in Figure 9.17b, the net force is negative and the unsupported object sinks.

Case II: A Floating Object. Now consider an object in static equilibrium floating on a fluid — that is, a partially submerged object. In this case, the upward buoyant force is balanced by the downward force of gravity acting on the object. If V_f is the volume of the fluid displaced by the object (which corresponds to the volume of the part of the object beneath the fluid level), then the magnitude of the buoyant force is given by $B = \rho_f V_f g$. Because the weight of the object is $w = mg = \rho_0 V_0 g$, and $w = B$, we see that $\rho_f V_f g = \rho_0 V_0 g$, or

$$\frac{\rho_0}{\rho_f} = \frac{V_f}{V_0} \qquad [9.12]$$

APPLICATION

Swim Bladders in Fish.

APPLICATION

Cerebrospinal Fluid.

APPLICATION

Testing Your Car's Antifreeze.

Under normal conditions, the average density of a fish is slightly greater than the density of water. This being the case, a fish would sink if it did not have a mechanism for adjusting its density: the internal regulation of the size of the swim bladder. In this manner, fish maintain neutral buoyancy as they swim to various depths.

The human brain is immersed in a fluid (the cerebrospinal fluid) of density 1007 kg/m³, which is slightly less than the average density of the brain, 1040 kg/m³. Consequently, most of the weight of the brain is supported by the buoyant force of the surrounding fluid. In some clinical procedures, it is necessary to remove a portion of this fluid for diagnostic purposes. During such procedures, the nerves and blood vessels in the brain are placed under great strain, which in turn can cause extreme discomfort and pain. Great care must be exercised with such patients until the initial brain fluid volume has been restored by the body.

When service station attendants check the antifreeze in your car or the condition of your battery, they often use devices that apply Archimedes's principle. Figure 9.18 shows a common device that is used to check the antifreeze in a car radiator. The small balls in the enclosed tube vary in density so that all of them float when the tube is filled with pure water, none float in pure antifreeze, one floats in a 5% mixture, two in a 10% mixture, and so forth. The number of balls that float thus serves as a measure of the percentage of antifreeze in the mixture, which in turn

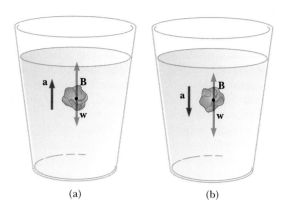

Figure 9.17 (a) A totally submerged object that is less dense than the fluid in which it is submerged experiences a net upward force. (b) A totally submerged object that is denser than the fluid sinks.

(a) (b)

is used to determine the lowest temperature the mixture can withstand without freezing.

Similarly, the charge of some newer car batteries can be determined with a so-called "magic-dot" process that is built into the battery (Fig. 9.19). When one looks down into a viewing port in the top of the battery, a red dot indicates that the battery is sufficiently charged; a black dot indicates that the battery has lost its charge. This is because, if the battery has sufficient charge, the density of the battery fluid is high enough to cause the red ball to float. As the battery loses its charge, the density of the battery fluid decreases and the ball sinks beneath the surface of the fluid, where the dot appears black.

APPLICATION

Checking the Battery Charge.

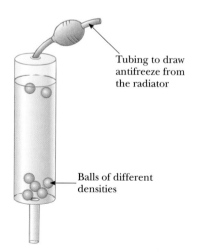

Tubing to draw antifreeze from the radiator

Balls of different densities

Figure 9.18 The number of balls that float in this device is a measure of the density of the antifreeze solution in a vehicle's radiator, and consequently a measure of the temperature at which freezing will occur.

Battery fluid

(a) Charged battery (b) Discharged battery

Figure 9.19 The red ball in the plastic tube inside the battery serves as an indicator of whether the battery is charged or discharged. As the battery loses its charge, the density of the battery fluid decreases, and the ball sinks out of sight.

EXAMPLE 9.6   A Red-Tag Special on Crowns

A bargain hunter purchases a "gold" crown at a flea market. After she gets home, she hangs it from a scale and finds its weight to be 7.84 N (Fig. 9.20a). She then weighs the crown while it is immersed in water of density 1000 kg/m³, as in Figure 9.20b, and now the scale reads 6.86 N. Is the crown made of pure gold?

(a)   (b)

Figure 9.20   (Example 9.6) (a) When the crown is suspended in air, the scale reads the true weight, **w.** (b) When the crown is immersed in water, the buoyant force, **B,** reduces the scale reading to the apparent weight, $T_2 = w - B$.

Reasoning   When the crown is suspended in air, the scale reads the true weight, w (neglecting the buoyancy of air). When it is immersed in water, the buoyant force, **B,** reduces the scale reading to an apparent weight of $T_2 = w - B$. Hence, the buoyant force on the crown is the difference between its weight in air and its weight in water:

$$B = w - T_2 = 7.84 \text{ N} - 6.86 \text{ N} = 0.98 \text{ N}$$

Because the buoyant force is equal in magnitude to the weight of the displaced fluid, w_f, we have

$$w_f = \rho_f g V_f = 0.98 \text{ N}$$

where V_f is the displaced fluid's volume, and ρ_f is its density (1000 kg/m³). Also, the volume of the crown, V_c, is equal to the volume of the displaced fluid (because the crown is completely submerged).

Solution   We find that

$$V_c = V_f = \frac{0.98 \text{ N}}{g\rho_f} = \frac{0.98 \text{ N}}{(9.8 \text{ m/s}^2)(1000 \text{ kg/m}^3)} = 1.0 \times 10^{-4} \text{ m}^3$$

Finally, the density of the crown is

$$\rho_c = \frac{m_c}{V_c} = \frac{w_c}{gV_c} = \frac{7.84 \text{ N}}{(9.8 \text{ m/s}^2)(1.0 \times 10^{-4} \text{ m}^3)} = 8.0 \times 10^3 \text{ kg/m}^3$$

From Table 9.2 we see that the density of gold is 19.3×10^3 kg/m³. Thus, either the crown is hollow or it is made of an alloy. This was not a good day for the bargain hunter!

EXAMPLE 9.7 Floating Down the River

A raft is constructed of wood having a density of 600 kg/m^3. Its surface area is 5.7 m^2 and its volume is 0.60 m^3. When the raft is placed in fresh water of density 1000 kg/m^3, as in Figure 9.21, how much of it is below water level?

Figure 9.21 (Example 9.7) A raft partially submerged in water.

Reasoning The magnitude of the upward buoyant force acting on the raft must equal the weight of the raft if the raft is to float. In addition, from Archimedes's principle the magnitude of the buoyant force is equal to the weight of the displaced water.

Solution The weight of the raft is

$$w_r = \rho_r g V_r = (600 \text{ kg/m}^3)(9.8 \text{ m/s}^2)(0.60 \text{ m}^3) = 3.5 \times 10^3 \text{ N}$$

The magnitude of the upward buoyant force acting on the raft equals the weight of the displaced water, which in turn must equal the weight of the raft:

$$B = \rho_w g V_w = \rho_w g A h = 3.5 \times 10^3 \text{ N}$$

Because the area A and density ρ_w are known, we can find the depth, h, at which the raft sinks into the water:

$$h = \frac{B}{\rho_w g A} = \frac{3.5 \times 10^3 \text{ N}}{(1000 \text{ kg/m}^3)(9.8 \text{ m/s}^2)(5.7 \text{ m}^2)} = \boxed{0.06 \text{ m}}$$

9.7 FLUIDS IN MOTION

When a fluid is in motion, its flow can be characterized in one of two ways. The flow is said to be **streamline,** or **laminar,** if every particle that passes a particular point moves exactly along the smooth path followed by particles that passed that point earlier. The path is called a *streamline* (Fig. 9.22a). Different streamlines cannot cross each other under this steady-flow condition, and the streamline at any point coincides with the direction of fluid velocity at that point.

In contrast, the flow of a fluid becomes irregular, or **turbulent,** above a certain velocity or under any conditions that can cause abrupt changes in velocity. Irregular motions of the fluid, called *eddy currents,* are characteristic in turbulent flow, as shown in Figure 9.22b.

In discussions of fluid flow, the term *viscosity* is used for the degree of internal friction in the fluid. This internal friction is associated with the resistance between two adjacent layers of the fluid moving relative to each other. A fluid such as kerosene has a lower viscosity than crude oil or molasses.

Most of the volume of this iceberg is beneath the water. Can you determine what fraction of the total volume is underwater? *(Geraldine Prentice/ Tony Stone Images)*

(a)

(b)

Figure 9.22 (a) This photograph was taken in a water tunnel using hydrogen bubbles to visualize the flow pattern around a cylinder. The flow was started from rest, and at this instant the pattern shows the development of a complex wake structure on the downstream side of the cylinder. *(Courtesy of Dr. Kenneth W. McAlister, Department of the Army)* (b) Hot gases from a cigarette made visible by smoke particles. The smoke first moves in streamline flow at the bottom and then in turbulent flow above. *(Werner Wolff/Black Star)*

Many features of fluid motion can be understood by considering the behavior of an **ideal fluid,** which satisfies the following conditions:

1. *The fluid is nonviscous;* that is, there is no internal friction force between adjacent layers.
2. *The fluid is incompressible,* which means that its density is constant.
3. *The fluid motion is steady,* meaning that the velocity, density, and pressure at each point in the fluid do not change in time.
4. *The fluid moves without turbulence.* This implies that each element of the fluid has zero angular velocity about its center; that is, there can be no eddy currents present in the moving fluid.

Equation of Continuity

Figure 9.23 represents a fluid flowing through a pipe of nonuniform size. The particles in the fluid move along the streamlines in steady-state flow. In a small time interval, Δt, the fluid entering the bottom end of the pipe moves a distance of $\Delta x_1 = v_1 \Delta t$, where v_1 is the speed of the fluid at this location. If A_1 is the cross-sectional area in this region, then the mass contained in the bottom blue region is $\Delta M_1 = \rho A_1 \Delta x_1 = \rho A_1 v_1 \Delta t$, where ρ is the density of the fluid. Similarly, the fluid that moves out of the upper end of the pipe in the same interval, Δt, has a mass of $\Delta M_2 = \rho A_2 v_2 \Delta t$. However, because mass is conserved and because the flow is steady, the mass that flows into the bottom of the pipe through A_1 in the time Δt must equal the mass that flows out through A_2 in the same interval. Therefore, $\Delta M_1 = \Delta M_2$, or

$$\rho A_1 v_1 = \rho A_2 v_2 \qquad [9.13]$$

Equation 9.13 reduces to

$$A_1 v_1 = A_2 v_2 \qquad [9.14]$$

This expression is called the **equation of continuity.** From this result, we see that **the product of any cross-sectional area of the pipe and the fluid speed at that cross-section is a constant.** Therefore, the speed is high where the tube is constricted and low where the tube has a larger diameter. The product Av, which has dimensions of volume per unit time, is called the **flow rate. The condition $Av =$ constant is equivalent to the fact that the amount of fluid that enters one end of the tube in a given time interval equals the amount of fluid leaving the tube in the same interval, assuming the absence of leaks.**

There are many instances in your everyday experiences when you have seen the equation of continuity in action. For example, you place your thumb over the open end of a garden hose to make the water spray our farther and with greater speed. As you reduce the cross-sectional area of the nozzle, the water drops leave the nozzle at a higher speed, causing them to move a longer distance. Similar reasoning explains why rising smoke from a smoldering piece of wood behaves as it does. The smoke first rises in a streamline pattern, getting thinner as it rises, and eventually breaks up into a swirling, turbulent pattern. The smoke rises because it is less dense than air, and the buoyant force of the air accelerates it upward. As the speed of the smoke stream increases, the cross-sectional area of the stream decreases, as seen from the equation of continuity. However, the stream soon reaches a speed so great that streamline flow is not possible. We shall study the relationship between speed of fluid flow and turbulence in a later discussion on the Reynolds number.

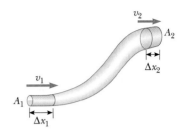

EXAMPLE 9.8 Filling a Water Bucket

A water hose 1.00 cm in radius is used to fill a 20.0-liter bucket. If it takes 1.00 min to fill the bucket, what is the speed, v, at which the water leaves the hose? (1 liter $= 10^3$ cm^3.)

Solution The cross-sectional area of the hose is

$$A = \pi r^2 = \pi(1.00 \text{ cm})^2 = \pi \text{ cm}^2$$

The flow rate is equal to the product Av. Thus,

$$Av = 20.0 \frac{\text{liters}}{\text{min}} = \frac{20.0 \times 10^3 \text{ cm}^3}{60.0 \text{ s}}$$

$$v = \frac{20.0 \times 10^3 \text{ cm}^3}{(\pi \text{ cm}^2)(60.0 \text{ s})} = \boxed{106 \text{ cm/s}}$$

Exercise If the radius of the hose is reduced to 0.500 cm, what is the speed of the water as it leaves the hose, assuming the same flow rate?

Answer 424 cm/s

Figure 9.23 A fluid moving with streamline flow through a pipe of varying cross-sectional area. The volume of fluid flowing through A_1 in a time interval of Δt must equal the volume flowing through A_2 in the same time interval. Therefore, $A_1 v_1 = A_2 v_2$.

Bernoulli's Equation

As a fluid moves through a pipe of varying cross-section and elevation, the pressure changes along the pipe. In 1738 the Swiss physicist Daniel Bernoulli (1700–1782) derived a fundamental expression that relates pressure to fluid speed and elevation. Bernoulli's equation is not a freestanding law of physics. It is, instead, a consequence of energy conservation as applied to the ideal fluid.

In deriving Bernoulli's equation, we again assume that the fluid is incompressible and nonviscous and flows in a nonturbulent, steady-state manner. Consider the flow through a nonuniform pipe in the time Δt, as illustrated in Figure 9.24. The force on the lower end of the fluid is $P_1 A_1$, where P_1 is the pressure at the lower end. The work done on the lower end of the fluid by the fluid behind it is

$$W_1 = F_1 \, \Delta x_1 = P_1 A_1 \, \Delta x_1 = P_1 V$$

where V is the volume of the lower blue region in Figure 9.24. In a similar manner, the work done on the fluid on the upper portion in the time Δt is

$$W_2 = -P_2 A_2 \, \Delta x_2 = -P_2 V$$

(Remember that the volume of fluid that passes through A_1 in the time Δt equals the volume that passes through A_2 in the same interval.) The work W_2 is negative because the force on the fluid at the top is opposite its displacement. Thus, the net work done by these forces in the time Δt is

$$W = P_1 V - P_2 V$$

Figure 9.24 A fluid flowing through a constricted pipe with streamline flow. The fluid in the section with a length of Δx_1 moves to the section with a length of Δx_2. The volumes of fluid in the two sections are equal.

Daniel Bernoulli, Swiss physicist and mathematician (1700–1782)

Bernoulli made important discoveries in hydrodynamics. In his most famous work, *Hydrodynamica*, Bernoulli showed that, as the velocity of fluid flow increases, its pressure decreases. In this same publication, Bernoulli also attempted the first explanation of the behavior of gases with changing pressure and temperature; this was the beginning of kinetic theory of gases. *(North Wind Picture Archives)*

Part of this work goes into changing the fluid's kinetic energy, and part goes into changing its gravitational potential energy. If m is the mass of the fluid passing through the pipe in the time interval Δt, then the change in kinetic energy of the volume of fluid is

$$\Delta KE = \tfrac{1}{2}mv_2^2 - \tfrac{1}{2}mv_1^2$$

The change in its gravitational potential energy is

$$\Delta PE = mgy_2 - mgy_1$$

We can apply the work-energy theorem in the form $W = \Delta KE + \Delta PE$ (Chapter 5) to this volume of fluid to give

$$P_1 V - P_2 V = \tfrac{1}{2}mv_2^2 - \tfrac{1}{2}mv_1^2 + mgy_2 - mgy_1$$

If we divide each term by V and recall that $\rho = m/V$, this expression reduces to

$$P_1 - P_2 = \tfrac{1}{2}\rho v_2^2 - \tfrac{1}{2}\rho v_1^2 + \rho gy_2 - \rho gy_1$$

Let us move those terms that refer to point 1 to one side of the equation and those that refer to point 2 to the other side:

$$P_1 + \tfrac{1}{2}\rho v_1^2 + \rho gy_1 = P_2 + \tfrac{1}{2}\rho v_2^2 + \rho gy_2 \qquad [9.15]$$

This is **Bernoulli's equation.** It is often expressed as

$$P + \tfrac{1}{2}\rho v^2 + \rho gy = \text{constant} \qquad [9.16]$$

> Bernoulli's equation states that the sum of the pressure (P), the kinetic energy per unit volume ($\tfrac{1}{2}\rho v^2$), and the potential energy per unit volume (ρgy) has the same value at all points along a streamline.

An important consequence of Bernoulli's equation can be demonstrated by considering Figure 9.25, which shows water flowing through a horizontal constricted pipe from a region of large cross-sectional area into a region of smaller cross-sectional area. This device, called a *Venturi tube,* can be used to measure the speed of fluid flow. Let us compare the pressure at point 1 to the pressure at point 2. Because the pipe is horizontal, $y_1 = y_2$ and Equation 9.15 applied to points 1 and 2 gives

$$P_1 + \tfrac{1}{2}\rho v_1^2 = P_2 + \tfrac{1}{2}\rho v_2^2 \qquad [9.17]$$

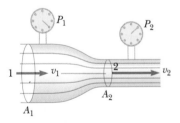

Figure 9.25 The pressure P_1 is greater than the pressure P_2, because $v_1 < v_2$. This device can be used to measure the speed of fluid flow.

Because the water is not backing up in the pipe, its speed in the constriction, v_2, must be greater than the speed v_1. From Equation 9.17, $v_2 > v_1$ means that P_2 must be less than P_1. This result is often expressed by the statement that **swiftly moving fluids exert less pressure than do slowly moving fluids.** As we shall see in the next section, this important result enables us to understand a wide range of everyday phenomena.

EXAMPLE 9.9   Shoot-Out at the Old Water Tank

A nearsighted sheriff fires at a cattle rustler with his trusty six-shooter. Fortunately for the cattle rustler, the bullet misses him and penetrates the town water tank and causes a leak (Fig. 9.26). If the top of the tank is open to the atmosphere, determine the speed at which the water leaves the hole when the water level is 0.500 m above the hole.

Reasoning   If we assume that the cross-sectional area of the tank is large relative to that of the hole ($A_2 \gg A_1$), then the water level drops very slowly and we can assume $v_2 \approx 0$. Let us apply Bernoulli's equation to points 1 and 2. If we note that $P_1 = P_0$ at the hole, we get

$$P_0 + \tfrac{1}{2}\rho v_1^2 + \rho g y_1 = P_0 + \rho g y_2$$

$$v_1 = \sqrt{2g(y_2 - y_1)} = \sqrt{2gh}$$

Solution   This says that the speed of the water emerging from the hole is equal to the speed acquired by a body falling freely through the vertical distance h. This is known as **Torricelli's law.** If the height h is 0.500 m, for example, the speed of the stream is

$$v = \sqrt{2(9.80 \text{ m/s}^2)(0.500 \text{ m})} = \quad 3.13 \text{ m/s}$$

Exercise   If the head of the cattle rustler is 3.00 m below the level of the hole in the tank, where must he stand to get doused with water?

Answer   2.45 m from the base of the tank.

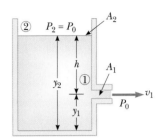

Figure 9.26   (Example 9.9) The water speed, v_1, from the hole in the side of the container is given by $v_1 = \sqrt{2gh}$.

9.8   OTHER APPLICATIONS OF BERNOULLI'S EQUATION

In this section we describe some common phenomena that can be explained, at least in part, by Bernoulli's equation.

First, consider the circulation of air around a thrown baseball. If the ball is not spinning (Fig. 9.27a), the motion of the airstream past the ball is nearly streamline.

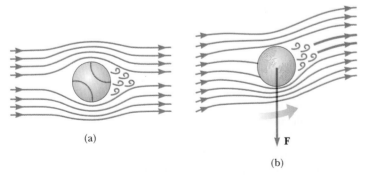

(a)

F

(b)

Figure 9.27   (a) The airstream around a nonrotating baseball moving from right to left. The streamlines represent the flow relative to the baseball. Note the symmetric region of turbulence behind the ball. (b) The airstream around a spinning baseball. The ball experiences a deflecting force because of the Bernoulli effect.

APPLICATION

Curve Balls and Sinkers.

APPLICATION

"Atomizers" in Perfume
Bottles and Paint Sprayers.

APPLICATION

Vascular Flutter and
Aneurysms.

Figure 9.28 A stream of air passing over a tube dipped in a liquid causes the liquid to rise in the tube as shown. This effect is used in perfume bottles and paint sprayers.

In this figure the ball is moving from right to left. Hence, from its point of view, the airstream is moving from left to right. A symmetric region of turbulence occurs behind the ball as shown. When the ball is spinning counterclockwise (Fig. 9.27b), layers of air near its surface are carried in the direction of spin because of viscosity. The combined effect of the steady flow of air and the air dragged along due to the spinning motion produces the streamlines shown in 9.24b and a corresponding turbulence pattern. The speed of the air below the ball is greater than the speed above the ball. Thus, from Bernoulli's equation, the air pressure above the ball is greater than the air pressure below the ball, and the ball experiences a downward deflecting force. When a pitcher wishes to throw a curve ball that deflects sideways, the spin axis should be vertical (perpendicular to the page in Fig. 9.27b). In contrast, if he wishes to throw a "sinker," the spin axis should be horizontal. Tennis balls and golf balls with spin also exhibit dynamic lift.

Many devices operate in the manner illustrated in Figure 9.28. A stream of air passing over an open tube reduces the pressure above the tube. This causes the liquid to rise into the airstream. The liquid is then dispersed into a fine spray of droplets. You might recognize that this so-called atomizer is used in perfume bottles and paint sprayers. The same principle is used in the carburetor of a gasoline engine. In that case, the low-pressure region in the carburetor is produced by air drawn in by the piston through the air filter. The gasoline vaporizes, mixes with the air, and enters the cylinder of the engine for combustion.

In a person with advanced arteriosclerosis, the Bernoulli effect produces a symptom called *vascular flutter*. In this situation, the artery is constricted as a result of accumulated plaque on its inner walls, as in Figure 9.29. To maintain a constant flow rate, the blood must travel faster than normal through the constriction. If the blood speed is sufficiently high in the constricted region, the artery may collapse under external pressure, causing a momentary interruption in blood flow. At this moment there is no Bernoulli effect, and the vessel reopens under arterial pressure. As the blood rushes through the constricted artery, the internal pressure drops and again the artery closes. Such variations in blood flow can be heard with a stethoscope. If the plaque becomes dislodged and ends up in a smaller vessel that delivers blood to the heart, the person can suffer a heart attack.

An aneurysm is a weakened spot on an artery where the artery walls have ballooned outward. Blood flows more slowly though this region, as can be seen from the equation of continuity, resulting in an increase in pressure in the vicinity of the aneurysm relative to the pressure in other parts of the artery. This condition is dangerous because the excess pressure can cause the artery to rupture.

The lift on an aircraft wing can also be explained, in part, by the Bernoulli effect. Airplane wings are designed so that the air speed above the wing is greater than that below. As a result, the air pressure above the wing is less than the pressure below, and there is a net upward force on the wing, called the "lift." Another factor influencing the lift on a wing is shown in Figure 9.30. The wing has a slight upward tilt that causes air molecules striking the bottom to be deflected downward. The air molecules bouncing off the wing at the bottom produce an upward force on the wing and a significant lift on the aircraft. Finally, turbulence also has an effect. If the wing is tilted too much, the flow of air across the upper surface becomes turbulent, and the pressure difference across the wing is not as great as that predicted by Bernoulli's equation. In an extreme case, this turbulence may cause the aircraft to stall.

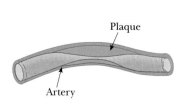

Figure 9.29 The flow of blood through a constricted artery.

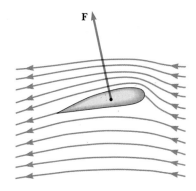

Figure 9.30 Streamline flow around an airplane wing. The pressure above is less than the pressure below, and there is a dynamic lift force upward.

Blow here

Applying Physics 5

The Bernoulli effect can be used to partially explain how a sailboat can accomplish the seemingly impossible task of sailing into the wind. How can this be done?

Explanation As shown in Figure 9.31, the wind blowing in the direction of the arrow causes the sail to billow out, taking on a shape similar to that of an airplane wing. The sail is oriented at an angle that is approximately halfway between the boat's axis (the keel axis) and the wind direction. The force of the wind on the sail, $\mathbf{F}_{wind}$, is due to the combined effect of the change in momentum of the wind as it bounces off the sail and the Bernoulli effect. The direction of this force is approximately perpendicular to the sail and tends to cause the boat to move sideways in the water. However, the keel (a broad vertical projection beneath the water) prevents this from happening because the water exerts a second force on the keel, $\mathbf{F}_{water}$, that is perpendicular to the keel axis. The resultant of the two forces, $\mathbf{F}_R$, is approximately in the forward direction as shown. The word almost is used here because a sailboat can move forward only when the wind direction is about 10 to 15° with respect to the forward direction. This means that in order to sail directly against the wind, a boat must follow a zigzag path, a procedure called tacking, so that the wind is always at some angle with respect to the direction of travel.

APPLICATION

Sailing into the Wind.

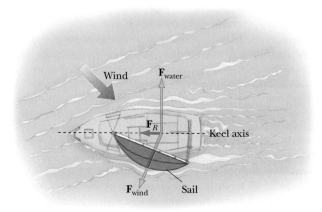

Wind $\mathbf{F}_{water}$

$\mathbf{F}_R$ ---- Keel axis

$\mathbf{F}_{wind}$ Sail

Figure 9.31
(Applying Physics 5)

Applying Physics 6

Consider the portion of a home plumbing system shown Figure 9.32. The water trap in the pipe below the sink captures a plug of water that prevents sewer gas from finding its way from the sewer pipe, up the sink drain, and into the home. Suppose the dishwasher is draining, so that water is moving to the right in the sewer pipe. What is the purpose of the vent, which is open to the air above the roof of the house? In which direction is air moving at the opening of the vent, upward or downward?

Figure 9.32   (Applying Physics 6)

Explanation   Let us imagine that the vent is not present, so that the drain pipe for the sink is simply connected through the trap to the sewer pipe. As water from the dishwasher moves to the right in the sewer pipe, the pressure in the sewer pipe is reduced below atmospheric pressure, according to Bernoulli's principle. The pressure at the drain in the sink is still at atmospheric pressure. Thus, this pressure differential can push the plug of water in the water trap of the sink down the drain pipe into the sewer pipe, removing it as a barrier to sewer gas. With the addition of the vent to the roof, the reduced pressure of the dishwasher water will result in air entering the vent pipe at the roof. This will keep the pressure in the vent pipe and the right-hand side of the sink drain pipe at a pressure close to atmospheric, so that the plug of water in the water trap will remain in the trap.

Optional Section

9.9   SURFACE TENSION, CAPILLARY ACTION, AND VISCOSITY

If you look closely at a dewdrop sparkling in the morning sunlight, you will find that the drop is spherical. The drop takes this shape because of a property of liquid surfaces called **surface tension.** In order to understand the origin of surface tension, consider a molecule at point A in a container of water, as in Figure 9.33. Although nearby molecules exert forces on this molecule, the net force on it is zero because it is completely surrounded by other molecules and hence is attracted equally in all directions. The molecule at B, however, is not attracted equally in all directions. Because there are no molecules above it to exert upward forces, the molecule at B is pulled toward the interior of the liquid. The contraction at the surface of the liquid ceases when the inward pull exerted on the surface molecules is balanced by the outward repulsive forces that arise from collisions with molecules in the interior of the liquid. **The net effect of this pull on all the surface molecules is to make the surface of the liquid contract and consequently to make the surface area of the liquid as small as possible.** Drops of water take on a spherical shape because a sphere has the smallest surface area for a given volume.

If you place a sewing needle very carefully on the surface of a bowl of water, you will find that the needle floats even though the density of steel is about eight times that of water. This also can be explained by surface tension. A close examination of the needle shows that it actually rests in a depression in the liquid surface, as shown in Figure 9.34. The water surface acts like an elastic membrane under tension. The weight of the needle produces a depression, thus increasing the surface area of the film. Molecular forces now act at all points along the depression in an attempt to restore the surface to its original horizontal position. The vertical components of these forces act to balance the force of gravity acting on the needle.

The **surface tension,** γ, in a film of liquid is defined as the ratio of the magnitude of the surface tension force, **F,** to the length along which the force acts:

$$\gamma \equiv \frac{F}{L} \qquad \text{[9.18]}$$

The SI units of surface tension are newtons per meter, and values for a few representative materials are given in Table 9.3.

The concept of surface tension can be thought of as the energy content of the fluid at its surface per unit surface area. To see that this is reasonable, we can manipulate the units of surface tension as

$$[\gamma] = \frac{N}{m} = \frac{N \cdot m}{m^2} = \frac{J}{m^2}$$

In general, **any equilibrium configuration of an object is one in which the energy is a minimum.** Consequently, a fluid will take on a shape such that its surface area is small as possible. For a given volume, a spherical shape is the one that has the smallest surface area; therefore, a drop of water takes on a spherical shape.

An apparatus used to measure the surface tension of liquids is shown in Figure 9.35. A circular wire with a circumference L is lifted from a body of liquid. The surface film clings to the inside and outside edges of the wire, holding back the wire and causing the spring to stretch. If the spring is calibrated, one can measure the force required to overcome the surface tension of the liquid. In this case, the surface tension is given by

$$\gamma = \frac{F}{2L}$$

Figure 9.33 The net force on a molecule at A is zero because such a molecule is completely surrounded by other molecules. The net force on a surface molecule at B is downward because it is not completely surrounded by other molecules.

Figure 9.34 End view of a needle resting on the surface of water. The components of surface tension balance the weight force.

Figure 9.35 An apparatus for measuring the surface tension of liquids. The force on the wire ring is measured just before it breaks free of the liquid.

TABLE 9.3 Surface Tensions for Various Liquids		
Liquid	T(°C)	**Surface Tension (N/m)**
Ethyl alcohol	20	0.022
Mercury	20	0.465
Soapy water	20	0.025
Water	20	0.073
Water	100	0.059

We must use $2L$ for the length because the surface film exerts forces on the inside and outside of the ring.

The surface tension of liquids decreases with increasing temperature. This occurs because the faster moving molecules of a hot liquid are not bound together as strongly as are those in a cooler liquid. Furthermore, certain ingredients added to liquids decrease surface tension. For example, soap or detergent decreases the surface tension of water. This reduction in surface tension makes it easier for soapy water to penetrate the cracks and crevices of your clothes to clean them better than plain water. An effect similar to this occurs in the lungs. The surface tissue of the air sacs in the lungs contains a fluid that has a surface tension of about 0.050 N/m. A liquid with a surface tension this high would make it very difficult for the lungs to expand as one inhales. However, as the area of the lungs increases with inhalation, the body secretes into the tissue a substance that gradually reduces the surface tension of the liquid. At full expansion, the surface tension of the lung fluid can drop to as low as 0.005 N/m.

APPLICATION

Surfactants.

EXAMPLE 9.10   Walking on Water

In this example, we shall illustrate how an insect is supported on the surface of water by surface tension. Let us assume that the insect's "foot" is spherical. When the insect steps onto the water with all six legs, a depression is formed in the water around each foot, as shown in Figure 9.36a. The surface tension of the water produces upward forces on the water that tend to restore the water surface to its normally flat shape. If the insect has a mass of 2.0×10^{-5} kg and if the radius of each foot is 1.5×10^{-4} m, find the angle θ.

Figure 9.36   (Example 9.10) (a) One foot of an insect resting on the surface of water. (b) This water-strider resting on the surface of a lake is able to remain on the surface, rather than sink, because an upward surface tension force acts on each leg, which balances the weight of the insect. *(Hermann Eisenbeiss/Photo Researchers, Inc.)*

Solution   From the definition of surface tension, we can find the net force F directed tangential to the depressed part of the water surface:

$$F = \gamma L$$

The length L along which this force acts is equal to the distance around the insect's foot, $2\pi r$. (It is assumed that the insect depresses the water surface such that the radius of the depression is equal to the radius of the foot.) Thus,

$$F = \gamma 2\pi r$$

and the net vertical force is

$$F_v = \gamma 2\pi r \cos\theta$$

Because the insect has six legs, this upward force must equal one sixth the weight of the insect, assuming its weight is equally distributed on all six feet. Thus,

$$\gamma 2\pi r \cos\theta = \tfrac{1}{6} w = \tfrac{1}{6} mg$$

$$(1) \qquad \cos\theta = \frac{mg}{12\pi r\gamma} = \frac{(2.0 \times 10^{-5}\text{ kg})(9.80\text{ m/s}^2)}{12\pi(1.5 \times 10^{-4}\text{ m})(0.073\text{ N/m})}$$

$$\theta = \boxed{62°}$$

Note that if the weight of the insect were great enough to make the right side of (1) greater than unity, a solution for θ would be impossible because the cosine can never be greater than unity. Under these conditions, the insect would sink.

The Surface of Liquids

If you have ever closely examined the surface of water in a glass container, you may have noticed that the surface of the liquid near the walls of the glass curves upward as you move from the center to the edge, as shown in Figure 9.37a. However, if mercury is placed in a glass container, the mercury surface curves downward, as in Figure 9.37b. These surface effects can be explained by considering the forces between molecules. In particular, we must consider the forces that the molecules of the liquid exert on one another and the forces that the molecules of the glass surface exert on those of the liquid. In general terms, forces between like molecules, such as the forces between water molecules, are called *cohesive forces* and forces between unlike molecules, such as those of glass on water, are called *adhesive forces*.

Water tends to cling to the walls of the glass because the adhesive forces between the liquid molecules and the glass molecules are *greater* than the cohesive forces between the liquid molecules. In effect, the liquid molecules cling to the

The surface of mercury *(left)* curves downward in a glass container, while the surface of water *(right)* curves upward as you move from the center to the edge. *(Charles D. Winters)*

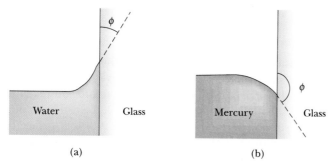

(a)

(b)

Figure 9.37 A liquid in contact with a solid surface. (a) For water, the adhesive force is greater than the cohesive force. (b) For mercury, the adhesive force is less than the cohesive force.

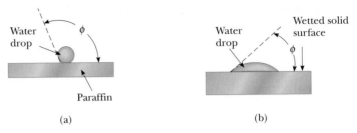

(a)   (b)

Figure 9.38 (a) The contact angle between water and paraffin is about 107°. In this case, the cohesive force is greater than the adhesive force. (b) When a chemical called a wetting agent is added to the water, it wets the paraffin surface, and $\phi < 90°$. In this case the adhesive force is greater than the cohesive force.

APPLICATION

Detergents and Waterproofing Agents.

Figure 9.39 A liquid rises in a narrow tube because of capillary action, a result of surface tension and adhesive forces.

surface of the glass rather than fall back into the bulk of the liquid. When this condition prevails, the liquid is said to wet the glass surface. The surface of the mercury curves downward near the walls of the container because the cohesive forces between the mercury atoms are greater than the adhesive forces between mercury and glass. That is, a mercury atom near the surface is pulled more strongly toward other mercury atoms than toward the glass surface; hence mercury does not wet the glass surface.

The angle ϕ between the solid surface and a line drawn tangent to the liquid at the surface is called *the angle of contact* (Fig. 9.38a and 9.38b). Note that ϕ is less than 90° for any substance in which adhesive forces are stronger than cohesive forces and greater than 90° if cohesive forces predominate. For example, if a drop of water is placed on paraffin, the contact angle is approximately 107° (Fig. 9.38a). If certain chemicals, called wetting agents or detergents, are added to the water, the contact angle becomes less than 90°, as shown in Figure 9.38b. The addition of such substances to water is of value when one wants to ensure that water makes intimate contact with a surface and penetrates it. For this reason, detergents are added to water to wash clothes or dishes. On the other hand, it is often necessary to keep water from making intimate contact with a surface, as in waterproofing clothing, where a situation somewhat the reverse of that shown in Figure 9.38 is called for. The clothing is sprayed with a waterproofing agent, which changes ϕ from less than 90° to greater than 90°. Thus, the water beads up on the surface and does not easily penetrate the clothing.

Capillary Action

Capillary tubes are tubes in which the diameter of the opening is very small. In fact, the word *capillary* means "hair-like." If such a tube is inserted into a fluid for which adhesive forces dominate over cohesive forces, the liquid will rise into the tube, as shown in Figure 9.39. The rising of the liquid in the tube can be explained in terms of the shape of the surface of the liquid and in terms of the surface tension effects in the liquid. At the point of contact between liquid and solid, the upward force of surface tension is directed as shown in Figure 9.39. From Equation 9.18, the magnitude of this force is

$$F = \gamma L = \gamma(2\pi r)$$

We use $L = 2\pi r$ here because the liquid is in contact with the surface of the tube at all points around its circumference. The vertical component of this force due to surface tension is

$$F_v = \gamma(2\pi r)(\cos \phi) \qquad \text{[9.19]}$$

In order for the liquid in the capillary tube to be in equilibrium, this upward force must be equal to the weight of the cylinder of water of height h inside the capillary tube. The weight of this water is

$$w = Mg = \rho Vg = \rho g \pi r^2 h \qquad \text{[9.20]}$$

Equating F_v in Equation 9.19 to w in Equation 9.20, we have

$$\gamma(2\pi r)(\cos \phi) = \rho g \pi r^2 h$$

Thus, the height to which water is drawn into the tube is

$$h = \frac{2\gamma}{\rho g r} \cos \phi \qquad \text{[9.21]}$$

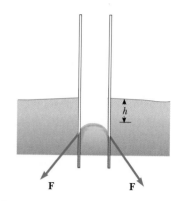

Figure 9.40 When cohesive forces between molecules of the liquid exceed adhesive forces, the liquid level in the capillary tube is below the surface of the surrounding fluid.

If a capillary tube is inserted into a liquid in which cohesive forces dominate over adhesive forces, the level of the liquid in the capillary tube will be below the surface of the surrounding fluid, as shown in Figure 9.40. An analysis similar to that done above would show that the distance h the surface is depressed is given by Equation 9.21.

Capillary tubes are often used to draw small samples of blood from a needle prick in the skin. Capillary action must also be considered in the construction of concrete-block buildings because water seepage through capillary pores in the blocks or the mortar may cause damage to the inside of the building. To prevent this, the blocks are usually coated with a waterproofing agent either outside or inside the building. Water seepage through a wall is an undesirable effect of capillary action, but paper towels employ capillary action in a useful manner to absorb spilled fluids.

EXAMPLE 9.11 How High Does the Water Rise?

Find the height to which water would rise in a capillary tube with a radius equal to 5.0×10^{-5} m. Assume that the angle of contact between the water and the material of the tube is small enough to be considered zero.

Solution The surface tension of water is 0.073 N/m. For a contact angle of 0°, we have $\cos \phi = \cos 0° = 1$, so that Equation 9.21 gives

$$h = \frac{2\gamma}{\rho g r} = \frac{2(0.073 \text{ N/m})}{(1.00 \times 10^3 \text{ kg/m}^3)(9.80 \text{ m/s}^2)(5.0 \times 10^{-5} \text{ m})} = \boxed{0.29 \text{ m}}$$

Viscosity

It is considerably easier to pour water out of a container than to pour syrup. This is because syrup has a higher viscosity than water. In a general sense, *viscosity refers to the internal friction of a fluid.* It is very difficult for layers of a viscous fluid to slide

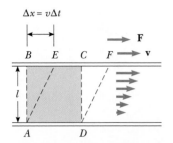

$\Delta x = v\Delta t$

Figure 9.41 A layer of liquid between two solid surfaces in which the lower surface is fixed and the upper surface moves to the right with a velocity **v.**

Coefficient of viscosity ▶

past one another. Likewise, it is difficult for one solid surface to slide past another if there is a highly viscous fluid, such as soft tar, between them.

To better understand the concept of viscosity, consider a liquid layer placed between two solid surfaces, as in Figure 9.41. The lower surface is fixed in position, and the top surface moves to the right with a velocity **v** under the action of an external force **F.** Because of this motion, a portion of the liquid is distorted from its original shape, *ABCD,* at one instant to the shape *AEFD* a moment later. You will recognize that the liquid has undergone a constantly increasing shear strain. Ideal fluids that have no internal friction forces between adjacent layers could not have their shape distorted. However, in viscous fluids, there are cohesive forces between molecules in the various layers that can lead to strains that change with time. By definition, the shear stress on the liquid is

$$\text{Shear stress} \equiv \frac{F}{A}$$

where A is the area of the top plate. Furthermore, the shear strain is defined as

$$\text{Shear strain} \equiv \frac{\Delta x}{L}$$

The velocity of the fluid changes from zero at the lower plate to **v** at the upper. Thus, in a time Δt, the fluid at the upper plate moves a distance $\Delta x = v \Delta t$. Therefore,

$$\frac{\text{Shear strain}}{\Delta t} = \frac{\Delta x/L}{\Delta t} = \frac{v}{L}$$

This equation states that the rate of change of the shearing strain is v/L.

The **coefficient of viscosity,** η (lowercase Greek letter eta), for the fluid is defined as the ratio of the shearing stress to the rate of change of the shear strain:

$$\eta \equiv \frac{FL}{Av} \qquad [9.22]$$

The SI units of viscosity are $\text{N} \cdot \text{s/m}^2$. You should note that the units of viscosity in many reference sources are often expressed in $\text{dyne} \cdot \text{s/cm}^2$, called 1 **poise** in honor of the French scientist J. L. Poiseuille (1799–1869). The relationship between the SI unit of viscosity and the poise is

$$1 \text{ poise} = 10^{-1} \text{ N} \cdot \text{s/m}^2 \qquad [9.23]$$

Small viscosities are often expressed in centipoise (cp), where $1 \text{ cp} = 10^{-2}$ poise. The coefficients of viscosity for some common substances are listed in Table 9.4.

TABLE 9.4 The Viscosities of Various Fluids

Fluid	$T(°C)$	Viscosity $\eta(\text{N} \cdot \text{s/m}^2)$
Water	20	1.0×10^{-3}
Water	100	0.3×10^{-3}
Whole blood	37	2.7×10^{-3}
Glycerin	20	1500×10^{-3}
10-wt motor oil	30	250×10^{-3}

Poiseuille's Law

Figure 9.42 shows a section of a tube containing a fluid under a pressure P_1 at the left end and a pressure P_2 at the right. Because of this pressure difference, the fluid will flow through the tube. The rate of flow (volume per unit time) depends on the pressure difference $(P_1 - P_2)$, the dimensions of the tube, and the viscosity of the fluid. The result, known as **Poiseuille's law,** is

$$\text{Rate of flow} = \frac{\Delta V}{\Delta t} = \frac{(P_1 - P_2)(\pi R^4)}{8L\eta} \qquad [9.24]$$

 ◀ Poiseuille's law

where R is the radius of the tube, L is its length, and η is the coefficient of viscosity. We shall not attempt to derive this equation here because the methods of integral calculus are required. However, you should note that the equation does agree with common sense. That is, it is reasonable that the rate of flow should increase if the pressure difference across the tube or the tube radius increases. Likewise, the flow rate should decrease if the viscosity of the fluid or the length of the tube increases. Thus, the presence of R and the pressure difference in the numerator of Equation 9.24 and of L and η in the denominator makes sense.

From Poiseuille's law, we see that in order to maintain a constant flow rate, the pressure difference across the tube has to increase if the viscosity of the fluid increases. This is important when one considers the flow of blood through the circulatory system. The viscosity of blood increases as the number of red blood cells rises. Thus, blood with a high concentration of red blood cells requires greater pumping pressure from the heart to keep it circulating than does blood of lower red blood cell concentration.

Note that the flow rate varies as the radius of the tube raised to the fourth power. Consequently, if a constriction occurs in a vein or artery, the heart will have to work considerably harder in order to produce a higher pressure drop and hence to maintain the required flow rate.

Figure 9.42 Velocity profile of a fluid flowing through a uniform pipe of circular cross-section. The rate of flow is given by Poiseuille's law. Note that the fluid velocity is greatest at the middle of the pipe.

EXAMPLE 9.12 A Blood Transfusion

A patient receives a blood transfusion through a needle of radius 0.20 mm and length 2.0 cm. The density of blood is 1050 kg/m³. The bottle supplying the blood is 0.50 m above the patient's arm. What is the rate of flow through the needle?

Solution The pressure differential between the level of the blood and the patient's arm is

$$P_1 - P_2 = \rho gh = (1050 \text{ kg/m}^3)(9.80 \text{ m/s}^2)(0.50 \text{ m}) = 5.15 \times 10^3 \text{ Pa}$$

Thus, the rate of flow, from Poiseuille's law, is

$$\frac{\Delta V}{\Delta t} = \frac{(P_1 - P_2)(\pi R^4)}{8L\eta}$$

$$= \frac{(5.15 \times 10^3 \text{ Pa})(\pi)(2.0 \times 10^{-4} \text{ m})^4}{8(2.0 \times 10^{-2} \text{ m})(2.7 \times 10^{-3} \text{ N·s/m}^2)} = 6.0 \times 10^{-8} \text{ m}^3/\text{s}$$

Exercise How long will it take to inject 1 pint (500 cm³) of blood into the patient?

Answer 140 min

Reynolds Number

As we mentioned earlier, at sufficiently high velocities, fluid flow changes from simple streamline flow to turbulent flow — that is, flow characterized by a highly irregular motion of the fluid. It is found experimentally that the onset of turbulence in a tube is determined by a dimensionless factor called the **Reynolds number,** given by

Reynolds number ▶

$$RN = \frac{\rho v d}{\eta} \qquad \text{[9.25]}$$

where ρ is the density of the fluid, v is the average speed of the fluid along the direction of flow, d is the diameter of the tube, and η is the viscosity of the fluid. If RN is below about 2000, the flow of fluid through a tube is streamline; turbulence occurs if RN is above 3000. In the region between 2000 and 3000, the flow is unstable, meaning that the fluid can move in streamline flow but any small disturbance will cause its motion to change to turbulent flow.

EXAMPLE 9.13   Turbulent Flow of Blood

Determine the speed at which blood flowing through an artery of diameter 0.20 cm would become turbulent. Assume that the density of blood is 1.05×10^3 kg/m³ and that its viscosity is 2.7×10^{-3} N·s/m².

Solution   At the onset of turbulence, the Reynolds number is 3000. Thus, the speed of the blood would have to be

$$v = \frac{\eta(RN)}{\rho d} = \frac{(2.7 \times 10^{-3} \text{ N·s/m}^2)(3000)}{(1.05 \times 10^3 \text{ kg/m}^3)(0.20 \times 10^{-2} \text{ m})} = \boxed{3.9 \text{ m/s}}$$

Optional Section

9.10   TRANSPORT PHENOMENA

When a fluid flows through a tube, the basic mechanism that produces the flow is a difference in pressure across the ends of the tube. This pressure difference is responsible for the transport of a mass of fluid from one location to another. The fluid may also move from place to place because of a second mechanism, one that depends on a concentration difference between two points in the fluid, as opposed to a pressure difference. When the concentration (the number of molecules per unit volume) is higher at one location than at another, molecules will flow from the point where the concentration is high to the point where it is lower. The two fundamental processes involved in fluid transport resulting from concentration differences are called *diffusion* and *osmosis*. The following sections examine the nature and importance of these processes.

Diffusion

You can imagine what happens when someone wearing a strong shaving lotion or perfume strolls into a crowded room. All eyes turn to seek out the source of the delightful smell. The aroma spreads through the room by a process called diffusion.

> In a diffusion process, molecules move from a region where their concentration is high to a region where their concentration is lower.

Figure 9.43 When the concentration of gas molecules on the left side of the container exceeds the concentration on the right side, there will be a net motion (diffusion) of molecules from left to right.

That is, the molecules of the lotion or perfume move from the source (near the person's face), where there are many molecules per unit volume, throughout the room, to regions where the concentration of these molecules is lower. Although the example used is one of diffusion in air, the process also occurs in liquids and, to a lesser extent, in solids. For example, if a drop of food coloring is placed in a glass of water, the coloring soon spreads throughout the liquid by diffusion. In either case, diffusion ceases when there is a uniform concentration at all locations in the fluid.

To understand why diffusion occurs, consider Figure 9.43, which represents a container in which a high concentration of molecules has been introduced into the left side. For example, this could be accomplished by releasing a few drops of perfume into the left side of the container. The dashed line in Figure 9.43 represents an imaginary barrier separating the region of high concentration from the region of lower concentration. Because the molecules are moving with high speeds in random directions, many of them will cross the imaginary barrier moving from left to right. Very few molecules of perfume will pass through this area moving from right to left simply because there are very few of them on the right side of the container at any instant. Thus, there will always be a *net* movement from the region where there are many molecules to the region where there are fewer molecules. For this reason, the concentration on the left side of the container will decrease in time and that on the right side will increase. Once a concentration equilibrium has been reached, there will be no *net* movement across the cross-sectional area. That is, when the concentration is the same on both sides, the number of molecules diffusing from right to left in a given time interval will equal the number moving from left to right in the same time interval.

The basic equation for diffusion is **Fick's law,** which in equation form is

$$\text{Diffusion rate} = \frac{\text{mass}}{\text{time}} = \frac{\Delta M}{\Delta t} = DA \left(\frac{C_2 - C_1}{L} \right) \qquad [9.26]$$

◀ Fick's law

where D is a constant of proportionality. The left side of this equation is called the diffusion rate and is a measure of the mass being transported per unit time. This equation says that

> the rate of diffusion is proportional to the cross-sectional area A and to the change in concentration per unit distance, $(C_2 - C_1)/L$, which is called the concentration gradient.

The concentrations C_1 and C_2 are measured in kilograms per cubic meter. The proportionality constant D is called the **diffusion coefficient** and has units of square meters per second. Table 9.5 lists diffusion coefficients for a few substances.

TABLE 9.5   Diffusion Coefficients for Various Substances at 20°C

Substance	$D(m^2/s)$
Oxygen through air	6.4×10^{-5}
Oxygen through tissue	1×10^{-11}
Oxygen through water	1×10^{-9}
Sucrose through water	5×10^{-10}
Hemoglobin through water	76×10^{-11}

The Size of Cells and Osmosis

Diffusion through cell membranes is extremely vital in carrying oxygen to the cells of the body and in removing carbon dioxide and other waste products from them. Oxygen is required by the cells for those metabolic processes in which substances are either synthesized or broken down. In such metabolic processes, the cell uses up oxygen and produces carbon dioxide as a by-product. A fresh supply of oxygen diffuses from the blood, where its concentration is high, into the cell, where its concentration is low. Likewise, carbon dioxide diffuses from the cell into the blood, where it is in lower concentration. Water, ions, and other nutrients also pass into and out of cells by diffusion.

A common characteristic of cells in all plants and animals is their extremely small size. The adult human body contains literally trillions of cells. In order to understand why cells are so small, we must consider the relationship between the surface area of an object and its volume.

Let us consider a cube 2 cm on a side. The area of one of its faces is 2 cm × 2 cm = 4 cm², and because a cube has six sides, the total surface area is 24 cm². Its volume is 2 cm × 2 cm × 2 cm = 8 cm³. Hence, the ratio of surface area to volume is 24/8 = 3. Now consider a larger cube, one measuring 3 cm on a side. Repeating the calculations gives us a surface area of 54 cm² and a volume of 27 cm³. In this case, the ratio of surface area to volume is 54/27 = 2. Thus, we see that as the size of an object decreases, the ratio of its surface area to its volume increases. This, of course, says that a small cell has a larger surface-area-to-volume ratio than a large cell. But how does this pertain to the operation of a cell?

A cell can function properly only if it can (a) rapidly receive vital substances such as oxygen and (b) rapidly eliminate waste products. If such substances are to readily move into and out of cells, the cells should have a large surface area. However, if the volume of the cell is too large, it could take a considerable period of time for the nutrients to diffuse into the interior of the cell where they are needed. Under optimum conditions, the surface area of the cell should be large enough so that the exposed membrane area can exchange materials effectively while at the same time the volume should be small enough so that materials can reach or leave particular locations rapidly. To reach these optimum conditions, a small cell with its high surface-area-to-volume ratio is necessary.

As we have seen, the movement of material through cell membranes is necessary for the efficient functioning of cells. The diffusion of material through a membrane is partially determined by the size of the pores (holes) in the membrane wall. That is, small molecules, such as water, may pass through the pores easily and larger molecules, such as sugar, may pass through only with difficulty or not at all. A membrane that allows passage of some molecules but not others is called a selectively permeable membrane.

> **Osmosis** is defined as the movement of water from a region where its concentration is high, across a selectively permeable membrane, into a region where its concentration is lower.

As in the case of diffusion, osmosis continues until the concentrations on the two sides of the membrane are equal. Osmosis is often described simply as the diffusion of water across a membrane.

To understand the effect of osmosis on living cells, let us consider a particular cell in the body that contains a sugar concentration of 1%. (That is, 1 g of sugar is dissolved in enough water to make 100 ml of solution.) Now assume that this cell is immersed in a 5% sugar solution (5 g of sugar dissolved in enough water to make 100 ml). In such a situation, water would diffuse from inside the cell, where its concentration is higher, across the cell wall membrane, to the outside solution, where the concentration of water is lower. This loss of water from the cell would cause it to shrink and perhaps become damaged through dehydration. If the concentrations were reversed, water would diffuse into the cell, causing it to swell and perhaps burst. It should be obvious from this description that normal osmotic relationships must be maintained in the body. If solutions are introduced into the body intravenously, care must be taken to ensure that these solutions do not disturb the osmotic balance of the body because such a disturbance could lead to cell damage. For example, if 9% saline solution surrounds a red blood cell, the cell will shrink. On the other hand, if the saline solution is about 1%, the cell will eventually burst.

In the body, blood is cleansed of impurities by osmosis as it flows through the kidneys (see Fig. 9.44a). Arterial blood first passes through a bundle of capillaries known as a glomerulus where most of the waste products along with some essential salts and minerals are removed from the blood. From the glomerulus, a narrow tube emerges that is in intimate contact with other capillaries throughout its length. In its passage through the tubules, most of the essential elements are returned to the blood, and the waste products are not allowed to re-enter and are eventually removed as urine.

If the kidneys fail, an artificial kidney or a dialysis machine can filter the blood. Figure 9.44b shows how this is done. Blood from an artery in the arm is mixed with heparin, a blood thinner, and allowed to pass through a tube covered with a semipermeable membrane. This tubing is immersed in a bath of a dialysate fluid with the same chemical composition as purified blood. Waste products from the blood enter the dialysate by diffusion through the membrane walls. The filtered blood is then returned to a vein.

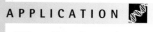

APPLICATION

Kidney Function and Dialysis.

Motion Through a Viscous Medium

When an object falls through air, its motion is impeded by the force of air resistance. In general, this force is dependent on the shape of the falling object and on its velocity. This viscous drag acts on all falling objects, but the exact details of the

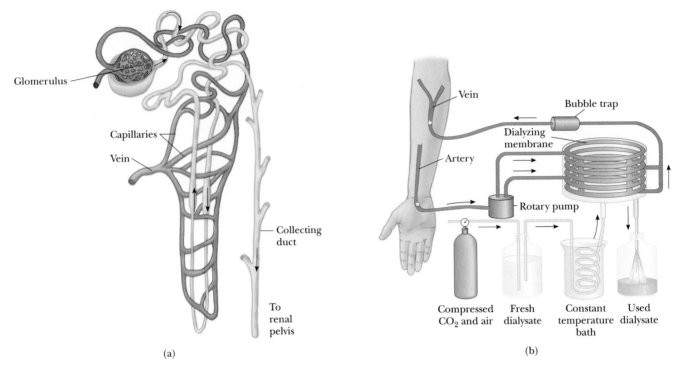

(a)

(b)

Figure 9.44 (a) Diagram of a single nephron in the human kidney system. (b) An artificial kidney.

motion can be calculated only for a few cases in which the object has a simple shape, such as a sphere. In this section, we shall examine the motion of a tiny spherical object falling slowly through a viscous medium.

In 1845 a scientist named George Stokes found that the magnitude of the resistive force on a very small spherical object of radius r falling slowly through a fluid of viscosity η with speed v is given by

$$F_r = 6\pi\eta r v \qquad \text{[9.27]}$$

This equation, called **Stokes's law,** has many important applications. For example, it describes the sedimentation of particulate matter in blood samples. It was used by Robert Millikan (1886–1953) to calculate the radius of charged oil droplets falling through air. From this, Millikan was ultimately able to determine the smallest known unit of electric charge. Millikan was awarded the Nobel prize in 1923 for this pioneering work on elemental charge.

As a sphere falls through a viscous medium, three forces act on it, as shown in Figure 9.45: $\mathbf{F}_r$ is the force of frictional resistance, $\mathbf{B}$ is the buoyant force of the fluid, and $\mathbf{w}$ is the force of gravity acting on the sphere, whose magnitude is given by

$$w = \rho g V = \rho g(\tfrac{4}{3}\pi r^3)$$

Figure 9.45 A sphere falling through a viscous medium. The forces acting on the sphere are the resistive frictional force $\mathbf{F}_r$, the buoyant force $\mathbf{B}$, and the weight of the sphere $\mathbf{w}$.

where ρ is the density of the sphere and $\tfrac{4}{3}\pi r^3$ is its volume. According to Archimedes's principle, the magnitude of the buoyant force is equal to the weight of the fluid displaced by the sphere:

$$B = \rho_f gV = \rho_f g(\tfrac{4}{3}\pi r^3)$$

where ρ_f is the density of the fluid.

At the instant the sphere begins to fall, the force of frictional resistance is zero because the speed of the sphere is zero. As it accelerates, the speed increases and so does $\mathbf{F}_r$. Finally, at a speed called the **terminal speed** v_t, *the net force goes to zero.* This occurs when the net upward force balances the downward force of gravity. Hence, the sphere reaches terminal speed when

◀ Terminal speed

$$F_r + B = w$$

or

$$6\pi\eta r v_t + \rho_f g(\tfrac{4}{3}\pi r^3) = \rho g(\tfrac{4}{3}\pi r^3)$$

When this is solved for v_t, we get

$$v_t = \frac{2r^2 g}{9\eta}(\rho - \rho_f)$$ [9.28]

EXAMPLE 9.14 A Falling Pearl

A pearl of density 2.0×10^3 kg/m^3 and radius 2.0 mm falls through a liquid shampoo of density 1.4×10^3 km/m^3 and viscosity 0.50×10^{-3} N·s/m^2. Find the terminal speed of the pearl.

Solution Substituting the given values into Equation 9.28, we have

$$v_t = \frac{2r^2 g}{9\eta}(\rho - \rho_f)$$

$$= \frac{2(2.0 \times 10^{-3}\text{ m})^2(9.80\text{ m/s}^2)}{9(0.50 \times 10^{-3}\text{ N·s/m}^2)}(2.0 \times 10^3\text{ kg/m}^3 - 1.4 \times 10^3\text{ kg/m}^3)$$

$$= 1.1 \times 10^{-2}\text{ m/s}$$

By spreading their arms and legs out from their bodies while keeping the planes of their bodies parallel to the ground, skydivers experience maximum air drag. The forces acting on this skydiver are his weight (the force of gravity) and the force of air resistance. When these two forces balance each other, the skydiver has zero acceleration and reaches a terminal speed of about 60 m/s. *(Heinz Fischer/The Image Bank)*

Sedimentation and Centrifugation

If an object is not spherical, we can still use the basic approach just described to determine its terminal speed. The only difference will be that we shall not be able to use Stokes's law for the resistive force. Instead, let us assume that the resistive force has a magnitude given by $F_r = kv$, where k is a coefficient of frictional resistance that must be determined experimentally. As we discussed previously, the object reaches its terminal speed when the downward force of gravity is balanced by the net upward force or

$$w = B + F_r \qquad [9.29]$$

where B is the buoyant force, given by $B = \rho_f g V$.

We can use the fact that volume, V, of the displaced fluid is related to the density of the falling object, ρ, by $V = m/\rho$. Hence, we can express the buoyant force as

$$B = \frac{\rho_f}{\rho} mg$$

Let us substitute this expression for B and $F_r = kv_t$ into Equation 9.29 (terminal speed condition):

$$mg = \frac{\rho_f}{\rho} mg + kv_t$$

or

$$v_t = \frac{mg}{k}\left(1 - \frac{\rho_f}{\rho}\right) \qquad [9.30]$$

ω

Figure 9.46 Simplified diagram of a centrifuge (top view).

The terminal speed for particles in biological samples is usually quite small. For example, the terminal speed for blood cells falling through plasma is about 5 cm/h in the gravitational field of the Earth. The terminal speeds for the molecules that make up a cell are many orders of magnitude smaller than this because of their much smaller mass. The speed at which materials fall through a fluid is called the *sedimentation rate*. This number is often important in clinical analysis.

It is often desired to increase the sedimentation rate in a fluid. A common method used to accomplish this is to increase the effective acceleration g that appears in Equation 9.30. A fluid containing various biological molecules is placed in a centrifuge and whirled at very high angular speeds (Fig. 9.46). Under these conditions, the particles experience a large radial acceleration, $a_c = v^2/r = \omega^2 r$, which is much greater than the free-fall acceleration, and so we can replace g in Equation 9.30 by $\omega^2 r$:

$$v_t = \frac{m\omega^2 r}{k}\left(1 - \frac{\rho_f}{\rho}\right) \qquad [9.31]$$

This equation indicates that those particles having the greatest mass will have the largest terminal speed. Therefore, the most massive particles will settle out on the bottom of a test tube first.

EXAMPLE 9.15 The Spinning Test Tube

A centrifuge rotates at 50 000 rev/min, which corresponds to an angular frequency of 5240 rad/s (a typical speed). A test tube placed in this device has its top 5.0 cm from the axis of rotation and its bottom 13 cm from this axis. Find the effective value of g at the midpoint of the test tube, which corresponds to a distance 9.0 cm from the axis of rotation.

Solution The acceleration experienced by the particles of the tube at a distance $r = 9.0$ cm from the axis of rotation is given by

$$a_c = \omega^2 r = \left(5240 \, \frac{\text{rad}}{\text{s}}\right)^2 (9.0 \times 10^{-2} \text{ m}) = \boxed{2.5 \times 10^6 \text{ m/s}^2}$$

Exercise If the mass of the contents of the test tube is 15 g, find the force that the bottom of the tube must exert on the contents of the tube to provide this centripetal acceleration. Assume a centripetal acceleration equal to that found at the midpoint of the tube.

Answer 3.7×10^4 N, or about 8000 lb! (Because of such large forces, the base of the tube in a centrifuge must be rigidly supported to keep the glass from shattering.)

SUMMARY

Matter is normally classified as being in one of three states: solid, liquid, or gaseous.

The elastic properties of a solid can be described using the concepts of stress and strain. **Stress** is related to the force producing a deformation; **strain** is a measure of the degree of deformation. Stress is proportional to strain, and the constant of proportionality is the **elastic modulus:**

$$\text{Elastic modulus} \equiv \frac{\text{stress}}{\text{strain}} \qquad [9.1]$$

Three common types of deformation are, (1) the resistance of a solid to elongation or compression, characterized by **Young's modulus, Y;** (2) the resistance to displacement of the faces of a solid sliding past each other, characterized by the **shear modulus, S;** (3) the resistance of a solid or liquid to a volume change, characterized by the **bulk modulus, B.**

In the SI system, pressure is expressed in pascals (Pa), where 1 Pa $\equiv$ 1 N/m^2.

The **density, ρ,** of a substance of uniform composition is its mass per unit volume — kilograms per cubic meter (kg/m^3) in the SI system:

$$\rho \equiv \frac{M}{V} \qquad [9.6]$$

The **pressure, P,** in a fluid is the force per unit area that the fluid exerts on an object immersed in it:

$$P \equiv \frac{F}{A} \qquad [9.7]$$

The pressure in a fluid varies with depth, h, according to the expression

$$P = P_0 + \rho g h \qquad [9.10]$$

where P_0 is atmospheric pressure (1.01×10^5 Pa) and ρ is the density of the fluid.

Pascal's principle states that, when pressure is applied to an enclosed fluid, the pressure is transmitted undiminished to every point of the fluid and to the walls of the containing vessel.

When an object is partially or fully submerged in a fluid, the fluid exerts an upward force, called the **buoyant force,** on the object. According to **Archimedes's principle,** the magnitude of the buoyant force is equal to the weight of the fluid displaced by the object.

Certain aspects of a fluid in motion can be understood by assuming that the fluid is nonviscous and incompressible and that its motion is in a steady state with no turbulence.

1. The flow rate through the pipe is a constant, which is equivalent to stating that the product of the cross-sectional area, A, and the speed, v, at any point is constant.

$$A_1 v_1 = A_2 v_2 \qquad [9.14]$$

This relation is referred to as the **equation of continuity.**

2. The sum of the pressure, the kinetic energy per unit volume, and the potential energy per unit volume has the same value at all points along a streamline:

$$P + \tfrac{1}{2}\rho v^2 + \rho g y = \text{constant} \qquad [9.16]$$

This is known as **Bernoulli's equation.**

MULTIPLE-CHOICE QUESTIONS

1. A test is performed on copper wire of radius 1.00 mm to determine how it will withstand the buildup of ice on it when used as a power line. It is found that a force of 300 N will cause it to stretch from a length of 100.00 m to a length of 100.11 m, and that when the weight is removed, the wire snaps back to its original length. The stress applied to the wire is _____, the strain is _____ and the elastic modulus is _____, in SI units.
   (a) 8.68×10^{10}, 0.11×10^{-2}, 9.55×10^7
   (b) 9.55×10^7, 0.11×10^{-2}, 8.68×10^{10}
   (c) 9.55×10^7, 8.68×10^{10}, 0.11×10^{-2}
   (d) 0.11×10^{-2}, 8.68×10^{10}, 9.55×10^7
   (e) 8.68×10^{10}, 9.55×10^7, 0.11×10^{-2}

2. As the experiment of Question 1 continues, it is found out that when a certain weight is added to the wire, it ceases to snap back to its original length. In this case, one has exceeded the _____.
   (a) breaking point   (b) stress limit   (c) strain limit
   (d) elastic limit   (e) elasticity

3. According to Poiseuille's law, the most effective way to increase the rate of flow of a transfusing fluid into a patient is to
   (a) increase the pressure at the level of the arm by raising the height of the bag.
   (b) dissolve the fluid in a less viscous material.
   (c) increase the radius of the needle.
   (d) increase the length of tubing from bag to needle.

4. The distance from the feet to the heart for an individual is 1.20 m, and the density of blood is 1.06×10^3 kg/m^3. Find the difference in blood pressure between the level of the heart and the level of the feet.
   (a) 1270 Pa   (b) 1.06×10^3 Pa   (c) 1.25×10^4 Pa
   (d) 1.00 atm

5. A solid rock, suspended in air by a spring scale, has a measured mass of 0.80 kg. When the rock is submerged in water, the scale reads 0.70 kg. What is the density of the rock?
   (a) 4.5×10^3 kg/m^3 (b) 3.5×10^3 kg/m^3
   (c) 1.2×10^3 kg/m^3 (d) 2.7×10^3 kg/m^3
   (e) 8.0×10^3 kg/m^3

CONCEPTUAL QUESTIONS

1. A woman wearing high-heeled shoes is invited into a home in which the kitchen has vinyl floor covering. Why should the homeowner be concerned?

2. Figure Q9.2 shows aerial views from directly above two dams. Both dams are equally long (the vertical dimension in the diagram) and equally deep (into the page in the diagram). The dam on the left holds back a very large lake, and the dam on the right holds back a narrow river. Which dam has to be built more strongly?

Figure Q9.2

3. A typical silo on a farm has many bands wrapped around its perimeter, as shown in Figure Q9.3. Why is the spacing between successive bands smaller at the lower portions of the silo?

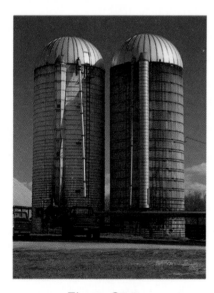

Figure Q9.3

4. During inhalation, the pressure in the lungs is slightly less than external pressure and the muscles controlling exhalation are relaxed. Underwater, the body equalizes internal and external pressures. Discuss the condition of the muscles if a person underwater is breathing through a snorkel. Would a snorkel work in deep water?

5. Atmospheric pressure varies from day to day. Does a ship float higher in the water on a high pressure day compared to a low pressure day?

6. Lead has a greater density than iron, and both are denser than water. Is the buoyant force on a solid lead object greater than, less than, or equal to the buoyant force on a solid iron object of the same dimensions?

7. Suppose a damaged ship just barely floats in the ocean after a hole in its hull has been sealed. It is pulled by a tugboat toward shore and into a river, heading toward a drydock for repair. As it is pulled up the river, it sinks. Why?

8. Will an ice cube float higher in water or in an alcoholic beverage?

9. A pound of styrofoam and a pound of lead have the same weight. If they are placed on an equal arm balance, will it balance?

10. An ice cube is placed in a glass of water. What happens to the level of the water as the ice melts?

11. A person in a boat floating in a small pond throws an anchor overboard. Does the level of the pond rise, fall, or remain the same?

12. Prairie dogs live in underground burrows with at least two entrances. They ventilate their burrows by building a mound over one entrance, which is open to a stream of air, as in the photograph (Figure Q9.12). A second entrance at ground level is open to almost stagnant air. How does this construction create an air flow through the burrow?

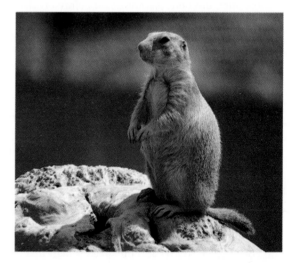

Figure Q9.12   *(Pamela Zilly/The Image Bank)*

13. Will a ship ride higher in an inland lake or in the ocean? Why?

14. A barge is carrying a load of gravel along a river. It approaches a low bridge and the captain realizes that the top of the pile of gravel is not going to make it under. The captain orders the crew to quickly shovel gravel from the pile into the water. Is this a good decision?

15. When you are driving a small car on the freeway and a truck passes you at high speed, you feel pulled toward the truck. Why?

16. Tornadoes and hurricanes often lift the roofs of houses. Use the Bernoulli effect to explain why. Why should you keep your windows open under these conditions?

PROBLEMS

1, 2, 3 = straightforward, intermediate, challenging   □ = full solution available in Study Guide/Student Solutions Manual
WEB = solution posted at **http://www.saunderscollege.com/physics/college**   ▨ = biomedical application   ▨ = Interactive Physics

Section 9.2 The Deformation of Solids

1. The heels on a pair of women's shoes have radii of 0.50 cm at the bottom. If 30% of the weight of a woman weighing 480 N is supported by each heel, find the stress on each heel.

2. Find the minimum diameter of an 18.0-m-long steel wire that will stretch no more than 9.00 mm when a load of 380 kg is hung on the lower end.

3. For safety in climbing, a mountaineer uses a nylon rope that is 50 m long and 1.0 cm in diameter. When supporting a 90-kg climber, the rope elongates 1.6 m. Find its Young's modulus.

4. If the elastic limit of steel is 5.0×10^8 Pa, determine the minimum diameter a steel wire can have if it is to support a 70-kg circus performer without its elastic limit being exceeded.

5. Bone has a Young's modulus of about 14.5×10^9 Pa. Under compression, it can withstand a stress of about 160×10^6 Pa before breaking. Assume that a femur (thigh bone) is 0.50 m long and calculate the amount of compression this bone can withstand before breaking.

6. The distortion of the Earth's crustal plates is an example of shear on a large scale. A particular crustal rock has a shear modulus of 1.5×10^{10} Pa. What shear stress is involved when a 10-km layer of this rock is sheared through a distance of 5.0 m?

7. A child slides across a floor in a pair of rubber-soled shoes. The friction force acting on each foot is 20 N, the cross-sectional area of each foot is 14 cm², and the height of the soles is 5.0 mm. Find the horizontal distance traveled by the sheared face of the sole. The shear modulus of the rubber is 3.0×10^6 Pa.

8. A high-speed lifting mechanism supports an 800-kg mass with a steel cable 25.0 m long and 4.00 cm² in cross-sectional area. (a) Determine the elongation of the cable. (b) By what additional amount does the cable increase in length if the mass is accelerated upward at a rate of 3.0 m/s²? (c) What is the greatest mass that can be accelerated upward at 3.0 m/s² if the stress in the cable is not to exceed the elastic limit of the cable, 2.2×10^8 Pa?

Section 9.3 Density and Pressure

9. A 50.0-kg ballet dancer stands on her toes during a performance with four square inches (26.0 cm²) in contact with the floor. What is the pressure exerted by the floor over the area of contact (a) if the dancer is stationary, and (b) if the dancer is leaping upwards with an acceleration of 4.00 m/s²?

10. The four tires of an automobile are inflated to a gauge pressure of 2.0×10^5 Pa. Each tire has an area of 0.024 m² in contact with the ground. Determine the weight of the automobile.

11. A 70-kg man in a 5.0-kg chair tilts back so that all the weight is balanced on two legs of the chair. Assume that each leg makes contact with the floor over a circular area with a radius of 1.0 cm, and find the pressure exerted on the floor by each leg.

12. If 1.0 m³ of concrete weighs 5.0×10^4 N, what is the height of the tallest cylindrical concrete pillar that will not collapse under its own weight? The compression strength of concrete (the maximum pressure that can be exerted on the base of the structure) is 1.7×10^7 Pa.

Section 9.4 Variation of Pressure with Depth

Section 9.5 Pressure Measurements

13. Water is to be pumped to the top of the Empire State Building, which is 1200 ft high. What gauge pressure is needed in the water line at the base of the building to raise the water to this height?

14. A collapsible plastic bag (Fig. P9.14) contains a glucose solution. If the average gauge pressure in the artery is 1.33×10^4 Pa, what must be the minimum height, h, of the bag in order to infuse glucose into the artery? Assume that the specific gravity of the solution is 1.02.

Figure P9.14

15. Air is trapped above liquid ethanol in a rigid container, as shown in Figure P9.15. If the air pressure above the liquid is 1.10 atm, determine the pressure inside a bubble 4.0 m below the surface of the ethanol.

Figure P9.15

16. One method of measuring the density of a liquid is illustrated in Figure P9.16. One side of the U-tube is in the liquid being tested; the other side is in water of density ρ_w. When the air is partially removed at the upper part of the tube, show that the density of the liquid on the left is given by $\rho = (h_w/h)\rho_w$.

WEB **17.** A container is filled to a depth of 20.0 cm with water. On top of the water floats a 30.0-cm-thick layer of oil with specific gravity 0.700. What is the absolute pressure at the bottom of the container?

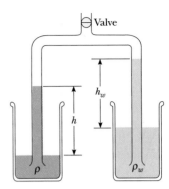

Figure P9.16

18. Piston 1 in Figure P9.18 has a diameter of 0.25 in.; piston 2 has a diameter of 1.5 in. In the absence of friction, determine the force, **F,** necessary to support the 500-lb weight.

Figure P9.18

19. Figure P9.19 shows the essential parts of a hydraulic brake system. The area of the piston in the master cylin-

Figure P9.19

der is 6.4 cm² and that of the piston in the brake cylinder is 1.75 cm². The coefficient of friction between shoe and wheel drum is 0.50. If the wheel has a radius of 34 cm, determine the frictional torque about the axle when a force of 44 N is exerted on the brake pedal.

Section 9.6   Buoyant Forces and Archimedes's Principle

20. A frog in a hemispherical pod finds that he just floats without sinking in a fluid of density 1.35 g/cm³. If the pod has a radius of 6.00 cm and negligible mass, what is the mass of the frog? (See Fig. P9.20.)

Figure P9.20

21. A small ferry boat is 4.00 m wide and 6.00 m long. When a loaded truck pulls onto it, the boat sinks an additional 4.00 cm into the water. What is the weight of the truck?

22. The density of ice is 920 kg/m³, and that of seawater is 1030 kg/m³. What fraction of the total volume of an iceberg is exposed?

23. An empty rubber balloon has a mass of 0.0120 kg. The balloon is filled with helium at a density of 0.181 kg/m³. At this density the balloon is spherical with a radius of 0.500 m. If the filled balloon is fastened to a vertical line, what is the tension in the line?

24. A light spring of constant $k = 160$ N/m rests vertically on the bottom of a large beaker of water (Fig. P9.24a). A 5.00-kg block of wood (density = 650 kg/m³) is connected to the spring and the mass-spring system is allowed to come to static equilibrium (Fig. P9.24b). What is the elongation, ΔL, of the spring?

25. A sample of an unknown material weighs 300 N in air and 200 N when immersed in alcohol of specific gravity 0.700. What are (a) the volume and (b) the density of the material?

26. An object weighing 300 N in air is immersed in water after being tied to a string connected to a balance. The scale now reads 265 N. Immersed in oil, the object weighs 275 N. Find (a) the density of the object and (b) the density of the oil.

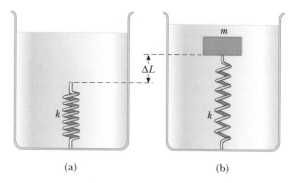

(a)   (b)

Figure P9.24

27. A thin spherical shell of mass 0.400 kg and diameter 0.200 m is filled with alcohol ($\rho = 806$ kg/m³). It is then released from rest on the bottom of a pool of water. Find the acceleration of the alcohol filled shell as it rises toward the surface of the water.

28. A hollow brass tube (diam. = 4.00 cm) is sealed at one end and loaded with lead shot to give a total mass of 0.200 kg. When the tube is floated in pure water, what is the depth, z, of its bottom end?

29. A rectangular air mattress is 2.0 m long, 0.50 m wide, and 0.08 m thick. If it has a mass of 2.0 kg, what additional mass can it support in water?

30. A 1.00-kg beaker containing 2.00 kg of oil (density = 916 kg/m³) rests on a scale. A 2.00-kg block of iron is suspended from a spring scale and completely submerged in the oil (Fig. P9.30). Find the equilibrium readings of both scales.

Figure P9.30

Section 9.7 Fluids in Motion

Section 9.8 Other Applications of Bernoulli's Equation

31. Water is pumped into a storage tank from a well delivering 20.0 gallons of water in 30.0 seconds through a pipe of 1.00 in.² cross-sectional area. What is the average velocity of the water in the pipe as the water is pumped from the well?

32. A cowboy at a dude ranch fills a horse trough that is 1.5 m long, 60 cm wide, and 40 cm deep. He uses a 2.0-cm-diameter hose from which water emerges at 1.5 m/s. How long does it take him to fill the trough?

33. (a) Calculate the mass flow rate (in grams per second) of blood ($\rho = 1.0$ g/cm³) in an aorta with a cross-sectional area of 2.0 cm² if the flow speed is 40 cm/s. (b) Assume that the aorta branches to form a large number of capillaries with a combined cross-sectional area of 3.0×10^3 cm². What is the flow speed in the capillaries?

34. What is the net upward force on an airplane wing of area 20.0 m² if the speed of air flow is 300 m/s across the top of the wing and 280 m/s across the bottom?

35. A liquid ($\rho = 1.65$ g/cm³) flows through two horizontal sections of tubing joined end to end. In the first section the cross-sectional area is 10.0 cm², the flow speed is 275 cm/s, and the pressure is 1.20×10^5 Pa. In the second section the cross-sectional area is 2.50 cm². Calculate the smaller section's (a) flow speed and (b) pressure.

36. A hypodermic syringe contains a medicine with the density of water (Fig. P9.36). The barrel of the syringe has a cross-sectional area of 2.50×10^{-5} m². In the absence of a force on the plunger, the pressure everywhere is 1.00 atm. A force, **F,** of magnitude 2.00 N is exerted on the plunger, making medicine squirt from the needle. Determine the medicine's flow speed through the needle. Assume that the pressure in the needle remains equal to 1.00 atm and that the syringe is horizontal.

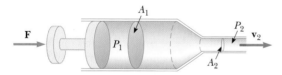

Figure P9.36

37. When a person inhales, air moves down the bronchus (windpipe) at 15 cm/s. The average flow speed of the air doubles through a constriction in the bronchus. Assuming incompressible flow, determine the pressure drop in the constriction.

38. A jet of water squirts out horizontally from a hole near the bottom of the tank in Figure P9.38. If the hole has a diameter of 3.50 mm, what is the height, h, of the water level in the tank?

Figure P9.38

39. A large storage tank, open to the atmosphere at the top and filled with water, develops a small hole in its side at a point 16.0 m below the water level. If the rate of flow from the leak is 2.50×10^{-3} m³/min, determine (a) the speed at which the water leaves the hole and (b) the diameter of the hole.

40. The inside diameters of the larger portions of the horizontal pipe in Figure P9.40 are 2.50 cm. Water flows to the right at a rate of 1.80×10^{-4} m³/s. Determine the inside diameter of the constriction.

Figure P9.40

41. The water supply of a building is fed through a main entrance pipe 6.0 cm in diameter. A 2.0-cm-diameter faucet tap positioned 2.0 m above the main pipe fills a 25-liter container in 30 s. (a) What is the speed at which the water leaves the faucet? (b) What is the gauge pressure in the main pipe? (Assume that the faucet is the only outlet in the system.)

42. A siphon is a device that allows a fluid to seemingly defy gravity (Fig. P9.42). The flow must be initiated by a partial vacuum in the tube, as in a drinking straw. (a) Show that the speed at which the water emerges from the siphon is given by $v = \sqrt{2gh}$. (b) For what values of y will the siphon work?

Figure P9.42

Section 9.9   Surface Tension, Capillary Action, and Viscosity (Optional)

43. In order to lift a wire ring of radius 1.75 cm from the surface of a container of blood plasma, a vertical force of 1.61×10^{-2} N greater than the weight of the ring is required. Calculate the surface tension of blood plasma from this information.

44. A square metal sheet 3.0 cm on a side and of negligible thickness is attached to a balance and inserted into a container of fluid. The contact angle is found to be zero, as shown in Figure P9.44a, and the balance to which the metal sheet is attached reads 0.40 N. A thin veneer of oil is then spread over the metal sheet and the contact angle becomes 180°, as shown in Figure P9.44b. The balance now reads 0.39 N. What is the surface tension of the fluid?

(a)   (b)

Figure P9.44

45. Whole blood has a surface tension of 0.058 N/m and a density of 1050 kg/m³. To what height can whole blood rise in a capillary blood vessel that has a radius of 2.0×10^{-6} m if the contact angle is zero?

46. A certain fluid has a density of 1080 kg/m³ and is observed to rise to a height of 2.1 cm in a 1.0-mm-diameter tube. The contact angle between the wall and the fluid is zero. Calculate the surface tension of the fluid.

47. A staining solution used in a microbiology laboratory has a surface tension of 0.088 N/m and a density 1.035 times the density of water. What must the diameter of a capillary tube be so that this solution will rise to a height of 5 cm? (Assume a contact angle of zero.)

48. The block of ice (temperature 0°C) shown in Figure P9.48 is drawn over a level surface lubricated by a layer of water 0.10 mm thick. Determine the magnitude of force **F** needed to pull the block with a constant speed of 0.50 m/s. At 0°C, the viscosity of water is equal to $\eta = 1.79 \times 10^{-3}$ N·s/m².

Figure P9.48

49. A 1.5-mm coating of glycerine has been placed between two microscope slides of width 1.0 cm and length 4.0 cm. Find the force required to pull one of the microscope slides at a constant speed of 0.30 m/s relative to the other.

50. A straight horizontal pipe with a diameter of 1.0 cm and a length of 50 m carries oil with a coefficient of viscosity of 0.12 Pa·s. At the output of the pipe, the flow rate is 8.6×10^{-5} m³/s and the pressure is 1.0 atmosphere. Find the gauge pressure at the pipe input.

51. The pulmonary artery, which connects the heart to the lungs, has an inner radius of 2.6 mm and is 8.4 cm long. If the pressure drop between the heart and lungs is 400 Pa, what is the average speed of blood in the pulmonary artery?

52. A hypodermic needle is 3.0 cm in length and 0.30 mm in diameter. What excess pressure is required along the needle so that the flow rate of water through it will be 1 g/s? (Use 1.0×10^{-3} Pa·s as the viscosity of water.)

WEB **53.** What diameter needle should be used to inject a volume of 500 cm³ of a solution into a patient in 30 min? Assume a needle length of 2.5 cm and that the solution is elevated 1.0 m above the point of injection. Furthermore, assume the viscosity and density of the solution are those of pure water and assume that the pressure inside the vein is atmospheric.

54. The aorta in humans has a diameter of about 2.0 cm and, at certain times, the blood speed through it is about 55 cm/s. Is the blood flow turbulent? The density of whole blood is 1050 kg/m^3, and its coefficient of viscosity is 2.7×10^{-3} N·s/m^2.

55. A pipe carrying 20°C water has a diameter of 2.5 cm. Estimate the maximum flow speed if the flow is to be laminar.

Section 9.10 Transport Phenomena (Optional)

56. Sucrose is allowed to diffuse along a 10-cm length of tubing filled with water. The tube is 6.0 cm^2 in cross-sectional area. The diffusion coefficient is equal to 5.0×10^{-10} m^2/s, and 8.0×10^{-14} kg is transported along the tube in 15 s. What is the difference in the concentration levels of sucrose at the two ends of the tube?

57. Glycerine in water diffuses along a horizontal column that has a cross-sectional area of 2.0 cm^2. The concentration gradient is 3.0×10^{-2} kg/m^4, and the diffusion rate is found to be 5.7×10^{-15} kg/s. Determine the diffusion coefficient.

58. The viscous force on an oil drop is measured to be 3×10^{-13} N when the drop is falling through air with a speed of 4.5×10^{-4} m/s. If the radius of the drop is 2.5×10^{-6} m, what is the viscosity of air?

59. Small spheres of diameter 1.00 mm fall through 20°C water with a terminal speed of 1.10 cm/s. Calculate the density of the spheres.

60. Spherical particles of a protein of density 1.8 g/cm^3 are shaken up in a solution of 20°C water. The solution is allowed to stand for 1.0 h. If the depth of water in the tube is 5.0 cm, find the radius of the largest particles that remain in solution at the end of the hour.

ADDITIONAL PROBLEMS

61. The approximate inside diameter of the aorta is 0.50 cm; that of a capillary is 10 μm. The approximate average blood flow speed is 1.0 m/s in the aorta and 1.0 cm/s in the capillaries. If all the blood in the aorta eventually flows through the capillaries, estimate the number of capillaries in the circulatory system.

62. Water at a pressure of 3.00×10^5 Pa flows through a horizontal pipe at a speed of 1.00 m/s. If the pipe narrows to one fourth its original diameter, find (a) the flow speed in the narrow section and (b) the pressure in the narrow section.

63. A block of wood weighs 50.0 N when weighed in air. A sinker is attached to the block, and the weight of the wood-sinker combination is 200 N when the sinker alone is immersed in water. Finally, the wood-sinker combination is completely immersed and the weight is 140 N. Find the density of the block.

64. A 600-kg weather balloon is designed to lift a 4000-kg package. What volume should the balloon have after being inflated with helium at standard temperature and pressure in order that the total load can be lifted?

65. A helium-filled balloon at atmospheric pressure is tied to a 2.0-m-long, 0.050-kg string. The balloon is spherical with a radius of 0.40 m. When released, it lifts a length (*h*) of the string and then remains in equilibrium, as in Figure P9.65. Determine the value of *h*. When deflated, the balloon has a mass of 0.25 kg. (*Hint:* Only that part of the string above the floor contributes to the weight of the system in equilibrium.)

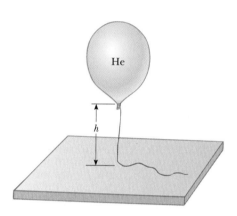

Figure P9.65

66. A stainless-steel orthodontic wire is applied to a tooth, as in Figure P9.66. The wire has an unstretched length of 3.1 cm and a diameter of 0.22 mm. If the wire is stretched 0.10 mm, find the magnitude and direction of the force on the tooth. Disregard the width of the tooth and assume that Young's modulus for stainless steel is 18×10^{10} Pa.

Figure P9.66

67. A small sphere 0.60 times as dense as water is dropped from a height of 10 m above the surface of a smooth lake. Determine the maximum depth to which the sphere will sink. Neglect any energy transferred to the water during impact and sinking.

68. A 2.0-cm-thick bar of soap is floating on a water surface so that 1.5 cm of the bar is underwater. Bath oil of specific gravity 0.60 is poured into the water and floats on top of the water. What is the depth of the oil layer when the top of the soap is just level with the upper surface of the oil?

69. A cube of ice whose edge is 20.0 mm is floating in a glass of ice-cold water with one of its faces parallel to the water surface. (a) How far below the water surface is the bottom face of the block? (b) Ice-cold ethyl alcohol is gently poured onto the water surface to form a layer 5.00 mm thick above the water. When the ice cube attains hydrostatic equilibrium again, what will be the distance from the top of the water to the bottom face of the block? (c) Additional cold ethyl alcohol is poured onto the water surface until the top surface of the alcohol coincides with the top surface of the ice cube (in hydrostatic equilibrium). How thick is the required layer of ethyl alcohol?

70. Water falls over a dam of height h meters at a rate of R kg/s. (a) Show that the power available from the water is given by

$$P = Rgh$$

where g is the acceleration of gravity. (b) Each hydroelectric unit at the Grand Coulee Dam discharges water at a rate of 8.50×10^5 kg/s from a height of 87.0 m. The power developed by the falling water is converted to electric power with an efficiency of 85.0%. How much electric power is produced by each hydroelectric unit?

71. A U-tube open at both ends is partially filled with water (Fig. 9.71a). Oil ($\rho = 750$ kg/m³) is then poured into the right arm and forms a column $L = 5.00$ cm high (Fig. 9.71b). (a) Determine the difference, h, in the heights of the two liquid surfaces. Assume that the density of air is 1.29 kg/m³, but be sure to include differences in the atmospheric pressure due to changes in al-

titude. (b) The right arm is then shielded from any air motion while air is blown across the top of the left arm until the surfaces of the two liquids are at the same height (Fig. 9.71c). Determine the speed of the air being blown across the left arm.

72. Figure P9.72 shows a water tank with a valve at the bottom. If this valve is opened, what is the maximum height attained by the water stream coming out of the right side of the tank? Assume that $h = 10$ m, $L = 2.0$ m, and $\theta = 30°$ and that the cross-sectional area at A is very large compared with that at B.

Figure P9.72

73. A solid copper ball with a diameter of 3.00 m at sea level is placed at the bottom of the ocean, at a depth of 10.0 km. If the density of the seawater is 1030 kg/m³, how much does the diameter of the ball decrease when it reaches bottom?

74. A 1.0-kg hollow ball with a radius of 0.10 m, filled with air, is released from rest at the bottom of a 2.0-m-deep pool of water. How high above the water does the ball shoot upward? Neglect all frictional effects, and neglect the ball's motion when it is only partially submerged.

75. In 1657 Otto von Guericke, inventor of the air pump, evacuated a sphere made of two brass hemispheres (Fig. P9.75). Two teams of eight horses each could pull the hemispheres apart only on some trials, and then with the greatest difficulty. (a) Show that the force required to pull the evacuated hemispheres apart is $\pi R^2(P_0 - P)$, where R is the radius of the hemispheres and P is the pressure inside the sphere, which is much less than atmospheric pressure, P_0. (b) Determine the required force if $P = 0.10 P_0$ and $R = 0.30$ m.

76. A water tank open to the atmosphere at the top has two holes punched in its side, one above the other. The holes are 5.00 cm and 12.0 cm above the floor. How high does water stand in the tank if the two streams of water hit the floor at the same place?

Figure P9.71

Figure P9.75   *(Courtesy of Henry Leap and Jim Lehman)*

WEB **77.** Oil having a density of 930 kg/m^3 floats on water. A rectangular block of wood 4.00 cm high and with a density of 960 kg/m^3 floats partly in the oil and partly in the water. The oil completely covers the block. How far below the interface between the two liquids is the bottom of the block?

78. A hollow object with an average density of 900 kg/m^3 floats in a pan containing 500 cm^3 of water. Ethanol is added to the water until the object is just on the verge of sinking. What volume of ethanol has been added? Disregard the loss of volume caused by mixing.

BIOLOGICAL PERSPECTIVE

Physics of the Human Circulatory System

WILLIAM G. BUCKMAN
Western Kentucky University

The circulatory system is an extremely complex and vital part of the human body. The blood supplies food and oxygen to the tissues, carries away waste products from the cells, distributes the heat generated by the cells to equalize body temperature, carries hormones that stimulate and coordinate the activity of organs, distributes antibodies to fight infection, and performs numerous other functions.

William Harvey (1579–1657), an English physician and physiologist, studied blood flow and the action of the heart. He established the essential mechanics of the heart and found that the blood flows from the arterial system through capillary beds and into the veins to be returned to the heart.

The Physical Properties of Blood

Blood is a liquid tissue consisting of two principal parts: the plasma, which is the intercellular fluid, and the cells, which are suspended in the plasma. Plasma is about 90% water, 9% proteins, and 1.0% salts, sugar, and traces of other materials. Blood contains white blood cells and red blood cells. The individual red blood cells are biconcave and have an average diameter of 7.5 μm. There are about 5×10^6 red blood cells per cubic millimeter of blood. The five types of white blood cells found in the blood have an average concentration of 8000 per cubic millimeter, with the concentration normally varying between 4500 and 11 000 per cubic millimeter. The density of blood is about 1.05×10^3 kg/m³, and its viscosity varies from 2.5 to 4 times that of water.

The Heart as a Pump

The heart can be considered a double pump, with each side consisting of an atrium and a ventricle (Fig. 1a). Oxygen-poor blood enters the right atrium, flows into the right ventricle, is pumped by the right ventricle to the lungs, and returns through the left atrium to the left ventricle. The left ventricle then pumps the oxygenated blood out through the aorta to the rest of the body. The heart has a system of one-way valves to ensure that the blood flows in the proper direction. The heart's pumping cycle has the two ventricles pumping at the same time, as shown in Figure 1b.

The pressure generated by the right ventricle is quite low (about 25 mm Hg), and the lungs offer a low resistance to blood flow. The left ventricle generates a larger pressure, typically greater than 120 mm Hg, at the peak (systole) of the pressure. During the resting stage (diastole) of the heartbeat, the pressure is typically about 80 mm Hg.

We shall now calculate the mechanical work done by the heart. Consider the fluid in the vessel shown in Figure 2. The net force on the fluid is equal to the product of the pressure drop across the fluid, ΔP, and the cross-sectional area, A. The power expended is equal to the net force times the average velocity: $(\Delta PA)(\overline{v})$. Because $A\overline{v} = AL/t =$ volume/time, which is the flow rate, we may now write for the power expended by the heart

$$\text{Power} = (\text{flow rate})(\Delta P)$$

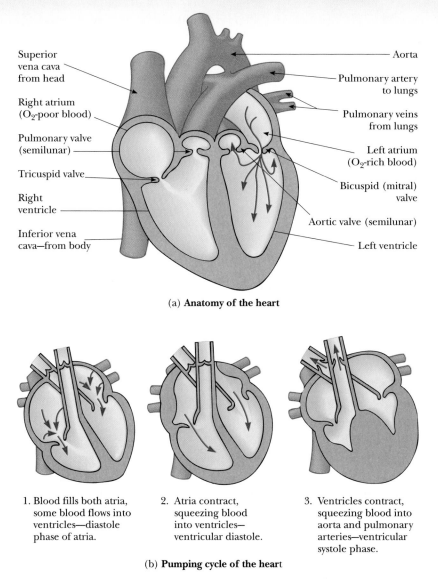

(a) **Anatomy of the heart**

1. Blood fills both atria, some blood flows into ventricles—diastole phase of atria.

2. Atria contract, squeezing blood into ventricles— ventricular diastole.

3. Ventricles contract, squeezing blood into aorta and pulmonary arteries—ventricular systole phase.

(b) **Pumping cycle of the heart**

Figure 1   (a) Anatomy of the heart. (b) Pumping cycle of the heart.

If a normal heart pumps blood at the rate of 97 cm³/s and the pressure drop from the arterial system to the venous system is 1.17×10^4 Pa, we then have

$$\text{Power} = (97 \text{ cm}^3/\text{s})(10^6 \text{ m}^3/\text{cm}^3)(1.17 \times 10^4 \text{ Pa}) = 1.1 \text{ W}$$

By measuring oxygen consumption, it is found that the heart of a 70-kg man at rest consumes about 10 W. In the previous calculation, it was determined that 1.1 W is required to do the mechanical work of pumping blood; hence, the heart is typically about 10% efficient. During strenuous exercise, the blood pressure may increase by 50% and the blood volume pumped may increase by a factor of 5 to yield an increase of 7.5 times in the power generated by the left ventricle. Because the right ventricle has a systolic pressure about

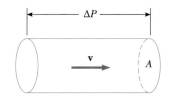

Figure 2   The power required to maintain blood flow against viscous forces.

307

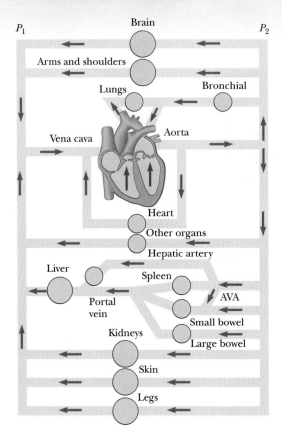

Figure 3 A diagram of the mammalian circulatory system. Pressure P_2 is that in the arterial system, and P_1 is that in the venous system. Arrows indicate the direction of blood flow.

one fifth that of the left ventricle, its power requirement is about one fifth that of the left ventricle.

When we listen to a heart with a stethoscope, we hear two sharp sounds. The first corresponds with the closing of the tricuspid and mitral valves, and the second corresponds with the closing of the aortic and pulmonary valves. Other sounds that are heard are those associated with the flow and turbulence of the blood.

The Cardiovascular System

The cardiovascular system includes the heart to pump the blood; arteries to carry the blood to the organs, muscle, and skin; and veins to return the blood to the heart (Fig. 3). The aorta branches to form smaller arteries, which in turn branch down to even smaller arteries, until finally the blood reaches the very small capillaries of the vascular bed. These capillaries are so small that the red blood cells must pass single file through them. After passing through the capillaries, where materials being carried by the blood are exchanged with the surrounding tissues, the blood flows to the veins and is returned to the heart.

The flow rate of the blood changes as the blood goes through this system. The cross-sectional area of the vascular bed, which is the product of the cross-sectional area and the number of capillaries, is much greater than the cross-sectional area of the aorta. Because the volume of the blood passing through a cross-sectional area per unit of time is Av, where v is the speed of the blood, we may express the volume flow rate of the blood as

$$\text{Flow rate} = A_{\text{aorta}} v_{\text{aorta}}$$

Furthermore, because the total average flow rate through the aorta and the capillaries must be the same, we have

$$\text{Flow rate} = A_{\text{aorta}}v_{\text{aorta}} = A_{\text{cap}}v_{\text{cap}}$$

EXAMPLE   Flow of Blood in the Aorta and Capillaries

The speed of blood in the aorta is 50 cm/s and this vessel has a radius of 1.0 cm. (a) What is the rate of flow of blood through this aorta? (b) If the capillaries have a total cross-sectional area of 3000 cm², what is the speed of the blood in them?

Solution

(a) The cross-sectional area of the aorta is

$$A = \pi r^2 = \pi (1.0 \text{ cm})^2 = 3.14 \text{ cm}^2$$

$$\text{Flow rate} = Av = (3.14 \text{ cm}^2)(50 \text{ cm/s}) = \boxed{160 \text{ cm}^3/\text{s}}$$

(b) The flow rate in the capillaries = 160 cm³/s = $A_c v_c$,

$$v_c = \frac{\text{flow rate}}{A_c} = \frac{160 \text{ cm}^3/\text{s}}{3000 \text{ cm}^2} = \boxed{0.053 \text{ cm/s}}$$

This low blood speed in the capillaries is necessary to enable the blood to exchange oxygen, carbon dioxide, and other nutrients with the surrounding tissues.

Questions and Problems

1. Explain why some individuals tend to black out when they stand up rapidly.
2. If a person is standing at rest, what is the relation between the blood pressure in the left arm and the left leg? What is the relation if the person is in a horizontal position?
3. At what upward acceleration would you expect the blood pressure in the brain to be zero? (Assume that no body mechanisms are operating to compensate for this condition.)
4. When a sphygmomanometer is used to measure blood pressure, will the blood pressure readings depend on the atmospheric pressure? If the atmospheric pressure decreases rapidly, how will this affect the blood pressure readings?
5. Why is it impractical to measure the pulse rate using a vein?
6. Assuming that an artery is clogged such that the effective radius is one half its normal radius, by what factor must the pressure differential be increased to obtain the normal flow rate through the clogged artery?
7. Determine the average speed of the blood in the aorta if it has a radius of 1.2 cm and the flow rate is 20 liters/min.
8. If the mean blood pressure in the aorta is 100 mm Hg, (a) determine the blood pressure in the artery located 6.5 cm above the heart. (b) One cannot apply, without significant error, Bernoulli's principle in the smaller arteries and the capillaries. Why not?
9. When the flow rate is 5.0 liters/min, the blood speed in the capillaries is 0.33 mm/s. Assuming the average diameter of a capillary to be 0.0080 mm, calculate the number of capillaries in the circulatory system.
10. An artery has a length of 20 cm and a radius of 0.50 cm, and blood is flowing at a rate of 6.0 liters/min. What is the difference in the pressure between the ends of the artery?

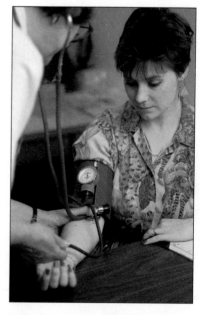

Figure 4   (Question 4) *(Charles Winters)*

Thermodynamics

As we saw in the first part of this book, Newtonian mechanics explains a wide range of phenomena, such as the motions of baseballs, rockets, and planets. We now turn to the study of thermodynamics, which is concerned with the concepts of heat and temperature. As we shall see, thermodynamics is very successful in explaining the bulk properties of matter and the correlation between those properties and the mechanics of atoms and molecules.

Historically, the development of thermodynamics paralleled the development of the atomic theory of matter. By the middle of the nineteenth century, chemical experiments provided solid evidence for the existence of atoms. At that time, scientists recognized that there must be a connection between the heat and temperature theory and the structure of matter. In 1827 the botanist Robert Brown reported that grains of pollen suspended in a liquid moved erratically from one place to another, as if under constant agitation. In 1905 Albert Einstein developed a theory about the cause of this erratic motion, which today is called *Brownian motion*. Einstein explained the phenomenon by assuming that the grains of pollen are under constant bombardment by "invisible" molecules in the liquid, which themselves are moving erratically. Einstein's insight gave scientists a means of discovering vital information concerning molecular motion. It also gave reality to the concept of the atomic constituents of matter.

Have you ever wondered how a refrigerator cools, what types of transformations occur in an automobile engine, or what happens to the kinetic energy of a falling object once the object comes to rest? The laws of thermodynamics and the concepts of heat and temperature enable us to answer such practical questions.

Many things can happen to an object when it is heated. Its size generally changes slightly, but it may also melt, boil, ignite, or even explode. The outcome depends on the composition of the object and the degree to which it is heated. In general, thermodynamics must concern itself with the physical and chemical transformations of matter in all of its forms: solid, liquid, and gas.

◀ *Paul Chesley/Tony Stone Images*

Thermal Physics

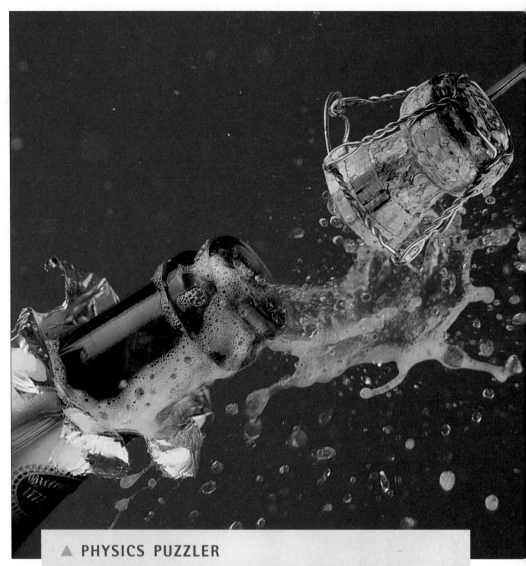

▲ **PHYSICS PUZZLER**

After the bottle is shaken, the cork is blown off. Contrary to common belief, shaking the champagne bottle does not increase the CO_2 pressure inside. Because the temperature of the bottle and its contents remain constant, the equilibrium pressure does not change, as can be shown by replacing the cork with a pressure gauge. Why, then, is champagne expelled when the cork pops? *(Steve Niedorf, The IMAGE Bank)*

Our study thus far has focused exclusively on mechanics. Such concepts as mass, force, and kinetic energy have been carefully defined in order to make the subject quantitative. We now move to a new branch of physics, **thermal physics.** Here we shall find that quantitative descriptions of thermal phenomena require careful definitions of the concepts of temperature, heat, and internal energy. We will seek answers to such questions as: What happens to an object when energy is added to it or removed from it by heat? What physical changes occur in an object when its temperature increases or decreases? One familiar outcome of a temperature increase or decrease for an object is a change in size. We shall examine the details of this process in our discussion of linear expansion.

This chapter concludes with a study of ideal gases. We approach this study on two levels. The first examines ideal gases on the macroscopic scale. Here we are concerned with the relationships among such quantities as pressure, volume, and temperature. On the second level we examine gases on a microscopic scale, using a model that pictures the components of a gas as small particles. This latter approach, called the kinetic theory of gases, helps us understand what happens on the atomic level to affect such macroscopic properties as pressure and temperature.

10.1 TEMPERATURE AND THE ZEROTH LAW OF THERMODYNAMICS

We often associate the concept of temperature with how hot or cold an object feels when we touch it. Thus, our senses provide us with qualitative indications of temperature. However, our senses are unreliable and often misleading. For example, if we remove a metal ice tray and a package of frozen vegetables from the freezer, the ice tray feels colder to the hand than the vegetables even though the two are at the same temperature. This is because metal is a better heat conductor than cardboard, and so the ice tray conducts energy from our hand more efficiently than does the cardboard package. What is needed is a reliable and reproducible method of making quantitative measurements to establish the relative "hotness" or "coldness" of objects. Scientists have developed a variety of thermometers to fulfill this purpose.

We are all familiar with the fact that two objects at different initial temperatures may eventually reach some intermediate temperature when placed in contact with each other. For example, if two soft drinks, one warm and the other cold, are placed in an insulated container, the two eventually reach an equilibrium temperature once the cold one warms up and the warm one cools off. Likewise, if a cup of hot coffee is cooled with an ice cube, the ice eventually melts and the coffee's temperature decreases.

In order to understand the concept of temperature, it is useful to first define two often-used phrases, *thermal contact* and *thermal equilibrium.* To grasp the meaning of thermal contact, imagine two objects placed in an insulated container so that they interact with each other but not with the rest of the world. If the objects are at different temperatures, energy is exchanged between them. The energy exchanged between objects because of a temperature difference between them is called **heat.** We shall examine the concept of heat in more detail in Chapter 11. For purposes of the current discussion, we shall assume that two objects are in **thermal contact** with each other if energy can be exchanged between them. **Ther-**

Molten lava flowing down a mountain in Kilauea, Hawaii. In this case, the hot lava flows smoothly out of a central crater until it cools and solidifies to form the mountains. However, violent eruptions sometimes occur, as in the case of Mount St. Helens in 1980, which can cause both local and global (atmospheric) damage. *(Ken Sakomoto/Black Star)*

mal equilibrium is the situation in which two objects in thermal contact with each other cease to have any exchange of energy.

Now consider two objects, A and B, that are not in thermal contact with each other, and a third object, C, that acts as a thermometer. We wish to determine whether or not A and B would be in thermal equilibrium with each other, once placed in thermal contact. The thermometer (object C) is first placed in thermal contact with A until thermal equilibrium is reached. At that point, the thermometer's reading remains constant, and we record it. The thermometer is then placed in thermal contact with B, and its reading is recorded after thermal equilibrium is reached. If the two readings are the same, then A and B will also be in thermal equilibrium with each other when they are placed in thermal contact.

We can summarize these results in a statement known as the **zeroth law of thermodynamics (the law of equilibrium):**

Zeroth law of thermodynamics ▶

> If bodies A and B are separately in thermal equilibrium with a third body, C, then A and B will be in thermal equilibrium with each other if placed in thermal contact.

This statement, insignificant and obvious as it may seem, is easily proved experimentally and is very important because it makes it possible to define temperature. We can think of **temperature** as the property that determines whether or not an object will be in thermal equilibrium with other objects. **Two objects in thermal equilibrium with each other are at the same temperature.**

Figure 10.1 Schematic diagram of a mercury thermometer. Because of thermal expansion, the level of the mercury rises as the mercury is heated from 0°C (the ice point) to 100°C (the steam point).

10.2   THERMOMETERS AND TEMPERATURE SCALES

Thermometers are devices used to measure the temperature of a system. All thermometers make use of a change in some physical property with temperature. One common thermometer in everyday use consists of a mass of liquid—usually mercury or alcohol—that expands into a glass capillary tube when heated (Fig. 10.1). In this case the physical property is the change in volume of a liquid, and one can define any temperature change to be proportional to the change in length of the liquid column in the capillary. The thermometer can be calibrated by placing it in thermal contact with some natural systems that remain at constant temperature. One such system is a mixture of water and ice in thermal equilibrium at atmospheric pressure. It is defined to have a temperature of zero degrees Celsius, written 0°C; this temperature is called **the ice point of water.** Another commonly used system is a mixture of water and steam in thermal equilibrium at atmospheric pressure; its temperature is 100°C, **the steam point of water.** Once the liquid levels in the thermometer at these two temperatures have been established, the column is divided into 100 equal segments, each corresponding to a change in temperature of one Celsius degree.

Thermometers calibrated in this way present problems when extremely accurate readings are needed. Because mercury and alcohol have different thermal expansion properties, when one indicates a temperature of 50°C, say, the other may indicate a slightly different value. In fact, an alcohol thermometer calibrated at the ice and steam points of water might agree with a mercury thermometer only at the calibration points. The discrepancies between different types of thermometers are especially large when the temperatures to be measured are far from the calibration points.

An additional practical disadvantage of any thermometer is its limited temperature range. A mercury thermometer, for example, cannot be used below the freezing point of mercury, $-39°C$. To surmount such problems, we need a universal thermometer whose readings are independent of the substance used. The gas thermometer approaches this requirement.

The Constant-Volume Gas Thermometer and the Kelvin Scale

In a **gas thermometer**, the temperature readings are nearly independent of the substance used in the thermometer. One type of gas thermometer is the constant-volume unit shown in Figure 10.2. The physical property used in this device is the pressure variation with temperature of a fixed volume of gas. When the constant-volume gas thermometer was developed, it was calibrated using the ice and steam points of water. (A different calibration procedure, to be discussed shortly, is used now.) The gas flask was inserted into an ice bath, and mercury reservoir *B* was raised or lowered until the volume of the confined gas was at some value, indicated by the zero point on the scale. The height h, the difference between the levels in the reservoir and column *A*, indicated the pressure in the flask at 0°C. The flask was inserted into water at the steam point, and reservoir *B* was readjusted until the height in column *A* was again brought to zero on the scale, ensuring that the gas volume was the same as it had been in the ice bath (hence the designation "constant-volume"). This gave a value for the pressure at 100°C.

The pressure and temperature values were then plotted on a graph, as in Figure 10.3. The line connecting the two points serves as a calibration curve for measuring unknown temperatures. If we wanted to measure the temperature of a substance, we would place the gas flask in thermal contact with the substance and adjust the column of mercury until the gas took on its specified volume. The height of the mercury column would tell us the pressure of the gas, and we could then find the temperature of the substance from the graph.

Experiments show that the thermometer readings are nearly independent of the type of gas used, so long as the gas pressure is low and the temperature is well above the point at which the gas liquifies (Fig. 10.4). The agreement among thermometers using different gases improves as the pressure is reduced.

If the curves in Figure 10.4 are extended back toward negative temperatures, in every case the pressure is zero when the temperature is $-273.15°C$. This significant temperature is used as the basis for the Kelvin temperature scale, which sets $-273.15°C$ as its zero point (0 K). The size of a Kelvin unit (called a kelvin) is identical to the size of a degree on the Celsius scale. Thus, the relationship of conversion between these temperatures is simply

$$T_C = T - 273.15 \qquad [10.1]$$

where T_C is the **Celsius temperature** and T is the **Kelvin temperature.**

Early gas thermometers made use of ice and steam points according to the procedure just described. However, these points are experimentally difficult to duplicate because they are pressure sensitive. Consequently, a new procedure based

Figure 10.2 A constant-volume gas thermometer measures the pressure of the gas contained in the flask on the left. The volume of gas in the flask is kept constant by raising or lowering the column of mercury such that the mercury level remains constant.

Figure 10.3 A typical graph of pressure versus temperature taken with a constant-volume gas thermometer. The dots represent known reference temperatures (the ice point and the steam point).

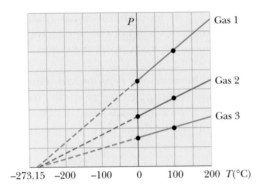

Figure 10.4 Pressure versus temperature for dilute gases. Note that, for all gases, the pressure extrapolates to zero at the unique temperature of −273.15°C.

on a single fixed point was adopted in 1954 by the International Committee on Weights and Measures. The **triple point of water, which is the single temperature and pressure at which water, water vapor, and ice can coexist in equilibrium,** was chosen as a convenient and reproducible reference temperature for the Kelvin scale. It occurs at a temperature of 0.01°C and a pressure of 4.58 mm of mercury. The temperature at the triple point of water on the Kelvin scale has been assigned a value of 273.16 kelvins (K). Thus, **the SI unit of temperature, the kelvin, is defined as 1/273.16 of the temperature of the triple point of water.**

▶ The kelvin

Figure 10.5 shows the Kelvin temperatures for various physical processes and structures. The temperature 0 K is often referred to as **absolute zero,** and, as Figure 10.5 shows, this temperature has never been achieved, although laboratory experiments have come close.

What would happen to a substance if its temperature could reach 0 K? As Figure 10.4 indicates, the pressure the substance exerted on the walls of its container would be zero (assuming that it remains a gas without liquefying or solidifying all the way to absolute zero, which, of course, is not the case). In Section 10.6 we shall show that the pressure of a gas is proportional to the kinetic energy of the molecules of that gas. Thus, according to classical physics, the kinetic energy of the gas would go to zero, and there would be no motion at all of the individual components of the gas; hence, the molecules would settle out on the bottom of the container. Quantum theory, to be discussed in Chapter 27, modifies this statement to indicate that there would be some residual energy, called the *zero-point energy,* at this low temperature.

Temperature (K)

10^9
10^8 ◀— Hydrogen bomb
10^7 ◀— Interior of the sun
10^6 ◀— Solar corona
10^5
10^4
        ◀— Surface of the sun
10^3 ◀— Copper melts
        ◀— Water freezes
10^2 ◀— Liquid nitrogen
        ◀— Liquid hydrogen
10
        ◀— Liquid helium
1

        Lowest temperature
        achieved ~10^{-6} K

Figure 10.5 Absolute temperature at which various selected physical processes take place. Note that the scale is logarithmic.

The Celsius, Kelvin, and Fahrenheit Temperature Scales

Equation 10.1 shows that the Celsius temperature, T_C, is shifted from the absolute (Kelvin) temperature, T, by 273.15. Because the size of a degree is the same on the two scales, a temperature difference of 5°C is equal to a temperature difference of 5 K. The two scales differ only in the choice of zero point. Thus, the ice point (273.15 K) corresponds to 0.00°C, and the steam point (373.15 K) is equivalent to 100.00°C.

The most common temperature scale in use in the United States is the Fahrenheit scale. It sets the temperature of the ice point at 32°F and the temperature

Figure 10.6 A comparison of the Celsius, Fahrenheit, and Kelvin temperature scales.

of the steam point at 212°F. The relationship between the Celsius and Fahrenheit temperature scales is

$$T_F = \tfrac{9}{5} T_C + 32 \qquad\qquad \text{[10.2]}$$

Equation 10.2 can easily be used to find a relationship between changes in temperature on the Celsius and Fahrenheit scales. In an end-of-chapter problem you will be asked to show that if the Celsius temperature changes by ΔT_C, the Fahrenheit temperature changes by the amount ΔT_F, given by

$$\Delta T_F = \tfrac{9}{5} \Delta T_C \qquad\qquad \text{[10.3]}$$

Figure 10.6 compares the Celsius, Fahrenheit and Kelvin temperature scales.

Thinking Physics 1

A group of future astronauts lands on an inhabited planet. They strike up a conversation with the life forms about temperature scales. It turns out that the inhabitants of this planet have a temperature scale based on the freezing and boiling points of water. Would these two temperatures on this planet be the same as those on Earth? Would the size of the planet inhabitants' degrees be the same as ours? Suppose that the inhabitants have also devised a scale similar to the Kelvin scale. Would their absolute zero be the same as ours?

Explanation The values of 0°C and 100°C for the freezing and boiling points of water are defined at atmospheric pressure. On another planet, it is unlikely that atmospheric pressure would be exactly the same as that on Earth. Thus, water would freeze and boil at different temperatures on the planet. The inhabitants may call these temperatures 0° and 100°, but their degrees would not be the same size as our Celsius degrees (unless their atmospheric pressure was the same as ours). For a version of the

Kelvin scale from this other planet, the absolute zero would be the same as ours, because it is based on a natural, universal definition, rather than being associated with a particular substance or a given atmospheric pressure.

EXAMPLE 10.1   Converting Temperatures

What is the temperature 50.0°F in degrees Celsius and in kelvins?

Solution   Let us solve Equation 10.2 for T_C and substitute $T_F = 50.0$°F:

$$T_C = \tfrac{5}{9}(T_F - 32.0) = \tfrac{5}{9}(50.0 - 32.0) = \boxed{10.0°C}$$

From Equation 10.1 we find that

$$T = T_C + 273.15 = \boxed{283 \text{ K}}$$

Exercise   On a hot summer day, the temperature is reported as 30.0°C. What is this temperature in Fahrenheit degrees and in kelvins?

Answer   86.0°F; 303 K

EXAMPLE 10.2   Heating a Pan of Water

A pan of water is heated from 25°C to 80°C. What is the change in its temperature on the Kelvin scale and on the Fahrenheit scale?

Solution   From Equation 10.1 we see that the change in temperature on the Celsius scale equals the change on the Kelvin scale. Therefore,

$$\Delta T = \Delta T_C = 80 - 25 = 55°C = \boxed{55 \text{ K}}$$

From Equation 10.3 we find that the change in temperature on the Fahrenheit scale is $\tfrac{9}{5}$ as great as the change on the Celsius scale. That is,

$$\Delta T_F = \tfrac{9}{5}\Delta T_C = \tfrac{9}{5}(80 - 25) = \boxed{99°F}$$

Thermal expansion joints are used to separate sections of roadways on bridges. Without these joints, the surfaces would buckle due to thermal expansion on very hot days or crack due to contraction on very cold days. (© *Frank Siteman, Stock/Boston*)

10.3   THERMAL EXPANSION OF SOLIDS AND LIQUIDS

Our discussion of the liquid thermometer made use of one of the best-known changes that occurs in a substance: As its temperature increases, its volume increases. (As we shall see, however, in some materials the volume *decreases* when the temperature increases.) This phenomenon, known as **thermal expansion,** plays an important role in numerous applications. For example, thermal expansion joints must be included in buildings, concrete highways, and bridges to compensate for changes in dimensions with temperature variations.

The overall thermal expansion of an object is a consequence of the change in the average separation between its constituent atoms or molecules. To understand this, consider how the atoms in a solid substance behave. At ordinary temperatures, the atoms vibrate about their equilibrium positions with an amplitude of about 10^{-11} m, and the average spacing between the atoms is about 10^{-10} m. As the temperature of the solid increases, the atoms vibrate with greater amplitudes and the average separation between them increases. Consequently, the solid as a whole expands. If the thermal expansion of an object is sufficiently small compared with the object's initial dimensions, then the change in any dimension is, to a good approximation, proportional to the first power of the temperature change.

Suppose an object has an initial length of L_0 along some direction at some temperature. The length increases by ΔL for the change in temperature ΔT. Experiments show that when ΔT is small enough, ΔL is proportional to ΔT and to L_0:

$$\Delta L = \alpha L_0 \, \Delta T \qquad\qquad \text{[10.4]}$$

or

$$L - L_0 = \alpha L_0 (T - T_0)$$

where L is the final length, T is the final temperature, and the proportionality constant α is called the **average coefficient of linear expansion** for a given material and has units of $(°C)^{-1}$.

It may be helpful to picture a thermal expansion as a magnification or a photographic enlargement. For example, as a metal washer is heated (Fig. 10.7), all dimensions, including the radius of the hole, increase according to Equation 10.4. One practical application of thermal expansion is a common technique used to loosen a metal lid that is stuck on a glass jar. As you are aware, running hot water over the lid causes the lid to expand, thus enabling you to remove it from the jar. Table 10.1 lists the average coefficients of linear expansion for various materials. Note that for these materials α is positive, indicating an increase in length with increasing temperature.

Because the linear dimensions of an object change with temperature, it follows that surface area and volume also change with temperature. Consider a square

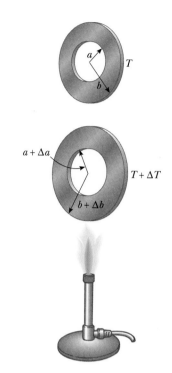

Figure 10.7 Thermal expansion of a homogeneous metal washer. As the washer is heated, all dimensions increase. (Note that the expansion is exaggerated in this figure.)

TABLE 10.1 Average Coefficients of Expansion for Some Materials Near Room Temperature

Material	Average Coefficient of Linear Expansion $[(°C)^{-1}]$	Material	Average Coefficient of Volume Expansion $[(°C)^{-1}]$
Aluminum	24×10^{-6}	Ethyl alcohol	1.12×10^{-4}
Brass and bronze	19×10^{-6}	Benzene	1.24×10^{-4}
Copper	17×10^{-6}	Acetone	1.5×10^{-4}
Glass (ordinary)	9×10^{-6}	Glycerin	4.85×10^{-4}
Glass (Pyrex®)	3.2×10^{-6}	Mercury	1.82×10^{-4}
Lead	29×10^{-6}	Turpentine	9.0×10^{-4}
Steel	11×10^{-6}	Gasoline	9.6×10^{-4}
Invar (Ni-Fe alloy)	0.9×10^{-6}	Air	3.67×10^{-3}
Concrete	12×10^{-6}	Helium	3.665×10^{-3}

Figure 10.8 (a) A bimetallic strip bends as the temperature changes because the two metals have different expansion coefficients. (b) A bimetallic strip used in a thermostat to break or make electrical contact. (c) The blade being heated in the photograph is a bimetallic strip that was straight before being heated. Note that the blade bends when heated. Which way would it bend if it were cooled? *(Courtesy of Central Scientific Company)*

having an initial length L_0 on a side and therefore an initial area of $A_0 = L_0{}^2$. As the temperature is increased, the length of each side increases to

$$L = L_0 + \alpha L_0 \, \Delta T$$

We are now able to calculate the change in the area of the square object as follows. The new area $A = L^2$ is

$$L^2 = (L_0 + \alpha L_0 \, \Delta T)(L_0 + \alpha L_0 \, \Delta T) = L_0{}^2 + 2\alpha L_0{}^2 \, \Delta T + \alpha^2 L_0{}^2 (\Delta T)^2$$

The last term in this expression contains the quantity $\alpha \, \Delta T$ raised to the second power. Because $\alpha \, \Delta T$ is much less than unity, squaring it makes it even smaller. Therefore, we can neglect this term to get a simpler expression:

$$A = L^2 = L_0{}^2 + 2\alpha L_0{}^2 \, \Delta T$$

$$A = A_0 + 2\alpha A_0 \, \Delta T$$

or

$$\Delta A = A - A_0 = \gamma A_0 \, \Delta T \qquad [10.5]$$

where $\gamma = 2\alpha$. The quantity γ (Greek letter gamma) is called the **average coefficient of area expansion.**

By a similar procedure we can show that the *increase in volume* of an object accompanying a change in temperature is

$$\Delta V = \beta V_0 \, \Delta T \qquad [10.6]$$

where β, the **average coefficient of volume expansion,** is equal to 3α.

As Table 10.1 indicates, each substance has its own characteristic coefficients of expansion. For example, when the temperatures of a brass rod and a steel rod of equal length are raised by the same amount from some common initial value, the brass rod expands more than the steel rod because brass has a larger coefficient of expansion than steel. A simple device that utilizes this principle, called a bimetallic strip, is found in practical devices such as thermostats. The strip is made by securely bonding two different metals together. As the temperature of the strip increases, the two metals expand by different amounts and the strip bends, as in Figure 10.8.

APPLICATION

Thermostats.

Applying Physics 1

A homeowner is painting the ceiling, and a drop of paint from the brush falls onto an operating incandescent light bulb. The bulb breaks. Why?

Explanation   The glass envelope of an incandescent light bulb receives energy on the inside surface by radiation from the very hot filament and convection of the gas filling the bulb. Thus, the glass can become very hot. If a drop of paint falls onto the glass, that portion of the glass envelope will suddenly become cold, and the contraction of this region could cause thermal stresses that could cause the glass to shatter.

Thermal expansion affects the choice of glassware used in kitchens and laboratories. If hot liquid is poured into a cold container made of ordinary glass, the container may well break due to thermal stress. The inside surface of the glass becomes hot and expands, while the outside surface is at room temperature, and ordinary glass may not withstand the differential expansion without breaking. Pyrex glass has a coefficient of linear expansion of about one third that of ordinary glass, so that the thermal stresses are smaller. Kitchen measuring cups and laboratory beakers are often made of Pyrex because they are likely to contain hot liquids.

APPLICATION

Pyrex Glass.

EXAMPLE 10.3   Expansion of a Railroad Track

A steel railroad track has a length of 30.000 m when the temperature is 0°C. What is its length on a hot day when the temperature is 40°C?

Solution   If we use Table 10.1 and Equation 10.4, and note that the change in temperature is 40°C, we find that the *increase* in length is

$$\Delta L = \alpha L_0 \, \Delta T = [11 \times 10^{-6} \, (°C)^{-1}] (30.000 \text{ m}) (40°C) = 0.013 \text{ m}$$

Therefore, the track's length at 40°C is

$$L = L_0 + \Delta L = \boxed{30.013 \text{ m}}$$

Exercise   What is the length of the same railroad track on a cold winter day when the temperature is 0°F?

Answer   29.994 m

Thermal expansion: The extreme heat of a July day in Asbury Park, New Jersey, caused these railroad tracks to buckle. *(Wide World Photos)*

EXAMPLE 10.4 Does the Hole Get Bigger or Smaller?

A hole of cross-sectional area 100.00 cm^2 is cut in a piece of steel at 20°C. What is the area of the hole if the steel is heated from 20°C to 100°C?

Solution A hole in a substance expands in exactly the same way as would a piece of the substance having the same shape as the hole (see Fig. 10.7). The change in area of the hole can be found by using Equation 10.5.

$$\Delta A = \gamma A_0 \, \Delta T = [22 \times 10^{-6} (°C)^{-1}](100.00 \text{ cm}^2)(80°C) = 0.18 \text{ cm}^2$$

Therefore, the area of the hole at 100°C is

$$A = A_0 + \Delta A = \boxed{100.18 \text{ cm}^2}$$

Exercise In the derivation of Eq. 10.5, the term $\alpha^2 L_0^2 (\Delta T)^2 = \alpha^2 A_0 (\Delta T)^2$ was neglected because it was said to be negligibly small. Use the values found in Example 10.4 to verify that this term is, indeed, extremely small.

The Unusual Behavior of Water

Liquids generally increase in volume with increasing temperature and have volume expansion coefficients about ten times greater than those of solids. Water is an exception to this rule, as we can see from its density-versus-temperature curve in Figure 10.9. As the temperature increases from 0°C to 4°C, water contracts and thus its density increases. Above 4°C, water expands with increasing temperature. In other words, the density of water reaches its maximum value (1000 kg/m^3) 4 degrees above the freezing point.

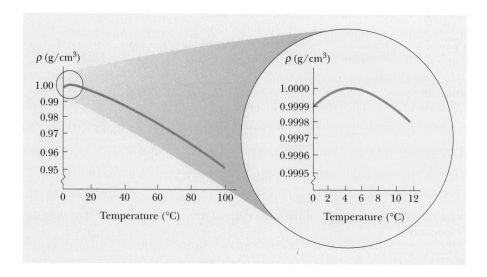

Figure 10.9 The density of water as a function of temperature. The inset at the right shows that the maximum density of water occurs at 4°C.

To understand why water behaves as it does, let us compare the volume of a given mass of water to the volume of an equal mass of ice. At a temperature of 0°C the ice occupies a larger volume than it does when in the form of liquid water at the same temperature. This is because solid ice has an open hexagonal crystal structure that results in relatively large amounts of empty space within its boundaries. On a microscopic level, the breakdown of all these intermolecular bonds is not complete until about 4°C. As liquid water is heated from 0°C to 4°C, its volume slowly decreases and, hence, its density slowly increases. Finally, at and above 4°C, random thermal motions lead to an increase in the volume of a given mass of water and a subsequent decrease in its density.

We can use this unusual thermal expansion behavior of water to explain why a pond freezes slowly from the top down. When the atmospheric temperature drops from, say, 7°C to 6°C, the water at the surface of the pond also cools and consequently decreases in volume. This means that the surface water is denser then the water below it, which has not cooled and decreased in volume. As a result, the surface water sinks and warmer water from below is buoyed up to the surface to be cooled. When the atmospheric temperature is between 4°C and 0°C, however, the surface water expands as it cools, becoming less dense than the water below it. The mixing process stops, and eventually the surface water freezes. As the water freezes, the ice remains on the surface because ice is less dense than water. The ice continues to build up on the surface, and water near the bottom of the pool remains at 4°C. If this did not happen, fish and other forms of marine life would not survive.

The same peculiar thermal expansion properties sometimes cause water pipes to burst in winter. As energy leaves the water through the pipe and is transferred to the outside cold air, the outer layers of water in the pipe freeze first. Continuing heat transfer causes ice formation to move ever closer to the center of the pipe. As long as there is still an opening through the ice, the water can expand as its temperature approaches 0°C or as it freezes into more ice, pushing itself into another part of the pipe. Eventually, however, the ice will freeze to the center somewhere along the pipe's length, forming a plug of ice there. If there is still liquid water between this plug and some other obstruction, such as another ice plug or a spigot, then there is no additional volume available for further expansion and freezing. It is possible for the pressure buildup due to this frustrated expansion to burst the pipe.

APPLICATION

Bursting Pipes in Winter.

Thinking Physics 2

Suppose a solid object is at the bottom of a container of water. Can the temperature of the system be changed in any way to make the object float to the surface?

Explanation Let us imagine lowering the temperature. Liquids tend to have larger coefficients of volume expansion than solids, so the density of water will increase more rapidly with the dropping temperature than that of the solid. If the density of the solid were initially only slightly larger than that of water, it would in principle be possible to make it float by lowering the temperature, although one faces the possible problem of reaching 4°C, the maximum density point, before this happens.

10.4   MACROSCOPIC DESCRIPTION OF AN IDEAL GAS

Figure 10.10 A gas confined to a cylinder whose volume can be varied with a movable piston.

In this section we are concerned with the properties of a gas of mass m confined to a container of volume V, pressure P, and temperature T. It is useful to know how these quantities are related. In general, the equation that interrelates them, called the *equation of state*, is very complicated. However, if the gas is maintained at a very low pressure (or low density), the equation of state is found experimentally to be quite simple. Such a low-density gas approximates what is called an **ideal gas.** Most gases at room temperature and atmospheric pressure behave as if they were ideal gases.

It is convenient to express the amount of gas in a given volume in terms of the number of moles, n. Recall that one mole of any substance is that mass of the substance that contains Avogadro's number, 6.022×10^{23}, of molecules. The number of moles, n, of a substance is related to its mass, m, by the expression

$$n = \frac{m}{M} \qquad [10.7]$$

where M is the molar mass, usually expressed in kilograms per mole. For example, the molar mass of molecular oxygen, O_2, is 32.0×10^{-3} kg/mol. Therefore, the mass of one mole of oxygen is 32.0×10^{-3} kg.

Now suppose an ideal gas is confined to a cylindrical container whose volume can be varied by means of a movable piston, as in Figure 10.10. We shall assume that the cylinder does not leak, and so the mass (or the number of moles) remains constant. For such a system, experiments provide the following information. First, when the gas is kept at a constant temperature, its pressure is inversely proportional to the volume (Boyle's law). Second, when the pressure of the gas is kept constant, the volume is directly proportional to the temperature (the law of Charles and Gay-Lussac). These observations can be summarized by the following **equation of state for an ideal gas:**

◀ Equation of state for an ideal gas

$$PV = nRT \qquad [10.8]$$

In this expression, called the **ideal gas law,** R is a constant for a specific gas that can be determined from experiments, and T is the temperature in kelvins. Experiments on several gases show that, as the pressure approaches zero, the quantity PV/nT approaches the same value of R for all gases. For this reason R is called the **universal gas constant.** In the SI system, where pressure is expressed in pascals and volume in cubic meters, the product PV has units of newton-meters, or joules, and R has the value

◀ The universal gas constant

$$R = 8.31 \text{ J/mol} \cdot \text{K} \qquad [10.9]$$

If the pressure is expressed in atmospheres and the volume is given in liters (1 L = 10^3 cm^3 = 10^{-3} m^3), then R has the value

$$R = 0.0821 \text{ L} \cdot \text{atm/mol} \cdot \text{K}$$

Using this value of R and Equation 10.8, one finds that the volume occupied by 1 mol of any gas at atmospheric pressure and 0°C (273 K) is 22.4 L (see Example 10.5).

Applying Physics 2

On some canned foods, you can test the seal by observing the lid of the container when it is opened. When sealed, the lid is indented, but when opened, the lid springs back out. How does this work?

Explanation The food is sealed inside the container while it is still warm. As the food, and the air above it, cools, the pressure of the trapped air is lowered. Thus, the atmospheric pressure pressing against the outside of the lid causes it to indent inward. When you open the container, breaking the seal allows outside air to rush in, and the lid snaps back to its original shape.

EXAMPLE 10.5 The Volume of One Mole of Gas

Verify that one mole of oxygen occupies a volume of 22.4 L at 1 atm and 0°C.

Solution Let us solve the ideal gas equation for V:

$$V = \frac{nRT}{P}$$

In this problem the mass of the gas, m, is assumed to be that of one mole, M. Thus,

$$n = \frac{m}{M} = 1 \text{ mol}$$

Now let us convert the temperature to kelvins and substitute into the ideal gas equation:

$$V = \frac{nRT}{P} = \frac{(1 \text{ mol})(0.0821 \text{ L} \cdot \text{atm/mol} \cdot \text{K})(273 \text{ K})}{1 \text{ atm}} = \boxed{22.4 \text{ L}}$$

This answer has general validity. **One mole of any gas at standard temperature and pressure (STP) occupies a volume of 22.4 L.**

EXAMPLE 10.6 How Many Gas Molecules Are in a Container?

An ideal gas occupies a volume of 100 cm³ at 20°C and a pressure of 100 Pa. Determine the number of moles of gas in the container.

Solution The quantities given are volume, pressure, and temperature: $V = 100$ cm³ $= 1.00 \times 10^{-4}$ m³, $P = 100$ Pa, and $T = 20$°C $= 293$ K. Using Equation 10.8, we get

$$n = \frac{PV}{RT} = \frac{(100 \text{ Pa})(10^{-4} \text{ m}^3)}{(8.31 \text{ J/mol} \cdot \text{K})(293 \text{ K})} = \boxed{4.11 \times 10^{-6} \text{ mol}}$$

Q U I C K L A B

Tie and tape two inverted empty paper bags to the ends of a rod as in the figure. Balance the setup. Then place a candle under one of the bags and note what happens. Why does this system become unbalanced? What do your results tell you concerning the density of warm air versus the density of cold air?

Rod

Tape

Candle

Exercise Show that the universal gas constant $R = 0.0821$ L·atm/mol·K is also equal to the value $R = 8.31$ J/mol·K.

EXAMPLE 10.7   Squeezing a Tank of Gas

Pure helium gas is admitted into a leakproof cylinder containing a movable piston. The initial volume, pressure, and temperature of the gas are 15 L, 2.0 atm, and 300 K. If the volume is decreased to 12 L and the pressure increased to 3.5 atm, find the final temperature of the gas. (Assume that helium behaves as an ideal gas.)

Solution Because no gas escapes from the cylinder, the number of moles remains constant; therefore, use of $PV = nRT$ at the initial and final points gives

$$\frac{P_i V_i}{T_i} = \frac{P_f V_f}{T_f}$$

where i and f refer to the initial and final values. Solving for T_f, we get

$$T_f = \left(\frac{P_f V_f}{P_i V_i}\right)(T_i) = \frac{(3.5 \text{ atm})(12 \text{ L})}{(2.0 \text{ atm})(15 \text{ L})}(300 \text{ K}) = \boxed{420 \text{ K}}$$

EXAMPLE 10.8   Heating a Bottle of Air

A sealed glass bottle at 27°C contains air at atmospheric pressure and has a volume of 30 cm³. The bottle is then tossed into an open fire. When the temperature of the air in the bottle reaches 200°C, what is the pressure inside the bottle? Assume that any volume changes of the bottle are small enough to be negligible.

Solution We start with the expression

$$\frac{P_i V_i}{T_i} = \frac{P_f V_f}{T_f}$$

Because the initial and final volumes of the gas are assumed equal, this expression reduces to

$$\frac{P_i}{T_i} = \frac{P_f}{T_f}$$

Before evaluating the final pressure, we must convert the given temperatures to kelvins: $T_i = 27°C = 300$ K and $T_f = 200°C = 473$ K. Thus,

$$P_f = \left(\frac{T_f}{T_i}\right)(P_i) = \left(\frac{473 \text{ K}}{300 \text{ K}}\right)(1.0 \text{ atm}) = \boxed{1.6 \text{ atm}}$$

Obviously, the higher the temperature, the higher the pressure exerted by the trapped air. Of course, if the pressure rises high enough, the bottle will shatter.

Exercise In this example we neglected the change in volume of the bottle. If the coefficient of volume expansion for glass is 27×10^{-6} (°C)$^{-1}$, find the magnitude of this volume change.

Answer 0.14 cm³

10.5 AVOGADRO'S NUMBER AND THE IDEAL GAS LAW

In the early 1800s an important field of experimental investigation was created to determine the relative masses of molecules. The equipment used in one such investigation is shown in Figure 10.11. The lower section of the glass vessel contains two electrodes that are connected to a battery so that an electric current can be passed through lightly salted water. (The purpose of the salt is to improve the electrical conductivity of the water.) Bubbles of gas are produced at each electrode and become trapped in the column above it. An analytical examination of the bubbles reveals that the column above the positive electrode contains oxygen gas and the column above the negative electrode contains hydrogen gas. Obviously, the current decomposes the water into its constituent parts.

Experimentally it is found that the volume of hydrogen gas collected is always exactly twice the volume of oxygen gas collected, as we now know in light of the chemical composition of water, H_2O. In addition, it is found that if 9 g of water are decomposed, 8 g of oxygen and 1 g of hydrogen are collected. From this information it is possible to determine the relative masses of oxygen and hydrogen molecules.

Following such an experimental investigation, Amedeo Avogadro in 1811 stated the following hypothesis:

Figure 10.11 When water is decomposed by an electric current, the volume of hydrogen gas collected is twice that of oxygen.

> **Equal volumes of gas at the same temperature and pressure contain the same numbers of molecules.**

Based on this hypothesis and the fact that the ratio of hydrogen to oxygen molecules collected in the experiment is 2:1, the mass of an oxygen molecule must be 16 times that of a hydrogen molecule.

In Example 10.8, we noted that single moles of all gases occupy the same volume at 1 atm and 0°C. Thus, a corollary to Avogadro's hypothesis is as follows:

> **One mole quantities of all gases at standard temperature and pressure contain the same numbers of molecules.**

Specifically, the number of molecules contained in one mole of any gas is Avogadro's number, 6.02×10^{23} molecules/mol, given the symbol N_A:

$$N_A = 6.02 \times 10^{23} \text{ molecules/mol} \qquad \text{[10.10]}$$

◀ Avogadro's number

It is of historical interest that Avogadro never knew, even approximately, the value for this number.

An alternative method for calculating the number of moles of any gas in a container is to divide the total number of molecules present, N, by Avogadro's number:

$$n = \frac{N}{N_A} \qquad \text{[10.11]}$$

With this expression we can rewrite the ideal gas law in the alternative form

$$PV = nRT = \frac{N}{N_A} RT$$

or

◁ Ideal gas law

$$PV = Nk_B T \qquad [10.12]$$

where k_B is **Boltzmann's constant** and has the value

◁ Boltzmann's constant

$$k_B = \frac{R}{N_A} = 1.38 \times 10^{-23} \, \text{J/K} \qquad [10.13]$$

EXAMPLE 10.9   What Is Avogadro's Number?

One mole of hydrogen atoms has a mass of 1.0078×10^{-3} kg. In the early twentieth century it was found that the mass of a hydrogen atom is approximately 1.673×10^{-27} kg. Use these values to find Avogadro's number.

Solution   The number of molecules in 1 mol of hydrogen can be found by dividing the mass of 1 mol by the mass per atom. Hence, we find that

$$N_A = \frac{1.0078 \times 10^{-3} \, \text{kg/mol}}{1.673 \times 10^{-27} \, \text{kg/atom}} = 6.024 \times 10^{23} \, \text{atoms/mol}$$

10.6   THE KINETIC THEORY OF GASES

In Section 10.4 we discussed the properties of an ideal gas, using such quantities as pressure, volume, number of moles, and temperature. In this section we find that pressure and temperature can be understood on the basis of what is happening on the atomic scale. In addition, we reexamine the ideal gas law in terms of the behavior of the individual molecules that make up the gas. Our discussion is restricted to the behavior of gases because the molecular interactions in a gas are much weaker than those in solids and liquids.

Molecular Model for the Pressure of an Ideal Gas

Let us first use the kinetic theory of gases to show that the pressure a gas exerts on the walls of its container is a consequence of the collisions of the gas molecules with the walls. We make the following assumptions:

1. **The number of molecules is large, and the average separation between them is large compared with their dimensions.** This means that the molecules occupy a negligible volume in the container.

◁ Assumptions of kinetic theory for an ideal gas

2. **The molecules obey Newton's laws of motion, but as a whole they move randomly.** By "randomly" we mean that any molecule can move equally in any direction.

3. **The molecules undergo elastic collisions with each other and with the walls of the container.** Thus, in the collisions kinetic energy is constant.
4. **The forces between molecules are negligible except during a collision.** The forces between molecules are short-range, so the molecules interact with each other only during collisions.
5. **The gas under consideration is a pure substance;** that is, all molecules are identical.

Although we often picture an ideal gas as consisting of single atoms, molecular gases at low pressures provide approximations equally as good for ideal gases. Molecular rotations or vibrations have no average effect on the motions considered.

Now let us derive an expression for the pressure of N molecules of an ideal gas in a container of volume V. The container is a cube with edges of length d (Fig. 10.12). We shall focus our attention on one of these molecules, of mass m and assumed to be moving so that its component of velocity in the x direction is v_x (Fig. 10.13). As the molecule collides elastically with any wall, its velocity is reversed. Because the momentum, p, of the molecule is mv_x before the collision and $-mv_x$ after the collision, the change in momentum of the molecule is

$$\Delta p_x = mv_f - mv_i = -mv_x - mv_x = -2mv_x$$

Thus, the change in momentum of the wall is $2mv_x$. Applying the impulse-momentum theorem (Eq. 6.4) to the wall gives

$$F_1 \Delta t = \Delta p = 2mv_x$$

where F_1 is the average force on the molecule (or wall) and Δt is the interval between collisions.

In order for the molecule to make another collision with the same wall, it must travel a distance of $2d$ in the x direction. Therefore, the time interval between two collisions with the same wall is

$$\Delta t = \frac{2d}{v_x}$$

The substitution of this result into the impulse-momentum equation enables us to obtain the average force exerted by a molecule on the wall:

$$F_1 = \frac{2mv_x}{\Delta t} = \frac{2mv_x^2}{2d} = \frac{mv_x^2}{d}$$

The total force exerted on the wall by all the molecules is found by adding the average forces exerted by all the individual molecules:

$$F = \frac{m}{d}(v_{x1}^2 + v_{x2}^2 + \cdots + v_{xN}^2)$$

In this equation, v_{x1} is the x component of velocity of molecule 1, v_{x2} is the x component of velocity of molecule 2, and so on. The summation terminates when we reach molecule N because there are N molecules in the container.

To proceed further, note that the average value of the square of the velocity in the x direction for the N molecules is

$$\overline{v_x^2} = \frac{v_{x1}^2 + v_{x2}^2 + \cdots + v_{xN}^2}{N}$$

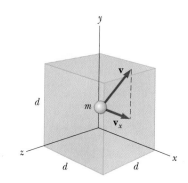

Figure 10.12 A cubical box with sides of length d containing an ideal gas. The molecule shown moves with velocity **v**.

Before collision

After collision

Figure 10.13 A molecule makes an elastic collision with the wall of the container. Its x component of momentum is reversed, and momentum is imparted to the wall.

The glass vessel contains dry ice (solid carbon dioxide). The white cloud is carbon dioxide vapor, which is denser than air and hence falls from the vessel as shown. *(© R. Folwell/Science Photo Library)*

Hence, the total force on the wall can be written

$$F = \frac{m}{d} N \overline{v_x^2}$$

Now let us focus on one molecule in the container and say that this molecule has velocity components v_x, v_y, and v_z. The Pythagorean theorem relates the square of the velocity to the square of these components:

$$v^2 = v_x^2 + v_y^2 + v_z^2$$

Hence, the average value of v^2 for all the molecules in the container is related to the average values of v_x^2, v_y^2, and v_z^2 according to the expression

$$\overline{v^2} = \overline{v_x^2} + \overline{v_y^2} + \overline{v_z^2}$$

Because all directions of motion are equivalent, the average velocity is the same in any direction. Therefore,

$$\overline{v_x^2} = \overline{v_y^2} = \overline{v_z^2}$$

and we have

$$\overline{v^2} = 3\overline{v_x^2}$$

Thus, the total force on the wall is

$$F = \frac{N}{3}\left(\frac{m\overline{v^2}}{d}\right)$$

This expression allows us to find the pressure exerted on the wall:

$$P = \frac{F}{A} = \frac{F}{d^2} = \tfrac{1}{3}\left(\frac{N}{d^3}\, m\overline{v^2}\right) = \tfrac{1}{3}\left(\frac{N}{V}\right)(m\overline{v^2})$$

[10.14]

$$P = \tfrac{2}{3}\left(\frac{N}{V}\right)(\tfrac{1}{2}m\overline{v^2})$$

Pressure of an ideal gas ▶

This result shows that **the pressure is proportional to the number of molecules per unit volume and to the average translational kinetic energy of the molecules, $\tfrac{1}{2}m\overline{v^2}$.** With this simplified model of an ideal gas, we have arrived at an important result that relates the macroscopic quantity of pressure to an atomic quantity, the average value of the square of the molecular speed. Thus, we have a key link between the atomic world and the large-scale world.

Note that Equation 10.14 verifies some features of pressure that are probably familiar to you. One way to increase the pressure inside a container is to increase the number of molecules per unit volume in the container. You do this when you add air to a tire. The pressure in the tire can also be raised by increasing the average translational kinetic energy of the molecules in the tire. As we shall see shortly, this can be accomplished by increasing the temperature of the gas inside the tire. That is why the pressure inside a tire increases as the tire heats up during long trips. The continuous flexing of the tire as it moves along the road surface represents work being done which increases internal energy in the rubber. By heat, energy is partially transferred to the air inside, increasing the air's temperature, which in turn produces an increase in pressure.

Molecular Interpretation of Temperature

We can gain some insight into the meaning of temperature by first writing Equation 10.14 in the form

$$PV = \tfrac{2}{3} N(\tfrac{1}{2} m\overline{v^2})$$

Let us compare this with the equation of state for an ideal gas:

$$PV = Nk_\text{B} T$$

Recall that the equation of state is based on experimental facts concerning the macroscopic behavior of gases. Equating the right sides of these expressions, we find that

$$T = \frac{2}{3k_\text{B}} (\tfrac{1}{2} m\overline{v^2}) \qquad [10.15]$$

◀ Temperature is proportional to average kinetic energy.

That is, **temperature is a direct measure of average molecular kinetic energy.** This result, together with Equation 10.14, implies that the pressure exerted by an ideal gas depends only on the number of molecules per unit volume and the temperature.

By rearranging Equation 10.15, we can relate the translational molecular kinetic energy to the temperature:

$$\tfrac{1}{2} m\overline{v^2} = \tfrac{3}{2} k_\text{B} T \qquad [10.16]$$

◀ Average kinetic energy per molecule

That is, the average translational kinetic energy per molecule is $\tfrac{3}{2} k_\text{B} T$. Because $\overline{v_x^2} = \tfrac{1}{3} \overline{v^2}$, it follows that

$$\tfrac{1}{2} m\overline{v_x^2} = \tfrac{1}{2} k_\text{B} T \qquad [10.17]$$

In a similar manner, for the y and z motions it follows that

$$\tfrac{1}{2} m\overline{v_y^2} = \tfrac{1}{2} k_\text{B} T \qquad \text{and} \qquad \tfrac{1}{2} m\overline{v_z^2} = \tfrac{1}{2} k_\text{B} T$$

Thus, each translational degree of freedom contributes an equal amount of energy to the gas, namely $\tfrac{1}{2} k_\text{B} T$. (In general, "degrees of freedom" refers to the number of independent means by which a molecule can possess energy.) A generalization of this result, known as the **theorem of equipartition of energy,** says that the energy of a system in thermal equilibrium is equally divided among all degrees of freedom.

◀ Theorem of equipartition of energy

The total translational kinetic energy of N molecules of gas is simply N times the average energy per molecule, which is given by Equation 10.16:

$$E = N(\tfrac{1}{2} m\overline{v^2}) = \tfrac{3}{2} Nk_\text{B} T = \tfrac{3}{2} nRT \qquad [10.18]$$

◀ Total kinetic energy of N molecules

where we have used $k_\text{B} = R/N_\text{A}$ for Boltzmann's constant and $n = N/N_\text{A}$ for the number of moles of gas. Equation 10.18 is an expression for the internal energy of the gas, and shows that the internal energy depends on the temperature.

The square root of $\overline{v^2}$ is called the **root-mean-square (rms) speed** of the molecules. From Equation 10.16 we get, for the rms speed,

$$v_\text{rms} = \sqrt{\overline{v^2}} = \sqrt{\frac{3k_\text{B} T}{m}} = \sqrt{\frac{3RT}{M}} \qquad [10.19]$$

◀ Root-mean-square speed

where M is the molar mass in kilograms per mole. This expression shows that, at a given temperature, lighter molecules move faster, on the average, than heavier

TABLE 10.2   Some rms Speeds

Gas	Molar Mass (kg/mol)	v_{rms} at 20°C (m/s)
H_2	2.02×10^{-3}	1902
He	4.0×10^{-3}	1352
H_2O	18×10^{-3}	637
Ne	20.1×10^{-3}	603
N_2 and CO	28×10^{-3}	511
NO	30×10^{-3}	494
CO_2	44×10^{-3}	408
SO_2	48×10^{-3}	390

molecules. For example, hydrogen, with a molar mass of 2×10^{-3} kg/mol, has molecules moving four times as fast as oxygen, whose molar mass is 32×10^{-3} kg/mol. The rms speed is not the speed at which a gas molecule moves across a room, because such a molecule undergoes several billion collisions per second with other molecules under standard conditions.

Table 10.2 lists the rms speeds for various molecules at 20°C.

Thinking Physics 3

Imagine a gas in an insulated cylinder with a movable piston. The piston has been pushed inward, compressing the gas, and is now released. As the molecules of the gas strike the piston, they move it outward. From the point of view of the kinetic theory, explain how the expansion of this gas causes its temperature to drop.

Explanation   From the point of view of kinetic theory, a molecule colliding with the piston causes the piston to rebound with some velocity. According to the conservation of momentum then, the molecule must rebound with less velocity than it had before the collision. Thus, as these collisions occur, the average velocity of the collection of molecules is reduced. Because temperature is related to the average velocity of the molecules, the temperature of the gas drops.

EXAMPLE 10.10   A Tank of Helium

A tank contains 2.0 mol of helium gas at 20°C. Assume that the helium behaves as an ideal gas.
(a) Find the total internal energy of the system.

Solution   Using Equation 10.18 with $n = 2.0$ and $T = 293$ K, we get

$$E = \tfrac{3}{2}nRT = \tfrac{3}{2}(2.0 \text{ mol})(8.31 \text{ J/mol·K})(293 \text{ K}) = \boxed{7.3 \times 10^3 \text{ J}}$$

(b) What is the average kinetic energy per molecule?

Solution   From Equation 10.16 we see that the average kinetic energy per molecule is

$$\tfrac{1}{2}m\overline{v^2} = \tfrac{3}{2}k_B T = \tfrac{3}{2}(1.38 \times 10^{-23} \text{ J/K})(293 \text{ K}) = \boxed{6.1 \times 10^{-21} \text{ J}}$$

Exercise Using the fact that the molar mass of helium is 4.0×10^{-3} kg/mol, determine the rms speed of the atoms at 20°C.

Answer 1.4×10^3 m/s

SUMMARY

The **zeroth law of thermodynamics** states that if two objects, A and B, are separately in thermal equilibrium with a third object, then A and B are in thermal equilibrium with each other.

The relationship between T_C, the **Celsius** temperature, and T, the **Kelvin (absolute)** temperature, is

$$T_C = T - 273.15 \qquad [10.1]$$

The relationship between the **Fahrenheit** and **Celsius** temperatures is

$$T_F = \tfrac{9}{5} T_C + 32 \qquad [10.2]$$

Generally, when a substance is heated, it expands. If an object has an initial length of L_0 at some temperature and undergoes a change in temperature of ΔT, this linear dimension changes by the amount ΔL, which is proportional to the object's initial length and the temperature change:

$$\Delta L = \alpha L_0 \, \Delta T \qquad [10.4]$$

The parameter α is called the **average coefficient of linear expansion.**

The change in area of a substance is given by

$$\Delta A = \gamma A_0 \, \Delta T \qquad [10.5]$$

where γ is the **average coefficient of area expansion** and is equal to 2α.

The change in volume of most substances is proportional to the initial volume, V_0, and the temperature change, ΔT:

$$\Delta V = \beta V_0 \, \Delta T \qquad [10.6]$$

where β is the **average coefficient of volume expansion** and is equal to 3α.

An **ideal gas** is one that obeys the equation

$$PV = nRT \qquad [10.8]$$

where P is the pressure of the gas, V is its volume, n is the number of moles of gas, R is the universal gas constant (8.31 J/mol·K), and T is the absolute temperature in kelvins. A real gas at very low pressures behaves approximately as an ideal gas.

The **pressure** of N molecules of an ideal gas contained in a volume V is given by

$$P = \tfrac{2}{3} \left(\frac{N}{V} \right) \left(\tfrac{1}{2} m \overline{v^2} \right) \qquad [10.14]$$

where $\tfrac{1}{2} m \overline{v^2}$ is the **average kinetic energy per molecule.**

The average kinetic energy of the molecules of a gas is directly proportional to the absolute temperature of the gas:

$$\tfrac{1}{2}m\overline{v^2} = \tfrac{3}{2}k_B T \qquad\qquad [10.16]$$

where k_B is **Boltzmann's constant** $(1.38 \times 10^{-23}\,\text{J/K})$.
The **root-mean-square (rms) speed** of the molecules of gas is

$$v_{\text{rms}} = \sqrt{\frac{3k_B T}{m}} = \sqrt{\frac{3RT}{M}} \qquad\qquad [10.19]$$

MULTIPLE-CHOICE QUESTIONS

1. Given that the molar mass of He is 4×10^{-3} kg/mol and for water vapor it is 18×10^{-3} kg/mol, at room temperature the ratio of the rms speed of He to water vapor molecules is approximately
   (a) 1 to 2   (b) 2 to 1   (c) 4.5 to 1
   (d) 1 to 4.5

2. One way to cool a gas is to let it expand. If a gas under 50 atm of pressure at 25°C is allowed to expand to 15 times its original volume, it reaches 1 atm pressure. What is its new temperature?
   (a) 177°C   (b) 0°C   (c) 200.7°C   (d) 89.4 K

3. The Statue of Liberty is 93.000 m tall and made of copper plates ($\alpha = 17 \times 10^{-6}$ C°$^{-1}$). If this is its height on a 20°C day, what is its height when the temperature is 35°C?
   (a) 92.764 m   (b) 93.024 m   (c) 95.46 m   (d) 108 m

4. An assumption below that is not a part of those made for the kinetic theory of gases is
   (a) The number of molecules is small.
   (b) The molecules obey Newton's laws of motion.
   (c) Collisions between molecules are elastic.
   (d) The gas is a pure substance.
   (e) The average separation between molecules is large compared to their dimensions.

5. Suppose for a brief moment that the gas molecules hitting a wall stuck to the wall instead of bouncing off the wall. How would the pressure on the wall be affected during that brief time?
   (a) the pressure would be zero.
   (b) the pressure would be halved.
   (c) the pressure would remain unchanged.
   (d) the pressure would be doubled.

CONCEPTUAL QUESTIONS

1. In an astronomy class, the temperature at the core of a star is given by the teacher as 1.5×10^7 degrees. A student asks if this is Kelvin or Celsius. How would you respond?

2. A piece of copper is dropped into a beaker of water. If the water's temperature rises, what happens to the temperature of the copper? When will the water and copper be in thermal equilibrium?

3. Common thermometers are made of a mercury column in a glass tube. Based on the operation of these common thermometers, which has the larger coefficient of linear expansion—glass or mercury? (Don't answer this by looking in a table.)

4. A steel wheel bearing is 1 mm smaller in diameter than an axle. How can the bearing be fit onto the axle without removing any material?

5. The boiling point of water is said to be 100°C in the text, but with a little luck and a very clean glass, you can often heat water in a microwave oven to 110°C for a few moments before it begins to boil. How might this occur?

6. Why is a power line more likely to break in winter than in summer even if it is loaded with the same weight?

7. Two spheres are made of the same metal and have the same radius, but one is hollow and the other is solid. The spheres are taken through the same temperature increase. Which sphere expands more?

8. After food is cooked in a pressure cooker, why is it very important to cool the container with cold water before attempting to remove the lid?

9. Some picnickers stop at a store to buy food, including bags of potato chips. They drive up into the mountains to their picnic site. When they unload the food, they no-

tice that the chip bags are puffed up like balloons. Why did this happen?

10. During normal breathing the lungs expand on inhalation and contract on exhalation. Does the ideal gas law apply to the volume of the lungs under these circumstances? Explain.

11. A common material for cushioning objects in packages is made by trapping small bubbles of air between sheets of plastic. Many individuals find satisfaction in popping these bubbles after receiving a package. (Other individuals who are nearby and listening to the sound may not be so satisfied.) Is this material more effective at cushioning the package contents on a hot day or a cold day?

12. Moving upward in the atmosphere, should you expect to find the ratio of nitrogen molecules relative to oxygen molecules increasing or decreasing? Explain.

13. Objects deep beneath the surface of the ocean are subjected to extremely high pressures, as we saw in Chapter 9. For example, some bacteria in these environments have adapted to pressures as much as a thousand times atmospheric. How might these bacteria be affected if they were rapidly moved to the surface of the ocean?

14. Two identical cylinders at the same temperature contain the same kind of gas. If cylinder *A* contains three times as much gas as cylinder *B*, what can you say about the relative pressures in the cylinders?

15. Small planets tend to have little or no atmosphere. Why is this?

16. Why do vapor bubbles in a pot of boiling water get larger as they approach the surface?

17. Markings to indicate length are placed on a steel tape in a room that is at a temperature of 22°C. Measurements are then made with the same tape on a day when the temperature is 27°C. Are the measurements too long, too short, or accurate?

18. Although the average speed of gas molecules in thermal equilibrium at some temperature is greater than zero, the average velocity is zero. Explain.

19. Explain why a column of mercury in a thermometer first descends slightly and then rises when placed in hot water.

20. One container is filled with helium gas and another with argon gas. If both containers are at the same temperature, which molecules have the higher rms speed?

21. What happens to a helium-filled balloon released into the air? Will it expand or contract? Will it stop rising at some height?

PROBLEMS

1, 2, 3 = straightforward, intermediate, challenging ☐ = full solution available in Study Guide/Student Solutions Manual
WEB = solution posted at **http://www.saunderscollege.com/physics/college** 🖼 = biomedical application 📱 = Interactive Physics

Section 10.2 Thermometers and Temperature Scales

1. For each of the following temperatures, find the equivalent temperature on the indicated temperature scale: (a) −273.15°C on the Fahrenheit scale, (b) 98.6°F on the Celsius scale, and (c) 100 K on the Fahrenheit scale.

2. The highest recorded temperature on Earth was 136°F, at Azizia, Libya, in 1922. The lowest recorded temperature was −127°F, at Vostok Station, Antarctica, in 1960. Express these temperature extremes in degrees Celsius.

3. Convert the following temperatures to Fahrenheit and kelvins: (a) the boiling point of liquid hydrogen −252.87°C; (b) the temperature of a room at 20°C.

4. Show that the temperature −40° is unique in that it has the same numerical value on the Celsius and Fahrenheit scales.

WEB 5. A constant-volume gas thermometer is calibrated in dry ice (−80.0°C) and in boiling ethyl alcohol (78.0°C). The two pressures are 0.900 atm and 1.635 atm. (a) What value of absolute zero does the calibration yield? (b) What pressures would be found at the freezing and boiling points of water?

6. The pressure in a constant-volume gas thermometer is 0.700 atm at 100°C and 0.512 atm at 0°C. (a) What is the temperature when the pressure is 0.0400 atm? (b) What is the pressure at 450°C?

7. Show that if the temperature on the Celsius scale changes by ΔT_C, the Fahrenheit temperature changes by the amount $\Delta T_F = (9/5)\Delta T_C$.

Section 10.3 Thermal Expansion of Solids and Liquids

8. The New River Gorge bridge in West Virginia is a 518-m-long steel arch. How much will its length change between temperature extremes −20°C and 35°C?

9. A gold ring has an inner diameter of 2.168 cm at a temperature of 15.0°C. Determine its inner diameter at 100°C. ($\alpha_{gold} = 1.42 \times 10^{-5}$°C^{-1}.)

10. A grandfather clock is controlled by a swinging brass pendulum that is 1.3000 m long at a temperature of 20.0°C. (a) What is the length of the pendulum rod when the temperature drops to 0.0°C? (b) If a pendulum's period is given by $T = 2\pi\sqrt{L/g}$, where *L* is its

length, does the change in length of the rod cause the clock to run fast or slow?

11. A pair of eyeglass frames are made of epoxy plastic (coefficient of linear expansion = $130 \times 10^{-6}°C^{-1}$). At room temperature (assume 20.0°C) the frames have circular lens holes 2.20 cm in radius. To what temperature must the frames be heated in order to insert lenses 2.21 cm in radius?

12. A cube of solid aluminum has a volume of $1.00 \ m^3$ at 20°C. What temperature change is required to produce a $100 \ cm^3$ increase in the volume of the cube?

13. A cylindrical brass sleeve is to be shrink-fitted over a brass shaft whose diameter is 3.212 cm at 0°C. The diameter of the sleeve is 3.196 cm at 0°C. (a) To what temperature must the sleeve be heated before it will slip over the shaft? (b) Alternatively, to what temperature must the shaft be cooled before it will slip into the sleeve?

14. Show that the coefficient of volume expansion, β, is related to the coefficient of linear expansion, α, through the expression $\beta = 3\alpha$.

15. At 20.000°C, an aluminum ring has an inner diameter of 5.000 cm, and a brass rod has a diameter of 5.050 cm. (a) To what temperature must the aluminum ring be heated so that it will just slip over the brass rod? (b) To what temperature must both be heated so the aluminum ring will slip off the brass rod? Would this work?

16. A construction worker uses a steel tape to measure the length of an aluminum support column. If the measured length is 18.7 m when the temperature is 21.2°C, what is the measured length when the temperature rises to 29.4°C? (*Note*: Do not neglect the expansion of the steel tape.)

17. The band in Figure P10.17 is stainless steel (coefficient of linear expansion = $17.3 \times 10^{-6}°C^{-1}$, Young's modulus = $18 \times 10^{10} \ N/m^2$). It is essentially circular with an initial mean radius of 5.0 mm, a height of 4.0 mm, and a thickness of 0.50 mm. If the band just fits snugly over the tooth when heated to a temperature of 80°C, what is the tension in the band when it cools to a temperature of 37°C?

Figure P10.17

18. An automobile fuel tank is filled to the brim with 45 L (12 gal) of gasoline at 10°C. Immediately afterward, the vehicle is parked in the Sun, where the temperature is 35°C. How much gasoline overflows from the tank as a result of the expansion? (Neglect the expansion of the tank.)

19. An underground gasoline tank at 54°F can hold 1000 gallons of gasoline. If the driver of a tanker truck fills the underground tank on a day when the temperature is 90°F, how many gallons, according to his measure on the truck, can he pour in? Assume that the temperature of the gasoline cools to 54°F on entering the tank.

Section 10.4 Macroscopic Description of an Ideal Gas

20. The ideal gas constant in the SI system is 8.315 J/mol·K. In chemistry classes, the mixed units L·atm/mol·K are used for this constant. Find the ideal gas constant in the mixed units to 4 significant figures.

21. One mole of oxygen gas is at a pressure of 6.00 atm and a temperature of 27.0°C. (a) If the gas is heated at constant volume until the pressure triples, what is the final temperature? (b) If the gas is heated so that both the pressure and volume are doubled, what is the final temperature?

22. Gas is contained in an 8.0-L vessel at a temperature of 20°C and a pressure of 9.0 atm. (a) Determine the number of moles of gas in the vessel. (b) How many molecules are in the vessel?

23. (a) An ideal gas occupies a volume of $1.0 \ cm^3$ at 20°C and atmospheric pressure. Determine the number of molecules of gas in the container. (b) If the pressure of the $1.0 \ cm^3$ volume is reduced to 1.0×10^{-11} Pa (an extremely good vacuum) while the temperature remains constant, how many moles of gas remain in the container?

24. The pressure on an ideal gas is cut in half, resulting in a decrease in temperature to three fourths of the original value. Calculate the ratio of the final volume to the original volume of the gas.

25. A cylinder with a movable piston contains gas at a temperature of 27.0°C, a volume of $1.50 \ m^3$, and an absolute pressure of 0.200×10^5 Pa. What will be its final temperature if the gas is compressed to $0.700 \ m^3$ and the absolute pressure increases to 0.800×10^5 Pa?

26. Gas is confined in a tank at a pressure of 10.0 atm and a temperature of 15.0°C. If half of the gas is withdrawn and the temperature is raised to 65.0°C, what is the new pressure in the tank?

27. A rigid tank contains 0.40 moles of oxygen (O_2). Determine the mass (in kg) of oxygen that must be withdrawn from the tank to lower the pressure of the gas from 40 atm to 25 atm. Assume that the volume of the tank and the temperature of the oxygen are constant during this operation.

28. A weather balloon is designed to expand to a maximum radius of 20 m when in flight at its working altitude, where the air pressure is 0.030 atm and the temperature is 200 K. If the balloon is filled at atmospheric pressure and 300 K, what is its radius at lift-off?

WEB **29.** A cylindrical diving bell, 3.00 m in diameter and 4.00 m tall with an open bottom, is submerged to a depth of 220 m in the ocean. The surface temperature is 25.0°C, and the temperature 220 m down is 5.00°C. The density of seawater is 1025 kg/m³. How high does the seawater rise in the bell when it is submerged?

30. An air bubble has a volume of 1.50 cm³ when it is released by a submarine 100 m below the surface of a lake. What is the volume of the bubble when it reaches the surface? Assume that the temperature of the air in the bubble remains constant during ascent.

31. The density of helium gas at $T = 0°C$ and atmospheric pressure is $\rho_0 = 0.179$ kg/m³. The temperature is then raised to $T = 100°C$, but the pressure is kept constant. Assuming that helium is ideal, calculate the new density, ρ_f, of the gas.

Section 10.5 Avogadro's Number and the Ideal Gas Law

Section 10.6 The Kinetic Theory of Gases

32. A sealed cubical container 20.0 cm on a side contains three times Avogadro's number of molecules at a temperature of 20.0°C. Find the force exerted by the gas on one of the walls of the container.

33. What is the average kinetic energy of oxygen molecules at a temperature of 300 K?

34. (a) What is the total random kinetic energy of all the molecules in one mole of hydrogen at a temperature of 300 K? (b) With what speed would a mole of hydrogen have to move so that the kinetic energy of the mass as a whole would be equal to the total random kinetic energy of its molecules?

35. Use Avogadro's number to find the mass of a helium atom.

36. Three moles of nitrogen gas, N_2, at 27.0°C are contained in a 22.4-L cylinder. Find the pressure the gas exerts on the cylinder walls.

WEB **37.** The temperature near the top of the atmosphere on Venus is 240 K. (a) Find the rms speed of hydrogen (H_2) at this point in the atmosphere. (b) Repeat for carbon dioxide (CO_2). (c) It has been found that if the rms speed exceeds one sixth of the planet's escape velocity, the gas eventually leaks out of the atmosphere and into outer space. If the escape velocity on Venus is 10.3 km/s, does hydrogen escape? Does carbon dioxide?

38. A cylinder contains a mixture of helium and argon gas in equilibrium at a temperature of 150°C. (a) What is the average kinetic energy of each type of molecule? (b) What is the rms speed of each type of molecule?

39. Superman leaps in front of Lois Lane to save her from a volley of bullets. In a one-minute interval, an automatic weapon fires 150 bullets, each of mass 8.0 g, at 400 m/s. The bullets strike his mighty chest, which has an area of 0.75 m². Find the average force exerted on Superman's chest if the bullets bounce back after an elastic, head-on collision.

40. In a period of 1.0 s, 5.0×10^{23} nitrogen molecules strike a wall of area 8.0 cm². If the molecules move at 300 m/s and strike the wall head on in a perfectly elastic collision, find the pressure exerted on the wall. (The mass of one N_2 molecule is 4.68×10^{-26} kg.)

41. If 2.0 mol of a gas are confined to a 5.0-L vessel at a pressure of 8.0 atm, what is the average kinetic energy of a gas molecule?

ADDITIONAL PROBLEMS

42. The active element of a certain laser is an ordinary glass rod 20 cm long and 1.0 cm in diameter. If the temperature of the rod increases by 75°C, find its increases in (a) length, (b) diameter, and (c) volume.

43. A 1.5-m-long glass tube, closed at one end, is weighted and lowered to the bottom of a fresh-water lake. When the tube is recovered, an indicator mark shows that water rose to within 0.40 m of the closed end. Determine the depth of the lake. Assume constant temperature.

44. The density of gasoline is 730 kg/m³ at 0°C. Its volume expansion coefficient is $9.6 \times 10^{-4}°C^{-1}$. If one gallon of gasoline occupies 0.0038 m³, how many extra kilograms of gasoline are obtained when 10 gallons of gasoline are bought at 0°C rather than at 20°C?

45. A vertical cylinder of cross-sectional area 0.050 m² is fitted with a tight-fitting, frictionless piston of mass 5.0 kg (Fig. P10.45). If there are 3.0 mol of an ideal gas in the cylinder at 500 K, determine the height, h, at which the piston will be in equilibrium under its own weight.

Figure P10.45

46. A liquid with coefficient of volume expansion β just fills a spherical container of volume V_0 at temperature T

(Fig. P10.46). The container is made of a material that has a coefficient of linear expansion of α. The liquid is free to expand into a capillary of cross-sectional area A at the top. (a) If the temperature increases by ΔT, show that the liquid rises in the capillary by the amount $\Delta h = (V_0/A)(\beta - 3\alpha)\Delta T$. (b) For a typical system, such as a mercury thermometer, why is it a good approximation to neglect the expansion of the container?

Figure P10.46 (Problems 46 and 57)

47. A hollow aluminum cylinder is to be fitted over a steel piston. At 20°C the inside diameter of the cylinder is 99% of the outside diameter of the piston. To what common temperature should the two pieces be heated in order that the cylinder just fits over the piston?

48. A steel measuring tape was designed to read correctly at 20°C. A parent uses the tape to measure the height of a 1.1-m-tall child. If the measurement is made on a day when the temperature is 25°C, is the tape reading longer or shorter than the actual height, and by how much?

49. Before beginning a long trip on a hot day, a driver inflates an automobile tire to a gauge pressure of 1.80 atm at 300 K. At the end of the trip the gauge pressure has increased to 2.20 atm. (a) Assuming the volume has remained constant, what is the temperature of the air inside the tire? (b) What percentage of the original mass of air in the tire should be released so the pressure returns to the original value? Assume the temperature remains at the value found in (a), and the volume of the tire remains constant as air is released.

50. Two small containers of equal volume, 100 cm³, each contain helium gas at 0°C and 1.00 atm pressure. The two containers are joined by a small open tube of negligible volume, allowing gas to flow from one container to the other. What common pressure will exist in the two containers if the temperature of one container is raised to 100°C while the other container is kept at 0°C?

WEB 51. A copper rod and steel rod are heated. At 0°C the copper rod has a length of L_C, the steel one has a length

L_S. When the rods are being heated or cooled, a difference of 5.00 cm is maintained between their lengths. Determine the values of L_C and L_S.

52. Nine grams of water are placed in a 2.00-L pressure cooker and heated to 500°C. What is the pressure inside the container?

53. An expandable cylinder has its top connected to a spring of constant 2.0×10^3 N/m (see Fig. P10.53). The cylinder is filled with 5.00 L of gas with the spring relaxed at a pressure of 1.00 atm and a temperature of 20.0°C. (a) If the lid has a cross-sectional area of 0.0100 m² and negligible mass, how high will the lid rise when the temperature is raised to 250°C? (b) What is the pressure of the gas at 250°C?

Figure P10.53

54. A 250-m-long bridge is improperly designed so that it cannot expand with temperature. It is made of concrete with $\alpha = 12 \times 10^{-6}$°C^{-1}. (a) Assuming that the maximum change in temperature at the site is expected to be 20°C, find the change in length the span would undergo if it were free to expand. (b) Show that the stress on an object with Young's modulus Y when raised by ΔT with its ends firmly fixed is given by $\alpha Y \Delta T$. (c) If the maximum stress the bridge can withstand without crumbling is 2.0×10^7 Pa, will it crumble because of this temperature increase? Young's modulus for concrete is about 2.0×10^{10} Pa.

55. A swimmer has 0.820 L of dry air in her lungs when she dives into a lake. Assuming the pressure of the dry air is 95% of the external pressure at all times, what is the volume of the dry air at a depth of 10.0 m? Assume that atmospheric pressure at the surface is 1.013×10^5 Pa.

56. Use the ideal gas equation and the relationship between moles and the mass of a gas to find an expression for the density of a gas.

57. A mercury thermometer is constructed as in Figure P10.46. The capillary tube has a diameter of 0.0050 cm, and the bulb has a diameter of 0.30 cm. Neglecting the expansion of the glass, find the change in height of the mercury column for a temperature change of 25°C.

58. A bimetallic bar is made of two thin strips of dissimilar metals bonded together. As they are heated, the one with the larger average coefficient of expansion expands more than the other, forcing the bar into an arc, with the outer radius having a larger circumference (see Fig. P10.58). (a) Derive an expression for the angle of bending θ as a function of the initial length of the strips, their average coefficients of linear expansion, the change in temperature, and the separation of the centers of the strips ($\Delta r = r_2 - r_1$). (b) Show that the angle of bending goes to zero when ΔT goes to zero or when the two coefficients of expansion become equal. (c) What happens if the bar is cooled?

Figure P10.58

11

Heat

Chapter Outline

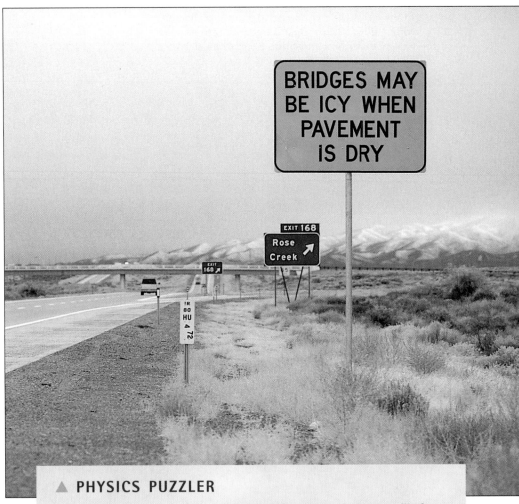

▲ PHYSICS PUZZLER

In regions where the temperature drops below freezing in winter, you will often see a road sign when approaching a bridge, warning "Bridge Freezes Before Roadway." Why does the bridge become icy before the roadway? *(© David R. Frazier Photolibrary)*

It is an experimentally established fact that when two objects at different temperatures are placed in thermal contact with each other, the temperature of the warmer object decreases and the temperature of the cooler object increases. If the two are left in contact for some time, they eventually reach a common equilibrium temperature that is intermediate between the two initial temperatures. When such a process occurs, we say that heat is transferred from the object at the higher temperature to the one at the lower temperature.

Up until about 1850, the subjects of heat and mechanics were considered to be two distinct branches of science, and the principle of conservation of energy seemed to be a rather specialized result that could be used to describe certain kinds of mechanical systems. After the two disciplines were shown to be related, the principle of conservation of energy emerged as a universal principle of nature. From this new perspective, heat and work are both treated as modes of energy transfer. Experiments performed by the Englishman James Joule (1818–1889) and his contemporaries demonstrated that whenever heat is gained or lost by a system during some process, the gain or loss can be accounted for by an equivalent quantity of mechanical work done on the system. Thus, with the broadening of the concept of energy to include heat as a path of energy transfer, the principle of energy conservation was extended.

The focus of this chapter is to introduce the concept of heat and some of the processes that enable heat to be transferred between a system and its surroundings.

James Prescott Joule, British physicist (1818–1889)

Joule received some formal education in mathematics, philosophy, and chemistry from John Dalton but was in large part self-educated. Joule's most active research period, from 1837 through 1847, led to the establishment of the principle of conservation of energy and the equivalence of heat and other forms of energy. His study of the quantitative relationship among electrical, mechanical, and chemical effects of heat culminated in his announcement in 1843 of the amount of work required to produce a unit of heat, called the mechanical equivalent of heat. *(By kind permission of the President and Council of the Royal Society)*

11.1 THE MECHANICAL EQUIVALENT OF HEAT

Heat (or **thermal energy**) is now defined as energy that is transferred between a system and its environment because of a temperature difference between them. Before scientists arrived at a correct understanding of heat, the units in which heat was measured had already been developed. They were chosen because of early misunderstandings about heat. These unusual units are still widely used in many applications, and so we shall discuss them briefly here. One of the most widely used is the **calorie (cal),** defined as **the heat required to raise the temperature of 1 g of water from 14.5°C to 15.5°C.** A related unit is the **kilocalorie (kcal),** 1 kcal = 10^3 cal. The Calorie with a capital C, used in describing the energy equivalent of foods, is equal to 1 kcal. The unit of heat in the British engineering system is the **British thermal unit** (Btu), defined as **the heat required to raise the temperature of 1 lb of water from 63°F to 64°F.**

◀ Definition of the calorie

Because heat is now recognized as energy being transferred, scientists are increasingly using the SI unit of energy, the **joule** (J), for quantities of heat. In this book, heat is most often measured in joules.

When the concept of mechanical energy was introduced in Chapter 5, we found that whenever friction is present in a mechanical system, some mechanical energy is lost. Experiments of various sorts show that this lost mechanical energy does not simply disappear; instead, it is transformed into thermal energy. Joule (1818–1889) was the first to establish the equivalence of the two forms of energy. He found that

$$1 \text{ cal} = 4.186 \text{ J} \qquad\qquad [11.1]$$

The ratio 4.186 J/cal is known, for purely historical reasons, as the **mechanical equivalent of heat.**

EXAMPLE 11.1   Losing Weight the Hard Way

A student eats a dinner rated at 2000 (food) Calories. He wishes to do an equivalent amount of work in the gymnasium by lifting a 50.0-kg mass. How many times must he raise the weight to expend this much energy? Assume that he raises the weight a distance of 2.00 m each time and that no work is done when the weight is dropped to the floor.

Solution   Because 1 Calorie = 10^3 cal, the work required is 2.00×10^6 cal. Converting this to joules, we have, for the total work required,

$$W = (2.00 \times 10^6 \text{ cal})(4.186 \text{ J/cal}) = 8.37 \times 10^6 \text{ J}$$

The work done in lifting the weight once through the distance h is equal to mgh, and the work done in lifting the weight n times is $nmgh$. If we set $nmgh$ equal to the total work required, we have

$$W = nmgh = 8.37 \times 10^6 \text{ J}$$

Because $m = 50.0$ kg and $h = 2.00$ m, we get

$$n = \frac{8.37 \times 10^6 \text{ J}}{(50.0 \text{ kg})(9.80 \text{ m/s}^2)(2.00 \text{ m})} = 8.54 \times 10^3 \text{ times}$$

If the student is in good shape and lifts the weight, say, once every 5.0 s, it will take him about 12 h to perform this feat. Clearly, it is much easier to lose weight by dieting.

This problem is somewhat misleading in that it assumes perfect conversion of chemical energy into mechanical energy. In actual practice, a more realistic assessment of the number of repetitions can be found by dividing the given answer by 6. Thus, about 1400 lifts are required to burn off the calories.

TABLE 11.1
Specific Heats of Some Materials at Atmospheric Pressure

Substance	J/kg·°C	cal/g·°C
Aluminum	900	0.215
Beryllium	1820	0.436
Cadmium	230	0.055
Copper	387	0.0924
Germanium	322	0.077
Glass	837	0.200
Gold	129	0.0308
Ice	2090	0.500
Iron	448	0.107
Lead	128	0.0305
Mercury	138	0.033
Silicon	703	0.168
Silver	234	0.056
Steam	2010	0.480
Water	4186	1.00

11.2   SPECIFIC HEAT

The quantity of heat energy required to raise the temperature of a given mass of a substance by some amount varies from one substance to another. For example, the heat required to raise the temperature of 1 kg of water by 1°C is 4186 J, but the heat required to raise the temperature of 1 kg of copper by 1°C is only 387 J. Every substance has a unique value for the amount of heat required to change the temperature of 1 kg of it by 1°C, and this number is referred to as the specific heat of the substance. Table 11.1 lists specific heats for a few substances. Note that specific heats do vary with temperature. The values given in Table 11.1 represent average values in the range of "ordinary" temperatures where they are approximately constant.

Suppose that a quantity, Q, of heat is transferred to a substance of mass m, thereby changing its temperature by ΔT. The **specific heat,** c, of the substance is defined as

Specific heat ▶

$$c \equiv \frac{Q}{m\,\Delta T} \qquad\qquad [11.2]$$

From this definition we can express the heat transferred between a system of mass *m* and its surroundings for a temperature change of ΔT as

$$Q = mc\,\Delta T \qquad\qquad [11.3]$$

For example, the heat required to raise the temperature of 0.500 kg of water by 3.00°C is equal to $(0.500 \text{ kg})(4186 \text{ J/kg·°C})(3.00°C) = 6280$ J. Note that when the temperature increases, ΔT and Q are taken to be *positive*, corresponding to heat flowing *into* the system. Likewise, when the temperature decreases, ΔT and Q are *negative* and heat flows *out* of the system.

Note from Table 11.1 that water has the highest specific heat of the substances we are likely to encounter on a routine basis. This high specific heat is responsible for the moderate temperatures found in regions near large bodies of water. As the temperature of a body of water decreases during winter, the water gives off heat to the air, which carries the heat landward when prevailing winds are favorable. For example, the prevailing winds off the western coast of the United States are toward the land, and the heat liberated by the Pacific Ocean as it cools keeps coastal areas much warmer than they would otherwise be. This explains why the western coastal states generally have more favorable winter weather than the eastern coastal states, where the winds do not carry the heat toward land. The same effect also keeps summers cooler in San Francisco than in a similar region on the East coast.

The fact that the specific heat of water is higher than that of land is responsible for the pattern of air flow at a beach. During the day, the Sun adds roughly equal amounts of energy to beach and water, but the lower specific heat of sand causes the beach to reach a higher temperature than the water. As a consequence, the air above the land reaches a higher temperature than that over the water, and cooler air from above the water is drawn in to displace this rising hot air, resulting in a breeze from ocean to land during the day. Because the hot air gradually cools as it rises, it subsequently sinks, setting up the circulating pattern shown in Figure 11.1a. During the night, the land cools more quickly than the water, and the circulating pattern reverses itself because the hotter air is now over the water (Fig. 11.1b). The offshore and onshore breezes are certainly well known to sailors.

APPLICATION

Sea Breezes and Thermals.

Day

Night

Water

Beach

Water

Beach

(a)

(b)

Figure 11.1 Circulation of air at the beach. (a) On a hot day, the air above the sand warms faster than the air above the cooler water. The cooler air over the water moves toward the beach, displacing the rising warmer air. (b) At night the sand cools more rapidly than the water, and hence the air currents reverse direction.

A similar effect produces rising layers of air, called *thermals,* that can help eagles to soar higher and hang gliders to stay in flight longer. A thermal is created when a portion of the Earth reaches a higher temperature than neighboring regions. This often happens to plowed fields, which are heated by the Sun to higher temperatures than nearby fields shaded by vegetation. The cooler, denser air over the vegetation-covered fields pushes the expanding air over the plowed field upward, and a thermal is formed.

11.3   CONSERVATION OF ENERGY: CALORIMETRY

Situations in which mechanical energy is converted to thermal energy occur frequently. We shall look at some of them in examples and in the problems at the end of the chapter, but most of our attention here is directed toward a particular kind of conservation-of-energy situation. In problems using the procedure called *calorimetry,* only the transfer of thermal energy between the system and its surroundings is considered.

One technique for measuring the specific heat of a solid or liquid is simply to heat the substance to some known temperature, place it in a vessel containing water of known mass and temperature, and measure the temperature of the water after equilibrium is reached. Because a negligible amount of mechanical work is done in the process, the law of conservation of energy requires that the heat that leaves the warmer substance (of unknown specific heat) equal the heat that enters the water. Devices in which this heat transfer occurs are called **calorimeters.**

Suppose that m_x is the mass of a substance whose specific heat we wish to measure, c_x its specific heat, and T_x its initial temperature. Let m_w, c_w, and T_w represent the corresponding values for the water. If T is the final equilibrium temperature after everything is mixed, then from Equation 11.2 we find that the heat gained by the water is $m_w c_w (T - T_w)$ and the heat lost by the substance of unknown specific heat, c_x, is $m_x c_x (T_x - T)$. Assuming that the system (water + unknown) neither loses nor gains heat, and if the cup has negligible mass, it follows that the heat gained by the water must equal the heat lost by the unknown (conservation of energy):

$$m_w c_w (T - T_w) = m_x c_x (T_x - T)$$

Solving for c_x gives

$$c_x = \frac{m_w c_w (T - T_w)}{m_x (T_x - T)}$$

Do not attempt to memorize this equation. Instead, always start from first principles in solving calorimetry problems. That is, determine which substances lose heat and which gain heat, and then equate heat loss to heat gain.

A word about sign conventions: In a later chapter we shall find it necessary to use a sign convention in which a positive sign for Q indicates heat gained by a substance and a negative sign indicates heat lost. However, for calorimetry problems it is less confusing if you ignore the positive and negative signs and instead equate heat loss to heat gain. That is, you should always write ΔT as a positive quantity. For example, if T_f is greater than T_i, let $\Delta T = T_f - T_i$. If T_f is less than T_i, let $\Delta T = T_i - T_f$.

EXAMPLE 11.2 Cooling a Hot Ingot

A 0.0500-kg ingot of metal is heated to 200.0°C and then dropped into a beaker containing 0.400 kg of water that is initially at 20.0°C. If the final equilibrium temperature of the mixed system is 22.4°C, find the specific heat of the metal.

Solution Because the heat lost by the ingot equals the heat gained by the water, we can write

$$m_x c_x (T_{ix} - T_{fx}) = m_w c_w (T_{fw} - T_{iw})$$

$$(0.0500 \text{ kg}) (c_x) (200.0°C - 22.4°C) = (0.400 \text{ kg}) (4186 \text{ J/kg} \cdot °C) (22.4°C - 20.0°C)$$

from which we find that

$$c_x = \quad 453 \text{ J/kg} \cdot °C$$

The ingot is most likely iron, as can be seen by comparing this result with the data in Table 11.1.

EXAMPLE 11.3 Fun Time for a Cowboy

A cowboy fires a 2.00-g silver bullet at a muzzle speed of 200 m/s into the pine wall of a saloon. Assume that all the internal energy generated by the impact remains with the bullet. What is the temperature change of the bullet?

Solution The kinetic energy of the bullet is

$$\tfrac{1}{2}mv^2 = \tfrac{1}{2}(2.00 \times 10^{-3} \text{ kg})(200 \text{ m/s})^2 = 40.0 \text{ J}$$

Nothing in the environment is hotter than the bullet, so the bullet gains no heat. Its temperature increases because the 40.0 J of kinetic energy becomes 40.0 J of extra internal energy. The temperature change would be the same as if 40.0 J of heat were transferred from a stove to the bullet, and we imagine this process to compute ΔT from

$$Q = mc \, \Delta T$$

Because the specific heat of silver is 234 J/kg·°C (Table 11.1), we get

$$\Delta T = \frac{Q}{mc} = \frac{40.0 \text{ J}}{(2.00 \times 10^{-3} \text{ kg})(234 \text{ J/kg} \cdot °C)} = \quad 85.5°C$$

Exercise Suppose the cowboy runs out of silver bullets and fires a lead bullet of the same mass and velocity into the wall. What is the temperature change of the bullet?

Answer 157°C

11.4 LATENT HEAT AND PHASE CHANGES

A substance usually undergoes a change in temperature when heat is transferred between it and its surroundings. There are situations, however, in which the flow of heat does not result in a change in temperature. This is the case whenever the substance undergoes a physical alteration from one form to another, referred to as

TABLE 11.2   Latent Heats of Fusion and Vaporization

Substance	Melting Point (°C)	Latent Heat of Fusion J/kg	(cal/g)	Boiling Point (°C)	Latent Heat of Vaporization J/kg	(cal/g)
Helium	−269.65	5.23×10^3	(1.25)	−268.93	2.09×10^4	(4.99)
Nitrogen	−209.97	2.55×10^4	(6.09)	−195.81	2.01×10^5	(48.0)
Oxygen	−218.79	1.38×10^4	(3.30)	−182.97	2.13×10^5	(50.9)
Ethyl alcohol	−114	1.04×10^5	(24.9)	78	8.54×10^5	(204)
Water	0.00	3.33×10^5	(79.7)	100.00	2.26×10^6	(540)
Sulfur	119	3.81×10^4	(9.10)	444.60	3.26×10^5	(77.9)
Lead	327.3	2.45×10^4	(5.85)	1750	8.70×10^5	(208)
Aluminum	660	3.97×10^5	(94.8)	2450	1.14×10^7	(2720)
Silver	960.80	8.82×10^4	(21.1)	2193	2.33×10^6	(558)
Gold	1063.00	6.44×10^4	(15.4)	2660	1.58×10^6	(377)
Copper	1083	1.34×10^5	(32.0)	1187	5.06×10^6	(1210)

a **phase change.** Some common phase changes are solid to liquid (melting), liquid to gas (boiling), and a change in crystalline structure of a solid. Every phase change involves a change in internal energy.

The heat required to change the phase of a given mass, m, of a pure substance is

$$Q = mL \qquad\qquad\qquad [11.4]$$

where L is called the **latent heat** ("hidden" heat) of the substance and depends on the nature of the phase change as well as on the properties of the substance. **Latent heat of fusion, L_f,** is the term used when the phase change is from solid to liquid, and **latent heat of vaporization, L_v,** is the term used when the phase change is from liquid to gas.[1] For example, the latent heat of fusion for water at atmospheric pressure is 3.33×10^5 J/kg, and the latent heat of vaporization for water is 2.26×10^6 J/kg. The latent heats of different substances vary considerably, as Table 11.2 shows.

Consider, for example, the heat required to convert a 1.00-g block of ice at −30.0°C to steam (water vapor) at 120.0°C. Figure 11.2 indicates the experimental results obtained when heat is gradually added to the ice. Let us examine each portion of the curve separately.

Part A.   On this portion of the curve the temperature of the ice is changing from −30.0°C to 0.0°C. Because the specific heat of ice is 2090 J/kg·°C, we can calculate the amount of heat added from Equation 11.3:

$$Q = m_i c_i \, \Delta T = (1.00 \times 10^{-3} \text{ kg})(2090 \text{ J/kg} \cdot {}^\circ\text{C})(30.0{}^\circ\text{C}) = 62.7 \text{ J}$$

[1] When a gas gives up heat, it eventually returns to the liquid phase, or *condenses.* The heat per unit mass given up during the process is called the *heat of condensation,* and it equals the heat of vaporization. When a liquid loses heat, it eventually solidifies, and the *heat of solidification* equals the heat of fusion.

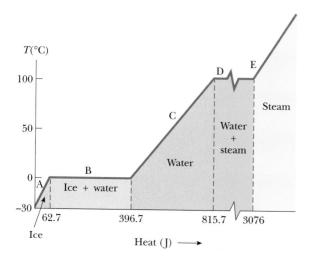

Figure 11.2 A plot of temperature versus heat added when 1.00 g of ice, initially at −30.0°C, is converted to steam at 120°C.

Part B. When the ice reaches 0°C, the ice–water mixture remains at this temperature — even though heat is being added — until all the ice melts to become water at 0°C. The heat required to melt 1.00 g of ice at 0°C is, from Equation 11.4,

$$Q = mL_f = (1.00 \times 10^{-3} \text{ kg})(3.33 \times 10^5 \text{ J/kg}) = 333 \text{ J}$$

Part C. Between 0°C and 100°C, nothing surprising happens. No phase change occurs in this region. The heat added to the water is being used to increase its temperature. The amount of heat necessary to increase the temperature from 0°C to 100°C is

$$Q = m_w c_w \, \Delta T = (1.00 \times 10^{-3} \text{ kg})(4.19 \times 10^3 \text{ J/kg} \cdot {}^\circ\text{C})(100^\circ\text{C})$$
$$= 4.19 \times 10^2 \text{ J}$$

Part D. At 100°C, another phase change occurs as the water changes from water at 100°C to steam at 100°C. Just as in Part B, the water–steam mixture remains at 100°C — even though heat is being added — until all the liquid has been converted to steam. The heat required to convert 1.00 g of water to steam at 100°C is

$$Q = mL_v = (1.00 \times 10^{-3} \text{ kg})(2.26 \times 10^6 \text{ J/kg}) = 2.26 \times 10^3 \text{ J}$$

Part E. On this portion of the curve, heat is being added to the steam with no phase change occurring. The heat that must be added to raise the temperature of the steam to 120°C is

$$Q = m_s c_s \, \Delta T = (1.00 \times 10^{-3} \text{ kg})(2.01 \times 10^3 \text{ J/kg} \cdot {}^\circ\text{C})(20^\circ\text{C}) = 40.2 \text{ J}$$

The *total amount of heat* that must be added to change one gram of ice at −30°C to steam at 120°C is therefore about 3.11×10^3 J. Conversely, to cool one gram of steam at 120°C down to the point at which we have ice at −30°C, we must remove 3.11×10^3 J of heat.

Phase changes can be described in terms of rearrangements of molecules when heat is added to or removed from a substance. Consider first the liquid-to-gas phase change. The molecules in a liquid are close together, and the forces between them are stronger than those between the more widely separated molecules of a gas. Therefore, work must be done on the liquid against these attractive molecular forces in order to separate the molecules. The latent heat of vaporization is the amount of energy that must be added to the liquid to accomplish this.

Similarly, at the melting point of a solid, we imagine that the amplitude of vibration of the atoms about their equilibrium positions becomes great enough to allow the atoms to pass the barriers of adjacent atoms and move to their new positions. The new locations are, on the average, less symmetrical and therefore have higher energy. The latent heat of fusion is equal to the work required at the molecular level to transform the mass from the ordered solid phase to the disordered liquid phase.

The average distance between atoms is much greater in the gas phase than in either the liquid or the solid phase. Each atom or molecule is removed from its neighbors, without the compensation of attractive forces to new neighbors. Therefore, it is not surprising that more work is required at the molecular level to vaporize a given mass of substance than to melt it; thus the latent heat of vaporization is much greater than the latent heat of fusion (Table 11.2).

Thinking Physics 1

The intuitive result of adding energy to a substance is to observe an increase in its temperature. But if energy is added to ice at 0°C, its temperature does not increase. Where does the energy go?

Explanation When energy is added to ice at, say, −10°C, much of the energy goes into increased molecular vibration, which has a physical manifestation of an increased temperature. When energy is added to ice at 0°C, the temperature of the ice remains constant and the ice melts. Much of the energy goes into increasing the electrical potential energy and breaking the bonds between molecules. This is similar to adding energy to a rock–Earth system by throwing the rock so fast that it escapes Earth — the gravitational bond between the rock and the Earth has been broken. Breaking the bonds between the water molecules in the ice allows the molecules to move with respect to each other, which we recognize as the behavior of the liquid that results from the melting.

Problem Solving Strategy

Calorimetry Problems

If you are having difficulty with calorimetry problems, make the following considerations.

1. Be sure your units are consistent throughout. That is, if you are using specific heats measured in cal/g·°C, be sure that masses are in grams and temperatures are in Celsius units throughout.
2. Losses and gains in heat are found by using $Q = mc\,\Delta T$ only for those intervals in which no phase changes occur. Likewise, the equations

$Q = mL_f$ and $Q = mL_v$ are to be used only when phase changes are taking place.

3. Often sign errors occur in heat loss = heat gain equations. One way to determine whether your equation is correct is to examine the signs of all ΔT's that appear in your equation. Every one should be a positive number.

EXAMPLE 11.4 Cooling the Steam

What mass of steam that is initially at 130°C is needed to warm 200 g of water in a 100-g glass container from 20.0°C to 50.0°C?

Solution This is a heat transfer problem in which we must equate the heat lost by the steam to the heat gained by the water and glass container. The steam loses heat in three stages. In the first, the steam is cooled to 100°C. The heat liberated in the process is

$$Q_1 = m_x c_s \, \Delta T = m_x (2.01 \times 10^3 \, \text{J/kg} \cdot {}^\circ\text{C})(30.0°\text{C}) = m_x (6.03 \times 10^4 \, \text{J/kg})$$

In the second stage, the steam is converted to water. In this case, to find the heat removed, we use the heat of vaporization and $Q = mL_v$:

$$Q_2 = m_x (2.26 \times 10^6 \, \text{J/kg})$$

In the last stage, the temperature of the water is reduced to 50.0°C. This liberates heat in the amount of

$$Q_3 = m_x c_w \, \Delta T = m_x (4.19 \times 10^3 \, \text{J/kg} \cdot {}^\circ\text{C})(50.0°\text{C}) = m_x (2.09 \times 10^5 \, \text{J/kg})$$

If we equate the heat lost by the steam to the heat gained by the water and glass, and use the given information, we find that

$$m_x (6.03 \times 10^4 \, \text{J/kg}) + m_x (2.26 \times 10^6 \, \text{J/kg}) + m_x (2.09 \times 10^5 \, \text{J/kg}) =$$
$$(0.200 \, \text{kg})(4.19 \times 10^3 \, \text{J/kg} \cdot {}^\circ\text{C})(30.0°\text{C}) + (0.100 \, \text{kg})(837 \, \text{J/kg} \cdot {}^\circ\text{C})(30.0°\text{C})$$
$$m_x = 1.09 \times 10^{-2} \, \text{kg} = \boxed{10.9 \, \text{g}}$$

EXAMPLE 11.5 Boiling Liquid Helium

Liquid helium has a very low boiling point, 4.2 K, and a very low latent heat of vaporization, $2.09 \times 10^4 \, \text{J/kg}$ (Table 11.2). A constant power of 10.0 W (1 W = 1 J/s) is transferred to a container of liquid helium from an immersed electric heater. At this rate, how long does it take to boil away 1.00 kg of liquid helium?

Solution Because $L_v = 2.09 \times 10^4 \, \text{J/kg}$, we must supply $2.09 \times 10^4 \, \text{J}$ of energy to boil away 1.00 kg. The power supplied to the helium is 10 W = 10 J/s. That is, in 1.00 s, 10.0 J of energy is transferred to the helium. Therefore, the time it takes to transfer $2.09 \times 10^4 \, \text{J}$ is

$$t = \frac{Q}{P} = \frac{2.09 \times 10^4 \, \text{J}}{10.0 \, \text{J/s}} = 2.09 \times 10^3 \, \text{s} \approx \boxed{35 \, \text{min}}$$

In contrast, 1.00 kg of liquid nitrogen ($L_v = 2.01 \times 10^5$ J/kg) would boil away in about 5.6 h with the same power input.

Exercise If 10.0 W of power is supplied to 1.00 kg of water at 100°C, how long will it take for the water to completely boil away?

Answer 62.8 h

11.5   HEAT TRANSFER BY CONDUCTION

There are three ways in which heat energy can be transferred from one location to another: conduction, convection, and radiation. Regardless of the process, however, no net heat transfer takes place between a system and its surroundings when the two are at the same temperature. In this section we discuss heat transfer by conduction. Convection and radiation will be discussed in Sections 11.6 and 11.7.

Each of the methods of heat transfer can be examined by considering the ways in which you can warm your hands over an open fire. If you insert a copper rod into the flame, as in Figure 11.3, the temperature of the metal in your hand increases rapidly. **Conduction,** the process by which heat is transferred from the flame through the copper rod to your hand, can be understood by examining what is happening to the atoms of the metal. Initially, before the rod is inserted into the flame, the copper atoms are vibrating about their equilibrium positions. As the flame heats the rod, the copper atoms near the flame begin to vibrate with greater and greater amplitudes. These wildly vibrating atoms collide with their neighbors and transfer some of their energy in the collisions. Gradually, copper atoms progressively farther up the rod increase their amplitudes of vibration until those at the held end are reached. This increased vibration results in an increase in temperature of the metal, and possibly a burned hand.

Although the transfer of heat through a metal can be partially explained by atomic collisions, the rate of heat conduction also depends on the properties of the substance being heated. For example, it is possible to hold a piece of asbestos in a flame indefinitely. This implies that very little heat is being conducted through the asbestos. In general, metals are good conductors of heat, and materials such as asbestos, cork, paper, and fiber glass are poor conductors; gases also are poor heat conductors because of their dilute nature. Metals are good conductors of heat because they contain large numbers of electrons that are relatively free to move through the metal and transport energy from one region to another. Thus, in a good conductor, such as copper, heat conduction takes place both via the vibration of atoms and via the motions of free electrons.

In this section we consider the rate at which heat is transferred from one location to another. If Q is the amount of heat transferred from one location on an object to another in the time Δt, the **heat transfer rate, H** (sometimes called the *heat current*), is defined as

$$H \equiv \frac{Q}{\Delta t} \qquad\qquad [11.5]$$

Figure 11.3   Heat reaches the hand by conduction through the copper rod.

The ceramic material being heated by a flame is a poor conductor of heat but is able to withstand a large temperature gradient; its left side is buried in ice (0°C) and its right side is extremely hot. *(Courtesy of Corning Glass Works)*

Note that H is expressed in watts when Q is in joules and Δt is in seconds (1 W = 1 J/s).

The conduction of heat occurs only if a difference in temperature exists between two parts of the conducting medium. Consider a slab of thickness L and cross-sectional area A, as in Figure 11.4. Suppose that one face is maintained at a temperature of T_2 and the other face is held at a lower temperature, T_1. Experimentally, one finds that the rate of heat flow—that is, the heat flow, Q, per unit time, Δt—is proportional to the difference in temperature $T_2 - T_1$ and the area A, and inversely proportional to the thickness of the slab. Specifically, the rate of flow of heat (or heat current) is given by

$$H = \frac{Q}{\Delta t} = kA\left(\frac{T_2 - T_1}{L}\right)$$ [11.6]

where k is a constant called the **thermal conductivity** of the material. This constant is a property of the material. Table 11.3 lists some values of k for metals, gases, and nonmetals. Note that k is large for metals, which are good heat conductors, and small for gases and nonmetals, which are poor heat conductors (good insulators).

The fact that different materials have different k values should help you understand the following phenomenon, first mentioned in Chapter 10. If you remove a metal ice tray and a package of frozen food from the freezer, which feels colder? Experience tells you that the metal tray feels colder even though it is at the same initial temperature as the cardboard package. This is explained by the fact that metal has a much higher thermal conductivity than cardboard, and so it conducts heat more rapidly and hence removes heat from your hand at a higher rate. Consequently, the metal tray *feels* colder than the carton even though it isn't. By use of a similar argument, you should be able to explain why a tile floor feels colder to bare feet than a carpeted floor does.

Figure 11.4 Heat transfer through a conducting slab of cross-sectional area A and thickness L. The opposite faces are at different temperatures, T_1 and T_2.

EXAMPLE 11.6 Heat Transfer Through a Concrete Wall

Find the amount of heat transferred in 1.00 h by conduction through a concrete wall 2.0 m high, 3.65 m long, and 0.20 m thick if one side of the wall is held at 20°C and the other side is at 5°C.

Solution Equation 11.6 gives the rate of heat transfer in joules per second. To find the amount of heat transferred in 1.00 h, we rewrite the equation as

$$Q = kA\,\Delta t\left(\frac{T_2 - T_1}{L}\right)$$

If we substitute the values given and consult Table 11.3, we find that

$$Q = (1.3\ \text{J/s}\cdot\text{m}\cdot°\text{C})(7.3\ \text{m}^2)(3600\ \text{s})\left(\frac{15°\text{C}}{0.20\ \text{m}}\right) = 2.6 \times 10^6\ \text{J}$$

Early houses were insulated by the material of which their walls were constructed—thick masonry blocks. Masonry restricts heat loss by conduction because its k is relatively low. The large thickness, L, also decreases heat loss, as shown by Equation 11.6.

TABLE 11.3
Thermal Conductivities

Substance	Thermal Conductivity (J/s·m·°C)
Metals (at 25°C)	
Aluminum	238
Copper	397
Gold	314
Iron	79.5
Lead	34.7
Silver	427
Gases (at 20°C)	
Air	0.0234
Helium	0.138
Hydrogen	0.172
Nitrogen	0.0234
Oxygen	0.0238
Nonmetals	
Asbestos	0.25
Concrete	1.3
Glass	0.84
Ice	1.6
Rubber	0.2
Water	0.60
Wood	0.10

Optional Section

Heat is conducted from the inside to the exterior more rapidly where the snow has melted. The dormer in this house appears to have been added—and insulated. The main roof does not appear to be well insulated. *(Courtesy of A. A. Bartlett, University of Colorado, Boulder)*

Home Insulation

If you would like to do some calculating to determine whether or not to add insulation to a ceiling or to some other portion of a building, you need to slightly modify what you have just learned about conduction, for two reasons.

1. The insulating properties of materials used in buildings are usually expressed in engineering rather than SI units. For example, measurements stamped on a package of fiber glass insulating board will be in units such as British thermal units, feet, and degrees Fahrenheit.
2. In dealing with the insulation of a building, we must consider heat conduction through a compound slab, with each portion of the slab having a different thickness and a different thermal conductivity. For example, a typical wall in a house consists of an array of materials, such as wood paneling, dry wall, insulation, sheathing, and wood siding.

It is found that the rate of heat transfer through a compound slab is

$$\frac{Q}{\Delta t} = \frac{A(T_2 - T_1)}{\sum_i L_i/k_i} \qquad [11.7]$$

where T_1 and T_2 are the temperatures of the *outer extremities* of the slab and the summation is over all portions of the slab. For example, if the slab consists of three different materials, the denominator is the sum of three terms. In engineering practice, the term L/k for a particular substance is referred to as the **R value** of the material. Thus, Equation 11.7 reduces to

TABLE 11.4	*R* Values for Some Common Building Materials
Material	**R value ($ft^2 \cdot °F \cdot h/Btu$)**
Hardwood siding (1.0 in. thick)	0.91
Wood shingles (lapped)	0.87
Brick (4.0 in. thick)	4.00
Concrete block (filled cores)	1.93
Styrofoam (1.0 in. thick)	5.0
Fiber-glass batting (3.5 in. thick)	10.90
Fiber-glass batting (6.0 in. thick)	18.80
Fiber-glass board (1.0 in. thick)	4.35
Cellulose fiber (1.0 in. thick)	3.70
Flat glass (0.125 in. thick)	0.89
Insulating glass (0.25-in. space)	1.54
Vertical air space (3.5 in. thick)	1.01
Air film	0.17
Dry wall (0.50 in. thick)	0.45
Sheathing (0.50 in. thick)	1.32

$$\frac{Q}{\Delta t} = \frac{A(T_2 - T_1)}{\displaystyle\sum_i R_i}$$ [11.8]

The R values for a few common building materials are listed in Table 11.4 (note the units).

Next to any vertical outside surface is a very thin, stagnant layer of air that must be considered when the total R value for a wall is figured. The thickness of this stagnant layer depends on the speed of the wind. As a result, heat loss from a house on a day when the wind is blowing hard is greater than heat loss on a day when the wind speed is zero. A representative R value for the stagnant layer of air is given in Table 11.4.

EXAMPLE 11.7 The R Value of a Typical Wall

Calculate the total R value for a wall constructed as shown in Figure 11.5a. Starting outside the house (to the left in the figure) and moving inward, the wall consists of brick, 0.50 in. of sheathing, a vertical air space 3.5 in. thick, and 0.50 in. of dry wall. Do not forget the dead-air layers inside and outside the house.

Solution   Referring to Table 11.4, we find the total R value for the wall as follows:

R_1(outside air film) $= 0.17 \text{ ft}^2 \cdot {}^\circ\text{F} \cdot \text{h/Btu}$

R_2(brick) $= 4.00$

R_3(sheathing) $= 1.32$

R_4(air space) $= 1.01$

R_5(dry wall) $= 0.45$

R_6(inside air film) $= 0.17$

R_{total} $= 7.12 \text{ ft}^2 \cdot {}^\circ\text{F} \cdot \text{h/Btu}$

Exercise   If a layer of fiber glass insulation 3.5 in. thick is placed inside the wall to replace the air space, as in Figure 11.5b, what is the total R value of the wall? By what factor is the heat loss reduced?

Answer $R = 17 \text{ ft}^2 \cdot {}^\circ\text{F} \cdot \text{h/Btu}$; a factor of 2.5.

Figure 11.5 (Example 11.7) A cross-sectional view of an exterior wall containing (a) an air space and (b) insulation.

11.6 CONVECTION

You have probably warmed your hands by holding them over an open flame, as illustrated in Figure 11.6. In this situation, the air directly above the flame is heated and expands. As a result, the density of the air decreases and the air rises. This warmed mass of air heats your hands as it flows by. **Heat transferred by the movement of a heated substance is said to have been transferred by convection.** When the movement results from differences in density, as it does in air around a fire, it is referred to as *natural convection*. When the heated substance is forced to move by a fan or pump, as in some hot-air and hot-water heating systems, the process is called *forced convection*.

Figure 11.6 Heating a hand by convection.

Schlieren photograph of a teakettle showing steam and turbulent convection air currents. *(Gary Settles/Science Source/Photo Researchers)*

APPLICATION

Cooling Automobile Engines.

APPLICATION

Algal Blooms in Ponds and Lakes.

Figure 11.7 Convection currents are set up in a room heated by a radiator.

The circulating pattern of air flow at a beach (see Fig. 11.1) is an example of convection. Likewise, the mixing that occurs as water is cooled and eventually freezes at its surface (Chapter 10) is an example of convection in nature. Recall that mixing by convection currents ceases when the water temperature reaches 4°C. Because the water in the pool cannot be cooled by convection below 4°C, and because water is a relatively poor conductor of heat (see Table 11.3), the water near the bottom remains near 4°C for a long time. As a result, fish live in water of a comfortable temperature even in periods of prolonged cold weather.

If it were not for convection currents, it would be very difficult to boil water. The lower layers of water in a teakettle are warmed first. These heated regions expand and are buoyed up to the top because their density is lowered. Meanwhile, denser cool water replaces the warm water at the bottom of the kettle so that it can be heated.

The same process occurs when a room is heated by a radiator. The hot radiator warms the air in the lower regions of the room. The warm air expands and rises to the ceiling because of its lower density. The denser regions of cooler air from above replace the warm air, setting up the continuous air current pattern shown in Figure 11.7.

Your automobile engine is maintained at a safe operating temperature by a combination of conduction and forced convection. Water (actually, a mixture of water and antifreeze) circulates in the interior of the engine. As the metal of the engine block increases in temperature, heat passes from the hot metal to the cooler water by conduction. The water pump forces water out of the engine and into the radiator, carrying heat along with it (forced convection). In the radiator, the hot water passes through metal pipes that are in contact with the cooler outside air, and heat passes into the air by conduction. The cooled water is then returned to the engine by the water pump to absorb more energy.

The algal blooms often seen in temperate lakes and ponds during spring or fall are caused by convection currents in the water. To understand this process, consider Figure 11.8. During the summer, bodies of water develop temperature gradients such that an upper warm layer of water is separated from a lower cold layer by a buffer zone called a thermocline. In the spring or fall, the temperature changes in the water break down this thermocline, setting up convection currents that mix the water. This mixing process transports nutrients from the bottom to the surface. The nutrient-rich water forming at the surface can cause a rapid, temporary increase in the population of algae.

Applying Physics 1

 The body temperature of mammals ranges from about 35°C to 38°C, a fairly narrow range, while that of birds ranges from about 40°C to 43°C. To maintain this temperature in cold weather, it is necessary to provide insulation to prevent heat loss from the surface of the skin. How is this accomplished?

Explanation   A natural method of accomplishing this is via layers of fat beneath the skin. Fat protects against both conduction and convection in that it has a low thermal conductivity, and there are few blood vessels in fat to carry blood to the surface where convective heat losses can occur. Birds ruffle their feathers in cold weather such that a layer of air with a low thermal conductivity is trapped between the feathers and the skin. Bristling the fur produces the same effect on fur-bearing animals.

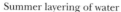

Warm Layer 25°–22°C

Thermocline 20°–10°C

Cool layer 5°–4°C

Summer layering of water

Fall and spring upwelling

Figure 11.8 (a) During the summer, the upper warm layer of water is separated from a cooler lower layer by a thermocline. (b) Convections currents during the spring or fall mix the water and can cause algal blooms.

11.7 RADIATION

The third way of transferring heat is through **radiation.** You have most likely experienced radiant heat when sitting in front of a fireplace. Figure 11.9 shows how you can warm your hands at an open flame by means of radiant heat. The hands are placed to one side of the flame; they are not in physical contact with the flame either directly or indirectly, and therefore conduction cannot account for the heat transfer. Furthermore, convection is not important in this situation because the hands are not above the flame in the path of convection currents. The important process in this case is the radiation of heat energy.

All objects continuously radiate energy in the form of electromagnetic waves, which we shall discuss in Chapter 21. Electromagnetic radiation associated with the loss of heat energy from an object at a temperature of a few hundred kelvins is referred to as *infrared* radiation.

The surface of the Sun is at a few thousand kelvins and most strongly radiates visible light. Approximately 1340 J of sunlight energy strikes 1 m² of the top of the Earth's atmosphere every second. Some of this energy is reflected back into space, and some is absorbed by the atmosphere, but enough arrives at the surface of the Earth each day to supply hundreds of times more energy than human technology requires—if it could be captured and used efficiently. The building of an increasing number of solar houses in this country is one attempt to make use of this free energy.

Radiant energy from the Sun affects our day-to-day existence in a number of ways. For example, consider what happens to the atmospheric temperature at night.

Figure 11.9 Warming hands by radiation.

If there is a cloud cover above the Earth, the water vapor in the clouds reflects back a part of the infrared radiation emitted by the Earth, and consequently the temperature remains moderate. In the absence of a cloud cover, however, there is nothing to prevent this radiation from escaping into space, and thus the temperature drops more on a clear night than when it is cloudy.

The rate at which an object emits radiant energy is proportional to the fourth power of its absolute temperature. This is known as **Stefan's law** and is expressed in equation form as

Stefan's law ▶

$$P = \sigma A e T^4 \qquad\qquad [11.9]$$

where P is the power radiated by the object in watts (or joules per second), σ is a constant equal to 5.6696×10^{-8} W/m$^2 \cdot$K^4, A is the surface area of the object in square meters, e is a constant called the **emissivity,** and T is the object's temperature in kelvins. The value of e can vary between zero and unity, depending on the properties of the surface.

An object radiates energy at a rate given by Equation 11.9. At the same time, the object also absorbs electromagnetic radiation. If the latter process did not occur, the object would eventually radiate all of its energy and its temperature would reach absolute zero. The absorbed energy comes from the body's surroundings, which consist of other objects that radiate energy. If an object is at temperature T and its surroundings are at temperature T_0, the net energy gained or lost each second by the object as a result of radiation is

$$P_{\text{net}} = \sigma A e (T^4 - T_0^4) \qquad\qquad [11.10]$$

When an object is in equilibrium with its surroundings, it radiates and absorbs energy at the same rate, and so its temperature remains constant. When an object is hotter than its surroundings, it radiates more energy than it absorbs, and so it cools. An *ideal absorber* is defined as an object that absorbs all of the energy incident on it; its emissivity is equal to unity. Such an object is often referred to as a **black body.** An ideal absorber is also an ideal radiator of energy. In contrast, an object with an emissivity equal to zero absorbs none of the energy incident on it. Such an object reflects all the incident energy and so is a *perfect reflector.*

A P P L I C A T I O N

Light-Colored Summer Clothing.

White clothing is more comfortable to wear in the heat of the summer than black clothing. Black fabric acts as a good absorber of incoming sunlight and as a good emitter of this absorbed energy. However, about half of the emitted energy travels toward the body, causing the person wearing the garment to feel uncomfortably warm. In contrast, white or light-colored clothing reflects away much of the incoming energy.

A P P L I C A T I O N

Thermography.

The amount of radiant energy emitted by a body can be measured with heat-sensitive recording equipment, using a technique called **thermography.** The radiation emitted by a body is greatest in the body's warmest regions. An image of the pattern formed by varying radiation levels, called a **thermogram,** is brightest in the warmest areas. Figure 11.10 reproduces a thermogram of a house. The center portions of the door and windows are yellow, signifying temperatures higher than those of surrounding areas. A higher temperature usually means that heat is escaping. Thermograms can be useful for purposes of energy conservation. For example, the owners of this house could conserve energy and reduce their heating costs by adding insulation to the attic area and by installing thermal draperies over the windows.

Figure 11.10 This thermogram of a home, made during cold weather, shows colors ranging from white and yellow (areas of greatest heat loss) to blue and purple (areas of least heat loss). *(Daedalus Enterprises, Inc./Peter Arnold, Inc.)*

Figure 11.11 shows a recently developed radiation thermometer that has removed most of the risk of taking the temperature of young children or the aged, risks such as bowel perforation or bacterial contamination. The instrument measures the intensity of the infrared radiation leaving the eardrum and surrounding tissues and converts this information to a standard numerical reading. The eardrum is a particularly good location to measure body temperature because it is near the hypothalamus—the body's temperature control center.

APPLICATION

Radiation Thermometers for Measuring Body Temperature.

Applying Physics 2

If you sit in front of a fire in a fireplace with your eyes closed, you can feel significant warmth in your eyelids. If you now put on a pair of eyeglasses and repeat this activity, your eyelids will not feel as warm. Why?

Explanation   Much of the warm feeling that one has while sitting in front of a fireplace is due to radiation from the fire. A large fraction of this radiation is in the infrared part of the spectrum. Your eyelids are particularly sensitive to infrared radiation. However, glass is partially opaque to infrared radiation. Thus, when you put on the glasses, you block much of the radiation from reaching your eyelids, and thus, they feel cooler.

Figure 11.11 A radiation thermometer measures a patient's temperature by monitoring the intensity of infrared radiation leaving the ear. *(© Blair Seitz/Photo Researchers, Inc.)*

 EXAMPLE 11.8   Who Turned Down the Thermostat?

A student is trying to decide what to wear. The air in his bedroom is at 20.0°C. If the skin temperature of the unclothed student is 37.0°C, how much heat is lost from his body in 10.0 min? Assume that the emissivity of skin is 0.900 and that the surface area of the student is 1.50 m².

Solution   Using Equation 11.10, the rate of heat loss from the skin is

$$P_{net} = \sigma Ae(T^4 - T_0^4)$$
$$= (5.67 \times 10^{-8} \text{ W/m}^2\cdot\text{K}^4)(1.50 \text{ m}^2)(0.90)[(310 \text{ K})^4 - (293 \text{ K})^4] = 143 \text{ J/s}$$

Note that it was necessary to change the temperature to kelvins. At this rate of heat loss, the total heat lost by the skin in 10.0 min is

$$Q = P_{net} \times \text{time} = (143 \text{ J/s})(600 \text{ s}) = \boxed{8.58 \times 10^4 \text{ J}}$$

11.8   HINDERING HEAT TRANSFER

The Thermos bottle, called a *Dewar flask* (after its inventor) in scientific applications, is designed to minimize heat transfer by conduction, convection, and radiation. It is used to store either cold or hot liquids for long periods of time. The standard vessel (Fig. 11.12) is double-walled Pyrex glass with a silvered inner wall. The space between the walls is evacuated to minimize heat transfer by conduction and convection. By reflecting most of the radiant heat, the silvered surface minimizes heat transfer by radiation. Very little heat is lost through the neck of the flask because Pyrex glass is a poor conductor. A further reduction in heat loss is achieved by reducing the size of the neck. A common scientific application of Dewar flasks is storage of liquid nitrogen (boiling point 77 K) and liquid oxygen (boiling point 90 K). For substances that have very low specific heats, such as liquid helium (boiling point 4.2 K), it is often necessary to use a double Dewar system in which the Dewar flask containing the liquid is surrounded by a second Dewar flask. The space between the two flasks is filled with liquid nitrogen.

Some of the principles of the Thermos bottle are used in the protection of sensitive electronic instruments in orbiting space satellites. In half of its orbit about the Earth, a satellite is exposed to intense radiation from the Sun, and in the other half it is in the Earth's cold shadow. Without protection, its interior would thus be subjected to tremendous extremes of heating and cooling. The interior of the satellite is wrapped with blankets of highly reflective aluminum foil. The foil's shiny surface reflects away much of the Sun's radiation while the satellite is in the unshaded part of the orbit, and helps retain interior heat while the satellite is in the Earth's shadow.

Wool sweaters and down jackets keep us warm by trapping the warmer air in regions close to our bodies and hence reducing heat loss by convection and conduction. In other words, what keeps us warm is not the clothing itself but the air trapped in the clothing.

Figure 11.12 A cross-sectional view of a Dewar vessel designed to store hot or cold liquids.

Vacuum

Silvered surfaces

Hot or cold substance

11.9   METABOLISM AND LOSING WEIGHT

In order to maintain life, humans must receive an input of energy from their surroundings. This is accomplished by an intake of food that supplies chemical energy inside the cells of the body. The **basal metabolism** is defined as the rate of energy consumption while an individual is awake and resting. For an average 20-year-old man this is approximately 1.2 W/kg (2.87×10^{-4} Cal/s·kg) and for a 20-year-old woman 1.1 W/kg (2.63×10^{-4} Cal/s·kg). Do the conversion yourself to show that this is equivalent to about 1700 Cal per day for a 70 kg man and 1400 Cal per day for a 60 kg woman. For an average sedentary person the energy required for daily activity is about 2500 Cal per day. That is, if such an individual consumes 2500 Cal per day, weight will be maintained and eating more or less will lead to a gain or loss of body weight, respectively. Table 11.5 shows the metabolic rates in Cal/s·kg for various activities, and Table 11.6 shows the energy content per kg of food.

TABLE 11.5
Metabolic Rates for Various Activities

Activity	Cal/s·kg
Sleeping	2.63×10^{-4}
Awake but resting	2.87×10^{-4}
Sitting	3.58×10^{-4}
Standing	6.21×10^{-4}
Walking	1.00×10^{-3}
Biking	1.81×10^{-3}
Swimming	2.63×10^{-3}
Running	4.30×10^{-3}

Applying Physics 3

 A common "old wives' tale" is that a camel stores fat in the hump on its back and converts this fat to water when deprived in the desert. This cannot be the case, however, because converting fat to water would require a tremendous increase in oxygen intake by the camel. This would result in a tremendous loss of water from the lungs by respiration, more water in fact than would be converted. What other factor(s) could cause a camel to function so well in the desert?

Explanation The major factor that allows the camel to function so well in the desert is that its body temperature can vary from about 34°C during the cool night to about 41°C in the hottest part of the day. Because of this, the camel does not begin to sweat until its body temperature is about 41°C, and when it does sweat, very little of the water lost comes from the blood stream. In contrast, a human loses a considerable amount of water from blood plasma, making the blood much thicker and harder to pump than normal. The reduced circulation of blood makes it even more difficult for a human to dissipate heat from the body.

TABLE 11.6
Energy Content per Gram of Dry Food

Food	Cal/g
Carbohydrate	4.1
Protein	4.2
Fat	9.3

EXAMPLE 11.9   Trying to Burn the Weight Off

(a) How much energy is expended by a 70.00 kg man who exercises each morning by walking for one hour?

Solution From Table 11.5 we see that the energy consumed is

$$(1.00 \times 10^{-3} \text{ Cal/s·kg})(70.00 \text{ kg})(3600 \text{ s/h}) = \boxed{252.0 \text{ Cal/h}}$$

(b) If the man consumes body fat to produce this energy, how much mass will be lost?

Solution The energy equivalent of fat is 9.3 Cal/g from Table 11.6. Thus, the fat consumed is equal to 27.0 g. An obvious moral to this story is that the easiest way to lose weight is to limit food intake.

11.10   APPLICATION: GLOBAL WARMING AND GREENHOUSE GASES

Many of the principles of heat transfer, and its prevention, can be understood by studying the operation of a glass greenhouse. During the day, sunlight passes into the greenhouse and is absorbed by the walls, earth, plants, and so on. This absorbed visible light is subsequently re-radiated as infrared radiation, which causes the temperature of the interior to rise.

In addition, convection currents are inhibited in a greenhouse. As a result, heated air cannot rapidly pass over the surfaces of the greenhouse that are exposed to the outside air and thereby cause a heat loss through those surfaces. Most experts now consider this to be a more important warming effect than that of any trapped infrared radiation. In fact, experiments have shown that when the glass over a greenhouse is replaced by a special glass known to transmit infrared light, the temperature inside is lowered only slightly. Based on this evidence, the primary mechanism that heats a greenhouse is not the absorption of infrared radiation but the inhibition of air flow that occurs under any roof (in an attic, for example).

A phenomenon formerly known as the *greenhouse effect* can also play a major role in determining the Earth's temperature. First note that the Earth's atmosphere is a good transmitter (and hence a poor absorber) of visible radiation and a good absorber of infrared radiation. Carbon dioxide (CO_2) in the Earth's atmosphere allows incoming visible radiation from the Sun to pass through more easily than infrared radiation. The visible light that reaches the Earth's surface is absorbed and re-radiated as infrared light, which in turn is absorbed (trapped) by the Earth's atmosphere. An extreme case is the warmest planet, Venus, which has a carbon dioxide-rich atmosphere and temperatures approaching 850°F.

As fossil fuels (coal, oil, and natural gas) are burned, large amounts of carbon dioxide are released into the atmosphere, causing it to retain more heat. This is of great concern to scientists and governments throughout the world. Many scientists are convinced that the 10% increase in the amount of atmospheric carbon dioxide in the past 30 years could lead to drastic changes in world climate. According to one estimate, doubling the carbon dioxide content in the atmosphere will cause temperatures to increase by 2°C! In temperate regions, such as Europe and the United States, the temperature rise would save billions of dollars per year in fuel costs. Unfortunately, the temperature increase would also melt polar ice caps, which could cause flooding and destroy many coastal areas; increase the frequency of droughts; and consequently decrease already low crop yields in tropical and subtropical countries. Even slightly higher average temperatures might make it impossible for certain plants and animals to survive in their customary ranges.

At present, about 3.5×10^{11} tons of CO_2 are released into the atmosphere each year. Most of this gas results from human activities such as the burning of fossil fuels, the cutting of forests, and manufacturing processes. Other greenhouse gases are also increasing in concentration in the atmosphere. One of these is methane, CH_4, which is released in the digestive process of cows and other ruminants. This gas originates from that part of the animal's stomach called the *rumen* where cellulose is digested. Termites also are major producers of this gas. Finally, greenhouse gases such as nitrous oxide (N_2O) and sulfur dioxide (SO_2) are increasing due to automobile and industrial pollution.

Whether the increasing greenhouse gases are responsible or not, there is convincing evidence that global warming is certainly underway. The evidence comes from the melting of ice in Antarctica and the retreat of glaciers at widely scattered sites throughout the world. For example, satellite images of Antarctica show James Ross Island completely surrounded by water for the first time since maps of the area were made about 100 years ago. Previously the island was connected to the mainland by an ice bridge. In addition, at various places across the continent, ice shelves are retreating, some at a rapid rate.

Perhaps at no place in the world are glaciers monitored with greater interest than in Switzerland. There, it is found that the Alps have lost about 50% of their glacial ice compared to 130 years ago. As evidence of this, in 1818 the citizens of the town of Brig became concerned about the advance of the Aletsch glacier, the largest glacier in Switzerland. It was threatening their summer pastures and forests. A religious procession was held, and a cross was placed against the glacier as a symbolic blockade. Today, the glacier is about a mile farther back in the mountains away from the cross, and it continues to retreat.

The retreat of glaciers on high altitude peaks in the tropics is even more severe than in Switzerland. The Lewis glacier on Mount Kenya and the snows of Kilimanjaro are two examples. However, in certain regions of the planet where glaciers are near large bodies of water and are fed by large and frequent snows, glaciers continue to advance, so the overall picture of a catastrophic global warming scenario is premature. However, in about fifty years, the amount of carbon dioxide in the atmosphere is expected to be about twice what it was in the preindustrial era. Because of this, most scientists voice the concern that reductions in greenhouse gas emissions need to be made now.

SUMMARY

Heat is a transfer of energy that takes place as a consequence of a temperature difference between a system and its surroundings. The internal energy of a substance is a function of the state of the substance, and generally increases with increasing temperature.

The **calorie** is the amount of heat necessary to raise the temperature of 1 g of water from 14.5°C to 15.5°C. The **mechanical equivalent of heat** is 4.186 J/cal.

The heat required to change the temperature of a substance by ΔT is

$$Q = mc\,\Delta T \qquad\qquad\qquad [11.3]$$

where m is the mass of the substance and c is its **specific heat.**

The heat required to change the phase of a mass m of a pure substance is

$$Q = mL \qquad\qquad\qquad [11.4]$$

The parameter L is called the **latent heat** of the substance and depends on the nature of the phase change and the properties of the substance.

Heat may be transferred by three fundamentally distinct processes: *conduction,* *convection,* and *radiation.* **Conduction** can be viewed as an exchange of kinetic energy between colliding molecules or through the motion of electrons. The rate at which heat flows by conduction through a slab of area A and thickness L is

$$H = \frac{Q}{\Delta t} = kA\left(\frac{T_2 - T_1}{L}\right) \qquad\qquad [11.6]$$

where k is the **thermal conductivity** of the material making up the slab. Heat transferred by the movement of a heated substance is said to have been transferred by **convection.**

All objects **radiate** and absorb energy in the form of electromagnetic waves. An object that is hotter than its surroundings radiates more energy than it absorbs, whereas an object that is cooler than its surroundings absorbs more energy than it radiates. The rate at which an object emits radiant energy is given by **Stefan's law:**

$$P = \sigma A e T^4 \qquad\qquad [11.9]$$

where σ is a constant equal to 5.6696×10^{-8} W/m$^2\cdot$K^4, and e is a constant called the **emissivity.**

MULTIPLE-CHOICE QUESTIONS

1. The specific heat of substance A is greater than that for substance B. If equal amounts of heat are added to both these substances, the one reaching the higher temperature, assuming no phase change occurs for either, will be
   (a) substance A.
   (b) substance B.
   (c) There will be no difference in the final temperatures.
   (d) Could be either A or B.

2. An amount of heat is added to ice, raising its temperature from $-10°$C to $-5°$C. A larger amount of heat is added to the same mass of liquid water, raising its temperature from 15°C to 20°C. From these results, we can conclude that
   (a) overcoming the latent heat of fusion of ice requires an input of energy.
   (b) the latent heat of fusion of ice delivers some energy to the system.
   (c) the specific heat of ice is less than that of water.
   (d) the specific heat of ice is greater than that of water.

3. Star A has twice the radius and twice the absolute temperature of star B. What is the ratio of the power output of star A to that of star B? The emissivity of both stars can assumed to be 1.
   (a) 4   (b) 8   (c) 16   (d) 32   (e) 64

4. How long would it take for a 1000 W heating element to melt 1 kg of ice at 0°C?
   (a) 4.19 s   (b) 41.9 s   (c) 5.55 min
   (d) 1 hour   (e) 555 min

5. A 100-g piece of copper, initially at 95°C, is dropped into 200 g of water contained in a 280-g aluminum can; the water and can are initially at 15°C. What is the final temperature of the system? (Specific heats of copper and aluminum are 0.092 and 0.215 cal/g·°C, respectively.)
   (a) 16.8°C   (b) 17.7°C   (c) 23.7°C   (d) 25.0°C

CONCEPTUAL QUESTIONS

1. Rub the palm of your hand on a metal surface for 30 to 45 seconds. Place the palm of your other hand on a unrubbed portion of the surface and then the rubbed portion. The rubbed portion will feel warmer. Now repeat this process on a wooden surface. You should notice that the difference in temperature between the rubbed and unrubbed portions of the wood surface seems larger than that of the metal surface. Why is this?

2. Pioneers stored fruits and vegetables in underground cellars. Discuss fully this choice for a storage site.

3. In usually warm climates that experience an occasional hard freeze, fruit growers will spray the fruit trees with water, hoping that a layer of ice will form on the fruit. Why is this advantageous?

4. In winter, the pioneers mentioned in Question 2 stored an open barrel of water alongside their produce. Why?

5. Cups of water for coffee or tea can be warmed with an immersion coil, which is immersed in the water and raised to a high temperature by means of electricity. The instructions warn the user not to operate the coils when they are out of water. Why? Can the immersion coil be used to warm up a cup of stew?

6. The U.S. penny is now made of copper-coated zinc. Can a calorimetric experiment be devised to test for the metal content in a collection of pennies? If so, describe the procedure you would use.

7. On a clear, cold night, why does frost tend to form on the tops of mailboxes and cars rather than the sides?

8. Consider two small balloons, one filled with air and the other with water. If a lighted match is placed within a few centimeters underneath the air-filled balloon, it bursts. However, if a lighted match is placed under the

water-filled balloon, the balloon does not burst even if the flame touches the balloon. Explain this observation in terms of heat conduction processes and the heat capacity of water.

9. A tile floor may feel uncomfortably cold to your bare feet, but a carpeted floor in an adjoining room at the same temperature feels warm. Why?

10. On a very hot day it is possible to cook an egg on the hood of a car. Would you select a black car or a white car on which to cook your egg? Why?

11. Concrete has a higher specific heat than does soil. Use this fact to explain (partially) why a city has a higher average temperature than the surrounding countryside. Would you expect breezes to blow from city to country or from country to city? Explain.

12. You need to pick up a very hot cooking pot in your kitchen. You have a pair of hot pads. Should you soak them in cold water or keep them dry in order to pick up the pot most comfortably?

13. In a daring lecture demonstration, a professor dips her wetted fingers into molten lead (327°C) and withdraws them quickly without getting burned. How is this possible?

14. The air temperature above coastal areas is profoundly influenced by the large specific heat of water. One reason is that the heat released when 1 cubic meter of water cools by 1.0°C will raise the temperature of an enormously larger volume of air by 1.0°C. Estimate the volume of air. The specific heat of air is approximately 1.0 kJ/kg·°C. Take the density of air to be 1.3 kg/m³.

15. Ethyl alcohol has about one half the specific heat of water. If equal masses of alcohol and water in separate beakers are supplied with the same amount of heat, compare the temperature increases of the two liquids.

PROBLEMS

1, 2, 3 = straightforward, intermediate, challenging   ☐ = full solution available in Study Guide/Student Solutions Manual
WEB = solution posted at **http://www.saunderscollege.com/physics/college**   🔲 = biomedical application   🖳 = Interactive Physics

Review Problem

An aluminum rod is 20.000 cm long at 20°C and has a mass of 350 g. If 10,000 J of heat energy is added to the rod, what is its new length?

Section 11.1   The Mechanical Equivalent of Heat

Section 11.2   Specific Heat

1. As part of an exercise routine a 50.0-kg person climbs 10.0 meters up a vertical rope. How many (food) Calories are expended in a single climb up the rope? [1 (food) Calorie = 10³ calories]

2. A 75.0-kg weight-watcher wishes to climb a mountain to work off the equivalent of a large piece of chocolate cake rated at 500 (food) Calories. How high must the person climb? [1 (food) Calorie = 10³ calories]

3. How many joules of energy are required to raise the temperature of 100 g of gold from 20.0°C to 100°C?

4. A 50.0-g sample of copper is at 25°C. If 1200 J of heat energy is added to the copper, what is its final temperature?

5. A 5.00-g lead bullet traveling at 300 m/s is stopped by a large tree. If half the kinetic energy of the bullet is transformed into internal energy and remains with the bullet while the other half is transmitted to the tree, what is the increase in temperature of the bullet?

6. A 50.0-g piece of cadmium is at 20°C. If 400 cal of heat is added to the cadmium, what is its final temperature?

7. Water at the top of Niagara Falls has a temperature of 10.0°C. If it falls a distance of 50.0 m and all of its potential energy goes into heating the water, calculate the temperature of the water at the bottom of the falls.

8. A 1.5-kg copper block is given an initial speed of 3.0 m/s on a rough horizontal surface. Because of friction, the block finally comes to rest. (a) If the block absorbs 85% of its initial kinetic energy as internal energy, calculate its increase in temperature. (b) What happens to the remaining energy?

9. A 200-g aluminum cup contains 800 g of water in thermal equilibrium at 80°C. The combination of cup and water is cooled uniformly so that the temperature decreases by 1.5°C per minute. At what rate is heat energy being removed? Express your answer in watts.

10. The Btu is defined as the heat required to raise the temperature of 1 lbm (pound of mass) of water by 1°F. Determine the number of Joules in a Btu. [1 lbm = a quantity of material that weighs 1 lb = 1/32.174 slugs = 0.4536 kg]

Section 11.3   Conservation of Energy: Calorimetry

11. Lead pellets, each of mass 1.00 g, are heated to 200°C. How many pellets must be added to 500 g of water that is initially at 20.0°C to make the equilibrium temperature 25.0°C? Neglect any heat transfer to or from the container.

12. A 0.40-kg iron horseshoe that is initially at 500°C is dropped into a bucket containing 20 kg of water at

22°C. What is the final equilibrium temperature? Neglect any heat transfer to or from the surroundings.

13. What mass of water at 25.0°C must be allowed to come to thermal equilibrium with a 3.00-kg gold bar at 100°C in order to lower the temperature of the bar to 50.0°C?

14. An aluminum cup contains 225 g of water at 27°C. A 400-g sample of silver at an initial temperature of 87°C is placed in the water. A 40-g copper stirrer is used to stir the mixture until it reaches its final equilibrium temperature of 32°C. Calculate the mass of the aluminum cup.

WEB 15. If 200 g of water is contained in a 300-g aluminum vessel at 10°C and an additional 100 g of water at 100°C is poured into the container, what is the final equilibrium temperature of the mixture?

16. A 100-g aluminum calorimeter contains 250 g of water. The two substances are in thermal equilibrium at 10°C. Two metallic blocks are placed in the water. One is a 50-g piece of copper at 80°C. The other sample has a mass of 70 g and is originally at a temperature of 100°C. The entire system stabilizes at a final temperature of 20°C. Determine the specific heat of the unknown second sample.

17. It is desired to cool iron parts from 500°F to 100°F by dropping them into water that is initially at 75°F. Assuming that all the heat from the iron is transferred to the water and that none of the water evaporates, how many kilograms of water are needed per kilogram of iron?

18. A combination of 0.250 kg of water at 20.0°C, 0.400 kg of aluminum at 26.0°C, and 0.100 kg of copper at 100°C is mixed in an insulated container and allowed to come to thermal equilibrium. Neglect any heat transfer to or from the container and determine the final temperature of the mixture.

19. A student drops two metallic objects into a 120-g steel container holding 150 g of water at 25°C. One object is a 200-g cube of copper that is initially at 85°C, and the other is a chunk of aluminum that is initially at 5°C. To the surprise of the student, the water reaches a final temperature of 25°C, exactly where it started. What is the mass of the aluminum chunk?

Section 11.4   Latent Heat and Phase Changes

20. A 50-g ice cube at 0°C is heated until 45 g has become water at 100°C and 5.0 g has become steam at 100°C. How much heat was added to accomplish this?

21. A 100-g cube of ice at 0°C is dropped into 1.0 kg of water that is originally at 80°C. What is the final temperature of the water after the ice has melted?

22. How much heat is required to change a 40-g ice cube from ice at −10°C to steam at 110°C?

23. What mass of steam that is initially at 120°C is needed to warm 350 g of water and its 300-g aluminum container from 20°C to 50°C?

24. Assuming no heat loss, determine the mass of water that boils away when 1.94 kg of mercury at a temperature of 200°C is added to 0.050 kg of water at a temperature of 80.0°C.

25. A 75-kg cross-country skier moves across snow such that the coefficient of friction between skis and snow is 0.20. Assume all the snow beneath his skis is at 0°C and that all the internal energy generated by friction is added to snow which sticks to his skis until melted. How far would he have to ski to melt 1.0 kg of snow?

Figure P11.25   A cross-country skier.
(Nathan Bilow, Leo de Wys, Inc.)

26. A 100-g ice cube at 0°C is placed in 650 g of water at 25°C. What is the final temperature of the mixture?

WEB 27. A 40-g block of ice is cooled to −78°C. It is added to 560 g of water in an 80-g copper calorimeter at a temperature of 25°C. Determine the final temperature. (If not all the ice melts, determine how much ice is left.) Remember that the ice must first warm to 0°C, melt, and then continue warming as water. The specific heat of ice is 0.500 cal/g·°C = 2090 J/kg·°C.

28. A beaker of water sits in the sun until it reaches an equilibrium temperature of 30°C. The beaker is made of 100 g of aluminum and contains 180 g of water. In an attempt to cool this system, 100 g of ice at 0°C is added to the water. (a) Determine the final temperature. If $T_f = 0°C$, determine how much ice remains. (b) Repeat this for 50 g of ice.

29. Steam at 100°C is added to ice at 0°C. (a) Find the amount of ice melted and the final temperature when the mass of steam is 10 g and the mass of ice is 50 g. (b) Repeat with steam of mass 1.0 g and ice of mass 50 g.

Section 11.5   Heat Transfer by Conduction

30. (a) Find the rate of heat flow through a copper block of cross-sectional area 15 cm² and length 8.0 cm when a temperature difference of 30°C is established across the block. Repeat the calculation assuming the material is

(b) a block of air with these dimensions; (c) a block of wood with these dimensions.

31. A window has a glass surface of 1.6×10^3 cm^2 and a thickness of 3.0 mm. (a) Find the rate of heat transfer by conduction through this pane when the temperature of the inside surface of the glass is 70°F and the outside temperature is 90°F. (b) Repeat for the same inside temperature and an outside temperature of 0°F.

32. Determine the R value for a wall constructed as follows: The outside of the house consists of lapped wood shingles placed over 0.50-in.-thick sheathing, over 3.0 in. of cellulose fiber, over 0.50 in. of dry wall.

33. A Thermopane window consists of two glass panes, each 0.50 cm thick, with a 1.0-cm-thick sealed layer of air between. If the inside temperature is 23.0°C and the outside temperature is 0.0°C, determine the rate of heat transfer through 1.0 m^2 of the window. Compare this with the rate of heat transfer through 1.0 m^2 of a single 1.0-cm-thick pane of glass.

34. The average thermal conductivity of the walls (including windows) and roof of the house in Figure P11.34 is 4.8×10^{-4} kW/m·°C, and their average thickness is 21.0 cm. The house is heated with natural gas, with a heat of combustion (heat given off per cubic meter of gas burned) of 9300 kcal/m^3. How many cubic meters of gas must be burned each day to maintain an inside temperature of 25.0°C if the outside temperature is 0.0°C? Disregard radiation and heat loss through the ground.

Figure P11.34

35. A copper rod and an aluminum rod of equal diameter are joined end to end in good thermal contact. The temperature of the free end of the copper rod is held constant at 100°C, and that of the far end of the aluminum rod is held at 0°C. If the copper rod is 0.15 m long, what must be the length of the aluminum rod so that the temperature at the junction is 50°C?

36. A Styrofoam box has a surface area of 0.80 m^2 and a wall thickness of 2.0 cm. The temperature of the inner surface is 5°C, and that outside is 25°C. If it takes 8.0 h for 5.0 kg of ice to melt in the container, determine the thermal conductivity of the Styrofoam.

Section 11.7   Radiation

WEB 37. A sphere that is to be considered as a perfect black-body radiator has a radius of 0.060 m and is at 200°C in a room where the temperature is 22°C. Calculate the net rate at which the sphere radiates energy.

38. Two identical objects are in the same surroundings at 0°C. One is at a temperature of 1200 K, and the other is at 1100 K. Find the ratio of the net power emitted by the hotter object to the net power emitted by the cooler object.

39. Calculate the temperature at which a tungsten filament that has an emissivity of 0.25 and a surface area of 2.5×10^{-5} m^2 will radiate energy at the rate of 25 W in a room where the temperature is 22°C.

40. Measurements on two stars indicate that Star X has a surface temperature of 6000°C and Star Y has a surface temperature of 12000°C. If both stars have the same radius, what is the ratio of the luminosity (total power output) of Star Y to the luminosity of Star X? Both stars can be considered to have an emissivity of 1.

ADDITIONAL PROBLEMS

41. The bottom of a copper kettle has a 10.0-cm radius and is 2.0 mm thick. The temperature of the outside surface is 102°C, and the water inside the kettle is boiling at 1 atm of pressure. Find the rate at which heat is being transferred through the bottom of the kettle.

42. A 1.00-m^2 solar collector collects radiation from the Sun and focuses it on 250 g of water that is initially at 23°C. The average thermal energy arriving from the Sun at the surface of the Earth at this location is 550 W/m^2, and we assume that this is collected with 100% efficiency. Find the time required for the collector to raise the temperature of the water to 100°C.

43. In a showdown on the streets of Laredo, the good guy drops a 5.0-g silver bullet, at a temperature of 20°C, into a 100-cm^3 cup of water at 90°C. Simultaneously, the bad guy drops a 5.0-g copper bullet, at the same initial temperature, into an identical cup of water. Which one ends the showdown with the coolest cup of water in the West? Neglect any heat transfer into or away from the container.

44. A brass statue (60.0% copper, 40.0% zinc) has a mass of 50.0 kg. If its temperature increases by 20.0°C, what is the change of internal energy of the statue? (Specific heat of brass = 380 J/kg·°C = 0.092 cal/g·°C)

45. A 40-g ice cube floats in 200 g of water in a 100-g copper cup; all are at a temperature of 0°C. A piece of lead at 98°C is dropped into the cup, and the final equilibrium temperature is 12°C. What is the mass of the lead?

46. A family comes home from a long vacation with laundry to do and showers to take. The water heater has been turned off during the vacation. If the water heater has a capacity of 50.0 gallons and a 4800-W heating element,

how much time is required to raise the temperature of the water from 20.0°C to 60.0°C? Assume that the heater is well insulated and no water is withdrawn from the tank during this time.

47. At time $t = 0$, a vessel contains a mixture of 10 kg of water and an unknown mass of ice in equilibrium at 0°C. The temperature of the mixture is measured versus time, with the following results. During the first 50 min, the mixture remains at 0°C. From 50 min to 60 min, the temperature increases to 2°C. Neglecting the heat capacity of the vessel, determine the mass of ice that was initially placed in the vessel. Assume a constant power input to the container.

48. A 200-g block of copper at a temperature of 90°C is dropped into 400 g of water at 27°C. The water is contained in a 300-g glass container. What is the final temperature of the mixture?

49. A "solar cooker" consists of a curved reflecting mirror that focuses sunlight onto the object to be heated (Fig. P11.49). The solar power per unit area reaching the Earth at the location of a 0.50-m-diameter solar cooker is 600 W/m². Assuming that 50% of the incident energy is converted to heat energy, how long would it take to boil away 1.0 L of water initially at 20°C? (Neglect the specific heat of the container.)

Figure P11.49

50. A class of 10 students taking an exam has a power output per student of about 200 W. Assume that the initial temperature of the room is 20°C and that its dimensions are 6.0 m by 15.0 m by 3.0 m. What is the temperature of the room at the end of 1.0 h if all the heat remains in the air in the room and none is added by an outside source? The specific heat of air is 837 J/kg·°C, and its density is about 1.3×10^{-3} g/cm³.

WEB 51. An aluminum rod and an iron rod are joined end to end in a good thermal contact. The two rods have equal lengths and radii. The free end of the aluminum rod is maintained at a temperature of 100°C, and the free end

of the iron rod is maintained at 0°C. **(a)** Determine the temperature of the interface between the two rods. **(b)** If each rod is 15 cm long and each has a cross-sectional area of 5.0 cm², what quantity of heat energy is conducted across the combination in 30 min?

52. Water is being boiled in an open kettle that has a 0.500-cm-thick circular aluminum bottom with a radius of 12.0 cm. If the water boils away at rate of 0.500 kg/min, what is the temperature of the lower surface of the bottom of the kettle? Assume that the top surface of the bottom of the kettle is at 100°C.

53. A solar collector has an effective collecting area of 12 m². The collector is thermally insulated, so conduction is negligible in comparison with radiation. On a cold but sunny day the temperature outside is −20.0°C, and the Sun irradiates the collector with a power per unit area of 300 W/m². Treating the collector as a black body (i.e., emissivity = 1.0), determine its interior temperature after the collector has achieved a steady-state condition (radiating energy as fast as it is received).

54. A *flow calorimeter* is an apparatus used to measure the specific heat of a liquid. The technique is to measure the temperature difference between the input and output points of a flowing stream of the liquid while adding heat at a known rate. **(a)** Start with the equations $Q = mc(\Delta T)$ and $m = \rho V$, and show that the rate at which heat is added to the liquid is given by the expression $\Delta Q/\Delta t = \rho c (\Delta T)(\Delta V/\Delta t)$. **(b)** In a particular experiment, a liquid of density 0.72 g/cm³ flows through the calorimeter at the rate of 3.5 cm³/s. At steady state, a temperature difference of 5.8°C is established between the input and output points when heat is supplied at the rate of 40 J/s. What is the specific heat of the liquid?

55. Three liquids are at temperatures of 10°C, 20°C, and 30°C, respectively. Equal masses of the first two liquids are mixed, and the equilibrium temperature is 17°C. Equal masses of the second and third are then mixed, and the equilibrium temperature is 28°C. Find the equilibrium temperature when equal masses of the first and third are mixed.

56. Lake Erie contains roughly 4.00×10^{11} m³ of water. **(a)** How much heat is required to raise the temperature of that volume of water from 11.0°C to 12.0°C? **(b)** Approximately how many years would it take to supply this amount of heat by using the full output of a 1000-MW electric power plant?

57. A 3.00-g copper penny at 25.0°C drops 50.0 m to the ground. **(a)** If 60.0% of its initial potential energy goes into increasing the internal energy, determine its final temperature. **(b)** Does the result depend on the mass of the penny? Explain.

58. A 3.00-g lead bullet at 30.0°C is fired at a speed of 240 m/s into a large block of ice at 0°C, in which it embeds itself. What quantity of ice melts?

59. A bar of gold is in thermal contact with a bar of silver of the same length and area (Fig. P11.59). One end of the compound bar is maintained at 80.0°C while the opposite end is at 30.0°C. When the heat flow reaches steady state, find the temperature at the junction.

Figure P11.59

60. A box with a total surface area of 1.20 m^2 and a wall thickness of 4.00 cm is made of an insulating material. A 10.0-W electric heater inside the box maintains the inside temperature at 15.0°C above the outside temperature. Find the thermal conductivity k of the insulating material.

61. A 60-kg runner dissipates 300 W of power while running a marathon. Assuming that 10% of the runner's energy is dissipated in the muscle tissue and that the excess heat is removed from the body primarily by sweating, determine the volume of bodily fluid (assume it is water) lost per hour. (At 37°C the latent heat of vaporization of water is 575 kcal/kg.)

62. An iron plate is held against an iron wheel so that a sliding frictional force of 50 N acts between the two pieces of metal. The relative speed at which the two surfaces slide over each other is 40 m/s. **(a)** Calculate the rate at which mechanical energy is converted to internal energy. **(b)** The plate and the wheel have masses of 5.0 kg each, and each receives 50% of the frictional work. If the system is run as described for 10 s and each object is then allowed to reach a uniform internal temperature, what is the resultant temperature increase?

63. An automobile has a mass of 1500 kg, and its aluminum brakes have an overall mass of 60 kg. **(a)** Assuming that all of the frictional work produced when the car stops is deposited in the brakes, and neglecting heat transfer, how many times could the car be braked to rest starting from 25 m/s (56 mph) before the brakes would begin to melt? (Assume an initial temperature of 20°C.) **(b)** Identify some effects that are neglected in part **(a)** but are likely to be important in a more realistic assessment of the heating of brakes.

64. A 1.0-m-long aluminum rod of cross-sectional area 2.0 cm^2 is inserted vertically into a thermally insulated vessel containing liquid helium at 4.2 K. The rod is initially at 300 K. If half of the rod is inserted into the helium, how many liters of helium boil off in the very short time while the inserted half cools to 4.2 K?

BIOLOGICAL PERSPECTIVE

Energy Management in the Human Body

DAVID GRIFFING

Miami University

Why can sprinters run at 10 m/s whereas distance runners can do only a little under 6 m/s? Why does a marathon runner's body temperature often increase by 3°F or more? Why is dieting a better strategy than exercise for losing weight? These and many other questions are readily answered if we examine energy management within the body.

Humans use energy to move, breathe, pump blood, and so forth. Chemical reactions within the body are responsible for storing, releasing, absorbing, and transferring this energy. Some of these reactions *require* energy (endothermic reactions), whereas others *release* energy (exothermic reactions).

The energy source our bodies use to produce endothermic reactions is the food we eat. Exothermic reactions release the energy needed to run, walk, build new cells, and so forth. As an example, locomotion in the human body depends on the cooperation of agonist/antagonist pairs of muscles that alternately contract and relax. Muscle fibers pull but cannot push. With skeletal support, the agonist pulls while the antagonist relaxes, to produce motion. Figure 1 shows the biceps and triceps muscles cooperating to bend the arm at the elbow.

The energy required for muscle contraction is made available in the muscle cells through a network of chemical reactions. Within this network, there are two input and two output ingredients. The input ingredients come from the air we breathe and the food we

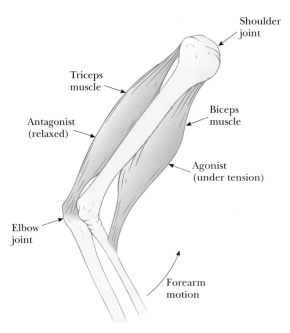

Figure 1 To bend the elbow, the biceps pulls (as agonist) while the triceps relaxes (as antagonist). To straighten the elbow, the roles of biceps and triceps are reversed. The bone is necessary because muscle fibers pull but cannot push.

eat, and the output ingredients are the carbon dioxide we exhale and the water produced as a by-product. In the lungs, the body removes oxygen from the inhaled air and transports it to muscle cells via hemoglobin in the bloodstream. In the mouth, stomach, and intestines, the body digests food, processing some of it into glucose ($C_6H_{12}O_6$). Some of this glucose is transported to the muscle cells. There, oxygen and glucose combine to form water and carbon dioxide in an exothermic reaction:

$$C_6H_{12}O_6 + 6O_2 \Rightarrow 6CO_2 + 6H_2O + E_{out}$$

Because this released energy is linked directly to the oxygen inhaled and carbon dioxide exhaled, it is theoretically simple to measure the energy a person generates. One measures either the net oxygen inhaled or the net carbon dioxide exhaled, and the energy released follows from the glucose oxidation reaction. Such measurements are routinely performed in hospitals and research laboratories.

Energy Production in Muscle Cells

The human body must take time to breathe, transport oxygen, and deliver glucose to the muscle cells. However, muscle fibers need not wait until glucose oxidation supplies the demanded energy because the body has the ability to store energy for future needs. Four processes are dedicated to energy production in a muscle cell; all use a high-energy molecule, adenosine triphosphate (ATP), to store and deliver energy in muscle cells. As long as ATP is available in the cell, immediate muscle contraction can occur without the benefit of breathing.

When the power demand in muscle fibers increases abruptly, the body needs time to adjust to the new level. Energy needed for muscle contraction during this transition time cannot be supplied by glucose oxidation because of the time needed to change to the new equilibrium state. To provide for this temporary energy gap, some "start-up" energy is stored in muscle cells in the form of existing ATP, and in the form of ingredients for anaerobic (without oxygen) production of ATP. After the transition time, respiration will gradually increase ATP production until a steady state is attained. During ordinary exercise, therefore, energy is supplied anaerobically for approximately the first two minutes and aerobically thereafter, via respiration.

Energy and Exercise

If either glucose or oxygen is absent, the production of energy in muscle cells stops. If either is in short supply, exercise is limited. Normally, enough glucose is stored in the muscle to supply ordinary energy needs, including vigorous exercise, for a few hours. Enough anaerobic energy is available for the body to function without oxygen for a couple of minutes. As an extreme example, Houdini, the celebrated escape artist of the 1920s, could function under water for 3 or 4 min without breathing.

When the body uses energy faster than respiration can support ATP production, ATP becomes scarce and the available energy declines. This condition is described as oxygen debt. At the end of vigorous exercise, when energy demand is back to normal, the debt is "repaid" as the exerciser continues to breathe rapidly, as when a sprinter gasps for air after the race.

Experiment has shown that the power, P, required to run a given distance is nearly directly proportional to the running speed v, that is, $P \propto v$. However, the power required to walk a given distance is $P \propto v^x$, with $x > 1$. So to walk a given distance the required energy increases with the walking speed, but to run the same distance the required energy is approximately independent of speed.[1] As a rule of thumb:

$$\text{Running energy/distance} \cong 1.5 \text{ kcal/mi} \cong 0.93 \text{ kcal/km}$$

Using this figure of merit, we can compare the energy expended running with the energy content of food. For example, an 80-kg person burns about 120 kcal running 1 mi at any speed. Burning 1 lb of fat generates about 4200 kcal. Thus, if the goal is to lose 1 lb of fat in 35 days, one could (1) run an extra mile a day, or (2) eat two fewer slices of bread a day.[2] Exercise probably stimulates the appetite enough so that the extra food one eats more than compensates for the extra energy needed for the exercise. Exercise is a poor substitute for dieting as a method to lose weight!

When the body is constructively stressed with exercise, it responds by improving its ability to cope with that stress. In weight lifting, for example, the muscles used to lift weights get stronger, while in training for distance running, the body's ability to ingest and process oxygen improves. In weight lifting, sudden bursts of energy are needed for short times. The exercise is largely anaerobic, and the body's ability to produce anaerobic bursts of energy improves. In distance running, however, a high level of steady-state energy production is needed. The exercise is largely aerobic, and the body responds by improving its capacity for aerobic energy production.

The primary goal of aerobic exercise is to attain good cardiovascular condition. Such exercise works constructively, provided the aerobic stress is regular and not harmfully intensive. Some forms of efficient aerobic activity are potentially harmful to the body in other ways. For example, running puts nonconstructive stress on the skeletal system, and joint problems are not uncommon among long-term distance runners and joggers. When a runner stops a regular exercise routine so that the stress is removed, the body quickly loses its high level of cardiovascular fitness. The time needed to achieve good fitness for distance running is several months or even several years. Other things being equal, runners who stop training during the summer cannot compete with those who train all year long.

To prevent glycogen (a polymer form of glucose) depletion during a marathon, glycogen loading has proven effective. Whereas long training times are needed for the body to develop excellent oxygen-processing in cells, times as short as a week can increase the glycogen stored in the muscle. The body quickly digests carbohydrates and converts them to glycogen. To increase the glycogen stored in muscle cells beyond the normal concentration, the athlete eats food containing no carbohydrates for a week or so and then eats food overconcentrated with carbohydrates for a few days before the race. In effect, the body is tricked into overstocking muscle cells with glycogen, which can provide extra fuel late in the race. Glycogen loading has been effective in helping marathon runners avoid "hitting the wall" after running about 20 mi.

Extreme energy demand, such as that required to swim the English Channel, to compete in the triathalon, or to cycle across the country, pushes the body to its limit both aerobically and anaerobically. Training for such high-stress events normally involves special high-energy diets or sleep deprivation.

Energy production in the body requires both oxygen and glucose. Oxygen cannot be stored, and so the steady-state rate of oxygen ingestion limits the long steady-state power production in muscles. Because glucose can be stored in the muscle cells and liver, the ultimate amount of energy that can be produced in the body without eating depends on

---

[1] Gilbert W. Fellingham et al., "Caloric Cost of Walking and Running," *Medicine and Science in Sports* 1978, 132–136.

[2] At 60 kcal/slice, 70 slices of bread correspond to 4200 kcal. At 2 slices a day, 35 days are needed.

how much glucose is stored in the form of glycogen. Beyond this point, the body will support itself by metabolizing body fat.

Work and Heat

A 70-kg person requires, depending on metabolism rate, about 2400 kcal a day for normal bodily functions. The power output of the body rises during exercise and falls during rest. What happens to the energy produced in the body? Some of this energy is used to perform work, and some appears as heat.

When a muscle contracts, a force is exerted through a distance, and so mechanical work is performed. Using only the upper body muscles, a skilled rope climber can climb 5 m in 2 s. For an 80-kg climber, about 2 kW of power is developed. Using leg muscles, up to 5–6 kW may be developed for similar short bursts of activity, such as in stair climbing or jumping. In extended activity, however, the power developed is considerably reduced. For example, in 1932 five men on the Polish Olympic Team climbed 362 m from the 5th to the 102nd floor of the Empire State Building in about 21 min. Again assuming a mass of 80 kg, this works out to a 225-W power output. The longer a person works, the lower the rate at which chemical energy is converted into mechanical work.

In a classic paper[3] the Nobel laureate A. V. Hill described measurements in which the work performed during muscle contraction is accompanied by a temperature increase of the muscle fiber. The energy delivered by ATP in the muscle is converted partly to thermal energy. To prevent the overall body temperature from rising when muscles contract, this thermal energy must be transferred from the body to the environment. A small amount is conducted away, and some is radiated away. Breathing also cools the body because of the water vapor exhaled as a by-product of respiration, and convection is an effective cooling process. Sweating is triggered to provide additional cooling when these cooling processes are saturated.

The body maintains a near-constant temperature by eliminating excess thermal energy as rapidly as possible. Blood vessels dilate so that blood flow increases, pores open so that sweating increases, and respiration increases. In hot weather, however, these mechanisms may be insufficient for extreme aerobic activities such as marathon running, and performance suffers. Athletes can pour water over themselves to increase cooling a little but may have to reduce their activity level to reduce their thermal energy production rate. It is not unusual for marathon runners to experience a rise in body temperature of 3 to 4°F during the race.

Joggers know that fewer clothes are needed to keep warm during exercise than when walking or standing. It is not unusual to observe joggers running in shorts and T-shirts in freezing weather. The heat developed during exercise keeps them warm. Surface blood vessels contract to inhibit heat loss. If any protection is needed against the cold, the most essential garment is a warm hat because up to 40% of the thermal energy lost by the body exits through the head, where a copious supply of blood vessels receives thermal energy from the rest of the body. (In bed in a cold room, the most efficient protection is a nightcap, as our ancestors well knew.)

The efficiency of converting energy to mechanical work is defined to be the ratio of the work output to the energy input. When viewed as an engine the human body's efficiency is probably less than 25%, depending on the basal metabolism rate of the individual. For comparison, this efficiency is about the same as that of an automobile engine.

---

[3] A. V. Hill, "The Heat of Shortening and the Dynamic Constants of Muscle," *Proc. Roy. Soc.* 1938, 136–195.

The Laws of Thermodynamics

Chapter Outline

▲ PHYSICS PUZZLER

Rockets, jet planes, gasoline automobiles, coal-fired and nuclear electric generating stations, and old-time steam locomotives all have the same basic limit to their efficiency. What is this limit? *(Phil Degginger/Tony Stone Images)*

The first law of thermodynamics is essentially the principle of conservation of energy generalized to include heat as a mode of energy transfer. According to the first law, the internal energy of an object (a concept to be discussed shortly) can be increased either by heat added to the object or by work done on it. The law places no restrictions on the types of energy conversions that occur. It tells us that the internal energy of a system can change as a result of two separate types of energy transfer across the boundary of the system.

The second law of thermodynamics, which can be stated in many equivalent ways, establishes which processes can occur in nature and which cannot. For example, the second law tells us that heat never flows spontaneously from a cold body to a hot body. One important application of this law is in the study of heat engines, such as the internal combustion engine, and the principles that limit their efficiency.

12.1 HEAT AND INTERNAL ENERGY

A major distinction must be made between internal energy and heat. **Internal energy** is all of the energy belonging to a system while it is stationary (neither translating nor rotating), including heat as well as nuclear energy, chemical energy, and strain energy (as in a compressed or stretched spring). **Thermal energy** is the portion of internal energy that changes when the temperature of the system changes. **Heat transfer** is caused by a temperature difference between the system and its surroundings.

In Chapter 10 we showed that the thermal energy of a monatomic ideal gas is associated with the motion of its atoms. In this special case, the thermal energy is simply kinetic energy on a microscopic scale; the higher the temperature of the gas, the greater the kinetic energy of the atoms and the greater the thermal energy of the gas. In general, however, thermal energy includes other forms of molecular energy, such as rotational energy and vibrational kinetic and potential energy (Fig. 12.1).

As an analogy, consider the distinction between work and energy that was discussed in Chapter 5. The work done on (or by) a system is a measure of the energy transferred between the system and its surroundings, whereas the system's mechanical energy (kinetic or potential) is a consequence of its motion and coordinates. Thus, when a person does work on a system, energy is transferred from the person to the system. It makes no sense to talk about the work *of* a system — one should refer only to the *work done on or by a system* when some process has occurred in which energy has been transferred to or from the system. Likewise, it makes no sense to use the term *heat* unless energy has been transferred as a result of a temperature difference.

It is also important to recognize that energy can be transferred between two systems even when no *thermal* energy transfer occurs. For example, when a piston quickly compresses a gas, the gas is warmed and its thermal energy increases, but no transfer of heat takes place; if the gas then expands rapidly, it cools and its thermal energy decreases, but there is no transfer of thermal energy to its surroundings. In each case, energy is transferred to or from the system as work but appears within the system as an increase or decrease in thermal energy. The change in

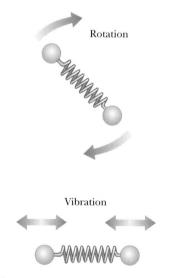

Rotation

Vibration

Figure 12.1 The rotational and vibrational energy of a diatomic gas contributes to the internal energy of the gas.

internal energy is equal to the change in thermal energy and is measured by a corresponding change in temperature.

12.2 WORK AND HEAT

In the macroscopic approach to thermodynamics, we describe the *state* of a system with the use of such variables as pressure, volume, temperature, and internal energy. The number of macroscopic variables needed to characterize a system depends on the system's nature. For a homogeneous system, such as a gas containing only one type of molecule, usually only two variables are needed. It is important to note that a *macroscopic state* of an isolated system can be specified only if the system is in thermal equilibrium internally. In the case of a gas in a container, internal thermal equilibrium requires that every part of the container be at the same pressure and temperature.

Consider a gas contained by a cylinder fitted with a movable piston (Fig. 12.2). In equilibrium, the gas occupies a volume of V and exerts a uniform pressure, P, on the cylinder walls and piston. If the piston has cross-sectional area A, the force exerted by the gas on the piston is $F = PA$. Now let us assume that the gas expands slowly enough to allow the system to remain essentially in thermodynamic equilibrium at all times. As the piston moves up a distance of Δy, the work done on the piston by the gas is

$$W = F \Delta y = PA \, \Delta y$$

(a)   (b)

Figure 12.2 (a) A gas in a cylinder occupying a volume V at a pressure P. (b) As the gas expands at constant pressure and the volume increases by the amount ΔV, the work done by the gas is $P \Delta V$.

Because $A \Delta y$ is the increase in volume of the gas (that is, $\Delta V = A \Delta y$), we can express the work done as

$$W = P \Delta V \qquad\qquad [12.1]$$

If the gas expands, as in Figure 12.2b, ΔV is positive and the work done by the gas is positive. If the gas is compressed, ΔV is negative and the work done by the gas is negative. (In the latter case, negative work can be interpreted as work being done *on* the system.) Clearly, the work done by (or on) the system is zero when the volume remains constant.

Equation 12.1 can be used to calculate the work done on or by the system only when the pressure of the gas remains constant during the expansion or compression. If the pressure changes, calculus is required to determine the work done. We do not attempt such calculations here.

Consider the process represented by the pressure-volume diagram in Figure 12.3. The gas has expanded from an initial volume, V_i, to a final volume, V_f, at a constant pressure of P. From Equation 12.1 we see that the work done by the gas in this case is $P(V_f - V_i)$. Note that this is just the area under the pressure–volume curve. **In general, the work done in an expansion from some initial state to some final state is the area under the curve on a *PV* diagram.** This statement is true whether or not the pressure remains constant during the process.

As Figure 12.3 shows, the work done in the expansion from the initial state to the final state depends on the path taken between the two states. To illustrate this important point, consider several different paths connecting i and f (Fig. 12.4). In the process depicted in Figure 12.4a, the pressure of the gas is reduced from P_i to P_f by cooling at constant volume, V_i, and then the gas expands from V_i to V_f at constant pressure, P_f. The work done along this path is $P_f(V_f - V_i)$. In Figure 12.4b, the gas expands from V_i to V_f at constant pressure, P_i, and then its pressure is reduced to P_f at constant volume, V_f. The work done along this path is $P_i(V_f - V_i)$, which is greater than the work done in the process of Figure 12.4a. Finally, in the process depicted in Figure 12.4c, where both P and V change continuously, the work done has some value intermediate between the values obtained in the first two processes. To evaluate the work in this case, the shape of the *PV* curve must be known. Thus, the work done by a system depends on the process by which the system goes from the initial to the final state. In other words, the work done depends on the initial, final, and intermediate states of the system.

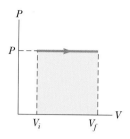

Figure 12.3 The *PV* diagram for a gas expanding at constant pressure. The shaded area represents the work done by the gas.

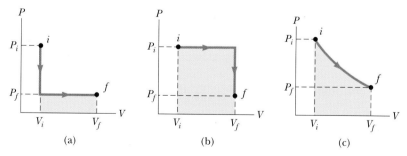

(a)   (b)   (c)

Figure 12.4 The work done by a gas as it is taken from an initial state to a final state depends on the path taken between these states.

Figure 12.5 (a) A gas at temperature T_i expands slowly by absorbing heat from a reservoir at the same temperature. (b) A gas expands rapidly into an evacuated region after a membrane separating it from that region is broken.

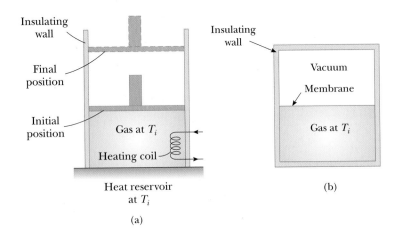

In a similar manner, the heat transferred into or out of a system is also found to depend on the process. This can be demonstrated by the situations depicted in Figure 12.5. In each case, the gas has the same initial volume, temperature, and pressure and is assumed to be ideal. In Figure 12.5a, the gas is in thermal contact with a heat reservoir, and a small heating coil allows energy to be transferred to the gas by heat from the surface of the coil. (A heat reservoir is a body whose heat capacity is so large that its temperature remains unchanged when heat is added or extracted from the body.) As an infinitesimal amount of energy enters the gas from the coil, the pressure of the gas becomes infinitesimally greater than atmospheric pressure, and the gas expands, causing the piston to rise. During this expansion to some final volume, V_f, and some final pressure, P_f, sufficient heat to maintain a constant temperature of T_i from the reservoir to the gas.

Now consider the thermally insulated system in Figure 12.5b. When the membrane is broken, the gas expands rapidly into the vacuum until it occupies a volume of V_f and is at a pressure of P_f. In this case the gas does no work, because there is no movable piston. Furthermore, no heat is transferred through the thermally insulated wall, which we call an *adiabatic wall*, and so the temperature remains at T_i. This process is often referred to as **adiabatic free expansion** or simply *free expansion*. In general, an adiabatic process is one in which no heat is transferred between the system and its surroundings.

The initial and final states of the ideal gas in Figure 12.5a are identical to the initial and final states in Figure 12.5b, but the paths are different. In the first case, heat is transferred slowly to the gas, and the gas does work on the piston. In the second case, no heat is transferred, and the work done is zero. Therefore, we conclude that **heat transfer, like work, depends on the initial, final, and intermediate states of the system.**

Heat reservoir ▷

Free expansion of a gas ▷

Heat transfer and work done depend ▷ on the path between the initial and final states.

Applying Physics 1

Warm air rises, so why is mountain air cold?

Explanation Consider what happens as warm air rises up a mountain. A certain volume of air expands as it rises because the pressure pushing against it is reduced. As the air expands, however, its temperature drops. By contrast, if prevailing winds drive air down a mountain, the falling air is compressed and, therefore, warmed. In Austria, a warm wind called the Foehn can blow from the Alps for several days at a time, proving very disconcerting for the Austrians. At one time, erratic behavior was excused, even by the courts, as being caused by the Foehn.

The usual situation, however, is for air to be pushed by prevailing winds from a valley to the top of a mountain. Air is a poor thermal conductor, so the process that occurs is one in which there is no heat transfer to or from the air. As the parcel of air rises, it moves into regions of lower atmospheric pressure. As a result, the parcel expands. By expanding, it is doing work on the surrounding air. This represents a transfer of energy out of the system, so that the internal energy drops and therefore the temperature drops. If the parcel of air contains water vapor, the drop in temperature as it rises up the mountain can result in condensation into raindrops. This is why the rain tends to fall on the windward side of mountains. On the leeward side, much of the water vapor has already condensed out as rain, and the air parcel is warming up again as it descends the mountain. Thus, there is little rain on the leeward side of the mountain.

An interesting example of this occurs in Hawaii. The prevailing winds are often from the northeast. The island of Kauai has a very tall mountain (Mount Waialeale, 1548 m), and it is one of the wettest places on Earth with an average rainfall of 1170 cm. Just to the southwest of Kauai is the island of Niihau, which is very dry and requires a system of irrigation for its farming. The leeward position of Niihau relative to Kauai is the reason.

APPLICATION

Rainfall Pattern Near Mountains.

Thinking Physics 1

Imagine a gas in an insulated cylinder with a movable piston. The piston has been pushed inward, compressing the gas, and is now released. As the molecules of the gas strike the piston, they move it outward. From the point of view of energy principles, explain how this expansion causes the temperature of the gas to drop.

Explanation   From the point of view of energy principles, the molecules strike the piston and move it through a distance. Thus, the molecules do work on the piston, which represents a transfer of energy out of the gas. As a result, the internal energy of the gas drops. Because the temperature is related to internal energy, the temperature of the gas drops.

EXAMPLE 12.1   Work Done by an Expanding Gas

In the system shown in Figure 12.2, the gas in the cylinder is at a pressure of 8000 Pa and the piston has an area of 0.10 m^2. As heat is slowly added to the gas, the piston is pushed up a distance of 4.0 cm. Calculate the work done on the surroundings by the expanding gas. Assume that the pressure remains constant.

Solution   The change in volume of the gas is

$$\Delta V = A \, \Delta y = (0.10 \text{ m}^2)(4.0 \times 10^{-2} \text{ m}) = 4.0 \times 10^{-3} \text{ m}^3$$

and from Equation 12.1, the work done is

$$W = P \, \Delta V = (8000 \text{ Pa})(4.0 \times 10^{-3} \text{ m}^3) = \boxed{32 \text{ J}}$$

12.3   THE FIRST LAW OF THERMODYNAMICS

When the principle of conservation of energy was first introduced in Chapter 5, it was stated that the mechanical energy of a system is constant in the absence of nonconservative forces, such as friction. That mechanical model did not encompass changes in the internal energy of the system. We now broaden our scope to use the term *principle of conservation of energy* for a generalization encompassing possible changes in internal energy. This is a universally valid law that can be applied to all kinds of processes. Furthermore, it provides a connection between the microscopic and macroscopic worlds. The result will be the **first law of thermodynamics.**

Change in internal energy ▶   We have seen that energy can be transferred between a system and its surroundings in two ways: via work done by (or on) the system, which requires a macroscopic displacement of the point of application of a force (or pressure), and via heat transfer, which occurs through random molecular collisions. Each of these represents a change of energy of the system, and therefore usually results in measurable changes in the macroscopic variables of the system, such as the pressure, temperature, and volume of a gas.

To express these ideas more quantitatively, suppose a system undergoes a change from an initial state to a final state. During this change, positive Q is the heat transferred *to* the system, and positive W is the work done *by* the system. For example, suppose the system is a gas whose pressure and volume change from P_i and V_i to P_f and V_f. If the quantity $Q - W$ is measured for various paths connecting the initial and final equilibrium states (that is, for various *processes*), one finds that it is the same for *all* paths connecting the initial and final states. We conclude that **the quantity $Q - W$ is determined completely by the initial and final states of the system, and we call it the change in the internal energy of the system.** Although Q and W both depend on the path, **$Q - W$ is independent of the path.** If we represent the internal energy function with the letter U, then the *change* in internal energy, $\Delta U = U_f - U_i$, can be expressed as

First-law equation ▶
$$\Delta U = U_f - U_i = Q - W \qquad [12.2]$$

where all quantities must have the same energy units. Equation 12.2 is known as the first law of thermodynamics. This law applies universally to all systems.

Let us examine some special cases in which the only changes in energy are changes in internal energy. First consider an *isolated system*—that is, one that does not interact with its surroundings. In this case, no heat transfer takes place and the work done is zero; hence, the internal energy remains constant. That is, because

For isolated systems, U remains ▶
constant.
$Q = W = 0$, $\Delta U = 0$, and so $U_i = U_f$. We conclude that **the internal energy of an isolated system remains constant.**

Next consider the case in which a system (not isolated from its surroundings) is taken through a **cyclic process**—that is, a process that originates and ends at the same state. In this case, the change in the internal energy must again be *zero*, and therefore the heat added to the system must equal the work done during the cycle. That is, in a cyclic process,

Cyclic process ▶
$$\Delta U = 0 \qquad \text{and} \qquad Q = W$$

Note that **the net work done per cycle equals the area enclosed by the path representing the process on a *PV* diagram.** As we shall see in the next section, cyclic processes are very important for describing the thermodynamics of *heat* engines—devices in which some part of the thermal energy input is converted into mechanical work.

If a process occurs in which the work done is zero, then the change in internal energy equals the heat entering or leaving the system. If heat enters the system, Q is positive and the internal energy increases. For a gas, we can associate this increase in internal energy with an increase in the kinetic energy of the molecules. If a process occurs in which the heat transferred is zero and work is done by the system, however, then the magnitude of the change in internal energy equals the negative of the work done by the system. That is, the internal energy of the system decreases. For example, if a gas is compressed with no heat transferred (by a moving piston, say), the work done by the gas is negative and the internal energy again increases. This is because kinetic energy is transferred from the moving piston to the gas molecules.

On a microscopic scale, no practical distinction exists between heat transfer and work. Each can produce a change in the internal energy of a system. Although the macroscopic quantities Q and W are *not* properties of a system, they are related to changes of the internal energy of a stationary system through the first law of thermodynamics. Once a process, or path, is defined, Q and W can be either calculated or measured, and the change the internal energy can be found from the first law of thermodynamics.

EXAMPLE 12.2   An Isobaric Process

A gas is enclosed in a container fitted with a piston of cross-sectional area 0.10 m². The pressure of the gas is maintained at 8000 Pa while heat is slowly added; as a result, the piston is pushed up a distance of 4.0 cm. (Any process in which the pressure remains constant is called an **isobaric process.**) If 42 J of heat is added to the system during the expansion, what is the change in internal energy of the system?

Solution   The work done by the gas is

$$W = P \Delta V = (8000 \text{ Pa})(0.10 \text{ m}^2)(4.0 \times 10^{-2} \text{ m}) = 32 \text{ N·m} = 32 \text{ J}$$

The change in internal energy is found from the first law of thermodynamics:

$$\Delta U = Q - W = 42 \text{ J} - 32 \text{ J} = \boxed{10 \text{ J}}$$

We see that **in an isobaric process the work done and the heat transferred are both nonzero.**

◀ An isobaric process is one that occurs at constant pressure.

Exercise   If 42 J of heat is added to the system with the piston clamped in a *fixed* position, what is the work done by the gas? What is the change in its internal energy?

Answer   No work is done; $\Delta U = 42$ J.

EXAMPLE 12.3   An Isovolumetric Process

Water with a mass of 2.0 kg is held at constant volume in a container while 10 000 J of heat is slowly added by a flame. The container is not well insulated, and as a result 2000 J of heat leaks out to the surroundings. What is the temperature increase of the water?

An isovolumetric process is one ▶
that takes place at constant
volume.

Solution   A process that takes place at constant volume is called an **isovolumetric process.** In such a process the work is equal to zero. Thus, the first law of thermodynamics gives

$$\Delta U = Q$$

This indicates that the net heat added to the water goes into increasing the internal energy of the water. The net heat added to the water is

$$Q = 10\ 000\ \text{J} - 2000\ \text{J} = 8000\ \text{J}$$

Because $Q = mc\,\Delta T$, the temperature increase of the water is

$$\Delta T = \frac{Q}{mc} = \frac{8000\ \text{J}}{(2.0\ \text{kg})(4.186 \times 10^3\ \text{J/kg} \cdot {}^\circ\text{C})} = \boxed{0.96^\circ\text{C}}$$

EXAMPLE 12.4   Boiling Water

One gram of water at atmospheric pressure (1.013×10^5 Pa) occupies a volume of 1.0 cm^3. When this water is boiled, it becomes 1671 cm^3 of steam. Calculate the change in internal energy for this process.

Solution   Because the latent heat of vaporization of water is 2.26×10^6 J/kg at atmospheric pressure, the heat required to boil 1.0 g is

$$Q = mL_v = (1.0 \times 10^{-3}\ \text{kg})(2.26 \times 10^6\ \text{J/kg}) = 2260\ \text{J}$$

The work done by the system is positive and equal to

$$W = P(V_{\text{steam}} - V_{\text{water}})$$
$$= (1.013 \times 10^5\ \text{Pa})[(1671 - 1.0) \times 10^{-6}\ \text{m}^3] = 169\ \text{J}$$

Hence, the change in internal energy is

$$\Delta U = Q - W = 2260\ \text{J} - 169\ \text{J} = \boxed{2.1 \times 10^3\ \text{J}}$$

The positive ΔU tells us that the internal energy of the system has increased. We see that most of the heat (93%) transferred to the liquid goes into increasing the internal energy. Only a small fraction (7%) goes into external work.

EXAMPLE 12.5   Heat Transferred to a Solid

The internal energy of a solid also increases when heat is transferred to it from its surroundings. To illustrate this point, suppose a 1.0-kg bar of copper is heated at atmospheric pressure. Its temperature increases from 20°C to 50°C.
(a) Find the work done by the copper.

Solution   The change in volume of the copper can be calculated using Equation 10.6 and the linear expansion coefficient for copper taken from Table 10.1 (remembering that $\beta = 3\alpha$):

$$\Delta V = \beta V \Delta T = [5.1 \times 10^{-5}(^\circ\text{C})^{-1}](50^\circ\text{C} - 20^\circ\text{C})V = (1.5 \times 10^{-3})V$$

The volume is equal to m/ρ, and the density of copper is 8.92×10^3 kg/m³. Hence,

$$\Delta V = (1.5 \times 10^{-3}) \left(\frac{m}{\rho}\right) = (1.5 \times 10^{-3}) \left(\frac{1.0 \text{ kg}}{8.92 \times 10^3 \text{ kg/m}^3}\right)$$

$$= 1.7 \times 10^{-7} \text{ m}^3$$

Because the expansion takes place at constant pressure (equal to normal atmospheric pressure), the work done is

$$W = P\Delta V = (1.013 \times 10^5 \text{ Pa})(1.7 \times 10^{-7} \text{ m}^3) = 1.9 \times 10^{-2} \text{ J}$$

(b) What quantity of heat is transferred to the copper?

Solution The specific heat of copper is given in Table 11.1, and from Equation 11.3 we find that the heat transferred is

$$Q = mc\,\Delta T = (1.0 \text{ kg})(387 \text{ J/kg}\cdot°\text{C})(30°\text{C}) = 1.2 \times 10^4 \text{ J}$$

Exercise What is the increase in internal energy of the copper?

Answer $\Delta U = 1.2 \times 10^4$ J

Applying Physics 2

In the late 1970s casino gambling was approved in Atlantic City, New Jersey, which can become quite cold in the winter. Energy projections that were performed on the design of the casinos showed that the air conditioning would need to operate in the casino even in the middle of a very cold January. Why?

Explanation If we consider the air in the casino to be a gas to which we apply the first law, imagine that there is no air conditioning and no ventilation, so that this gas simply stays in the room. There is no work being done by the gas, so we focus on the heat input. A casino consists of a large number of people, many of whom are active (throwing dice, cheering, etc.). As a result, these people have large rates of energy flow by heat from their bodies. This energy goes into the air of the casino, resulting in an increase in internal energy of the air. With the large number of excited people in a casino (along with the large number of incandescent lights and machines), the temperature of the gas can rise quickly and to a high value. In order to keep the temperature at reasonable values, there must be an energy transfer out of the air. Calculations show that conduction through the walls even in a 10°F January are not sufficient to maintain the required energy transfer, so the air conditioning system must be in almost continuous use throughout the year.

12.4 HEAT ENGINES AND THE SECOND LAW OF THERMODYNAMICS

A heat engine is a device that converts thermal energy to other useful forms, such as electrical and mechanical energy. In a typical process for producing electricity in a power plant, for instance, coal or some other fuel is burned, and the thermal energy produced is used to convert water to steam. This steam is directed at the

Figure 12.6 A schematic representation of a heat engine. The engine receives heat Q_h from the hot reservoir, expels heat Q_c to the cold reservoir, and does work W.

blades of a turbine, setting it in rotation. Finally, the mechanical energy associated with this rotation is used to drive an electric generator. Another heat engine, the automobile internal combustion engine, extracts heat from a burning fuel and converts a fraction of this energy to mechanical energy.

A heat engine carries some working substance through a cyclic process during which (1) heat is absorbed from a source at a high temperature, (2) work is done by the engine, and (3) heat is expelled by the engine to a reservoir at a lower temperature. As an example, consider the operation of a steam engine in which the working substance is water. The water is carried through a cycle in which it first evaporates into steam in a boiler and then expands against a piston. After the steam is condensed with cooling water, it is returned to the boiler, and the process is repeated.

It is useful to represent a heat engine schematically, as in Figure 12.6. The engine absorbs a quantity of heat, Q_h, from the hot reservoir, does work W, and then gives up heat Q_c to the cold reservoir. Because the working substance goes through a cycle, its initial and final internal energies are equal, so $\Delta U = 0$. Hence, from the first law of thermodynamics we see that **the net work, W, done by a heat engine equals the net heat flowing into it.** As we can see from Figure 12.6, $Q_{net} = Q_h - Q_c$; therefore,

$$W = Q_h - Q_c \qquad [12.3]$$

where Q_h and Q_c are taken to be positive quantities.

If the working substance is a gas, **the net work done for a cyclic process is the area enclosed by the curve representing the process on a PV diagram.** This is shown for an arbitrary cyclic process in Figure 12.7. **The thermal efficiency, e, of a heat engine is the ratio of the net work done to the heat absorbed at the higher temperature during one cycle:**

Thermal efficiency ▷

$$e \equiv \frac{W}{Q_h} = \frac{Q_h - Q_c}{Q_h} = 1 - \frac{Q_c}{Q_h} \qquad [12.4]$$

We can think of the efficiency as the ratio of what we get (mechanical energy) to what we give (thermal energy at the higher temperature). Equation 12.4 shows that a heat engine has 100% efficiency ($e = 1$) only if $Q_c = 0$—that is, if no heat is expelled to the cold reservoir. In other words, a heat engine with perfect efficiency would have to convert all of the absorbed heat to mechanical work. One of the consequences of the second law of thermodynamics is that this is impossible.

Second law ▷

The **second law of thermodynamics** can be stated as follows: **It is impossible to construct a heat engine that, operating in a cycle, produces no other effect than the absorption of heat from a reservoir and the performance of an equal amount of work.**

This form of the second law is useful for understanding the operation of heat engines. With reference to Equation 12.4, the second law says that, during the operation of a heat engine, W can never be equal to Q_h or, alternatively, that some heat, Q_c, must be rejected to the environment. As a result, it is theoretically impossible to construct an engine that works with 100% efficiency.

Heat engines can, however, operate in reverse—we send in energy, modeled as work, W, in Figure 12.6, resulting in heat being extracted from the cold reservoir and transferred to the hot reservoir. This system is now operating as a heat pump. Your kitchen refrigerator is a common heat pump. Energy (heat) is extracted from

APPLICATION

Heat Pumps and Refrigerators.

the cold food in the interior and delivered to the warmer air in the kitchen. The work is done in the compressor unit of the refrigerator, in which a piston compresses the refrigerant (freon, or other similar material) to a smaller volume, increasing its temperature. A household air conditioner is another example of a heat pump. Some homes are both heated and cooled by heat pumps. In the winter, the heat pump extracts energy from the outside air and delivers it to the inside, and the process is reversed for summer cooling. In a ground-water heat pump, energy is extracted in the winter from water deep in the ground rather than from the outside air, and energy is delivered to that water in the summer. This increases the year-round efficiency of the heating and cooling unit, because the ground water is at a higher temperature than the air in winter and is at a cooler temperature than the air in summer.

Our assessment of the first two laws of thermodynamics can be summed up as follows: the first law says **we cannot get a greater amount of energy out of a cyclic process than we put in,** and the second law says **we cannot break even.**

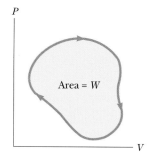

Figure 12.7 The *PV* diagram for an arbitrary cyclic process. The area enclosed by the curve equals the net work done.

EXAMPLE 12.6 The Efficiency of an Engine

Find the efficiency of an engine that introduces 2000 J of heat during the combustion phase and loses 1500 J at exhaust.

Solution The efficiency of the engine is given by Equation 12.4 as

$$e = 1 - \frac{Q_c}{Q_h} = 1 - \frac{1500 \text{ J}}{2000 \text{ J}} = \boxed{0.25 \text{ or } 25\%}$$

Exercise If an engine has an efficiency of 20% and loses 3000 J at exhaust and to the cooling water, how much work is done by the engine?

Answer 750 J

A model steam engine equipped with a built-in horizontal boiler. The water is heated electrically, generating steam that is used to power the electric generator at the left. *(Courtesy of Central Scientific Company)*

12.5 REVERSIBLE AND IRREVERSIBLE PROCESSES

In the next section we shall discuss a theoretical heat engine that is the most efficient engine possible. In order to understand its nature, we must first examine the meanings of reversible and irreversible processes. A **reversible** process is one in which every state between the initial and final states is an equilibrium state, and that can be reversed in order to be followed exactly from the final state back to the initial state. A process that does not satisfy these requirements is **irreversible.**

All natural processes are known to be irreversible. As an example, let us examine the free expansion of a gas (already discussed in Section 12.2) and show that it cannot be reversed. The gas is contained in an insulated container with a membrane separating it from a vacuum (Fig. 12.8). If the membrane is punctured, the gas expands freely into the vacuum. Because the gas does not exert a force on its surroundings through a distance, it does no work as it expands. In addition, no heat is transferred to or from the gas because the container is insulated from its surroundings. Thus, in this process the system has changed but the surroundings have not.

Figure 12.8 Free expansion of a gas.

Figure 12.9 A gas in thermal contact with a heat reservoir is compressed slowly by grains of sand dropped onto the piston. The compression is isothermal and reversible.

Now imagine that we try to reverse the process by first compressing the gas to its original volume. Let's say an engine is being used to force the piston inward. Note, however, that this action is changing both the system and surroundings. The surroundings are changing because work is being done by an outside agent on the system, and the system is changing because the compression is increasing the temperature of the gas. We can lower the temperature of the gas by allowing it to come in contact with an external heat reservoir. Although this second procedure returns the gas to its original state, the surroundings are again affected because thermal energy is added to the surroundings. If this heat could somehow be used to drive the engine and compress the gas, the system and its surroundings could be returned to their initial states. However, our statement of the second law says that this extracted heat cannot be completely converted to mechanical energy isothermally. We must conclude that a reversible process has not occurred.

Although real processes are always irreversible, some are *almost* reversible. If a real process occurs very slowly so that the system is virtually always in equilibrium, the process can be considered reversible. For example, imagine compressing a gas very slowly by dropping some grains of sand onto a frictionless piston, as in Figure 12.9. The compression process can be reversed by the placement of the gas in thermal contact with a heat reservoir. The pressure, volume, and temperature of the gas are well defined during this isothermal compression. Each added grain of sand represents a change to a new equilibrium state. The process can be reversed by the slow removal of grains of sand from the piston.

A general characteristic of a reversible process is that no dissipative effects that convert mechanical energy to thermal energy, such as turbulence or friction, can be present. In reality, such effects are impossible to eliminate completely, and hence it is not surprising that real processes in nature are irreversible.

Sadi Carnot, French engineer (1796–1832)

Carnot is considered to be the founder of the science of thermodynamics. Some of his notes found after his death indicate that he was the first to recognize the relationship between work and heat.

12.6   THE CARNOT ENGINE

In 1824 a French engineer named Sadi Carnot (1796–1832) described a theoretical engine, now called a *Carnot engine,* that is of great importance from both practical and theoretical viewpoints. He showed that a heat engine operating in an ideal, reversible cycle—called a Carnot cycle—between two reservoirs is the most efficient engine possible. **Carnot's theorem** can be stated as follows:

> No real engine operating between two heat reservoirs can be more efficient than a Carnot engine, operating between the same two reservoirs.

To describe the Carnot cycle, we assume that the substance working between temperatures T_c and T_h is an ideal gas contained in a cylinder with a movable piston at one end. The cylinder walls and the piston are thermally nonconducting. Figure 12.10 shows four stages of the Carnot cycle, and Figure 12.11 is the PV diagram for the cycle. The cycle consists of two adiabatic and two isothermal processes, all reversible.

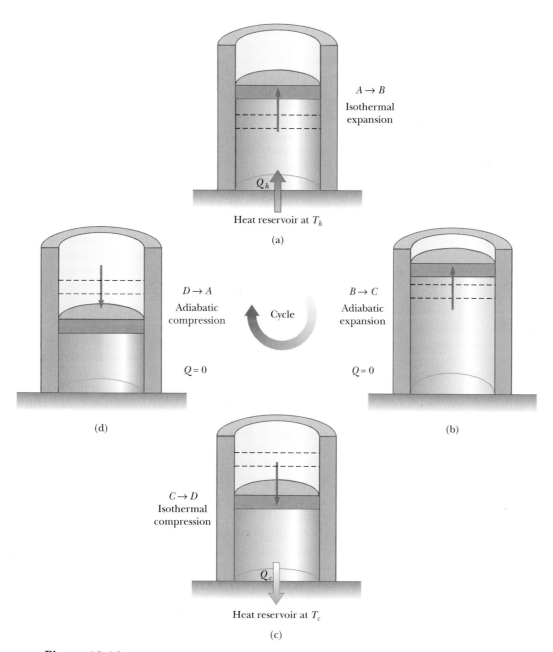

Figure 12.10 The Carnot cycle. In process $A \rightarrow B$, the gas expands isothermally while in contact with a reservoir at T_h. In process $B \rightarrow C$, the gas expands adiabatically ($Q = 0$). In process $C \rightarrow D$, the gas is compressed isothermally while in contact with a reservoir at $T_c < T_h$. In process $D \rightarrow A$, the gas is compressed adiabatically. The upward arrows on the piston indicate removal of sand during the expansions, and the downward arrows indicate addition of sand during the compressions.

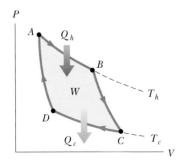

Figure 12.11 The *PV* diagram for the Carnot cycle. The net work done, *W*, equals the net heat received in one cycle, $Q_h - Q_c$.

1. The process $A \rightarrow B$ is an isothermal expansion at temperature T_h, in which the gas is placed in thermal contact with a heat reservoir at temperature T_h (Fig. 12.10a). During the process, the gas absorbs heat Q_h from the reservoir and does work W_{AB} in raising the piston.

2. In the process $B \rightarrow C$, the base of the cylinder is replaced by a thermally non-conducting wall and the gas expands adiabatically; that is, no heat enters or leaves the system (Fig. 12.10b). During the process, the temperature falls from T_h to T_c and the gas does work W_{BC} in raising the piston.

3. In the process $C \rightarrow D$, the gas is placed in thermal contact with a heat reservoir at temperature T_C (Fig. 12.10c) and is compressed isothermally at temperature T_c. During this time, the gas expels heat Q_c to the reservoir, and the work done on the gas is W_{CD}.

4. In the final stage, $D \rightarrow A$, the base of the cylinder is again replaced by a thermally nonconducting wall (Fig. 12.10d) and the gas is compressed adiabatically. The temperature of the gas increases to T_h, and the work done on the gas is W_{DA}.

Carnot showed that the thermal efficiency of a Carnot engine is

$$e_c = \frac{T_h - T_c}{T_h} = 1 - \frac{T_c}{T_h} \qquad [12.5]$$

where *T* must be in kelvins. From this result, we see that all Carnot engines operating reversibly between the same two temperatures have the same efficiency. Furthermore, the efficiency of any reversible engine operating in a cycle between two temperatures is greater than the efficiency of any irreversible (real) engine operating between the same two temperatures.

Equation 12.5 can be applied to any working substance operating in a Carnot cycle between two heat reservoirs. According to this result, the efficiency is zero if $T_c = T_h$. The efficiency increases as T_c is lowered and as T_h is increased. However, the efficiency can be unity (100%) only if $T_c = 0$ K. Such reservoirs are not available, and so the maximum efficiency is always less than unity. In most practical cases, the cold reservoir is near room temperature, about 300 K. Therefore, one usually strives to increase the efficiency by raising the temperature of the hot reservoir. **All real engines are less efficient than the Carnot engine because they are subject to practical difficulties, including friction, but especially the need to operate irreversibly to complete a cycle in a brief time period.**

Lord Kelvin, British physicist and mathematician (1824–1907)

Born William Thomson in Belfast, Kelvin was the first to propose the use of an absolute scale of temperature. His study of Carnot's theory led to the idea that heat cannot pass spontaneously from a colder body to a hotter body; this is known as the second law of thermodynamics.

(J. L. Charmet/SPL/Photo Researchers)

EXAMPLE 12.7   The Steam Engine

A steam engine has a boiler that operates at 500 K. The heat changes water to steam, which drives the piston. The temperature of the exhaust is that of the outside air, about 300 K. What is the maximum thermal efficiency of this steam engine?

Solution   From the expression for the efficiency of a Carnot engine, we find the maximum thermal efficiency for any engine operating between these temperatures:

$$e_c = 1 - \frac{T_c}{T_h} = 1 - \frac{300 \text{ K}}{500 \text{ K}} = \boxed{0.4, \text{ or } 40\%}$$

This is the highest theoretical efficiency of the engine. In practice, the efficiency is considerably lower.

PHYSICS IN ACTION

The device in the photograph on the left, called *Hero's engine*, was invented around 150 B.C. by Hero in Alexandria. When water is boiled in the flask, which is suspended by a cord, steam exits through two tubes at the sides of the flask (in opposite directions), creating a torque that rotates the flask.

The device shown in the center and right-hand photographs, called a *thermoelectric converter*, uses a series of semiconductor cells to convert thermal energy to electrical energy. In the center photograph, the two "legs" of the device are at the same temperature, and no electrical energy is produced. However, when one leg is at a higher temperature than the other, as in the photograph on the right, electrical energy is produced as the device extracts energy from the hot reservoir and drives a small electric motor. How does this intriguing experiment demonstrate the second law of thermodynamics?

Devices That Convert Thermal Energy into Other Forms of Energy

(Courtesy of Central Scientific Company)

(Courtesy of PASCO Scientific Company)

Exercise Determine the maximum work the engine can perform in each cycle of operation if it absorbs 200 J of thermal energy from the hot reservoir during each cycle.

Answer 80 J

EXAMPLE 12.8 The Carnot Efficiency

The highest theoretical efficiency of a gasoline engine, based on the Carnot cycle, is 30%. If this engine expels its gases into the atmosphere, which has a temperature of 300 K, what is the temperature in the cylinder immediately after combustion?

Solution   The Carnot efficiency is used to find T_h:

$$e_c = 1 - \frac{T_c}{T_h}$$

$$T_h = \frac{T_c}{1 - e_c} = \frac{300 \text{ K}}{1 - 0.30} = \boxed{430 \text{ K}}$$

Actual gasoline engines operate on a cycle significantly different from the Carnot cycle and therefore have lower maximum possible efficiencies.

Exercise   If the heat engine absorbs 837 J of heat from the hot reservoir during each cycle, how much work can it perform in each cycle?

Answer   251 J

12.7   ENTROPY

The concept of temperature is involved in the zeroth law of thermodynamics (Chapter 10), and the concept of internal energy is involved in the first law. Temperature and internal energy are both state functions; that is, they can be used to describe the thermodynamic state of a system. Another state function related to the second law of thermodynamics is the **entropy function, S.**

Consider a reversible process between two equilibrium states. If ΔQ_r is the heat absorbed or expelled by the system during some small interval of the path, **the change in entropy, ΔS, between two equilibrium states is given by the heat transferred, ΔQ_r, divided by the absolute temperature, T, of the system in this interval.** That is,

$$\Delta S \equiv \frac{\Delta Q_r}{T} \qquad [12.6]$$

The subscript r on the term ΔQ_r emphasizes that the definition applies only to reversible processes. When heat is absorbed by the system, ΔQ_r is positive and hence the entropy increases. When heat is expelled by the system, ΔQ_r is negative and the entropy decreases. Note that Equation 12.6 defines not entropy but the change in entropy. This is consistent with the fact that a change in state always accompanies heat transfer. Hence, the meaningful quantity in a description of a process is the change in entropy.

The concept of entropy was introduced into the study of thermodynamics by Rudolf Clausius in 1865. One reason it became useful and gained wide acceptance is because it provides another variable to describe the state of a system, to go along with pressure, volume, and temperature. The concept of entropy reached a position of even more significance when it was found that **the entropy of the Universe increases in all natural processes.** This is yet another way of stating the second law of thermodynamics.

The statement of the second law just given must be interpreted with care. Although it says that the entropy of the Universe always increases in all natural processes, there are processes in which the entropy of a system decreases. That is, there

Rudolf Clausius (1822–1888)

"I propose . . . to call S the entropy of a body, after the Greek word 'transformation.' I have designedly coined the word 'entropy' to be similar to energy, for these two quantities are so analogous in their physical significance, that an analogy of denominations seems to be helpful." *(AIP Niels Bohr Library, Lande Collection)* are situations in which the entropy of one system (system *A*) decreases but in correspondence with a net increase in entropy of some other system (system *B*). In all cases the change in entropy of system *B* is greater than the change in entropy of system *A*.

For centuries, individuals have attempted to build perpetual motion machines that would operate continuously without input of energy and without any net increase in entropy. Using the laws of thermodynamics, we can understand why a perpetual motion machine will *never* be invented. Perpetual motion machines of the first kind are those that violate the first law, putting out more energy than is put into the machine. Perpetual motion machines of the second kind are those that violate the second law. Imagine, for example, a waterfall in which water passes over a paddlewheel, which turns an electric generator providing electricity to a nearby motor. The temperature of the falling water has risen after it hits the paddlewheel and the pond below. The pond is used as the high temperature reservoir of a heat engine. The electric motor and the heat engine together are used to pump water from the pond back up to the top of the waterfall. This is a perpetual motion machine of both the first and second kinds. It violates the first law because it assumes all energy remains in the system so that it operates forever at the same rate. This would require no friction in any of the mechanisms, and it requires that the mechanisms make no sound, because sound would carry energy away from the system. The machine we have described also violates the second law because it assumes that there is no exhaust heat from the heat engine, which would result in energy loss from the system.

The concept of entropy is satisfying because it enables us to present the second law of thermodynamics in the form of a mathematical statement. In the next section we will find that entropy can also be interpreted in terms of probabilities, a relationship that has profound implications for our world.

► **APPLICATION**

"Perpetual Motion" Machines.

EXAMPLE 12.9   Melting a Piece of Lead

Calculate the change in entropy when 300 g of lead melts at 327°C (600 K). Lead has a latent heat of fusion of 2.45×10^4 J/kg.

Solution   The amount of heat added to the lead to melt it is

$$Q = mL_f = (0.300 \text{ kg})(2.45 \times 10^4 \text{ J/kg}) = 7.35 \times 10^3 \text{ J}$$

From Equation 12.6, the entropy change of the lead is

$$\Delta S = \frac{Q}{T} = \frac{7.35 \times 10^3 \text{ J}}{600 \text{ K}} = \boxed{12.3 \text{ J/K}}$$

EXAMPLE 12.10   Which Way Does the Heat Flow?

A large, cold object is at 273 K, and a large, hot object is at 373 K. Show that it is impossible for a small amount of heat energy, say 8.00 J, to be transferred from the cold object to the hot object without decreasing the entropy of the Universe and hence violating the second law.

Solution We assume here that during the heat transfer the two systems undergo no significant temperature change. This is not a necessary assumption; it is used to avoid a need for the techniques of integral calculus. The entropy change of the hot object is

$$\Delta S_h = \frac{Q}{T_h} = \frac{8.00 \text{ J}}{373 \text{ K}} = 0.0214 \text{ J/K}$$

The cold reservoir loses heat, and its entropy change is

$$\Delta S_c = \frac{Q}{T_c} = \frac{-8.00 \text{ J}}{273 \text{ K}} = -0.0293 \text{ J/K}$$

The net entropy change of the Universe is

$$\Delta S_U = \Delta S_c + \Delta S_h = -0.0079 \text{ J/K}$$

This is in violation of the law that the entropy of the Universe always increases in natural processes. That is, **the spontaneous transfer of heat from a cold object to a hot object cannot occur.**

Exercise In the preceding example, suppose that 8.00 J of heat were transferred from the hot to the cold object. What would be the net change in entropy of the Universe?

Answer $+0.0079 \text{ J/K}$

12.8 ENTROPY AND DISORDER

As you look around at the beauty of nature, it is easy to recognize that the events of natural processes have in them a large element of chance. For example, the spacing between trees in a natural forest is quite random; if you were to discover a forest where all the trees were equally spaced, you would conclude that it was a planted forest. Likewise, leaves fall to the ground with random arrangements. It would be highly unlikely to find the leaves laid out in perfectly straight rows. We can express the results of such observations by saying that **a disorderly arrangement is much more probable than an orderly one if the laws of nature are allowed to act without interference.**

Entropy originally found its place in thermodynamics, but its importance grew tremendously as the field of statistical mechanics developed. This analytical approach employs an alternative interpretation of entropy. In statistical mechanics, the behavior of a substance is described in terms of the statistical behavior of the atoms and molecules contained in the substance. One of the main products of this approach is the conclusion that **isolated systems tend toward greater disorder, and entropy is a measure of that disorder.**

In light of this new view of entropy, Boltzmann found an alternative method for calculating entropy through use of the relation

$$S = k_B \ln W \qquad\qquad [12.7]$$

where k_B is Boltzmann's constant ($k_B = 1.38 \times 10^{-23}$ J/K) and W is a number proportional to the probability that the system has a particular configuration. The symbol "ln" in Equation 12.7 is an abbreviation for the natural logarithm discussed in Appendix A.

TABLE 12.1 Possible Results of Drawing Four Marbles from a Bag		
End Result	**Possible Draws**	**Total Number of Same Results**
All R	RRRR	1
1G, 3R	RRRG, RRGR, RGRR, GRRR	4
2G, 2R	RRGG, RGRG, GRRG, RGGR, GRGR, GGRR	6
3G, 1R	GGGR, GGRG, GRGG, RGGG	4
All G	GGGG	1

Let us explore the meaning of this equation through a specific example. Imagine that you have a bag of 100 marbles, 50 red and 50 green. You are allowed to draw four marbles from the bag according to the following rules. Draw one marble, record its color, return it to the bag, and draw again. Continue this process until four marbles have been drawn. Note that because each marble is returned to the bag before the next one is drawn, the probability of drawing a red marble is always the same as the probability of drawing a green one.

The results of all possible drawing sequences are shown in Table 12.1. For example, the result RRGR means that a red marble was drawn first, a red one second, a green one third, and a red one fourth. This table indicates that there is only one possible way to draw four red marbles. There are four possible sequences that produce one green and three red marbles, six sequences that produce two green and two red, four sequences that produce three green and one red, and one sequence that produces all green. From Equation 12.7, we see that the state with the greatest disorder (two red and two green marbles) has the highest entropy because it is most probable. In contrast, the most ordered states (all red marbles and all green marbles) are least likely to occur and are states of lowest entropy.

The outcome of the draw can range between these highly ordered (lowest entropy) and highly disordered (highest entropy) states. Thus, entropy can be regarded as an index of how far a system has progressed from an ordered to a disordered state.

The second law of thermodynamics is really a statement of what is most probable rather than of what must be. Imagine placing an ice cube in contact with a hot piece of pizza. There is nothing in nature that absolutely forbids the transfer of heat from the ice to the much warmer pizza. Statistically, it is possible for a slow-moving molecule in the ice to collide with a faster-moving molecule in the pizza so that the slow one transfers some of its energy to the faster one. However, when the great number of molecules present in the ice and pizza are considered, the odds are overwhelmingly in favor of the transfer of energy from the faster-moving molecules to the slower-moving molecules. Furthermore, this example demonstrates that a system naturally tends to move from a state of order to a state of disorder. The initial state, in which all the pizza molecules have high kinetic energy and all the ice molecules have lower kinetic energy, is much more ordered than the final state after heat transfer has taken place and the ice has melted.

Even more generally, the second law of thermodynamics defines the direction of time for all events as the direction in which the entropy of the universe increases. Although conservation of energy is not violated if energy spontaneously flows from a cold object (the ice cube) to a hot object (the pizza slice), the event violates the

APPLICATION

The Direction of Time.

second law because it represents a spontaneous increase in order, and it also violates everyday experience. If the melting ice cube were filmed and the film speeded up, the difference between running the film in forward and reverse directions would be obvious to an audience. The same would be true of filming any event involving a large number of particles — such as a dish dropping to the floor and shattering.

Thinking Physics 2

Suppose you film the collision of two billiard balls on a pool table. Now suppose you run the film in reverse. Will the difference between forward and reverse directions be obvious? Why or why not?

Explanation   The film shows nothing about the molecular arrangement of the billiard balls, so there are effectively only two particles in the event — the two billiard balls. During the collision, there is no change from order to disorder or vice versa, so the film appears to be authentic (describing a valid situation) when run in either direction.

As another example, suppose you were able to measure the velocities of all the air molecules in a room at some instant. It is very unlikely that you would find all molecules moving in the same direction with the same speed — this would be a highly ordered state, indeed. The most probable situation is a system of molecules moving haphazardly in all directions with a distribution of speeds — a highly disordered state. Let us compare this case to that of drawing marbles from a bag. If a container held 10^{23} molecules of a gas, the probability of finding all of the molecules moving in the same direction with the same speed at some instant would be similar to that of drawing a marble from the bag 10^{23} times and getting a red marble on every draw. This is clearly an unlikely set of events.

The tendency of nature to move toward a state of disorder affects the ability of a system to do work. Consider a ball thrown toward a wall. The ball has kinetic energy, and its state is an ordered one; that is, all of the atoms and molecules of the ball move in unison at the same speed and in the same direction (apart from their random thermal motions). When the ball hits the wall, however, part of the ball's kinetic energy is transformed into the random, disordered, thermal motion of the molecules in the ball and the wall and the temperatures of the ball and the wall both increase slightly. Before the collision, the ball was capable of doing work. It could drive a nail into the wall, for example. With the transformation of part of the ordered energy into disordered thermal energy, this capability of doing work is reduced. That is, the ball rebounds with less kinetic energy than it originally had, because the collision is inelastic.

Various forms of energy can be converted to thermal energy, as in the collision between the ball and the wall, but the reverse transformation is never complete. In general, if two kinds of energy, A and B, can be completely interconverted, we say that they are the *same grade*. However, if form A can be completely converted to form B and the reverse is never complete, then form A is a *higher grade* of energy than form B. In the case of a ball hitting a wall, the kinetic energy of the ball is of a higher grade than the thermal energy contained in the ball and the wall after the collision. Therefore, when high-grade energy is converted to thermal energy, it can never be fully recovered as high-grade energy.

A full house is a very good hand in the game of poker. Can you calculate the probability of being dealt a full house (a pair and three of a kind) from a standard deck of 52 cards? *(Tom Mareschel, The IMAGE Bank)*

This conversion of high-grade energy to thermal energy is referred to as **degradation of energy.** The energy is said to be degraded because it takes on a form that is less useful for doing work. In other words, **in all real processes where heat transfer occurs, the energy available for doing work decreases.**

Finally, note once again that the statement that entropy must increase in all natural processes is true only for isolated systems. There are instances in which the entropy of some system decreases, but with a corresponding net increase in entropy for some other system. When all systems are taken together to form the Universe, *the entropy of the Universe always increases.*

Ultimately, the entropy of the Universe should reach a maximum. At this time the Universe will be in a state of uniform temperature and density. All physical, chemical, and biological processes will have ceased, because a state of perfect disorder implies no available energy for doing work. This gloomy state of affairs is sometimes referred to as an ultimate "heat death" of the Universe.

An illustration from Flammarion's novel *La Fin du Monde*, depicting the heat death of the Universe.

Thinking Physics 3

If you shake a jar of jelly beans of two different sizes, the larger jelly beans tend to appear near the top, while the smaller ones tend to settle at the bottom. Why does this occur? Does this process violate the second law?

Explanation   Shaking opens up spaces between the jelly beans. The smaller ones have a chance of falling down into spaces below them. The accumulation of larger ones on top and smaller ones on the bottom implies an increase in order and a decrease in one contribution to the total entropy. However, the second law is not violated and the total entropy of the system increases. The increase in the internal energy of the system comes from the work required to shake the jar of beans (that is, work your muscles must do, with an increase in entropy accompanying the biological processes) and also from the small loss of gravitational potential energy as the beans settle together more compactly.

SUMMARY

The work done as a gas expands or contracts at a constant pressure is

$$W = P \, \Delta V \qquad [12.1]$$

The work done is negative if the gas is compressed and positive if the gas expands. In general, the work done in an expansion from some initial state to some final state is the area under the curve on a *PV* diagram.

From the first law of thermodynamics, when a system undergoes a change from one state to another, the **change in its internal energy,** ΔU, is

$$\Delta U = U_f - U_i = Q - W \qquad [12.2]$$

where Q is the heat transferred into (or out of) the system and W is the work done by (or on) the system. Q is positive when heat enters the system and negative when the system loses heat.

The **first law of thermodynamics** is the generalization of the law of conservation of energy that includes heat transfer.

In a cyclic process (one in which the system returns to its initial state), $\Delta U = 0$ and therefore $Q = W$. That is, the heat transferred into the system equals the work done during the cycle.

An **adiabatic process** is one in which no heat is transferred between the system and its surroundings ($Q = 0$). In this case, the first law gives $\Delta U = -W$. That is, the internal energy changes as a consequence of work being done by (or on) the system.

An **isobaric process** is one that occurs at constant pressure. The work done in such a process is $P \Delta V$.

A **heat engine** is a device that converts thermal energy to other forms of energy, such as mechanical and electrical energy. The work done by a heat engine in carrying a substance through a cyclic process ($\Delta U = 0$) is

$$W = Q_h - Q_c \qquad \text{[12.3]}$$

where Q_h is the heat absorbed from a hot reservoir and Q_c is the heat expelled to a cold reservoir.

The **thermal efficiency** of a heat engine is defined as the ratio of the net work done to the heat absorbed per cycle:

$$e \equiv \frac{W}{Q_h} = 1 - \frac{Q_c}{Q_h} \qquad \text{[12.4]}$$

No real heat engine operating between the kelvin temperatures T_h and T_c can exceed the efficiency of an engine operating between the same two temperatures in a **Carnot cycle,** given by

$$e_c = 1 - \frac{T_c}{T_h} \qquad \text{[12.5]}$$

Real processes proceed in an order governed by the **second law of thermodynamics,** which can be stated in two ways:

1. Heat will not flow spontaneously from a cold object to a hot object.
2. No heat engine operating in a cycle can absorb thermal energy from a reservoir and perform an equal amount of work.

The second law can also be stated in terms of a quantity called **entropy.** The **change in entropy** of a system is equal to the heat flowing into (or out of) the system as the system changes from one state (A) to another (B), divided by the absolute temperature:

$$\Delta S \equiv \frac{\Delta Q_r}{T} \qquad \text{[12.6]}$$

One of the primary findings of statistical mechanics is that systems tend toward disorder and that entropy is a measure of this disorder. The entropy of the Universe increases in all natural processes; this is an alternative statement of the second law.

MULTIPLE-CHOICE QUESTIONS

1. If a process starts at (P_0, V_0) and doubles in volume, which of the following type involves the most work:
   (a) adiabatic   (b) isobaric   (c) isovolumetric
   (d) No answer is correct always.

2. An ideal gas is maintained at a constant pressure of 70 000 Pa during an isobaric process and its volume decreases by 0.2 m^3. What work is done by the system on its environment?
   (a) 14 000 J   (b) 350 000 J   (c) $-14\,000$ J
   (d) $-35\,000$ J

3. A 2-mole ideal-gas system is maintained at a constant volume of 4 liters; if 100 J of heat is added, what is the change in internal energy of the system?
   (a) zero   (b) 50 J   (c) 67 J   (d) 100 J

4. A 1-kg block of ice at 0°C and 1 atm melts completely to water. What is its change in entropy? (For ice, $L_f = 3.34 \times 10^5$ J/kg.)
   (a) 3340 J/K   (b) 2170 J/K   (c) 613 J/K
   (d) 1220 J/K

5. An steam turbine operates at a boiler temperature of 450 K and an exhaust temperature of 300 K. What is the maximum theoretical efficiency of this system?
   (a) 24%   (b) 50%   (c) 33%   (d) 67%

CONCEPTUAL QUESTIONS

1. For an ideal gas in an isothermal process, there is no change in internal energy. Suppose the gas does work W during such a process. How much energy was transferred by heat?

2. A designer of an electric heating unit describes her product as being 100% efficient. Is her claim correct?

3. Consider the human body performing a strenuous exercise, such as lifting weights or riding a bicycle. Work is being done by the body, and energy is leaving by conduction from the skin into the surrounding air. According to the first law of thermodynamics, the temperature of the body should be steadily decreasing during the exercise. This is not what happens, however. Is the first law invalid for this situation? Explain.

4. Clearly distinguish among temperature, thermal energy, heat, and internal energy.

5. What is wrong with the statement, "Given any two bodies, the one with the higher temperature contains more heat?"

6. A steam-driven turbine is one major component of an electric power plant. Why is it advantageous to increase the temperature of the steam as much as possible?

7. When a sealed Thermos bottle full of hot coffee is shaken, what changes, if any, take place in (a) the temperature of the coffee and (b) its internal energy?

8. In solar ponds constructed in Israel, the Sun's energy is concentrated near the bottom of a salty pond. With the proper layering of salt in the water, convection is prevented, and temperatures of 100°C may be reached. Can you guess the maximum efficiency with which useful energy can be extracted from the pond?

9. What are some factors that affect the efficiency of automobile engines?

10. Why does your automobile burn more gas in winter than in summer?

11. Is it possible to construct a heat engine that creates no thermal pollution?

12. Suppose your roommate is "Mr. Clean" and tidies up your messy room after a big party. Because more order is being created by your roommate, does this represent a violation of the second law of thermodynamics?

13. A thermodynamic process occurs in which the entropy of a system changes by -8.0 J/K. According to the second law of thermodynamics, what can you conclude about the entropy change of the environment?

14. If a supersaturated sugar solution is allowed to evaporate slowly, sugar crystals form in the container. Hence, sugar molecules go from a disordered form (in solution) to a highly ordered, crystalline form. Does this process violate the second law of thermodynamics? Explain.

15. The first law of thermodynamics says we cannot get more out of a process than we put in, but and the second law says that we cannot break even. Explain this statement.

16. Give some examples of irreversible processes that occur in nature. Give an example of a process in nature that is nearly reversible.

17. The thermoelectric converter shown on the right side photograph on page 387 uses semiconductor cells to convert thermal energy to electrical energy. (a) Why does a temperature difference between the two legs of the device produce electric energy? (b) In what sense does this intriguing device demonstrate the second law of thermodynamics?

PROBLEMS

1, 2, 3 = straightforward, intermediate, challenging ☐ = full solution available in Study Guide/Student Solutions Manual
WEB = solution posted at **http://www.saunderscollege.com/physics/college** 🖼 = biomedical application 🖼 = Interactive Physics

Review Problem

A 40.0-gram projectile is launched by the expansion of hot gas in an arrangement shown in part (a) of the figure. The cross-sectional area of the launch tube is 1.0 cm², and the length that the projectile travels down the launch tube after starting from rest is 32 cm. As the gas expands, the pressure varies as shown in part (b) of the figure. The values for the initial pressure and volume are $P_0 = 11 \times 10^5$ Pa and $V_0 = 8.0$ cm³ and the final values are $P_f = 1.0 \times 10^5$ Pa and $V_f = 40.0$ cm³. Friction between the projectile and the launch tube is negligible. **(a)** If the projectile is launched into a vacuum, what is the speed of the projectile as it leaves the launch tube? **(b)** If instead the projectile is launched into air at a pressure of 1.0×10^5 Pa, what fraction of the work done by the expanding gas in the tube is spent by the projectile pushing air out of the way as it proceeds down the tube?

(a)   (b)

a volume of 1.0 L to 3.0 L at a constant pressure of 3.0 atm. **(b)** The gas is then cooled at constant volume until the pressure falls to 2.0 atm. **(c)** The gas is then compressed at a constant pressure of 2.0 atm from a volume of 3.0 L to 1.0 L. (*Note:* Be careful of signs.) **(d)** The gas is heated until its pressure increases from 2.0 atm to 3.0 atm at a constant volume. **(e)** Find the net work done during the complete cycle.

5. A gas expands from I to F along the three paths indicated in Figure P12.5. Calculate the work done by the gas along paths **(a)** IAF, **(b)** IF, and **(c)** IBF.

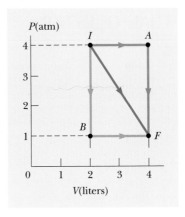

Figure P12.5

Section 12.1 Heat and Internal Energy

Section 12.2 Work and Heat

WEB ☐ **1.** The only form of energy possessed by molecules of a monatomic ideal gas is translational kinetic energy. Using the results from the discussion of kinetic theory in Section 10.6, show that the internal energy of a monatomic ideal gas at pressure P and occupying volume V may be written as $U = \frac{3}{2}PV$.

2. Steam moves into the cylinder of a steam engine at a constant pressure of 2.00×10^5 Pa. The diameter of the piston is 16.0 cm, and the piston travels 20.0 cm in one stroke. How much work is done during one stroke?

3. A container of volume 0.40 m³ contains 3.0 mol of argon gas at 30°C. Assuming argon behaves as an ideal gas, find the total internal energy of this gas. (*Hint:* See Problem 1.)

4. Sketch a *PV* diagram and find the work done by the gas during the following stages. **(a)** A gas is expanded from

6. Sketch a *PV* diagram of the following processes. **(a)** A gas expands at constant pressure P_1 from volume V_1 to volume V_2. It is then kept at constant volume while the pressure is reduced to P_2. **(b)** A gas is reduced in pressure from P_1 to P_2 while its volume is held constant at V_1. It is then expanded at constant pressure P_2 to a final volume, V_2. **(c)** In which of the processes is more work done? Why?

7. Gas in a container is at a pressure of 1.5 atm and a volume of 4.0 m³. What is the work done by the gas if **(a)** it expands at constant pressure to twice its initial volume? **(b)** it is compressed at constant pressure to one quarter its initial volume?

8. A movable piston having a mass of 8.00 kg and a cross-sectional area of 5.00 cm² traps 0.200 moles of an ideal gas in a vertical cylinder. If the piston slides without friction in the cylinder, how much work will the gas do when its temperature is increased from 20°C to 300°C?

9. One mole of an ideal gas initially at a temperature of $T_0 = 0°C$ undergoes an expansion, at a constant pressure of one atmosphere, to four times its original volume. (a) Calculate the new temperature of the gas, T_f. (b) Calculate the work done by the gas during the expansion.

Section 12.3 The First Law of Thermodynamics

10. A container is placed in a water bath and held at constant volume as a mixture of fuel and oxygen is burned inside it. The temperature of the water is observed to rise during the burning (the water is also held at constant volume). (a) Consider the burning mixture to be the system. What are the signs of Q, ΔU, and W? (b) What are the signs of these quantities if the water bath is considered to be the system?

11. A quantity of a monatomic ideal gas undergoes a process in which both its pressure and volume are doubled as shown in Figure P12.11. What is the heat absorbed by the gas during this process? (*Hint:* See Problem 1.)

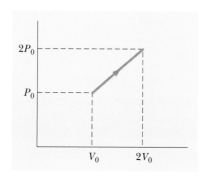

Figure P12.11

12. A monatomic ideal gas undergoes the thermodynamic process shown in the *PV* diagram of Figure P12.12. Determine whether each of the values ΔU, Q, and W for the gas are positive, negative, or zero. (*Hint:* See Problem 1.)

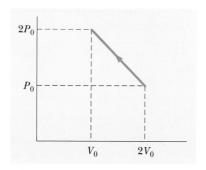

Figure P12.12

13. A gas is compressed at a constant pressure of 0.300 atm from a volume of 8.00 L to 3.00 L. In the process, 400 J of heat energy flows out of the gas. (a) What is the work done by the gas? (b) What is the change in its internal energy?

14. A gas expands from I to F in Figure P12.5. The heat added to the gas is 418 J when the gas goes from I to F along the diagonal path. (a) What is the change in internal energy of the gas? (b) How much heat must be added to the gas for the indirect path IAF to give the same change in internal energy?

15. A gas is taken through the cyclic process described by Figure P12.15. (a) Find the net heat transferred to the system during one complete cycle. (b) If the cycle is reversed—that is, the process follows the path $ACBA$—what is the net heat transferred per cycle?

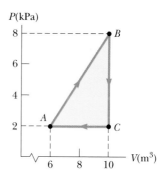

Figure P12.15 (Problems 15 and 18)

16. Two cm³ water is boiled at atmospheric pressure to become 3342 cm³ of steam, also at atmospheric pressure. (a) Calculate the work done by the gas during this process. (b) Find the amount of heat added to the water to accomplish this process. (c) From (a) and (b), find the change in internal energy.

17. A gas is enclosed in a container fitted with a piston of cross-sectional area 0.150 m². The pressure of the gas is maintained at 6000 Pa as the piston moves inward 20.0 cm. (a) Calculate the work done by the gas. (b) If the internal energy of the gas decreases by 8.00 J, find the amount of heat removed from the system during the compression.

18. Consider the cyclic process described by Figure P12.15. If Q is negative for the process BC and ΔU is negative for the process CA, determine the signs of Q, W, and ΔU associated with each process.

19. One gram of water changes from liquid to solid at a constant pressure of one atmosphere and a constant temperature of 0°C. In the process, the volume changes from 1.00 cm³ to 1.09 cm³. (a) Find the work done by the water, and (b) the change in the internal energy of the water.

20. A 5.0-kg block of aluminum is heated from 20°C to 90°C at atmospheric pressure. Find (a) the work done by the aluminum, (b) the amount of heat transferred to it, and (c) the increase in its internal energy.

21. One mole of gas is initially at a pressure of 2.00 atm, a volume of 0.300 L, and an internal energy equal to 91.0 J. In its final state the gas is at a pressure of 1.50 atm and a volume of 0.800 L, and its internal energy equals 180 J. For the paths *IAF*, *IBF*, and *IF* in Figure P12.21, calculate (a) the work done by the gas and (b) the net heat transferred to the gas in the process.

Figure P12.21

Section 12.4 Heat Engines and the Second Law of Thermodynamics

Section 12.5 Reversible and Irreversible Processes

Section 12.6 The Carnot Engine

22. A heat engine operates between two reservoirs at temperatures of 20°C and 300°C. What is the maximum efficiency possible for this engine?

WEB 23. The efficiency of a Carnot engine is 30%. The engine absorbs 800 J of heat per cycle from a hot reservoir at 500 K. Determine (a) the heat expelled per cycle and (b) the temperature of the cold reservoir.

24. The exhaust temperature of a Carnot heat engine is 300°C. What is the intake temperature if the efficiency of the engine is 30%?

25. A steam engine has a boiler that operates at 300°F, and the temperature of the exhaust is 150°F. Find the maximum efficiency of this engine.

26. A heat engine performs 200 J of work in each cycle and has an efficiency of 30%. For each cycle of operation, (a) how much heat is absorbed and (b) how much heat is expelled?

27. An engine absorbs 1700 J from a hot reservoir and expels 1200 J to a cold reservoir in each cycle. (a) What is the engine's efficiency? (b) How much work is done in

each cycle? (c) What is the power output of the engine if each cycle lasts for 0.300 s?

28. A particular engine has a power output of 5.00 kW and an efficiency of 25.0%. If the engine expels 8000 J of heat in each cycle, find (a) the heat absorbed in each cycle and (b) the time required for each cycle.

29. In one cycle, a heat engine absorbs 500 J from the high-temperature reservoir and expels 300 J to a low-temperature reservoir. If the efficiency of this engine is 60% of the efficiency of a Carnot engine, what is the ratio of the low temperature to the high temperature in the Carnot engine?

30. A heat engine operates in a Carnot cycle between 80.0°C and 350°C. It absorbs 21 000 J of heat per cycle from the hot reservoir. The duration of each cycle is 1.00 s. (a) What is the maximum power output of this engine? (b) How much heat does it expel in each cycle?

31. A nuclear power plant has a power output of 1000 MW and operates with an efficiency of 33%. If excess heat is carried away from the plant by a river with a flow rate of 1.0×10^6 kg/s, what is the rise in temperature of the flowing water?

Section 12.7 Entropy

Section 12.8 Entropy and Disorder

32. A freezer is used to freeze 1.0 L of water completely into ice. The water and the freezer remain at a constant temperature of $T = 0°C$. Determine (a) the change in the entropy of the water and (b) the change in the entropy of the freezer.

33. What is the change in entropy of 1.0 kg of liquid water at 100°C as it changes to steam at 100°C?

34. A 70-kg log falls from a height of 25 m into a lake. If the log, the lake, and the air are all at 300 K, find the change in entropy of the Universe for this process.

WEB 35. Two 2000-kg cars, both traveling at 20 m/s, undergo a head-on collision and stick together. Find the entropy change of the Universe during the collision if the temperature is 23°C.

36. Repeat the procedure used to construct Table 12.1 (a) for the case in which you draw three marbles rather than four from your bag and (b) for the case in which you draw five rather than four.

37. Prepare a table like Table 12.1 for the following occurrence. You toss four coins into the air simultaneously. Record all the possible results of the toss in terms of the numbers of heads and tails that can result. (For example, HHTH and HTHH are two possible ways in which three heads and one tail can be achieved.) (a) On the basis of your table, what is the most probable result of a toss? (b) In terms of entropy, what is the most ordered state and (c) what is the most disordered?

38. Consider a standard deck of 52 playing cards that has been thoroughly shuffled. (a) What is the probability of

drawing the ace of spaces in one draw? **(b)** What is the probability of drawing any ace? **(c)** What is the probability of drawing any spade?

ADDITIONAL PROBLEMS

39. A student claims that she has constructed a heat engine that operates between the temperatures of 200 K and 100 K with 60% efficiency. The professor does not give her credit for the project. Why not?

40. A power plant that uses the temperature gradient of the ocean has been proposed. The system is to operate between 20°C (surface temperature) and 5.0°C (water temperature at a depth of about 1 km). **(a)** What is the maximum efficiency of such a system? **(b)** If the power output of the plant is 75 MW, what is the minimum thermal energy absorbed per hour? **(c)** In view of your answer to part (a), do you think such a system is worthwhile?

41. A heat engine extracts heat Q_h from a hot reservoir at constant temperature T_h and rejects heat Q_c to a cold reservoir at constant temperature T_c. Find the entropy changes of the **(a)** hot reservoir, **(b)** the cold reservoir, **(c)** the engine, and **(d)** the complete system.

42. One end of a copper rod is in thermal contact with a hot reservoir at $T = 500$ K, and the other end is in thermal contact with a cooler reservoir at $T = 300$ K. If 8000 J of thermal energy is transferred from one end to the other, with no change in the temperature distribution, find the entropy change of each reservoir and the total entropy change of the Universe.

43. Suppose that a nuclear power plant operates in a Carnot cycle. Estimate the temperature change in Celsius of a river due to the exhausted heat from the plant. Assume that the input power to the boiler in the plant is 25×10^8 W, the efficiency of the use of this power is 30%, and the river flow rate is 9.0×10^6 kg/min.

44. Every second at Niagara Falls, some 5000 m³ of water falls a distance of 50 m. What is the increase in entropy per second due to the falling water? (Assume a 20°C environment.)

Figure P12.44 Niagara Falls. *(Jan Kopec/Tony Stone Images)*

45. One object is at a temperature of T_h and another is at a lower temperature, T_c. Use the second law of thermodynamics to show that heat transfer can only occur from the hotter to the colder object. Assume a constant-temperature process.

46. When a gas follows path 123 on the *PV* diagram in Figure P12.46, 418 J of heat flows into the system and 167 J of work is done. **(a)** What is the change in the internal energy of the system? **(b)** How much heat flows into the system if the process follows path 143? The work done by the gas along this path is 63.0 J. What net work would be done on or by the system if the system followed **(c)** path 12341? **(d)** path 14321? **(e)** What is the change in internal energy of the system in the processes described in parts (c) and (d)?

Figure P12.46

47. A substance undergoes the cyclic process shown in Figure P12.47. A work output occurs along path *AB*, while a work input is required along path *BC*, and no work is involved in the constant volume process *CA*. Heat transfers occur during each process involved in the cycle. **(a)** What is the work output during process *AB*? **(b)** How much work input is required during process *BC*? **(c)** What is the net heat input during this cycle?

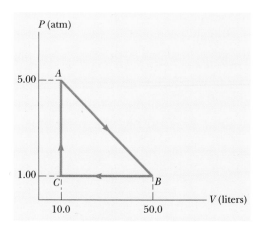

Figure P12.47

48. A 100-kg steel support rod in a building has a length of 2.0 m at a temperature of 20°C. The rod supports a load of 6000 kg. Find (a) the work done by the rod as the temperature increases to 40°C, (b) the heat added to the rod (assume the specific heat of steel is the same as that for iron), and (c) the change in internal energy of the rod.

49. A 1500-kW heat engine operates at 25% efficiency. The heat energy expelled at the low temperature is absorbed by a stream of water that enters the cooling coils at 20°C. If 60 L flows across the coils per second, determine the increase in temperature of the water.

50. An ideal gas initially at pressure P_0, volume V_0, and temperature T_0 is taken through the cycle described in Figure P12.50. (a) Find the net work done by the gas per cycle in terms of P_0 and V_0. (b) What is the net heat added to the system per cycle? (c) Obtain a numerical value for the net work done per cycle for 1.00 mol of gas initially at 0°C. (*Hint:* Recall that work equals the area under a *PV* curve.)

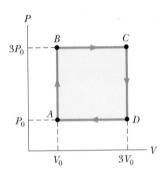

Figure P12.50

51. (a) Determine the work done by a fluid that expands from *i* to *f*, as indicated in Figure P12.51. (b) How much work is performed by the fluid if it is compressed from *f* to *i* along the same path?

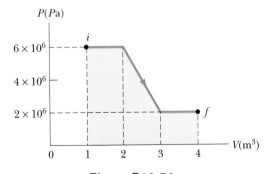

Figure P12.51

52. A 20.0%-efficient real engine is used to speed up a train from rest to 5.00 m/s. It is known that an ideal (Carnot) engine using the same cold and hot reservoirs would accelerate the same train from rest to a speed of 6.50 m/s using the same amount of fuel. If the engines use air at 300 K as a cold reservoir, find the temperature of the steam serving as the hot reservoir.

53. The surface of the Sun is approximately at 5700 K, and the temperature of the Earth's surface is approximately 290 K. What entropy change occurs when 1000 J of thermal energy is transferred from the Sun to the Earth?

54. A force, **F,** is applied to a metal wire, stretching it by the amount ΔL. Show that the work done on the wire is given by $W = \frac{1}{2}(\text{stress})(\text{strain})V$, where *V* is the volume of the wire.

55. A 1.0-kg block of aluminum is heated at atmospheric pressure so that its temperature increases from 22°C to 40°C. Find (a) the work done by the aluminum, (b) the heat added to the aluminum, and (c) the change in internal energy of the aluminum.

56. Suppose a heat engine is connected to two heat reservoirs, one a pool of molten aluminum (660°C) and the other a block of solid mercury (−38.9°C). The engine runs by freezing 1.00 g of aluminum and melting 15.0 g of mercury during each cycle. The latent heat of fusion of aluminum is 3.97×10^5 J/kg, and that of mercury is 1.18×10^4 J/kg. (a) What is the efficiency of this engine? (b) How does the efficiency compare with that of a Carnot engine?

WEB 57. One mole of neon gas is heated from 300 K to 420 K at constant pressure. Calculate (a) the heat energy transferred to the gas, (b) the change in the internal energy of the gas, and (c) the work done by the gas. Note that neon has a molar specific heat of $c = 20.79$ J/mol·K.

58. One mole of an ideal gas is taken through the reversible cycle shown in Figure P12.58. At point *A*, the pressure, volume, and temperature are P_0, V_0, and T_0. In terms of *R* and T_0, find (a) the total heat entering the system per cycle, (b) the total heat leaving the system per cycle,

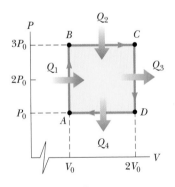

Figure P12.58

(c) the efficiency of an engine operating in this reversible cycle, and (d) the efficiency of an engine operating in a Carnot cycle between the temperature extremes for this process. (*Hint:* Recall that work equals the area under a *PV* curve.)

59. An electrical power plant has an overall efficiency of 15%. The plant is to deliver 150 MW of power to a city, and its turbines use coal as fuel. The burning coal produces steam at 190°C, which drives the turbines. This steam is then condensed into water at 25°C by passing through coils in contact with river water. (a) How many metric tons of coal does the plant consume each day (1 metric ton = 1×10^3 kg)? (b) What is the total cost of the fuel per year if the delivered price is $8 per metric ton? (c) If the river water is delivered at 20°C, at what minimum rate must it flow over the cooling coils in order that its temperature not exceed 25°C? (*Note:* The heat of combustion of coal is 7.8×10^3 cal/g.)

Vibrations and Wave Motion

A s we look around us, we find many examples of objects that vibrate or oscillate: a pendulum, the strings of a guitar, an object suspended on a spring, the piston of an engine, the head of a drum, the reed of a saxophone. Most elastic objects vibrate when an impulse is applied to them; that is, once they are distorted, their shape tends to be restored to some equilibrium configuration. Even at the atomic level, the atoms in a solid vibrate about some position as if they were connected to their neighbors by imaginary springs.

Wave motion is closely related to the phenomenon of vibration. Sound waves, earthquake waves, waves on stretched strings, and water waves are all produced by vibrations. As a sound wave travels through a medium (such as air), the molecules of the medium vibrate back and forth; as a water wave travels across a pond, the water molecules vibrate up and down. When any wave travels through a medium, the particles of the medium move in repetitive cycles. Therefore, the motion of the particles bears a strong resemblance to the periodic motion of a vibrating pendulum or a mass attached to a spring.

Many other natural phenomena occur whose explanations require an understanding of vibrations and waves. Although many large structures, such as skyscrapers and bridges, appear to be rigid, they actually vibrate — a fact that must be taken into account by the architects and engineers who design and build them. To understand how radio and television work, we must comprehend the origin and nature of electromagnetic waves and how they propagate through space. Finally, much of what scientists have learned about atomic structure has come from information carried by waves; therefore, to understand the concepts and theories of atomic physics, we must first study waves and vibrations.

◀ *Giraudon/Art Resource, NY*

403

Vibrations and Waves

▲ **PHYSICS PUZZLER**

On musical instruments such as guitars, the frequency of the note played by a string can be changed in two ways — by turning the tuning pegs or by pressing the string to the fingerboard. What is being changed in each of these cases and why does this cause the frequency of vibration of the string to change? *(Michael Tamborrino/FPG International)*

This chapter constitutes a brief return to the subject of mechanics as we examine various forms of periodic motion. We concentrate especially on motion that occurs when the force on an object is proportional to the displacement of the object from its equilibrium position. When such a force acts only toward the equilibrium position, the result is a back-and-forth motion called *simple harmonic motion*—oscillation, or vibration, between two extreme positions for an indefinite period of time with no loss of energy. The terms *harmonic motion* and *periodic motion* are used interchangeably in this chapter. Both refer to back-and-forth motion. You may be familiar with several types of periodic motion, such as the oscillations of a mass on a spring, the motion of a pendulum, and the vibrations of a stringed musical instrument.

Because vibrations can move through a medium, we also study wave motion in this chapter. Many kinds of waves occur in nature, including sound waves, seismic waves, and electromagnetic waves. We end this chapter with a brief discussion of some terms and concepts that are common to all types of waves, and in later chapters we shall focus our attention on specific categories of waves.

Figure 13.1 The force of a spring on a mass varies with the displacement of the mass from the equilibrium position, $x = 0$. (a) When x is positive (stretched spring), the spring force is to the left. (b) When x is zero (unstretched spring), the spring force is zero. (c) When x is negative (compressed spring), the spring force is to the right.

13.1   HOOKE'S LAW

One of the simplest types of vibrational motion is that of a mass attached to a spring, which we discussed briefly in Chapter 5. Let us assume that the mass moves on a frictionless horizontal surface. If the spring is stretched or compressed a small distance, x, from its unstretched, or equilibrium, position and then released, it exerts a force on the mass, as shown in Figure 13.1. From experiment this spring force is found to obey the equation

$$F_s = -kx \qquad\qquad [13.1]$$

where x is the displacement of the mass from its unstretched ($x = 0$) position and k is a positive constant called the **spring constant.** This force law for springs was discovered by Robert Hooke in 1678 and is known as **Hooke's Law.** The value of k is a measure of the stiffness of the spring. Stiff springs have large k values, and soft springs have small k values.

◀ Hooke's law

The negative sign in Equation 13.1 signifies that the force exerted by the spring is always directed *opposite* the displacement of the mass. When the mass is to the right of the equilibrium position, as in Figure 13.1a, x is positive and F_s is negative. This means that the force is in the negative direction, to the left. When the mass is to the left of the equilibrium position, as in Figure 13.1c, x is negative and F_s is positive, indicating that the direction of the force is to the right. Of course, when $x = 0$, as in Figure 13.1b, the spring is unstretched and $F_s = 0$. Because the spring force always acts toward the equilibrium position, it is sometimes called a *restoring force.* **The direction of the restoring force is such that the mass is being either pulled or pushed toward the equilibrium position.**

Let us examine the motion of the mass if it is initially pulled a distance of A to the right and released from rest. The force exerted on the mass by the spring pulls the mass back toward the equilibrium position. As the object moves toward $x = 0$, the magnitude of the force decreases (because x decreases) and reaches zero at $x = 0$. However, the mass gains speed as it moves toward the equilibrium position. In fact, the speed reaches its maximum value when $x = 0$. The momentum achieved by the mass causes it to overshoot the equilibrium position and to compress the spring. As the object moves to the left of the equilibrium position (negative x values), the force acting on it begins to increase to the right and the speed of the mass begins to decrease. The mass finally comes to a stop at $x = -A$. The process is then repeated, and the mass continues to oscillate back and forth over the same path. This type of motion is called *simple harmonic motion*. **Simple harmonic motion occurs when the net force along the direction of motion is a Hooke's law type of force—that is, when the net force is proportional to the displacement and in the opposite direction.**

Simple harmonic motion ▶

Not all repetitive motion over the same path can be classified as simple harmonic motion. For example, consider a ball being tossed back and forth between a parent and child. The ball moves repetitively, but the motion is not simple harmonic. Unless the force acting on an object along the direction of motion has the form of Equation 13.1, the object does not exhibit simple harmonic motion.

The motion of a mass suspended from a vertical spring is also simple harmonic. In this case, the force of gravity acting on the attached mass stretches the spring until equilibrium is reached. (The mass is suspended and at rest.) This position of equilibrium establishes the position of the mass for which $x = 0$. When the mass is stretched a distance of x beyond this equilibrium position and then released, a net force acts (in the form of Hooke's law) toward the equilibrium position. Because the net force is proportional to x, the motion is simple harmonic.

Before we can discuss simple harmonic motion in more detail, it is necessary to define a few terms.

- The **amplitude, A,** is the maximum distance traveled by an object away from its equilibrium position. In the absence of friction, an object continues in simple harmonic motion and reaches a maximum displacement equal to the amplitude on each side of the equilibrium position during each cycle.
- The **period, T,** is the time it takes the object to execute one complete cycle of motion.
- The **frequency, f,** is the number of cycles or vibrations per unit of time.

Figure 13.2 (Example 13.1) Determining the spring constant. The elongation, d, of the spring is due to the suspended weight, mg. Because the upward spring force balances the weight when the system is in equilibrium, it follows that $k = mg/d$.

EXAMPLE 13.1   Measuring the Spring Constant

A common technique used to evaluate the spring constant is illustrated in Figure 13.2. The spring is hung vertically (Fig. 13.2a), and a body of mass m is attached to the lower end of the spring (Fig. 13.2b). The spring stretches a distance of d from its initial position under the action of the "load" mg. Because the spring force is upward, it must balance the weight, mg, downward *when the system is at rest*. In this case, we can apply Hooke's law to give

$$F_s = kd = mg$$

$$k = \frac{mg}{d}$$

For example, if a spring is stretched 2.0 cm by a mass of 0.55 kg, the force constant is

$$k = \frac{mg}{d} = \frac{(0.55 \text{ kg})(9.80 \text{ m/s}^2)}{2.0 \times 10^{-2} \text{ m}} = 2.7 \times 10^2 \text{ N/m}$$

EXAMPLE 13.2 Simple Harmonic Motion on a Frictionless Surface

A 0.35-kg mass attached to a spring of spring constant 130 N/m is free to move on a frictionless horizontal surface, as in Figure 13.1. If the mass is released from rest at $x = 0.10$ m, find the force on it and its acceleration at (a) $x = 0.10$ m, (b) $x = 0.050$ m, (c) $x = 0$ m, and (d) $x = -0.050$ m.

Solution (a) The point at which the mass is released defines the amplitude of the motion. In this case, $A = 0.10$ m. The mass moves continuously between the limits 0.10 m and -0.10 m. When x is a maximum $(x = A)$, the force on the mass is a maximum and is calculated as follows:

$$F = -kx$$

$$F_{max} = -kA = -(130 \text{ N/m})(0.10 \text{ m}) = \boxed{-13 \text{ N}}$$

The negative sign indicates that the force acts to the left, in the negative x direction. We can use Newton's second law to calculate the acceleration at this position:

$$F_{max} = ma_{max}$$

$$a_{max} = \frac{F_{max}}{m} = -\frac{13 \text{ N}}{0.35 \text{ kg}} = \boxed{-37 \text{ m/s}^2}$$

Again, the negative sign indicates that the acceleration is to the left.

(b) We can use the same approach to find the force and acceleration at other positions. At $x = 0.05$ m, we have

$$F = -kx = -(130 \text{ N/m})(0.050 \text{ m}) = \boxed{-6.5 \text{ N}}$$

$$a = \frac{F}{m} = -\frac{6.5 \text{ N}}{0.35 \text{ kg}} = \boxed{-19 \text{ m/s}^2}$$

Note that the acceleration of an object moving with simple harmonic motion *is not constant* because F is not constant.

(c) At $x = 0$, the spring force is zero (because $F = -kx = 0$) and the acceleration is zero. In other words, when the spring is unstretched, it exerts no force on the mass attached to it.

(d) At $x = -0.050$ m, we have

$$F = -kx = -(130 \text{ N/m})(-0.050 \text{ m}) = \boxed{6.5 \text{ N}}$$

$$a = \frac{F}{m} = \frac{6.5 \text{ N}}{0.35 \text{ kg}} = \boxed{19 \text{ m/s}^2}$$

This result shows that the force and acceleration are positive when the mass is on the negative side of the equilibrium position. This indicates that the force of the spring on the mass is acting to the right as the spring is being compressed. At the same time, the mass is slowing down as it moves from $x = 0$ to $x = -0.050$ m.

Exercise Find the force and acceleration when $x = -0.10$ m.

Answer 13 N; 37 m/s^2

As indicated in Example 13.2, the acceleration of an object moving with simple harmonic motion can be found by using Hooke's law as the force in the equation for Newton's second law, $F = ma$. This gives

$$-kx = ma$$

$$a = -\frac{k}{m}x \qquad [13.2]$$

Because the maximum value of x is defined to be the amplitude, A, we see that the acceleration ranges over the values $-kA/m$ to $+kA/m$. Equation 13.2 enables us to find the acceleration of the object as a function of its position. In subsequent sections, we shall find equations for velocity as a function of position and position as a function of time.

Earlier we stated that an object moves with simple harmonic motion when the net force acting on it is proportional to its displacement from equilibrium and is directed toward the equilibrium position. Equation 13.2 provides an alternative definition of simple harmonic motion. An object moves with simple harmonic motion if its acceleration is proportional to its displacement and is in the opposite direction to it.

13.2   ELASTIC POTENTIAL ENERGY

For the most part, we have worked thus far with three types of mechanical energy: gravitational potential energy, translational kinetic energy, and rotational kinetic energy. In Chapter 5 we briefly discussed a fourth type of mechanical energy, elastic potential energy. We now consider this form of energy in more detail.

An object has potential energy by virtue of its shape or position. As we learned in Chapter 5, an object of mass m at height h above the ground has gravitational potential energy equal to mgh. This means that the object can do work after it is released. Likewise, a compressed spring has potential energy by virtue of its shape. In this case, the compressed spring, when allowed to expand, can move an object and thus do work on it. As an example, Figure 13.3 shows a ball being projected from a spring-loaded toy gun, where the spring is compressed a distance of x. As

Energy = elastic *PE*

Energy = *KE*

Figure 13.3   A ball projected from a spring-loaded gun. The elastic potential energy stored in the spring is transformed into the kinetic energy of the ball.

the gun is fired, the compressed spring does work on the ball and imparts kinetic energy to it. **The energy stored in a stretched or compressed spring or other elastic material is called elastic potential energy, PE_s, given by**

$$PE_s \equiv \tfrac{1}{2}kx^2 \qquad\qquad [13.3]$$ ◀ Elastic potential energy

Note that the elastic potential energy stored in a spring is zero when the spring is unstretched or uncompressed ($x = 0$). **Energy is stored in a spring only when it is either stretched or compressed. Furthermore, the elastic potential energy is a maximum when a spring has reached its maximum compression or extension.** Finally, the potential energy in a spring is always positive because it is proportional to x^2. We now include this new form of energy in our equation for conservation of mechanical energy:

$$(KE + PE_g + PE_s)_i = (KE + PE_g + PE_s)_f \qquad [13.4]$$

If nonconservative forces such as friction are present, then the final mechanical energy does not equal the initial mechanical energy. In this case, the difference in the two energies must equal the work done by the nonconservative force, W_{nc}. From the work-energy theorem,

$$W_{nc} = (KE + PE_g + PE_s)_f - (KE + PE_g + PE_s)_i \qquad [13.5]$$

As an example of the energy conversions that take place when a spring is included in the system, consider Figure 13.4. A block of mass m slides on a frictionless

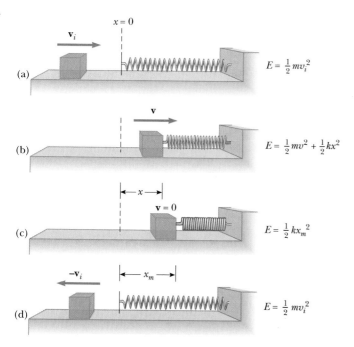

Figure 13.4 A block sliding on a frictionless horizontal surface collides with a light spring. (a) Initially, the mechanical energy is entirely the kinetic energy of the block. (b) The mechanical energy at some arbitrary position is the sum of the kinetic energy of the block and the elastic potential energy stored in the spring. (c) When the block comes to rest, the mechanical energy is entirely elastic potential energy stored in the compressed spring. (d) When the block leaves the spring, the mechanical energy is equal to the block's kinetic energy. The total energy remains constant.

Figure 13.5   Elastic potential energy is stored in this drawn bow. *(Eric Lars Baleke/Black Star)*

horizontal surface with constant velocity $\mathbf{v_i}$ and collides with a coiled spring. The description that follows is greatly simplified by assuming that the spring is very light and therefore has negligible kinetic energy. As the spring is compressed, it exerts a leftward force on the block. At maximum compression, the block momentarily stops (Fig. 13.4c). The initial total energy in the system before the collision (block plus spring) is the kinetic energy of the block. After the block collides with the spring and the spring is partially compressed, as in Figure 13.4b, the block has kinetic energy $\frac{1}{2}mv^2$ (where $v < v_i$) and the spring has potential energy $\frac{1}{2}kx^2$. When the block stops momentarily after colliding with the spring, the kinetic energy is zero. Because the spring force is conservative and because there are no external forces that can do work on the system, **the total mechanical energy of the system consisting of the block and spring remains constant.** Thus, energy is transformed from kinetic energy of the block into potential energy stored in the spring. As the spring expands, the block moves in the opposite direction and regains all of its initial kinetic energy, as in Figure 13.4d.

Elastic potential energy is stored in the bow of an archer when he or she pulls back the bowstring (Figure 13.5). This is a more complicated system than a single mass on a spring, but if we consider the essential features, there is a restoring force on the arrow as a result of the stretching of the string and the bending of the bow. The potential energy stored in the system is transformed into kinetic energy of the arrow as it is released. We could make similar statements about devices such as crossbows and slingshots.

APPLICATION

Archery.

EXAMPLE 13.3   Stop That Car!

A 13 000-N car starts at rest and rolls down a hill from a height of 10 m (Fig. 13.6). It then moves across a level surface and collides with a light spring-loaded guardrail. Neglecting any losses due to friction, find the maximum distance the spring is compressed. Assume a spring constant of 1.0×10^6 N/m.

Solution   The initial and final values of the kinetic energy of the car are zero. Thus, the initial potential energy of the car is completely converted to elastic potential energy in the spring at the end of the trip (assuming we neglect any energy losses due to friction during the collision). Conservation of energy gives

$$mgh = \tfrac{1}{2}kx^2$$

Figure 13.6 (Example 13.3) A car starts from rest on a hill at the position shown. When the car reaches the bottom of the hill, it collides with a spring-loaded guardrail.

Solving for x gives

$$x = \sqrt{\frac{2mgh}{k}} = \sqrt{\frac{2(13000 \text{ N})(10 \text{ m})}{1.0 \times 10^6 \text{ N/m}}} = 0.50 \text{ m}$$

Note that it was not necessary at any point to calculate the velocity of the car to obtain a solution. This demonstrates the power of the principle of conservation of energy. One works with the initial and final energy values only, without having to consider all the details in between.

Exercise What is the speed of the car just before it collides with the guardrail?

Answer 14 m/s

EXAMPLE 13.4 Motion With and Without Friction

A 1.6-kg block is attached to a spring with a spring constant of 1.0×10^3 N/m (Fig. 13.1). The spring is compressed a distance of 2.0 cm, and the block is released from rest. (a) Calculate the speed of the block as it passes through the equilibrium position, $x = 0$, if the surface is frictionless.

Solution By using Equation 13.3, we can find the initial elastic potential energy of the spring when $x = -2.0 \text{ cm} = -2.0 \times 10^{-2}$ m:

$$PE_s = \tfrac{1}{2}kx_i^2 = \tfrac{1}{2}(1.0 \times 10^3 \text{ N/m})(-2.0 \times 10^{-2} \text{ m})^2 = 0.20 \text{ J}$$

Because the block is always at the same height above the Earth's surface, its gravitational potential energy remains constant. Hence, the initial potential energy stored in the spring is converted to kinetic energy at $x = 0$. That is,

$$\tfrac{1}{2}kx_i^2 = \tfrac{1}{2}mv_f^2$$

$$0.20 \text{ J} = \tfrac{1}{2}(1.6 \text{ kg})(v_f^2)$$

$$v_f = 0.50 \text{ m/s}$$

(b) Calculate the speed of the block as it passes through the equilibrium position if a constant frictional force of 4.0 N retards its motion.

Solution   Because sliding friction is present in this situation, we know that the final mechanical energy is less than the initial mechanical energy. The work done by the frictional force for a displacement of 2.0×10^{-2} m is

$$W_{nc} = W_f = -fs = -(4.0 \text{ N})(2.0 \times 10^{-2} \text{ m}) = -0.080 \text{ J}$$

Applying Equation 13.5 to this situation gives

$$-0.080 \text{ J} = \tfrac{1}{2}(1.6 \text{ kg})(v_f^2) - 0.20 \text{ J}$$

$$v_f = \boxed{0.39 \text{ m/s}}$$

Note that this value for v_f is less than that obtained in the frictionless case. Does the result make sense?

Exercise   How far does the block travel before coming to rest? Assume a constant friction force of 4.0 N and the same initial conditions as before.

Answer   5.0 cm

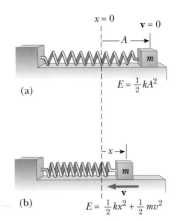

Figure 13.7   (a) A mass attached to a spring on a frictionless surface is released from rest with the spring extended a distance of A. Just before release, the total energy is elastic potential energy, $kA^2/2$. (b) When the mass reaches position x, it has kinetic energy $mv^2/2$ and the elastic potential energy has decreased to $kx^2/2$.

13.3   VELOCITY AS A FUNCTION OF POSITION

Conservation of energy provides a simple method of deriving an expression for the velocity of a mass undergoing periodic motion as a function of position. The mass in question is initially at its maximum extension, A (Fig. 13.7a), and is then released from rest. In this situation, the initial energy of the system is entirely elastic potential energy stored in the spring, $\tfrac{1}{2}kA^2$. As the mass moves toward the origin to some new position, x (Fig. 13.7b), part of this energy is transformed into kinetic energy, and the potential energy stored in the spring is reduced to $\tfrac{1}{2}kx^2$. Because the total energy of the system is equal to $\tfrac{1}{2}kA^2$ (the initial energy stored in the spring), we can equate this to the sum of the kinetic and potential energies at the final position:

$$\tfrac{1}{2}kA^2 = \tfrac{1}{2}mv^2 + \tfrac{1}{2}kx^2$$

Solving for v, we get

$$v = \pm\sqrt{\frac{k}{m}(A^2 - x^2)} \qquad [13.6]$$

This expression shows us that the speed is a maximum at $x = 0$ and zero at the extreme positions $x = \pm A$.

The right side of Equation 13.6 is preceded by the $\pm$ sign because the square root of a number can be either positive or negative. The sign of v that is selected depends on the circumstances of the motion. If the mass in Figure 13.7 is moving to the right, v must be positive; if the mass is moving to the left, v must be negative.

 EXAMPLE 13.5   The Mass-Spring System Revisited

A 0.50-kg mass connected to a light spring with a spring constant of 20 N/m oscillates on a frictionless horizontal surface.
(a) Calculate the total energy of the system and the maximum speed of the mass if the amplitude of the motion is 3.0 cm.

Solution From Equation 13.3, we have

$$E = PE_s = \tfrac{1}{2}kA^2 = \tfrac{1}{2}(20 \text{ N/m})(3.0 \times 10^{-2} \text{ m})^2 = \boxed{9.0 \times 10^{-3} \text{ J}}$$

When the mass is at $x = 0$, $PE_s = 0$ and $E = \tfrac{1}{2}mv_{max}^2$; therefore,

$$\tfrac{1}{2}mv_{max}^2 = 9.0 \times 10^{-3} \text{ J}$$

$$v_{max} = \sqrt{\frac{18 \times 10^{-3} \text{ J}}{0.50 \text{ kg}}} = \boxed{0.19 \text{ m/s}}$$

(b) What is the velocity of the mass when the displacement is 2.0 cm?

Solution We can apply Equation 13.6 directly:

$$v = \pm\sqrt{\frac{k}{m}(A^2 - x^2)} = \pm\sqrt{\frac{20}{0.50}(3.0^2 - 2.0^2) \times 10^{-4}} = \boxed{\pm 0.14 \text{ m/s}}$$

The $\pm$ sign indicates that the mass could be moving to the right or to the left at this instant.

(c) Compute the kinetic and potential energies of the system when the displacement is 2.0 cm.

Solution The results of (b) can be used to give

$$KE = \tfrac{1}{2}mv^2 = \tfrac{1}{2}(0.50 \text{ kg})(0.14 \text{ m/s})^2 = \boxed{5.0 \times 10^{-3} \text{ J}}$$

$$PE_s = \tfrac{1}{2}kx^2 = \tfrac{1}{2}(20 \text{ N/m})(2.0 \times 10^{-2} \text{ m})^2 = \boxed{4.0 \times 10^{-3} \text{ J}}$$

Note that the sum $KE + PE_s$ equals the total energy, E, found in part (a).

Exercise For what values of x is the speed of the mass 0.10 m/s?

Answer ± 2.6 cm

13.4 COMPARING SIMPLE HARMONIC MOTION WITH UNIFORM CIRCULAR MOTION

We can better understand and visualize many aspects of simple harmonic motion along a straight line by looking at their relationships to uniform circular motion. Figure 13.8 is a top view of an experimental arrangement that is useful for this purpose. A ball is attached to the rim of a phonograph turntable of radius A, illuminated from the side by a lamp. Rather than concentrating on the ball, let us focus our attention on the shadow that the ball casts on the screen. We find that **as the turntable rotates with constant angular speed, the shadow of the ball moves back and forth with simple harmonic motion.**

In order to understand why, let us examine Equation 13.6 more closely. This equation says that the velocity of an object moving with simple harmonic motion is related to the displacement by

$$v = C\sqrt{A^2 - x^2}$$

where C is a constant. To see that the shadow also obeys this relation, consider Figure 13.9, which shows the ball moving with a constant speed, v_0, in a direction

Figure 13.8 An experimental setup for demonstrating the connection between simple harmonic motion and uniform circular motion. As the ball rotates on the turntable with constant angular speed, its shadow on the screen moves back and forth with simple harmonic motion.

Figure 13.9 The ball rotates with constant speed v_0. The x component of the velocity of the ball equals the projection of $\mathbf{v}_0$ on the x axis.

tangent to the circular path. At this instant, the velocity of the ball in the x direction is given by $v = v_0 \sin \theta$, or

$$\sin \theta = \frac{v}{v_0}$$

Likewise, the larger triangle containing the angle θ in Figure 13.9 enables us to write

$$\sin \theta = \frac{\sqrt{A^2 - x^2}}{A}$$

Equating the right-hand sides of the last two expressions, we see that v is related to the displacement, x, as

$$\frac{v}{v_0} = \frac{\sqrt{A^2 - x^2}}{A}$$

or

$$v = \frac{v_0}{A} \sqrt{A^2 - x^2} = C\sqrt{A^2 - x^2}$$

The velocity of the ball in the x direction is related to the displacement, x, in exactly the same manner as the velocity of an object undergoing simple harmonic motion. Hence, the shadow moves with simple harmonic motion.

A valuable example of the relationship between simple harmonic motion and circular motion is seen in vehicles and machines that use the back-and-forth motion of a piston to create rotational motion of a wheel. For example, consider the drive wheels of a locomotive. In Figure 13.10, the black mechanism at the lower right contains a piston that moves back and forth in simple harmonic motion. The piston is connected to an arrangement of rods that transforms its back-and-forth motion into rotational motion of the wheels. A similar mechanism in an automobile engine transforms the back-and-forth motion of the pistons to rotational motion of the crankshaft.

APPLICATION

Pistons and Drive Wheels.

Figure 13.10 An old locomotive from the Strasburg Railroad, Pennsylvania. The piston housing is shown at the lower right. A piston moves back and forth with simple harmonic motion. *(Grant Heilman Photography)*

Period and Frequency

Note that the period, T, of the shadow in Figure 13.8, which represents the time required for one complete trip back and forth, is also the time it takes the ball to make one complete circular trip on the turntable. Because the ball moves through the distance $2\pi A$ (the circumference of the circle) in the time T, the speed, v_0, of the ball around the circular path is

$$v_0 = \frac{2\pi A}{T}$$

$$T = \frac{2\pi A}{v_0}$$

However, for our purposes, let us consider only a fraction of the complete trip. Imagine that the ball moves from P to Q, a quarter of a revolution, in Figure 13.8. This requires a time interval equal to one fourth of the period, and the distance traveled by the ball is $2\pi A/4$. Therefore,

$$\frac{T}{4} = \frac{2\pi A}{4v_0}$$ [13.7]

Now imagine that the motion of the shadow is equivalent to the horizontal motion of a mass on the end of a spring. During this quarter of a cycle, the shadow moves from a point where its energy is solely elastic potential energy to a point where its energy is solely kinetic energy. That is,

$$\tfrac{1}{2}kA^2 = \tfrac{1}{2}mv_0^2$$

$$\frac{A}{v_0} = \sqrt{\frac{m}{k}}$$

Substituting for A/v_0 in Equation 13.7, we find that the period is

$$T = 2\pi\sqrt{\frac{m}{k}}$$ [13.8]   ◀ The period of a mass-spring system moving with simple harmonic motion

This expression gives the time required for an object to make a complete cycle of its motion. Now recall that the frequency, f, is the number of cycles per unit of time. The symmetry in the units of period and frequency should lead you to see that the two must be related inversely as

$$f = \frac{1}{T}$$ [13.9]

Therefore, the **frequency** of the periodic motion is

$$f = \frac{1}{2\pi}\sqrt{\frac{k}{m}}$$ [13.10]   ◀ Frequency

The units of frequency are s^{-1}, or hertz (Hz). The **angular frequency, ω,** is

$$\omega = 2\pi f = \sqrt{\frac{k}{m}}$$ [13.11]   ◀ Angular frequency

Thinking Physics 1

We know that the period of oscillation of a mass hung on a spring is proportional to the square root of the mass. Thus, if we perform an experiment in which we hang a range of masses on the end of a spring and measure the period of oscillation, a graph of the square of the period on the vertical axis versus the suspended mass m on the horizontal axis will result in a straight line. But we would find that the line does not go through the origin. That is, for $m = 0$, T^2 has a finite value. How do you explain this result?

Explanation   The reason that the line does not go through the origin is that the spring itself has mass. Thus, the resistance to change in motion of the system is a combination of the mass hung on the end of the spring and the mass of the oscillating coils of the spring. The entire mass of the spring is not oscillating, however. The bottom-most coil is oscillating with the same amplitude as the mass, and the top-most coil is not oscillating at all. For a cylindrical spring, energy arguments can be used to show that the effective additional mass affecting the oscillations of a light spring is one third of the mass of the spring. The square of the period is proportional to the total oscillating mass. Therefore, a graph of period squared against total mass (mass hung on the spring plus the effective oscillating mass of the spring) would pass through the origin.

Thinking Physics 2

A mass oscillating on the end of a horizontal spring slides back and forth over a frictionless surface. During one oscillation, you set an identical mass on the first mass when the system is at the maximum displacement point where it is at rest. The two are connected by Velcro and continue their oscillation together. Does the period of the oscillation change? Does the amplitude of the oscillation change? Does the energy of the oscillation change?

Explanation   The period of the oscillation does change, because the period depends on the mass that is oscillating. The amplitude does not change. Because the new mass was added while the original mass was at rest, the combined masses are at rest at this point, also, defining the amplitude as the same as in the original oscillation. The energy does not change, either. At the maximum displacement point, the energy is all potential energy stored in the spring, which depends only on the spring constant and the amplitude, not the mass. The increased mass will pass through the equilibrium point with less velocity than in the original oscillation, but with the same kinetic energy. Another approach is to think about how energy could be transferred into the oscillating system—no work was done (nor was there any other form of energy transfer), so the energy in the system cannot change.

APPLICATION

Bungee Jumping.

Figure 13.11   (Applying Physics 1) Bungee jumping from a bridge. *(Telegraph Colour Library/FPG International)*

Applying Physics 1

A bungee cord can be modeled as a spring. If you go bungee jumping, you will bounce up and down at the end of the cord after your daring dive off a bridge (Fig. 13.11). Suppose you perform this dive and measure the frequency of your bouncing. You then move to another bridge. You discover that the bungee cord is too long for dives off this bridge; you will hit the ground. Thus, you fold the bungee cord in half and make the dive from the doubled bungee cord. How does the frequency of your bouncing at the end of this dive compare to the frequency after the dive from the first bridge?

Explanation   The force exerted by the bungee cord, modeled as a spring, is proportional to the separation of the coils as the spring is extended. Imagine that we extend a spring by a given distance and measure the distance between coils. We then cut the spring in half. If one of the half springs is now extended by the same distance, the coils will be twice as far apart as they were for the complete spring. Thus, it takes twice as much force to stretch the half-spring, from which we conclude that the half- spring has a spring constant that is twice that of the complete spring. Now consider the folded bungee cord that we model as two half-springs in parallel. Each half has a spring constant that is twice the original spring constant of the bungee cord. In addition, a given mass hanging on the folded bungee cord will experience two forces—one from each half-spring. As a result, the required force for a given extension will be four times as much as for the original bungee cord. The effective spring constant of the folded bungee cord is, therefore, four times as large as the original spring constant. Because the frequency of oscillation is proportional to the square root of the spring constant, your bouncing frequency on the folded cord will be twice that of the original cord.

EXAMPLE 13.6 That Car Needs a New Set of Shocks!

A 1300-kg car is constructed on a frame supported by four springs. Each spring has a spring constant of 20 000 N/m. If two people riding in the car have a combined mass of 160 kg, find the frequency of vibration of the car when it is driven over a pothole in the road.

Solution We assume that the weight is evenly distributed; thus each spring supports one fourth of the load. The total mass supported by the springs is 1460 kg, and therefore each spring supports 365 kg. Hence, the frequency of vibration is

$$f = \frac{1}{2\pi}\sqrt{\frac{k}{m}} = \frac{1}{2\pi}\sqrt{\frac{20\ 000\ \text{N/m}}{365\ \text{kg}}} = 1.18\ \text{Hz}$$

Exercise How long does it take the car to execute three complete vibrations?

Answer 2.54 s

13.5 POSITION, VELOCITY, AND ACCELERATION AS A FUNCTION OF TIME

We can obtain an expression for the position of an object moving with simple harmonic motion as a function of time by returning to the relationship between simple harmonic motion and uniform circular motion. Again, consider a ball on the rim of a rotating turntable of radius A, as in Figure 13.12. We shall refer to the circle made by the ball as the *reference circle* for the motion. Let us assume that the turntable revolves at a constant angular speed of ω. As the ball rotates on the reference circle, the angle, θ, made by the line OP with the x axis changes with time. Furthermore, as the ball rotates, the projection of P on the x axis, labeled point Q, moves back and forth along the axis with simple harmonic motion.

From the right triangle, OPQ, we see that $\cos\theta = x/A$. Therefore, the x coordinate of the ball is

$$x = A\cos\theta$$

Because the ball rotates with constant angular speed, it follows that $\theta = \omega t$ (see Chapter 7). Therefore,

$$x = A\cos(\omega t) \qquad [13.12]$$

In one complete revolution, the ball rotates through an angle of 2π rad in the period T. In other words, the motion repeats itself every T seconds. Therefore,

$$\omega = \frac{\Delta\theta}{\Delta t} = \frac{2\pi}{T} = 2\pi f \qquad [13.13]$$

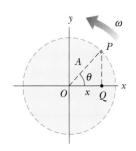

Figure 13.12 A reference circle. As the ball at P rotates in a circle with uniform angular speed, its projection, Q, along the x axis moves with simple harmonic motion.

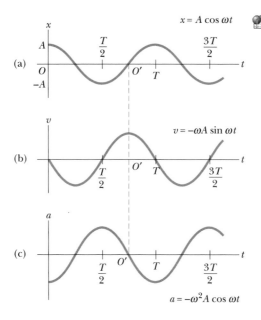

$x = A \cos \omega t$

(a)

$v = -\omega A \sin \omega t$

(b)

(c)

$a = -\omega^2 A \cos \omega t$

Figure 13.13   (a) Displacement, (b) velocity, and (c) acceleration versus time for an object moving with simple harmonic motion under the initial conditions that $x_0 = A$ and $v_0 = 0$ at $t = 0$.

where f is the frequency of the motion. The angular speed of the rotation on the reference circle is the same as the angular frequency of the projected simple harmonic motion. Consequently, Equation 13.12 can be written

$$x = A \cos(2\pi f t) \qquad [13.14]$$

This equation represents the position of an object moving with simple harmonic motion as a function of time; it is graphed in Figure 13.13a. The curve should be familiar to you from trigonometry. Its shape is called sinusoidal. Note that x varies between A and $-A$ because the cosine function varies between 1 and -1.

Figures 13.13b and 13.13c represent curves for velocity and acceleration as a function of time. An end-of-chapter problem will ask you to show that the velocity and acceleration are sinusoidal functions of time. Note that when x is a maximum or minimum, the velocity is zero, and when x is zero, the magnitude of the velocity is a maximum. Furthermore, when x has its maximum positive value, the acceleration is a maximum but in the negative x direction, and when x is at its maximum negative position, the acceleration has its maximum value in the positive direction. These curves are consistent with our earlier discussion of the points at which v and a reach their maximum, minimum, and zero values.

Figure 13.14 illustrates one experimental arrangement that demonstrates simple harmonic motion. A mass connected to a spring has a marking pen attached to it. While the mass vibrates vertically, a sheet of paper is moved horizontally with constant speed. The pen traces out a sinusoidal pattern.

Figure 13.14   An experimental apparatus for demonstrating simple harmonic motion. A pen attached to the oscillating mass traces out a sinusoidal wave on the moving chart paper.

Motion of paper

EXAMPLE 13.7   The Vibrating Mass-Spring System

(a) Find the amplitude, frequency, and period of motion for an object vibrating at the end of a spring if the equation for its position as a function of time is

$$x = (0.25 \text{ m}) \cos\left(\frac{\pi}{8.0} t\right)$$

Solution We can find two of our unknowns by comparing this equation with the general equation for such motion:

$$x = A \cos(2\pi f t)$$

By comparison, we see that

$$A = \boxed{0.25 \text{ m}}$$

$$2\pi f = \frac{\pi}{8.0} \text{ s}^{-1}$$

$$f = \boxed{\tfrac{1}{16} \text{ Hz}}$$

Because the period $T = 1/f$, if follows that $T = 1/f = \boxed{16 \text{ s}}$.

(b) What is the position of the object after 2.0 seconds have elapsed?

Solution Direct substitution for t in the expression for x gives

$$x = (0.25 \text{ m}) \cos\left(\frac{\pi}{4}\right) = \boxed{0.18 \text{ m}}$$

Note that in evaluating the cosine function, the angle is in radians. Thus, you should either set your calculator to evaluate trigonometric functions based on radian measure or convert from radians to degrees. The angle is called the *phase* of the oscillation at that instant.

13.6   MOTION OF A PENDULUM

A simple pendulum is another mechanical system that exhibits periodic motion. It consists of a small bob of mass m suspended by a light string of length L fixed at its upper end, as in Figure 13.15. (By a light string, we mean that the string's mass is assumed to be very small compared to the mass of the bob and hence can be ignored.) When released, the mass swings to and fro over the same path; but is its motion simple harmonic? In order to answer this question, we must examine the force that acts as the restoring force on the pendulum. If this force is proportional to the displacement, s, then the force is of the Hooke's law form, $F = -ks$, and hence the motion is simple harmonic. Furthermore, because $s = L\theta$ in this case, we see that the motion is simple harmonic if F is proportional to the angle θ.

The component of the force of gravity tangential to the circular path is the force that acts to restore pendulum to its equilibrium position. Thus, the restoring force is

$$F_t = -mg \sin\theta$$

From this equation, we see that the restoring force is proportional to $\sin\theta$ rather than to θ. Thus, in general, the motion of a pendulum is *not* simple harmonic. However, for small angles, less than about 15 degrees, the angle θ measured in radians and the sine of the angle are approximately equal. Therefore, if we restrict the motion to small angles, the restoring force can be written as

$$F_t = -mg\,\theta$$

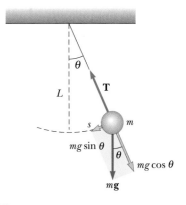

Figure 13.15 A simple pendulum consists of a bob of a mass m suspended by a light string of length L. (L is the distance from the pivot to the center of mass of the bob.) The restoring force that causes the pendulum to undergo simple harmonic motion is the component of weight tangent to the path of motion, $mg \sin\theta$.

Because $s = L\theta$, we have

$$F_t = -\left(\frac{mg}{L}\right)s$$

This equation follows the general form of the Hooke's law force, given by $F = -ks$, with $k = mg/L$. Thus, we are justified in saying that a pendulum undergoes simple harmonic motion only when it swings back and forth at very small amplitudes (or, in this case, small values of θ, so that $\sin\theta \cong \theta$).

We can find the period of a pendulum by first recalling that the period of a mass-spring system (Eq. 13.8) is

$$T = 2\pi\sqrt{\frac{m}{k}}$$

If we replace k with its equivalent, mg/L, we see that the period of a simple pendulum is

$$T = 2\pi\sqrt{\frac{m}{mg/L}}$$

$$T = 2\pi\sqrt{\frac{L}{g}} \qquad\qquad [13.15]$$

A multiflash photograph of a swinging pendulum. Is the oscillating motion simple harmonic in this case? *(Richard Megna/Fundamental Photographs)*

▷ The period of a simple pendulum depends only on *L* and *g*.

This equation reveals the somewhat surprising result that the period of a simple pendulum depends not on mass but only on the pendulum's length and on the free-fall acceleration. Furthermore, the amplitude of the motion is not a factor as long as it is relatively small. The analogy between the motion of a simple pendulum and the mass-spring system is illustrated in Figure 13.16.

It is of historical interest to point out that it was Galileo who first noted that the period of a pendulum was independent of its amplitude. He supposedly observed this while attending church services at a cathedral in Pisa. The pendulum he studied was a swinging chandelier that was set in motion when someone bumped it while lighting candles. Galileo was able to measure its frequency, and hence its period, by timing the swings with his pulse.

APPLICATION

Pendulum Clocks.

The dependence of the period of a pendulum on its length and on the free-fall acceleration allows us to use a pendulum as a timekeeper for a clock. A number of clock designs employ a pendulum, with the length adjusted so that its period serves as the basis for the rate at which the clock's hands turn. Of course, these clocks are used at different locations on the Earth, so there will be some variation of the free-fall acceleration. In order to compensate for this, the pendulum of a clock should have some movable mass so that the effective length can be adjusted for variations in *g*.

APPLICATION

Use of Pendulum in Prospecting.

Geologists often make use of the simple pendulum and Equation 13.15 when prospecting for oil or minerals. Deposits beneath the Earth's surface can produce irregularities in the free-fall acceleration over the region being studied. A specially designed pendulum of known length is used to measure the period, which in turn is used to calculate *g*. Although such a measurement in itself is inconclusive, it is an important tool for geological surveys.

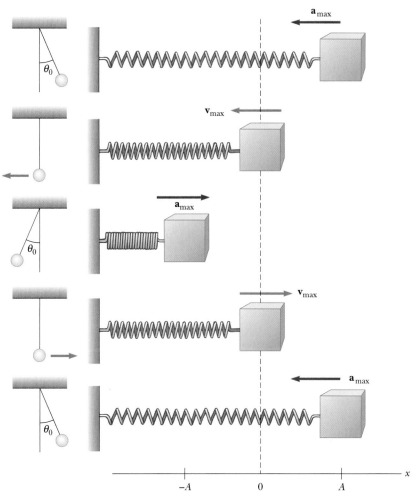

EXAMPLE 13.8 What Is the Height of That Tower?

A man needs to know the height of a tower, but darkness obscures the ceiling. He knows, however, that a long pendulum extends from the ceiling almost to the floor and that its period is 12.0 s. How tall is the tower?

Solution If we use $T = 2\pi\sqrt{L/g}$ and solve for L, we get

$$L = \frac{gT^2}{4\pi^2} = \frac{(9.80 \text{ m/s}^2)(12.0 \text{ s})^2}{4\pi^2} = 35.8 \text{ m}$$

Exercise If the length of the pendulum were halved, what would its period of vibration be?

Answer 8.49 s

Q U I C K L A B

Construct a simple pendulum by tying a metal bolt to one end of a string of length 1.00 m, and measure the period of your pendulum while fixing the opposite end. You should obtain the period by timing 25 complete oscillations, making sure that the string always makes small angles with the vertical. Repeat the measurement for string lengths ranging from 0.40 m to 1.60 m in increments of 0.20 m. Plot the square of the period versus the length of the string, and measure the slope of the line through your data points. Does your slope agree with that predicted by Equation 13.15? What value of g do you obtain from your data?

Figure 13.17 A cross-sectional view of a shock absorber connected to a spring in the suspension system of an automobile. The upper part of the shock absorber and the ends of the suspension spring are attached to the frame of the automobile (not shown).

Oil or other viscous fluid

Piston with holes

Suspension spring

13.7 DAMPED OSCILLATIONS

The vibrating motions we have discussed so far have taken place in ideal systems — that is, systems that *oscillate indefinitely* under the action of a linear restoring force. In many real systems, forces of friction retard the motion, and consequently the systems do not oscillate indefinitely. The friction reduces the mechanical energy of the system as time passes, and the motion is said to be **damped.**

Shock absorbers in automobiles (Fig. 13.17) make practical application of damped motion. A shock absorber consists of a piston moving through a liquid such as oil. The upper part of the shock absorber is firmly attached to the body of the car, and the piston is attached to a leaf spring that, with the other springs, acts as the main suspension for the car. When the car travels over a bump in the road, holes in the piston allow it to move up and down in the fluid in a damped fashion.

Damped motion varies depending on the fluid used. For example, if the fluid has a relatively low viscosity, the vibrating motion is preserved but the amplitude of vibration decreases in time and the motion ultimately ceases. This is known as *underdamped* oscillation, and its position-time curve appears in Figure 13.18. If the fluid viscosity is increased, the mass returns rapidly to equilibrium after it is released and does not oscillate. In this case, the system is said to be *critically damped* (Fig. 13.19a), and the piston returns to the equilibrium position in the shortest time possible without once overshooting the equilibrium position. If the viscosity is made greater still, the system is said to be *overdamped.* In this case, the piston returns to equilibrium without ever passing through the equilibrium point, but the time required to reach equilibrium is greater than at critical damping, as shown by Figure 13.19b.

To make automobiles more comfortable to ride in, shock absorbers are designed to be slightly underdamped. This can be demonstrated by a sharp downward

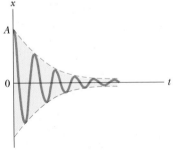

Figure 13.18 A graph of displacement versus time for an underdamped oscillator. Note the decrease in amplitude with time.

push on the hood of a car. After the applied force is removed, the body of the car oscillates a few times about the equilibrium position before returning to its fixed position.

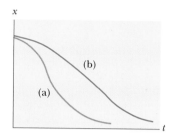

Figure 13.19 Plots of displacement versus time for (a) a critically damped oscillator and (b) an overdamped oscillator.

13.8 WAVE MOTION

Most of us first saw waves when, as children, we dropped a pebble into a pool of water. The disturbance created by the pebble generates water waves, which move outward until they finally reach the edge of the pool. There are a wide variety of physical phenomena that have wave-like characteristics. The world is full of waves: sound waves, waves on a string, earthquake waves, and electromagnetic waves, such as visible light, radio waves, television signals, and x-rays. All of these waves have as their source a vibrating object. Thus, we shall use the terminology and concepts of simple harmonic motion as we move into the study of wave motion.

In the case of sound waves, the vibrations that produce waves arise from such sources as a person's vocal cords or a plucked guitar string. The vibrations of electrons in an antenna produce radio or television waves, and the simple up-and-down motion of a hand can produce a wave on a string. Regardless of the type of wave under consideration, there are certain concepts common to all varieties. In the remainder of this chapter, we shall focus our attention on a general study of wave motion. In later chapters we shall study specific types of waves, such as sound and electromagnetic waves.

What Is a Wave?

As we said before, when you drop a pebble into a pool of water, the disturbance produced by the pebble excites water waves, which move away from the point at which the pebble entered the water. If you carefully examined the motion of a leaf floating near the disturbance, you would see that it moves up and down and back and forth about its original position but does not undergo any net displacement attributable to the disturbance. That is, the water wave (or disturbance) moves from one place to another *but the water is not carried with it.*

Einstein and Infeld made these remarks about wave phenomena:

A bit of gossip starting in Washington reaches New York very quickly, even though not a single individual who takes part in spreading it travels between these two cities. There are two quite different motions involved, that of the rumor, Washington to New York, and that of the persons who spread the rumor. The wind, passing over a field of grain, sets up a wave which spreads out across the whole field. Here again we must distinguish between the motion of the wave and the motion of the separate plants, which undergo only small oscillations. . . . The particles constituting the medium perform only small vibrations, but the whole motion is that of a progressive wave. The essentially new thing here is that for the first time we consider the motion of something which is not matter, but energy propagated through matter.[1]

A disturbance traveling from right to left on a stretched spring. *(Education Development Center, Newton, MA)*

[1] A. Einstein and L. Infeld, *The Evolution of Physics,* New York, Simon and Schuster, 1961.

When we observe what is called a *water wave*, what we see is a rearrangement of the water's surface. Without the water there would be no wave. A wave traveling on a string would not exist without the string. Sound waves travel through air as a result of pressure variations from point to point. (We shall discuss sound waves in Chapter 14.) The wave motion we consider in this chapter corresponds to the disturbance of a body or medium. Therefore, we can consider a wave to be *the motion of a disturbance*. In a later chapter we shall discuss electromagnetic waves, which do not require a medium.

The mechanical waves discussed in this chapter require (1) some source of disturbance, (2) a medium that can be disturbed, and (3) some physical connection or mechanism through which adjacent portions of the medium can influence each other. All waves carry energy and momentum. The amount of energy transmitted through a medium and the mechanism responsible for the transport of energy differ from case to case. For instance, the energy carried by ocean waves during a storm is much greater than that carried by a sound wave generated by a single human voice.

Thinking Physics 3

We have shown that waves carry energy—not only waves on strings, but all waves. Devise a demonstration that shows that waves transfer momentum as well as energy.

Explanation   Imagine that we cover the ends of a cardboard tube with rubber sheets secured with rubber bands, to form a crude drum with two heads. We now lay the tube horizontally on the table and suspend a Ping-Pong ball from a string so that the Ping-Pong ball is just touching one of the rubber sheets. If we now thump the other rubber sheet with a finger, the sound will travel down the tube, strike the first rubber sheet, and the Ping-Pong ball will bounce away from the rubber sheet. This demonstrates that the sound wave is transferring momentum and energy to the Ping-Pong ball.

Applying Physics 2

Why is it important to be quiet in avalanche country? In a spy movie, the bad guys try to stop the hero, who is escaping on skis, by firing a gun and causing an avalanche. Why did the avalanche happen?

Explanation   The essence of a wave is the propagation of a disturbance through a medium. An impulsive sound, like the gunshot, can cause an acoustical disturbance that propagates through the air and can impact a ledge of snow that is just ready to break free to begin an avalanche. Such a disastrous event occurred in 1916 during World War I, when Austrian soldiers in the Alps were smothered by an avalanche caused by cannon fire.

Figure 13.20 A wave pulse traveling along a stretched rope. The shape of the pulse is approximately unchanged as it travels.

13.9   TYPES OF WAVES

One of the simplest ways to demonstrate wave motion is to flip one end of a long rope that is under tension and has its opposite end fixed, as in Figure 13.20. The bump (called a *pulse*) travels to the right with a definite speed. A disturbance of this type is called a **traveling wave.** Figure 13.20 shows the shape of the rope at three closely spaced times.

As a traveling wave pulse travels along the rope, **each segment of the rope that is disturbed moves perpendicularly to the wave motion.** Figure 13.21 illustrates this point for a particular tiny segment, *P.* Never does the rope move in the direction of the wave. A traveling wave such as this, in which the particles of the disturbed medium move perpendicularly to the wave velocity, is called a **transverse wave.** Figure 13.22 illustrates the formation of transverse waves on a long spring.

In another class of waves, called **longitudinal waves, the particles of the medium undergo displacements parallel to the direction of wave motion.** Sound waves in air, for instance, are longitudinal. Their disturbance corresponds to a series of high- and low-pressure regions that may travel through air or through any material medium with a certain speed. A longitudinal pulse can be easily produced in a stretched spring, as in Figure 13.22b. The free end is pumped back and forth along the length of the spring. This action produces compressed and stretched regions of the coil that travel along the spring, parallel to the wave motion.

Figure 13.21 A pulse traveling on a stretched rope is a transverse wave. That is, any element *P* on the rope moves (blue arrows) perpendicularly to the wave motion (red arrows).

Picture of a Wave

Figure 13.23 shows the curved shape of a vibrating string. This pattern, sometimes called a *waveform,* should be familiar to you from our study of simple harmonic motion; it is a sinusoidal curve. The red curve can be thought of as a snapshot of a traveling wave taken at some instant of time, say *t* = 0; the blue curve, a snapshot of the same traveling wave at a later time. It is not difficult to imagine that this picture can as easily be used to represent a wave on water. In such a case, point *A* would correspond to the *crest* of the wave and point *B* to the low point, or *trough,* of the wave.

(a) Transverse wave

(b) Longitudinal wave

Figure 13.22 (a) A transverse wave is set up in a spring by moving one end of the spring perpendicularly to its length. (b) A longitudinal pulse along a stretched spring. The displacement of the coils is in the direction of the wave motion. For the starting motion described in the text, the compressed region is followed by a stretched region.

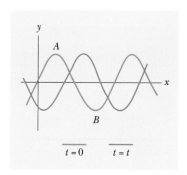

Figure 13.23 A one-dimensional sinusoidal wave traveling to the right with a speed of *v.* The red curve is a snapshot of the wave at *t* = 0, and the blue curve is another snapshot at some later time *t.*

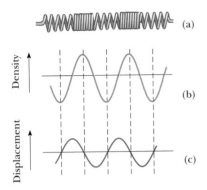

Figure 13.24 (a) A longitudinal wave on a spring. (b) The crests of the waveform correspond to compressed regions of the spring, and the troughs correspond to stretched regions of the spring. (c) The displacement wave.

This same waveform can be used to describe a longitudinal wave. To see how this is done, consider a longitudinal wave traveling on a spring. Figure 13.24a is a snapshot of this wave at some instant, and Figure 13.24b shows the sinusoidal curve that represents the wave. Points where the coils of the spring are compressed correspond to the crests of the waveform, and stretched regions correspond to troughs.

The type of wave represented by the curve in Figure 13.24b is often referred to as a *density* or *pressure wave*. This is because the crests, where the spring coils are compressed, are regions of high density, and the troughs, where the coils are stretched, are regions of low density.

An alternative method for representing wave motion along a spring is through the concept of a displacement wave, shown in Figure 13.24c. In this representation, coils that are displaced the greatest distance from equilibrium in one direction are indicated by crests, and coils that are displaced the greatest distance from equilibrium in the opposite direction are represented by troughs. Both of these wave representations will be used in future sections.

13.10   FREQUENCY, AMPLITUDE, AND WAVELENGTH

Figure 13.25 illustrates a method of producing a wave on a very long string. One end of the string is connected to a blade that is set in vibration. As the blade oscillates vertically with simple harmonic motion, a traveling wave moving to the right is set up in the string. Figure 13.25 consists of views of the wave at intervals of one quarter of a period. Note that **each particle of the string, such as *P*, oscillates vertically in the *y* direction with simple harmonic motion.** This must be the case, because each particle follows the simple harmonic motion of the blade. Therefore, every segment of the string can be treated as a simple harmonic oscillator vibrating with the same frequency as the blade that drives the string.

Figure 13.25 One method for producing traveling waves on a continuous string. The left end of the string is connected to a blade that is set in vibration. Note that every segment, such as *P*, oscillates vertically with simple harmonic motion.

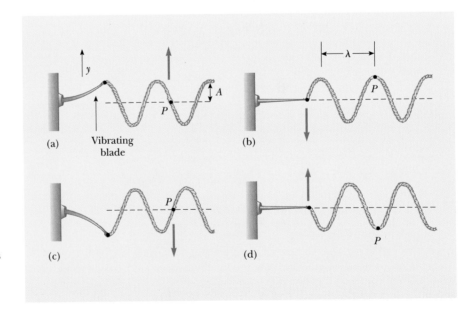

The frequencies of the waves we shall study will range from rather low values for waves on strings and waves on water to values between 20 Hz and 20 000 Hz (recall that 1 Hz = 1 s^{-1}) for sound waves, and much higher frequencies for electromagnetic waves.

The horizontal dashed line in Figure 13.25 represents the position of the string if no wave were present. The maximum distance the string moves above or below this equilibrium value is called the **amplitude, A,** of the wave. For the waves we work with, the amplitudes at the crest and the trough will be identical.

Figure 13.25b illustrates another characteristic of a wave. The horizontal arrows show the distance between two successive points that behave identically. This distance is called the **wavelength, λ** (Greek letter lambda).

We can use these definitions to derive an expression for the velocity of a wave. We start with the defining equation for velocity:

$$v = \frac{\Delta x}{\Delta t}$$

A little reflection should convince you that a wave will advance a distance of one wavelength in a time interval equal to one period of the vibration. Thus

$$v = \frac{\lambda}{T}$$

Because the frequency is equal to the reciprocal of the period, we have

$$v = f\lambda \qquad\qquad\qquad \text{[13.16]} \qquad \blacktriangleleft \text{ Wave velocity}$$

We shall apply this equation to many types of waves. For example, we shall use it often in our study of sound and electromagnetic waves.

EXAMPLE 13.9   A Traveling Wave

A wave traveling in the positive x direction is pictured in Figure 13.26. Find the amplitude, wavelength, period, and speed of the wave if it has a frequency of 8.0 Hz.

Solution   The amplitude and wavelength can be read directly from the figure:

$$A = \boxed{15 \text{ cm}} \qquad \lambda = 40 \text{ cm} = \boxed{0.40 \text{ m}}$$

The period of the wave is

$$T = \frac{1}{f} = \frac{1}{8.0} \text{ s} = \boxed{0.13 \text{ s}}$$

and the speed is

$$v = f\lambda = (8.0 \text{ s}^{-1})(0.40 \text{ m}) = \boxed{3.2 \text{ m/s}}$$

Figure 13.26 (Example 13.9) A harmonic wave of wavelength $\lambda = 40$ cm and amplitude $A = 15$ cm.

EXAMPLE 13.10   Give Me a "C" Note

The note middle C on a piano has a frequency of approximately 262 Hz and a wavelength in air of 1.31 m. Find the speed of sound in air.

Solution By direct substitution into Equation 13.16, we find that

$$v = f\lambda = (262 \text{ s}^{-1})(1.31 \text{ m}) = \boxed{343 \text{ m/s}}$$

EXAMPLE 13.11 The Speed of Radio Waves

An FM station broadcasts at a frequency of 100 MHz (M = mega = 10^6), with a radio wave having a wavelength of 3.00 m. Find the speed of the radio wave.

Solution As in the last example, we use Equation 13.16:

$$v = f\lambda = (100 \times 10^6 \text{ s}^{-1})(3.00 \text{ m}) = \boxed{3.00 \times 10^8 \text{ m/s}}$$

This, in fact, is the speed of *any* electromagnetic wave traveling through empty space.

Exercise Find the wavelength of an electromagnetic wave whose frequency is 9.0 GHz = 9.0×10^9 Hz (G = giga = 10^9), which is the microwave range.

Answer 0.033 m

13.11   THE SPEED OF WAVES ON STRINGS

In this section we focus our attention on the speed of a wave on a stretched string. Rather than deriving the equation, we use dimensional analysis to verify that the expression can be valid.

It is easy to understand why the wave speed depends on the tension in the string. If a string under tension is pulled sideways and released, the tension is responsible for accelerating a particular segment back toward its equilibrium position. The acceleration and wave speed increase with increasing tension in the string. Likewise, the wave speed is inversely dependent on the mass per unit length of the string. This is because it is more difficult to accelerate (and impart a large wave speed to) a massive string than a light string. Thus, wave speed is directly dependent on the tension and inversely dependent on the mass per unit length of the string. The exact relationship of the wave speed, v, the tension, F, and the mass per length, μ, is

$$v = \sqrt{\frac{F}{\mu}} \qquad\qquad [13.17]$$

From this expression we see that the speed of a mechanical wave, such as a wave on a string, depends only on the properties of the medium through which the disturbance travels.

Now let us verify that this expression is dimensionally correct. The dimensions of F are ML/T^2, and the dimensions of μ are M/L. Therefore, the dimensions of F/μ are L^2/T^2, and those of $\sqrt{F/\mu}$ are L/T, which are indeed the dimensions of speed. No other combination of F and μ is dimensionally correct, assuming that these are the only variables relevant to the situation.

Equation 13.17 indicates that we can increase the speed of a wave on a stretched string by increasing the tension in the string. It also shows that if we wrap a string with a metallic winding, as is done to the bass strings of pianos and guitars, we decrease the speed of a transmitted wave because the mass per unit length is increased.

EXAMPLE 13.12   A Pulse Traveling on a String

A uniform string has a mass, M, of 0.300 kg and a length, L, of 6.00 m. Tension is maintained in the string by suspending a 2.00-kg block from one end (Fig. 13.27). Find the speed of a pulse on this string.

Solution   The tension, F, in the string is equal to the mass, m, of the block multiplied by the free-fall acceleration:

$$F = mg = (2.00 \text{ kg})(9.80 \text{ m/s}^2) = 19.6 \text{ N}$$

For the string, the mass per unit length, μ, is

$$\mu = \frac{M}{L} = \frac{0.300 \text{ kg}}{6.00 \text{ m}} = 0.0500 \text{ kg/m}$$

Therefore, the wave speed is

$$v = \sqrt{\frac{F}{\mu}} = \sqrt{\frac{19.6 \text{ N}}{0.0500 \text{ kg/m}}} = 19.8 \text{ m/s}$$

Exercise   Find the time it takes the pulse to travel from the wall to the pulley.

Answer   0.253 s

Figure 13.27 (Example 13.12) The tension, F, in the string is maintained by the suspended block. The wave speed is given by the expression $v = \sqrt{F/\mu}$.

13.12   SUPERPOSITION AND INTERFERENCE OF WAVES

Many interesting wave phenomena in nature are impossible to describe with a single moving wave. Instead, one must analyze what happens when two or more waves pass through the same region of space. For such analyses one can use the **superposition principle:**

> If two or more traveling waves are moving through a medium, the resultant wave is found by adding together the displacements of the individual waves point by point.

Experiments show that the superposition principle is valid only when the individual waves have small amplitudes of displacement—an assumption we make in all our examples.

One consequence of the superposition principle is that **two traveling waves can pass through each other without being destroyed or even altered.** For instance, when two pebbles are thrown into a pond, the expanding circular waves do

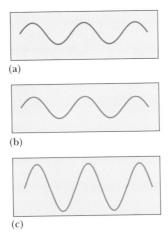

(a)

(b)

(c)

Figure 13.28 Constructive interference. If two waves having the same frequency and amplitude are in phase, as in (a) and (b), the resultant wave when they combine (c) has the same frequency as the individual waves but twice their amplitude.

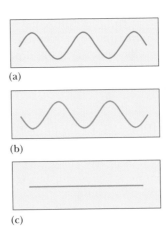

(a)

(b)

(c)

Figure 13.29 Destructive interference. When two waves with the same frequency and amplitude are 180° out of phase, as in (a) and (b), the result when they combine (c) is complete cancellation.

Figure 13.30 Interference patterns produced by outward-spreading waves from several drops of water falling into a pond. *(Martin Dohrn/SPL/Photo Researchers)*

not destroy each other. In fact, the ripples pass through each other. Likewise, when sound waves from two sources move through air, they pass through each other. The sound one hears at a given location is the result of both disturbances.

Figures 13.28a and 13.28b show two waves of the same amplitude and frequency. If at some instant of time these two waves attempted to travel through the same region of space, the resultant wave at that instant would have a shape like Figure 13.28c. For example, suppose these are water waves of amplitude 1 m. At the instant they overlap so that crest meets crest and trough meets trough, the resultant wave has an amplitude of 2 m. Waves coming together like this are said to be *in phase* and to undergo **constructive interference.**

Figures 13.29a and 13.29b show two similar waves. In this case, however, the crest of one coincides with the trough of the other; that is, one wave is *inverted* relative to the other. The resultant wave, shown in Figure 13.29c, is seen to be a state of complete cancellation. If these were water waves coming together, one of the waves would be trying to pull an individual drop of water upward at the same instant the other wave was trying to pull it downward. The result would be no motion of the water at all. In such a situation, the two waves are said to be 180° out of phase and to undergo **destructive interference.** Figure 13.30 illustrates the interference of water waves produced by drops of water falling into a pond.

Figure 13.31 shows constructive interference in two pulses moving toward each other along a stretched string; Figure 13.32 shows destructive interference in two pulses. Notice in each case that, when the two pulses separate, their shapes are unchanged, as if they had never met!

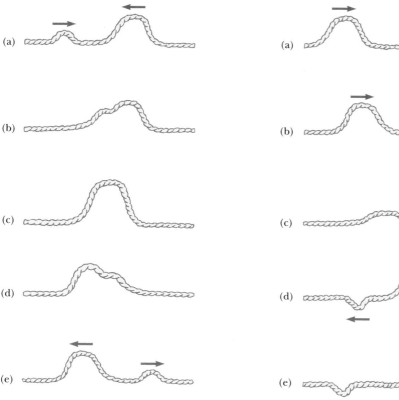

Figure 13.31 Two wave pulses traveling on a stretched string in op-posite directions pass through each other. When the pulses overlap, as in (b), (c), and (d), the net dis-placement of the string equals the sum of the displacements produced by each pulse. Because each pulse produces positive displacements of the string, we refer to their super-position as *constructive interference.*

Figure 13.32 Two wave pulses traveling in opposite directions with displacements that are inverted rela-tive to each other. When the two overlap as in (c), their displace-ments subtract from each other.

13.13 REFLECTION OF WAVES

In our discussion so far, we have assumed that the waves being analyzed could travel indefinitely without striking anything. Often, such conditions are not realized in practice. Whenever a traveling wave reaches a boundary, part or all of the wave is reflected. For example, consider a pulse traveling on a string that is fixed at one end (Fig. 13.33). When the pulse reaches the wall, it is reflected.

Note that the reflected pulse is inverted. This can be explained as follows. When the pulse meets the wall, the string exerts an upward force on the wall. According

Figure 13.33 The reflection of a traveling wave at the fixed end of a stretched string. Note that the re-flected pulse is inverted, but its shape remains the same.

(a)

(b)

(c)

Reflected pulse

(d)

Figure 13.34 The reflection of a traveling wave at the free end of a stretched string. In this case, the reflected pulse is not inverted.

to Newton's third law, the wall must exert an equal and opposite (downward) reaction force on the string. This downward force causes the pulse to invert on reflection.

Now consider a case in which the pulse arrives at the string's end, which is attached to a ring of negligible mass that is free to slide along the post without friction (Fig. 13.34). Again the pulse is reflected, but this time it is not inverted. On reaching the post, the pulse exerts a force on the ring, causing it to accelerate upward. The ring is then returned to its original position by the downward component of the tension.

An alternative method of showing that a pulse is reflected without inversion when it strikes a free end of a string is to send a pulse down a string hanging vertically. When the pulse hits the free end, it is reflected without inversion, similarly to the pulse in Figure 13.34.

SUMMARY

Simple harmonic motion occurs when the net force along the direction of motion is a **Hooke's law** type of force — that is, when the net force is proportional to the displacement and in the opposite direction:

$$F_s = -kx \qquad [13.1]$$

The time required for one complete vibration is called the **period** of the motion. The inverse of the period is the **frequency** of the motion, which is the number of oscillations per second.

When an object is moving with simple harmonic motion, its **acceleration** as a function of location is

$$a = -\frac{k}{m}x \qquad [13.2]$$

The energy stored in a stretched or compressed spring or other elastic material is called **elastic potential energy**:

$$PE_s \equiv \tfrac{1}{2}kx^2 \qquad [13.3]$$

The **velocity** of an object as a function of position, when the object is moving with simple harmonic motion, is

$$v = \pm\sqrt{\frac{k}{m}(A^2 - x^2)} \qquad [13.6]$$

The **period** of an object of mass m moving with simple harmonic motion while attached to a spring of spring constant k is

$$T = 2\pi\sqrt{\frac{m}{k}} \qquad [13.8]$$

Also, the **frequency** of a mass-spring system is $f = 1/T$.

The **position** of an object as a function of time, when the object is moving with simple harmonic motion, is

$$x = A\cos(2\pi ft) \qquad [13.14]$$

A **simple pendulum** of length L moves with simple harmonic motion for small angular displacements from the vertical, with a period of

$$T = 2\pi\sqrt{\frac{L}{g}} \qquad\qquad \text{[13.15]}$$

The period is independent of the suspended mass.

A **transverse wave** is one in which the particles of the medium move perpendicularly to the direction of the wave velocity. An example is a wave on a stretched string.

A **longitudinal wave** is one in which the particles of the medium move parallel to the direction of the wave velocity. An example is a sound wave.

The relationship of the velocity, wavelength, and frequency of a wave is

$$v = f\lambda \qquad\qquad \text{[13.16]}$$

The speed of a wave traveling on a stretched string of mass per unit length μ and under tension F is

$$v = \sqrt{\frac{F}{\mu}} \qquad\qquad \text{[13.17]}$$

The **superposition principle** states that if two or more traveling waves are moving through a medium, the resultant wave is found by adding the individual waves together point by point. When waves meet crest to crest and trough to trough, they undergo **constructive interference.** When crest meets trough, the waves undergo complete **destructive interference.**

When a wave pulse reflects from a rigid boundary, the pulse is inverted. When the boundary is free, the reflected pulse is not inverted.

MULTIPLE-CHOICE QUESTIONS

1. The distance between a crest of a sinusoidal water wave and the next trough is 2 m. If the frequency of this wave is 2 Hz, find the speed of the wave.
   (a) 4 m/s (b) 1 m/s (c) 8 m/s
   (d) Impossible to say from the information given.
2. Light waves and sound waves have the following in common.
   (a) Neither will travel in a vacuum.
   (b) Both are longitudinal waves.
   (c) They travel at the same speed.
   (d) None of the above is correct.
3. Replacing an object on a spring with one of nine times the mass will have the result of changing the frequency of the vibrating spring by what multiplying factor?
   (a) 1/9 (b) 1/3 (c) 3.0 (d) 9.0
4. A mass of 0.4 kg, hanging from a spring with a spring constant of 80 N/m, is set into an up-and-down simple harmonic motion. What is the speed of the mass when moving through the equilibrium point? The starting displacement is 0.1 m.
   (a) zero (b) 1.4 m/s (c) 2.0 m/s (d) 3.4 m/s
5. A mass of 0.4 kg, hanging from a spring with a spring constant of 80 N/m, is set into an up-and-down simple harmonic motion. What is the acceleration of the mass when at its maximum displacement of 0.1 m?
   (a) zero (b) 5 m/s^2 (c) 10 m/s^2 (d) 20 m/s^2
6. A runaway railroad car, with mass 30×10^4 kg, coasts across a level track at 2.0 m/s when it collides with a spring-loaded bumper at the end of the track. If the spring constant of the bumper is 2.0×10^6 N/m, what is the maximum compression of the spring during the collision? (Assume the collision is elastic.)
   (a) 0.77 m (b) 0.58 m
   (c) 0.34 m (d) 1.07 m

CONCEPTUAL QUESTIONS

1. A mass-spring system undergoes simple harmonic motion with an amplitude A. Does the total energy change if the mass is doubled but the amplitude is not changed? Are the kinetic and potential energies at a given point in its motion affected by the change in mass? Explain.

2. Does the acceleration of a simple harmonic oscillator remain constant during its motion? Is the acceleration ever zero? Explain.

3. A mass is hung on a spring and the frequency of the oscillation of the system, f, is measured. The mass, a second identical mass, and the spring are carried in the space shuttle to space. The two masses are attached to the ends of the spring, and the system is taken out on a space walk. The spring is extended, and the system is released to oscillate while floating in space. What is the frequency of oscillation for this system in terms of f?

4. If a mass-spring system is hung vertically and set into oscillation, why does the motion eventually stop?

5. Is a bouncing ball an example of simple harmonic motion? Is the daily movement of a student from home to school and back simple harmonic motion?

6. If a pendulum clock keeps perfect time at the base of a mountain, will it also keep perfect time when moved to the top of the mountain? Explain.

7. A simple pendulum is suspended from the ceiling of a stationary elevator, and the period is determined. Describe the changes, if any, in the period when the elevator (a) accelerates upward, (b) accelerates downward, and (c) moves with constant velocity.

8. If a grandfather clock were running slow, how could we adjust the length of the pendulum to correct the time?

9. A grandfather clock depends on the period of a pendulum to keep correct time. Suppose a grandfather clock is calibrated correctly and then the temperature of the room in which it resides increases. Does the grandfather clock run slow, fast, or correctly? (*Hint:* Remember that a metal expands when its temperature increases.)

10. The amplitude of a system moving in simple harmonic motion is doubled. Determine the change in (a) the total energy, (b) the maximum speed, (c) the maximum acceleration, and (d) the period.

11. In a long line of people waiting to buy tickets at a movie theater, when the first person leaves, a pulse of motion occurs as people step forward to fill in the gap. The gap moves through the line of people. What determines the speed of this pulse? Is it transverse or longitudinal? How about the "wave" at a baseball game, where people in the stands stand up and shout as the wave arrives at their location, and this pulse moves around the stadium —what determines the speed of this pulse? Is it transverse or longitudinal?

12. How would you create a longitudinal wave in a stretched spring? Would it be possible to create a transverse wave in a spring?

13. In mechanics, massless strings are often assumed. Why is this *not* a good assumption when discussing waves on strings?

14. What happens to the wavelength of a wave on a string when the frequency is doubled? Assume that the tension in the string remains the same.

15. Explain why the kinetic and potential energies of a mass-spring system can never be negative.

16. What happens to the speed of a wave on a string when the frequency is doubled? Assume that the tension in the string remains the same.

17. What is the total distance traveled by a body executing simple harmonic motion in a time equal to its period if its amplitude is A?

18. By what factor would you have to multiply the tension in a stretched spring in order to double the wave speed?

19. Determine whether or not the following quantities can be in the same direction for a simple harmonic oscillator: (a) displacement and velocity, (b) velocity and acceleration, (c) displacement and acceleration.

20. The motion of the Earth going around the Sun is periodic with a period of 1 year. Is this motion simple harmonic? Explain.

PROBLEMS

1, 2, 3 = straightforward, intermediate, challenging ☐ = full solution available in Study Guide/Student Solutions Manual
WEB = solution posted at **http://www.saunderscollege.com/physics/college** ▨ = biomedical application ▨ = Interactive Physics

Section 13.1 Hooke's Law

1. A 0.40-kg mass is attached to a spring with a spring constant 160 N/m so that the mass is allowed to move on a horizontal frictionless surface. The mass is released from rest when the spring is compressed 0.15 m. Find (a) the force on the mass and (b) its acceleration at this instant.

2. A load of 50 N attached to a spring hanging vertically stretches the spring 5.0 cm. The spring is now placed

horizontally on a table and stretched 11 cm. (a) What force is required to stretch the spring by this amount? (b) Plot a graph of force (on the *y* axis) versus spring displacement from the equilibrium position along the *x* axis.

3. A ball dropped from a height of 4.00 m makes a perfectly elastic collision with the ground. Assuming no energy lost due to air resistance, (a) show that the motion is periodic and (b) determine the period of the motion. (c) Is the motion simple harmonic? Explain.

4. A small ball is set in horizontal motion by rolling it with a speed of 3.00 m/s across a room 12.0 m long, between two walls. Assume that the collisions made with each wall are perfectly elastic and that the motion is perpendicular to the two walls. (a) Show that the motion is periodic, and determine its period. (b) Is this motion simple harmonic? Explain.

Section 13.2 Elastic Potential Energy

5. A slingshot consists of a light leather cup containing a stone that is pulled back against two rubber bands. It takes a force of 30 N to stretch the bands 1.0 cm. (a) What is the potential energy stored in the bands when a 50-g stone is placed in the cup and pulled back 0.20 m from the equilibrium position? (b) With what speed does the stone leave the slingshot?

6. An archer pulls her bow string back 0.400 m by exerting a force that increases uniformly from zero to 230 N. (a) What is the equivalent spring constant of the bow? (b) How much work is done in pulling the bow?

WEB **7.** A child's toy consists of a piece of plastic attached to a spring (Fig. P13.7). The spring is compressed against the floor a distance of 2.00 cm, and the toy is released. If the toy has a mass of 100 g and rises to a maximum height of 60.0 cm, estimate the force constant of the spring.

Figure P13.7

8. A child's pogo stick (Fig. P13.8) stores energy in a spring ($k = 2.50 \times 10^4$ N/m). At position *A* ($x_1 = -0.100$ m) the spring compression is a maximum, and the child momentarily stops. At position *B* ($x = 0$) the spring is relaxed, and the child is moving upward. At position *C* the spring remains relaxed and the child again stops at the top of the jump. Assume that the com-

bined mass of the child and pogo stick is 25.0 kg. (a) Calculate the total energy of the system if both the spring and the gravitational potential energies are zero at $x = 0$. (b) Determine x_2. (c) Calculate the speed of the child at $x = 0$. (d) Determine the acceleration of the child at $x = x_1$.

Figure P13.8

9. The spring constant of the spring in Figure P13.9 is 19.6 N/m, and the mass of the object is 1.5 kg. The spring is unstretched and the surface is frictionless. A constant 20-N force is applied horizontally to the object as shown. Find the speed of the object after it has moved a distance of 0.30 m.

Figure P13.9

10. A 10.0-g bullet is fired into and embeds in a 2.00-kg block attached to a spring with a spring constant of 19.6 N/m and whose mass is negligible. How far is the spring compressed if the bullet has a speed of 300 m/s just before it strikes the block, and the block slides on a frictionless surface? (*Note:* You must use conservation of momentum in this problem. Why?)

11. A 1.5-kg block is attached to a spring with a spring constant of 2000 N/m. The spring is then stretched a distance of 0.30 cm and the block is released from rest. **(a)** Calculate the speed of the block as it passes through the equilibrium position if no friction is present. **(b)** Calculate the speed of the block as it passes through the equilibrium position if a constant frictional force of 2.0 N retards its motion. **(c)** What would be the strength of the frictional force if the block reached the equilibrium position the first time with zero velocity?

12. A simple harmonic oscillator has a total energy of E. **(a)** Determine the kinetic and potential energies when the displacement is one half the amplitude. **(b)** For what value of the displacement does the kinetic energy equal the potential energy?

Section 13.3   Velocity as a Function of Position

13. A mass of 0.40 kg connected to a light spring with a spring constant of 19.6 N/m oscillates on a frictionless horizontal surface. If the spring is compressed 4.0 cm and released from rest, determine **(a)** the maximum speed of the mass, **(b)** the speed of the mass when the spring is compressed 1.5 cm, and **(c)** the speed of the mass when the spring is stretched 1.5 cm. **(d)** For what value of x does the speed equal one half the maximum speed?

14. A mass-spring system oscillates with an amplitude of 3.5 cm. If the spring constant is 250 N/m and the mass is 0.50 kg, determine **(a)** the mechanical energy of the system, **(b)** the maximum speed of the mass, and **(c)** the maximum acceleration.

15. At an outdoor market, a bunch of bananas is set into oscillatory motion with an amplitude of 20.0 cm on a spring with a spring constant of 16.0 N/m. It is observed that the maximum speed of the bunch of bananas is 40.0 cm/s. What is the weight of the bananas in newtons?

16. A 50.0-g mass is attached to a horizontal spring with a spring constant of 10.0 N/m and released from rest with an amplitude of 25.0 cm. What is the velocity of the mass when it is halfway to the equilibrium position if the surface is frictionless?

Section 13.4   Comparing Simple Harmonic Motion with Uniform Circular Motion

17. An object moves uniformly around a circular path of radius 20.0 cm, making one complete revolution every 2.00 s. What is the **(a)** translational speed of the object, **(b)** the frequency of motion in hertz, and **(c)** the angular speed of the object?

18. Consider the simplified single-piston engine in Figure P13.18. If the wheel rotates at a constant angular speed of ω, explain why the piston rod oscillates in simple harmonic motion.

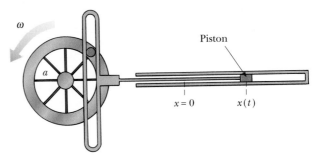

Figure P13.18

19. A 200-g mass is attached to a spring and executes simple harmonic motion with a period of 0.250 s. If the total energy of the system is 2.00 J, find **(a)** the force constant of the spring and **(b)** the amplitude of the motion.

20. The frequency of vibration of a mass-spring system is 5.00 Hz when a 4.00-g mass is attached to the spring. What is the force constant of the spring?

21. A spring stretches 3.9 cm when a 10-g mass is hung from it. If a total mass of 25 g attached to this spring oscillates in simple harmonic motion, calculate the period of motion.

22. When four people with a combined mass of 320 kg sit down in a car, they find that the car drops 0.80 cm lower on its springs. Then they get out of the car and bounce it up and down. What is the frequency of the car's vibration if its mass (empty) is 2.0×10^3 kg?

Section 13.5   Position, Velocity, and Acceleration as a Function of Time

23. The motion of an object is described by the equation

$$x = (0.30 \text{ m}) \cos(\pi t/3)$$

Find **(a)** the position of the object at $t = 0$ and $t = 0.60$ s, **(b)** the amplitude of the motion, **(c)** the frequency of the motion, and **(d)** the period of the motion.

24. A 2.00-kg mass on a frictionless horizontal track is attached to the end of a horizontal spring whose force constant is 5.00 N/m. The mass is displaced 3.00 m to the right from its equilibrium position and then released, which initiates simple harmonic motion. **(a)** What is the force (magnitude and direction) acting on the mass 3.50 s after it is released? **(b)** How many times does the mass oscillate in 3.50 s?

25. A spring of negligible mass stretches 3.00 cm from its relaxed length when a force of 7.50 N is applied. A 0.500-kg particle rests on a frictionless horizontal surface and is attached to the free end of the spring. The particle is pulled horizontally so that it stretches the spring 5.00 cm and is then released from rest at $t = 0$. **(a)** What is the force constant of the spring? **(b)** What are the period, frequency, and angular frequency (ω) of the mo-

tion? **(c)** What is the total energy of the system? **(d)** What is the amplitude of the motion? **(e)** What are the maximum velocity and the maximum acceleration of the particle? **(f)** Determine the displacement, *x*, of the particle from the equilibrium position at *t* = 0.500 s.

26. Given that $x = A\cos(\omega t)$ is a sinusoidal function of time, show that *v* (velocity) and *a* (acceleration) are also sinusoidal functions of time. (*Hint:* Use Equations 13.6 and 13.2.)

Section 13.6 Motion of a Pendulum

27. A simple 2.00-m-long pendulum oscillates in a location where $g = 9.80$ m/s². How many complete oscillations does it make in 5.00 min?

28. A visitor to a lighthouse wishes to determine the height of the tower. She ties a spool of thread to a small rock to make a simple pendulum, which she hangs down the center of a spiral staircase of the tower. The period of oscillation is 9.40 s. What is the height of the tower?

29. An aluminum clock pendulum having a period of 1.00 s keeps perfect time at 20.0°C. **(a)** When placed in a room at a temperature of −5.0°C, will it gain time or lose time? **(b)** How much time will it gain or lose every hour? (*Hint:* See Chapter 10.)

30. A pendulum clock that works perfectly on Earth is taken to the Moon. **(a)** Does it run fast or slow there? **(b)** If the clock is started at 12:00:00 A.M., what will it read after one Earth day (24.0 h)? Assume that the free-fall acceleration on the Moon is 1.63 m/s².

WEB 31. The acceleration due to gravity on Mars is 3.7 m/s². **(a)** What length pendulum has a period of one second on Earth? What length pendulum would have a one second period on Mars? **(b)** A mass is suspended from a spring with force constant 10 N/m. Find the mass suspended from this spring that would result in a one second period on Earth and on Mars.

Section 13.10 Frequency, Amplitude, and Wavelength

32. A wave traveling in the positive *x* direction has a frequency of 25.0 Hz, as in Figure P13.32. Find the **(a)** amplitude, **(b)** wavelength, **(c)** period, and **(d)** speed of the wave.

33. A bat can detect small objects such as an insect whose size is approximately equal to one wavelength of the sound the bat makes. If bats emit a chirp at a frequency of 60.0 kHz and if the speed of sound in air is 340 m/s, what is the smallest insect a bat can detect?

34. If the frequency of oscillation of the wave emitted by an FM radio station is 88.0 MHz, determine the wave's **(a)** period of vibration and **(b)** wavelength. (*Hint:* Radio waves travel at the speed of light, 3.00×10^8 m/s.)

35. The distance between two successive maxima of a transverse wave is 1.20 m. Eight crests, or maxima, pass a given point along the direction of travel every 12.0 s. Calculate the wave speed.

36. A piano emits frequencies that range from a low of about 28 Hz to a high of about 4200 Hz. Find the range of wavelengths spanned by this instrument. The speed of sound in air is approximately 343 m/s.

37. A harmonic wave is traveling along a rope. It is observed that the oscillator that generates the wave completes 40.0 vibrations in 30.0 s. Also, a given maximum travels 425 cm along the rope in 10.0 s. What is the wavelength?

38. Ocean waves are traveling to the east at 4 m/s with a distance of 20 m between crests. With what frequency do the waves hit the front of a boat **(a)** when the boat is at anchor and **(b)** when the boat is moving westward at 1 m/s?

Section 13.11 The Speed of Waves on Strings

39. A circus performer stretches a tightrope between two towers. He strikes one end of the rope and sends a wave along it toward the other tower. He notes that it takes the wave 0.800 s to reach the opposite tower, 20.0 m away. If one meter of the rope has a mass of 0.350 kg, find the tension in the tightrope.

40. Transverse waves with a speed of 50.0 m/s are to be produced on a stretched string. A 5.00-m length of string with a total mass of 0.0600 kg is used. **(a)** What is the required tension in the string? **(b)** Calculate the wave speed in the string if the tension is 8.00 N.

WEB 41. A string is 50.0 cm long and has a mass of 3.00 g. A wave travels at 5.00 m/s along this string. A second string has the same length but half the mass of the first. If the two strings are under the same tension, what is the speed of a wave along the second string?

42. Tension is maintained in a string, as in Figure P13.42.

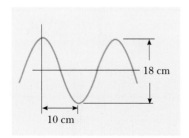

18 cm

10 cm

Figure P13.32

3.0 kg

Figure P13.42

The observed wave speed is 24 m/s when the suspended mass is 3.0 kg. **(a)** What is the mass per unit length of the string? **(b)** What is the wave speed when the suspended mass is 2.0 kg?

43. Transverse waves travel at 20.0 m/s on a string that is under a tension of 6.00 N. What tension is required for a wave speed of 30.0 m/s in the same string?

Section 13.12   Superposition and Interference of Waves

Section 13.13   Reflection of Waves

44. A series of pulses of amplitude 0.15 m are sent down a string that is attached to a post at one end. The pulses are reflected at the post and travel back along the string without loss of amplitude. What is the amplitude at a point on the string where two pulses are crossing, **(a)** if the string is rigidly attached to the post? **(b)** if the end at which reflection occurs is free to slide up and down?

45. A wave of amplitude 0.30 m interferes with a second wave of amplitude 0.20 m traveling in the same direction. What are the **(a)** largest and **(b)** smallest resultant amplitudes that can occur, and under what conditions will these maxima and minima occur?

ADDITIONAL PROBLEMS

46. The position of a 0.30-kg object attached to a spring is described by

$$x = (0.25 \text{ m}) \cos(0.4\pi t)$$

Find **(a)** the amplitude of the motion, **(b)** the spring constant, **(c)** the position at $t = 0.30$ s, and **(d)** the object's velocity at $t = 0.30$ s.

47. A spring with a spring constant of 30.0 N/m is stretched 0.200 m from its equilibrium position. How much work must be done to stretch it an additional 0.100 m?

48. A 500-g block is released from rest and slides down a frictionless track that begins 2.00 m above the horizontal, as shown in Figure P13.48. At the bottom of the track, where the surface is horizontal, the block strikes and sticks to a light spring with a spring constant of 20.0 N/m. Find the maximum distance the spring is compressed.

Figure P13.48

500 g

2.00 m

k

49. A 3.00-kg mass is fastened to a light spring that passes over a pulley (Fig. P13.49). The pulley is frictionless, and its inertia may be neglected. The mass is released from rest when the spring is unstretched. If the mass drops 10.0 cm before stopping, find **(a)** the spring constant of the spring and **(b)** the speed of the mass when it is 5.00 cm below its starting point.

3.00 kg

k

Figure P13.49

50. A 5.00-g bullet moving with an initial speed of 400 m/s is fired into and passes through a 1.00-kg block, as in Figure P13.50. The block, initially at rest on a frictionless horizontal surface, is connected to a spring with a spring constant of 900 N/m. If the block moves 5.00 cm to the right after impact, find **(a)** the speed at which the bullet emerges from the block and **(b)** the energy lost in the collision.

400 m/s

5.00 cm

v

Figure P13.50

51. A 25-kg block is connected to a 30-kg block by a light string that passes over a frictionless pulley. The 30-kg block is connected to a light spring of force constant

200 N/m, as in Figure P13.51. The spring is unstretched when the system is as shown in the figure, and the incline is smooth. The 25-kg block is pulled 20 cm down the incline (so that the 30-kg block is 40 cm above the floor) and is released from rest. Find the speed of each block when the 30-kg block is 20 cm above the floor (that is, when the spring is unstretched).

Figure P13.51

52. A spring in a toy gun has a spring constant of 9.80 N/m and can be compressed 20.0 cm beyond the equilibrium position. A 1.00-g pellet resting against the spring is propelled forward when the spring is released. (a) Find the muzzle speed of the pellet. (b) If the pellet is fired horizontally from a height of 1.00 m above the floor, what is its range?

53. A 2.0-kg block situated on a rough incline is connected to a light spring with a spring constant of 100 N/m (Fig. P13.53). The block is released from rest when the spring is unstretched, and the pulley is frictionless. The block moves 20 cm down the incline before stopping. Find the coefficient of kinetic friction between the block and the incline.

Figure P13.53

54. A 60.0-kg fire fighter slides down a pole while a constant frictional force of 300 N retards his motion. A horizontal 20.0-kg platform is supported by a spring at the bottom of the pole to cushion the fall. The fire fighter starts from rest 5.00 m above the platform, and the spring constant is 2500 N/m. Find (a) the fire fighter's speed just before he collides with the platform and

(b) the maximum distance the spring is compressed. Assume that the frictional force acts during the entire motion. (*Hint:* The collision between the fire fighter and the platform is perfectly inelastic.)

55. A mass, $m_1 = 9.0$ kg, is in equilibrium while connected to a light spring of constant $k = 100$ N/m that is fastened to a wall as in Figure P13.55a. A second mass, $m_2 = 7.0$ kg, is slowly pushed up against mass m_1, compressing the spring by the amount $A = 0.20$ m, as shown in Figure P13.55b. The system is then released, causing both masses to start moving to the right on the frictionless surface. (a) When m_1 reaches the equilibrium point, m_2 loses contact with m_1 (Fig. P13.55c) and moves to the right with speed v. Determine the value of v. (b) How far apart are the masses when the spring is fully stretched for the first time (Fig. P13.55d)? (*Hint:* First determine the period of oscillation and the amplitude of the m_1-spring system after m_2 loses contact with m_1.)

Figure P13.55

56. An astronaut on the Moon wishes to measure the local value of the free-fall acceleration g_{Moon} by timing pulses traveling down a wire that has a large mass suspended from it. Assume a wire of mass 4.00 g is 1.60 m long and has a 3.00 kg mass suspended from it. A pulse requires 36.1 ms to traverse the length of the wire. Calculate

g_{Moon} from these data. (You may neglect the mass of the wire when calculating the tension in it.)

57. A 2.00-kg block hangs without vibrating at the end of a spring ($k = 500$ N/m) that is attached to the ceiling of an elevator car. The car is rising with an upward acceleration of $g/3$ when the acceleration suddenly ceases (at $t = 0$). (a) What is the angular frequency of oscillation of the block after the acceleration ceases? (b) By what amount is the spring stretched during the time that the elevator car is accelerating?

58. A mass m is connected to two rubber bands of length L, each under tension F, as in Figure P13.58. The mass is displaced vertically by a small distance, y. Assuming the tension does not change, show that (a) the restoring force is $-(2F/L)y$ and (b) the system exhibits simple harmonic motion with an angular frequency $\omega = \sqrt{2F/mL}$.

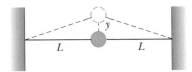

Figure P13.58

59. A light balloon filled with helium of density 0.180 kg/m³ is tied to a light string of length $L = 3.00$ m. The string is tied to the ground, forming an "inverted" simple pendulum (Fig. P13.59a). If the balloon is displaced slightly from equilibrium, as in Figure P13.59b, show that the motion is simple harmonic, and determine the period of the motion. Take the density of air to be 1.29 kg/m³. (*Hint:* Use an analogy with the simple pendulum discussed in the text, and see Chapter 9.)

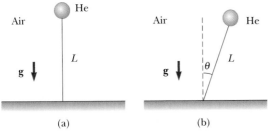

(a)   (b)

Figure P13.59

60. Two points, A and B, on the Earth are at the same longitude and 60.0° apart in latitude. An earthquake at point A sends two waves toward B. A transverse wave travels along the surface of the Earth at 4.50 km/s, and a longitudinal wave travels through the body of the Earth at 7.80 km/s. (a) Which wave arrives at B first? (b) What is the time difference between the arrivals of the two waves at B? Take the radius of the Earth to be 6.37×10^6 m.

WEB 61. A light string of mass 10.0 g and length $L = 3.00$ m has its ends tied to two walls that are separated by the distance $D = 2.00$ m. Two masses, each of mass $m = 2.00$ kg, are suspended from the string, as in Figure P13.61. If a wave pulse is sent from point A, how long does it take to travel to point B?

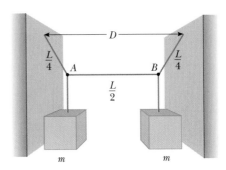

Figure P13.61

62. Figure P13.62 shows a crude model of an insect wing. The mass, m, represents the entire mass of the wing, which pivots about the fulcrum, F. The spring represents the surrounding connective tissue. Motion of the wing corresponds to vibration of the spring. Suppose the mass of the wing is 0.30 g and the effective spring constant of the tissue is 4.7×10^{-4} N/m. If the mass m moves up and down a distance of 2.0 mm from its position of equilibrium, what is the maximum speed of the outer tip of the wing?

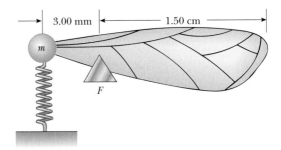

Figure P13.62

63. Assume that a hole is drilled through the center of the Earth. It can be shown that an object of mass m at a distance of r from the center of the Earth is pulled toward the center of the Earth only by the mass in the shaded portion of Figure P13.63. Write down Newton's law of gravitation for an object at distance r from the center of

the Earth, and show that the force on it is of Hooke's law form, $F = -kr$, with an effective force constant of $k = (4/3)\pi\rho Gm$, where ρ is the density of the Earth and G is the gravitational constant.

Earth

Figure P13.63

64. An 8.00-kg block travels on a rough horizontal surface and collides with a spring. The speed of the block just before the collision is 4.00 m/s. As it rebounds to the left with the spring uncompressed, the block travels at 3.00 m/s. If the coefficient of kinetic friction between the block and the surface is 0.400, determine (a) the loss in mechanical energy due to friction while the block is in contact with the spring and (b) the maximum distance the spring is compressed.

Sound

▲ **PHYSICS PUZZLER**

You may be familiar with a technique to determine how far away a lightning strike is. You count the seconds between the time when you see the lightning and when you hear the thunder. You divide that time interval by five, which gives you the distance to the lightning in miles. Why does this calculation give you this distance? *(Keith Kent/Peter Arnold, Inc.)*

S ound waves are the most important example of longitudinal waves. In this chapter we discuss the characteristics of sound waves—how they are produced, what they are, and how they travel through matter. We then investigate what happens when sound waves interfere with each other. The insights gained in this chapter will help you understand how we hear.

14.1 PRODUCING A SOUND WAVE

Whether it conveys the shrill whine of a jet engine or the soft melodies of a pop singer, any sound wave has its source in a vibrating object. Musical instruments produce sounds in a variety of ways. For example, the sound from a clarinet is produced by a vibrating reed, the sound from a drum by the vibration of the taut drum head, the sound from a piano by vibrating strings, and the sound from a singer by vibrating vocal folds.

Sound waves are longitudinal waves traveling through a medium, such as air. In order to investigate how sound waves are produced, we focus our attention on the tuning fork, a common device for producing pure musical notes. A tuning fork consists of two metal prongs, or tines, that vibrate when struck. Their vibration disturbs the air near them, as shown in Figure 14.1. (The amplitude of vibration of the tine in Figure 14.1 has been greatly exaggerated for clarity.) When a tine swings to the right, as in Figure 14.1a, the air molecules in front of its movement are forced closer together than normal. Such a region of high molecular density and high air pressure is called a **compression** or **condensation.** This compression moves away from the fork like a ripple on a pond. When the tine swings to the left, as in Figure 14.1b, the molecules to the right of the tine spread apart and the density and air pressure in this region are then lower than normal. Such a region of lower-than-normal density is called a **rarefaction.** Molecules to the right of the rarefaction in the figure move to the left. Hence, the rarefaction itself moves to the right, following the previously produced compression.

As the tuning fork continues to vibrate, a succession of condensations and rarefactions forms and spreads out from it. The resultant pattern in the air is somewhat like that pictured in Figure 14.2a. We can use a sinusoidal curve to represent a sound wave, as in Figure 14.2b. Notice that there are crests in the sinusoidal wave at the points where the sound wave has condensations and troughs where the sound

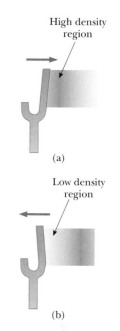

High density region

(a)

Low density region

(b)

(c)

Figure 14.1 A vibrating tuning fork. (a) As the right tine of the fork moves to the right, a high-density region (condensation) of air is formed in front of its movement. (b) As the right tine moves to the left, a low-density region (rarefaction) of air is formed behind it. (c) A set of tuning forks. *(Courtesy of Henry Leap and Jim Lehman)*

(a)

(b)

Figure 14.2 (a) As the tuning fork vibrates, a series of condensations and rarefactions moves outward, away from the fork. (b) The crests of the wave correspond to condensations, and the troughs correspond to rarefactions.

wave has rarefactions. The molecular motion of the sound waves is superposed on the random thermal motion of the atoms and molecules (discussed in Chapter 10).

14.2   CHARACTERISTICS OF SOUND WAVES

As already noted, the general motion of air molecules near a vibrating object is back and forth between regions of compression and rarefaction. Back-and-forth molecular motion in the direction of the disturbance is characteristic of a longitudinal wave. **The motion of the medium particles in a longitudinal sound wave is back and forth along the direction in which the wave travels.** In contrast, **in a transverse wave the vibrations of the medium are at right angles to the direction of travel of the wave.**

Categories of Sound Waves

Sound waves fall into three categories covering different ranges of frequencies. **Audible waves** are longitudinal waves that lie within the range of sensitivity of the human ear, approximately 20 Hz to 20 000 Hz. **Infrasonic waves** are longitudinal waves with frequencies below the audible range. Earthquake waves are an example. **Ultrasonic waves** are longitudinal waves with frequencies above the audible range for humans. For example, certain types of whistles produce ultrasonic waves. Some animals, such as dogs, can hear the waves emitted by these whistles, even though humans cannot.

Applications of Ultrasound

APPLICATION

Medical Use of Ultrasound.

Direction of vibration

Electrical connections

Crystal

Figure 14.3 An alternating voltage applied to the faces of a piezoelectric crystal causes the crystal to vibrate.

Ultrasonic waves are sound waves with frequencies from 20 kHz to 100 kHz, frequencies that are beyond the audible range. Because of their high frequency and corresponding short wavelengths, ultrasonic waves can be used to produce images of small objects and are currently in wide use in medical applications, both as a diagnostic tool and in certain treatments. Internal organs can be examined via the images produced by the reflection and absorption of ultrasonic waves. Although ultrasonic waves are far safer than x-rays, their images do not always have as much detail. Certain organs, however, such as the liver and the spleen, are invisible to x-rays but can be imaged with ultrasonic waves.

Medical workers can measure the speed of the blood flow in the body using a device called an *ultrasonic flow meter*, which makes use of the Doppler effect. The Doppler effect will be discussed in Section 14.6. By comparing the frequency of the waves scattered by the blood vessels with the incident frequency, one can obtain the flow speed.

Figure 14.3 illustrates the technique used to produce ultrasonic waves for clinical use. Electrical contacts are made to the opposite faces of a crystal, such as quartz or strontium titanate. If an alternating voltage of very high frequency is applied to these contacts, the crystal vibrates at the same frequency as the applied voltage, emitting a beam of ultrasonic waves. (At one time, this was how almost all headphones produced sound.) This method of transforming electrical energy into mechanical energy, called the **piezoelectric effect,** is also reversible: If some external source causes the crystal to vibrate, an alternating voltage is produced across the crystal. Hence, a single crystal can be used to both transmit and receive ultrasonic waves.

The primary physical principle that makes ultrasound imaging possible is the fact that a sound wave is partially reflected whenever it is incident on a boundary between two materials having different densities. If a sound wave is traveling in a material of density ρ_i, and strikes a material of density ρ_t, the percentage of the incident sound wave reflected, PR, is given by

$$PR = \left(\frac{\rho_i - \rho_t}{\rho_i + \rho_t}\right)^2 \times 100$$

This equation assumes that the incident sound wave travels perpendicularly to the boundary and that the speed of sound is approximately the same in the two materials. The latter assumption holds very well for the human body because the speed of sound does not vary much in the organs of the body.

Physicians commonly use ultrasonic waves to observe fetuses. This technique presents far less risk than x-rays, which can produce birth defects. First the abdomen of the mother is coated with a liquid, such as mineral oil. If this were not done, most of the incident ultrasonic waves from the piezoelectric source would be reflected at the boundary between the air and the mother's skin. Mineral oil has a density similar to that of skin, and a very small fraction of the incident ultrasonic wave is reflected when $\rho_i \approx \rho_t$. The ultrasound energy is emitted as pulses rather than as a continuous wave, so the same crystal can be used as a detector as well as a transmitter. The source-receiver is then passed over the mother's abdomen. The reflected sound waves picked up by the receiver are converted to an electric signal, which forms an image along a line on a fluorescent screen. The sound source is then moved a few centimeters on the mother's body, and the process is repeated. The reflected signal produces a second line on the fluorescent screen. In this fashion a complete scan of the fetus can be made. Difficulties such as the likelihood of spontaneous abortion or of breech birth are easily detected with this technique. Also, such fetal abnormalities as spina bifida and water on the brain are readily observable.

A relatively new medical application of ultrasonics is the cavitron ultrasonic surgical aspirator (CUSA). This device has made it possible to surgically remove brain tumors that were previously inoperable. It is a long needle that emits ultrasonic waves (about 23 kHz) at its tip. When the tip touches a tumor, the part of the tumor near the needle is shattered and the residue can be sucked up (aspirated) through the hollow needle.

Another interesting application of ultrasound is the ultrasonic ranging unit designed by the Polaroid Corporation and used in some of their cameras to provide an almost instantaneous measurement of the distance between the camera and object to be photographed. The principal component of this device is a crystal that acts as both a loudspeaker and a microphone. A pulse of ultrasonic waves is transmitted from the transducer to the object to be photographed. The object reflects part of the signal, producing an echo that is detected by the device. The time interval between the outgoing pulse and the detected echo is then electronically converted to a distance value, because the speed of sound is a known quantity.

An ultrasound image of a human fetus in the womb after 20 weeks of development, showing the head, body, arms, and legs in profile. (*U.H.B. Trust/Tony Stone Images*)

APPLICATION

Ultrasonic Range Unit for Cameras.

14.3   THE SPEED OF SOUND

The speed of a sound wave in a liquid or gas depends on the medium's compressibility and inertia. If the fluid has a bulk modulus of B and an equilibrium density of ρ, the speed of sound is

▶ Speed of sound in a liquid

$$v = \sqrt{\frac{B}{\rho}} \qquad [14.1]$$

Recall from Chapter 9 that bulk modulus is defined as the ratio of the change in pressure, ΔP, to the resulting fractional change in volume, $\Delta V/V$:

$$B \equiv -\frac{\Delta P}{\Delta V/V} \qquad [14.2]$$

Note that B is always positive because an increase in pressure (positive ΔP) results in a decrease in volume. Hence the ratio $\Delta P/\Delta V$ is always negative.

It is interesting to compare Equation 14.1 with Equation 13.17 for the speed of transverse waves on a string, $v = \sqrt{F/\mu}$, discussed in Chapter 13. In both cases, the wave speed depends on an elastic property of the medium (B or F) and on an inertial property of the medium (ρ or μ). In fact, the speed of all mechanical waves follows an expression of the general form

$$v = \sqrt{\frac{\text{elastic property}}{\text{inertial property}}}$$

Another example of this general form is the **speed of a longitudinal wave in a solid rod,** which is

$$v = \sqrt{\frac{Y}{\rho}} \qquad [14.3]$$

where Y is the Young's modulus of the solid, defined as the longitudinal stress divided by the longitudinal strain (Equation 9.3), and ρ is the density of the solid.

Table 14.1 lists the speeds of sound in various media. As you can see, the speed of sound is much higher in solids than in gases. This makes sense because the molecules in a solid are closer together than those in a gas and hence respond more rapidly to a disturbance. In general, sound travels more slowly in liquids than in solids because liquids are more compressible and hence have smaller bulk moduli. To show that sound travels faster in a solid than in air, suppose that a student places her ear against a long metal object, such as a long steel rail, and listens as her distant boyfriend strikes the rail with a hammer. The student will hear two sounds. The first is produced by the rapidly moving sound traveling through the rail, and the second is the sound taking the slower path through the air.

The speed of sound also depends on the temperature of the medium. For sound traveling through air, the relationship between the speed of sound and temperature is

$$v = (331 \text{ m/s})\sqrt{1 + \frac{T}{273}} \qquad [14.4]$$

TABLE 14.1
Speeds of Sound in Various Media

Medium	v (m/s)
Gases	
Air (0°C)	331
Air (100°C)	386
Hydrogen (0°C)	1290
Oxygen (0°C)	317
Helium (0°C)	972
Liquids at 25°C	
Water	1490
Methyl alcohol	1140
Seawater	1530
Solids	
Aluminum	5100
Copper	3560
Iron	5130
Lead	1320
Vulcanized rubber	54

where 331 m/s is the speed of sound in air at 0°C and T is the temperature in degrees Celsius. Using this equation, one finds that at 20°C the speed of sound in air is approximately 343 m/s.

Thinking Physics 1

Why does thunder produce an extended "rolling" sound? And how does lightning produce thunder in the first place?

Explanation Let us assume that we are at ground level and neglect ground reflections. When lightning strikes, a channel of ionized air carries a very large electric current from a cloud to the ground. This results in a rapid temperature increase of this channel of air as the current moves through it. The temperature increase causes a sudden expansion of the air. This expansion is so sudden and so intense that a tremendous disturbance is produced in the air—thunder. The thunder rolls due to the fact that the lightning channel is a long extended source—the entire length of the channel produces the sound at essentially the same instant of time. Sound produced at the bottom of the channel reaches you first, if you are on the ground, because that is the point closest to you, and then sounds from progressively higher portions of the channel reach you. If the lightning channel were a perfectly straight line, the resulting sound might be a steady roar, but the zig-zagged shape of the path results in the rolling variation in loudness, with sound from some portions of the channel arriving at your location simultaneously. The result is a particularly loud sound, interspersed with other instants when the net sound reaching you is low in intensity.

EXAMPLE 14.1   Speed of Sound in a Liquid

Find the speed of sound in water, which has a bulk modulus of about 2.1×10^9 Pa and a density of about 1.0×10^3 kg/m³.

Solution   From Equation 14.1, we find that

$$v_{water} = \sqrt{\frac{B}{\rho}} = \sqrt{\frac{2.1 \times 10^9 \text{ Pa}}{1.0 \times 10^3 \text{ kg/m}^3}} \approx 1500 \text{ m/s}$$

EXAMPLE 14.2   Sound Waves in a Solid Bar

If a solid bar is struck at one end with a hammer, a longitudinal pulse propagates down the bar. Find the speed of sound in a bar of aluminum, which has a Young's modulus of 7.0×10^{10} Pa and a density of 2.7×10^3 kg/m³.

Solution   From Equation 14.3 we find that

$$v_{Al} = \sqrt{\frac{Y}{\rho}} = \sqrt{\frac{7.0 \times 10^{10} \text{ Pa}}{2.7 \times 10^3 \text{ kg/m}^3}} \approx 5100 \text{ m/s}$$

This is a typical value for the speed of sound in solids (see Table 14.1).

14.4   ENERGY AND INTENSITY OF SOUND WAVES

As the tines of a tuning fork move back and forth through the air, they exert a force on a layer of air and cause it to move. In other words, the tines do work on the layer of air. The fact that the fork pours energy into the air as sound is one of the reasons that the vibration of the fork slowly dies out. (Other factors, such as the energy lost to friction as the tines bend, also are responsible for the diminution of movement.)

> We define the **intensity, *I*,** of a wave to be the rate at which energy flows through a unit area, *A*, perpendicularly to the direction of travel of the wave.

In equation form this is

$$I \equiv \frac{1}{A}\frac{\Delta E}{\Delta t} \qquad [14.5]$$

Equation 14.5 can be written in an alternative form if you recall that the rate of transfer of energy is defined as power. Thus,

Intensity of a wave ▶

$$I \equiv \frac{\text{power}}{\text{area}} = \frac{P}{A} \qquad [14.6]$$

where *P* is the sound power passing through *A*, measured in watts, and the intensity has units of watts per square meter.

The faintest sounds the human ear can detect at a frequency of 1000 Hz have an intensity of about 1×10^{-12} W/m^2. This intensity is called the **threshold of hearing.** The loudest sounds the ear can tolerate have an intensity of about 1 W/m^2 (the **threshold of pain**). At the threshold of hearing, the increase in pressure in the ear is approximately 3×10^{-5} Pa over normal atmospheric pressure. Because atmospheric pressure is about 1×10^5 Pa, this means the ear can detect pressure fluctuations as small as about 3 parts in 10^{10}! Also, at the threshold of hearing, the maximum displacement of an air molecule is about 1×10^{-11} m. This is a remarkably small number! If we compare this result with the diameter of a molecule (about 10^{-10} m), we see that the ear is an extremely sensitive detector of sound waves.

In a similar manner, one finds that the loudest sounds the human ear can tolerate correspond, at 1000 Hz, to a pressure variation of about 29 Pa away from normal atmospheric pressure, with a maximum displacement of air molecules of 1×10^{-5} m.

Intensity Levels in Decibels

As was just mentioned, the human ear can detect a wide range of intensities, with the loudest tolerable sounds having intensities about 1.0×10^{12} times greater than those of the faintest detectable sounds. However, the most intense sound is not perceived as being 1.0×10^{12} times louder than the faintest sound. This is because the sensation of loudness is approximately logarithmic in the human ear. The relative intensity of a sound is called the **intensity level** or **decibel level, *β*,** and is defined as

$$\beta \equiv 10 \log \left(\frac{I}{I_0} \right)$$ [14.7] ◀ Intensity level

The constant I_0 is the reference intensity, taken to be the sound intensity at the threshold of normal acute hearing ($I_0 = 1.0 \times 10^{-12}$ W/m²), I is any intensity, and β is the corresponding intensity level, measured in decibels (dB). (The word *decibel* comes from the name of the inventor of the telephone, Alexander Graham Bell, 1847–1922.) On this scale, the threshold of pain ($I = 1.0$ W/m²) corresponds to an intensity level of $\beta = 10 \log(1/1 \times 10^{-12}) = 10 \log(10^{12}) = 120$ dB. Nearby jet airplanes can create intensity levels of 150 dB, and subways and riveting machines have levels of 90 to 100 dB. The electronically amplified sound heard at rock concerts can be at levels of up to 120 dB, the threshold of pain. Exposure to such high intensity levels can seriously damage the ear. Earplugs are recommended whenever prolonged intensity levels exceed 90 dB. Recent evidence suggests that noise pollution, which is common in most large cities and in some industrial environments, may be a contributing factor to high blood pressure, anxiety, and nervousness. Table 14.2 gives some idea of the intensity levels of various sounds.

TABLE 14.2
Intensity Levels in Decibels for Different Sources

Source of Sound	β (dB)
Nearby jet airplane	150
Jackhammer, machine gun	130
Siren, rock concert	120
Subway, power mower	100
Busy traffic	80
Vacuum cleaner	70
Normal conversation	50
Mosquito buzzing	40
Whisper	30
Rustling leaves	10
Threshold of hearing	0

EXAMPLE 14.3 Intensity Levels of Sound

Calculate the intensity level of a sound wave having an intensity of (a) 1.0×10^{-12} W/m²; (b) 1.0×10^{-11} W/m²; (c) 1.0×10^{-10} W/m².

Solution (a) For an intensity of 1.0×10^{-12} W/m² the intensity level, in decibels, is

$$\beta = 10 \log \left(\frac{1.0 \times 10^{-12} \text{ W/m}^2}{1.0 \times 10^{-12} \text{ W/m}^2} \right) = 10 \log(1) = 0 \text{ dB}$$

This answer should have been obvious without calculation, because an intensity of 1.0×10^{-12} W/m² corresponds to the threshold of hearing.

(b) In this case, the intensity is exactly ten times that in part (a). The intensity level is

$$\beta = 10 \log \left(\frac{1.0 \times 10^{-11} \text{ W/m}^2}{1.0 \times 10^{-12} \text{ W/m}^2} \right) = 10 \log(10) = 10 \text{ dB}$$

(c) Here the intensity is 100 times greater than at the threshold of hearing, and the intensity level is

$$\beta = 10 \log \left(\frac{1.0 \times 10^{-10} \text{ W/m}^2}{1.0 \times 10^{-12} \text{ W/m}^2} \right) = 10 \log(100) = 20 \text{ dB}$$

Note the pattern in these answers. A sound with an intensity level of 10 dB is ten times more intense than the 0-dB sound, and a sound with an intensity level of 20 dB is 100 times more intense than a 0-dB sound. This pattern is continued throughout the decibel scale. In short, on the decibel scale **an increase of 10 dB means that the intensity of the sound is multiplied by a factor of 10.** For example, a 50-dB sound is 10 times as intense as a 40-dB sound and a 60-dB sound is 100 times as intense as a 40-dB sound.

Exercise Determine the intensity level of a sound wave with an intensity of 5.0×10^{-7} W/m².

Answer 57 dB

EXAMPLE 14.4 A Noisy Grinding Machine

A noisy grinding machine in a factory produces a sound intensity of 1.0×10^{-5} W/m^2. Find the decibel level of this machine, and calculate the new intensity level when a second, identical machine is added to the factory.

Solution   The intensity level of the single grinder is

$$\beta = 10 \log \left(\frac{1.0 \times 10^{-5} \text{ W/m}^2}{1.0 \times 10^{-12} \text{ W/m}^2} \right) = 10 \log(10^7) = \boxed{70 \text{ dB}}$$

Adding the second grinder doubles the energy input into sound and hence doubles the intensity. The new intensity level is

$$\beta = 10 \log \left(\frac{2.0 \times 10^{-5} \text{ W/m}^2}{1.0 \times 10^{-12} \text{ W/m}^2} \right) = \boxed{73 \text{ dB}}$$

APPLICATION

OSHA Noise Level Regulations.

Federal regulations now demand that no office or factory worker be exposed to noise levels that average more than 90 dB over an 8-hour day. The results in this example read like one of the old jokes that start "There is some good news and some bad news." First the good news. Imagine that you are a manager analyzing the noise conditions in your factory. One machine in the factory produces a noise level of 70 dB. When you add a second machine, the noise level increases by only 3 dB. Because of the logarithmic nature of intensity levels, doubling the intensity does not double the intensity level; in fact, it alters it by a surprisingly small amount. This means that additional equipment can be added to the factory without appreciably altering the intensity level of the environment.

Now the bad news. The results also work in reverse. As you remove noisy machinery, the intensity level is not lowered appreciably. For example, consider a factory with 60 machines producing a noise level of 93 dB, which is 3 dB above the maximum allowed. In order to reduce the noise level by 3 dB, half the machines would have to be removed! That is, you would have to remove 30 machines to reduce the noise level to 90 dB. To reduce the level another 3 dB, you would have to remove half of the remaining machines, and so on.

14.5   SPHERICAL AND PLANE WAVES

If a small spherical object oscillates so that its radius changes periodically with time, a spherical sound wave is produced (Fig. 14.4). The wave moves outward from the source at a constant speed.

Because all points on the vibrating sphere behave in the same way, we conclude that the energy in a spherical wave propagates equally in all directions. That is, no one direction is preferred over any other. If P_{av} is the average power emitted by the source, then at any distance r from the source, this power must be distributed over a spherical surface of area $4\pi r^2$, assuming no absorption in the medium. (Recall that $4\pi r^2$ is the surface area of a sphere.) Hence, the **intensity** of the sound at a distance of r from the source is

$$I = \frac{\text{average power}}{\text{area}} = \frac{P_{av}}{A} = \frac{P_{av}}{4\pi r^2} \qquad [14.8]$$

Spherical wave front

r_1

r_2

Figure 14.4 A spherical wave propagating radially outward from an oscillating sphere. The intensity of the spherical wave varies as $1/r^2$.

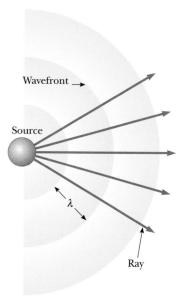

Wavefront →

Source

λ

Ray

Figure 14.5 Spherical waves emitted by a point source. The circular arcs represent the spherical wavefronts concentric with the source. The rays are radial lines pointing outward from the source, perpendicular to the wavefronts.

This shows that the intensity of a wave decreases with increasing distance from its source, as you might expect. The fact that I varies as $1/r^2$ is a result of the assumption that the small source (sometimes called a **point source**) emits a spherical wave. (In fact, light waves also obey this so-called inverse square relationship.) Because the average power is the same through any spherical surface centered at the source, we see that the intensities at distances r_1 and r_2 (Fig. 14.4) from the center of the source are

$$I_1 = \frac{P_{av}}{4\pi r_1{}^2} \qquad I_2 = \frac{P_{av}}{4\pi r_2{}^2}$$

Therefore, the ratio of intensities at these two spherical surfaces is

$$\frac{I_1}{I_2} = \frac{r_2{}^2}{r_1{}^2}$$

It is useful to represent spherical waves graphically with a series of circular arcs (lines of maximum intensity) concentric with the source representing part of a spherical surface, as in Figure 14.5. We call such an arc a **wavefront**. The distance between adjacent wavefronts equals the wavelength, λ. The radial lines pointing outward from the source and cutting the arcs perpendicularly are called **rays.**

Now consider a small portion of a wavefront that is at a *great* distance (great relative to λ) from the source, as in Figure 14.6. In this case, the rays are nearly parallel to each other and the wavefronts are very close to being planes. Therefore, at distances from the source that are great relative to the wavelength, we can approximate the wavefront with parallel planes. We call such waves plane waves. Any small portion of a spherical wave that is far from the source can be considered a **plane wave.** Figure 14.7 illustrates a plane wave propagating along the x axis. If x is taken to be the direction of the wave motion (or ray) in this figure, then the wavefronts are parallel to the plane containing the y and z axes.

Rays

Wavefronts

Figure 14.6 Far away from a point source, the wavefronts are nearly parallel planes and the rays are nearly parallel lines perpendicular to the planes. Hence, a small segment of a spherical wavefront is approximately a plane wave.

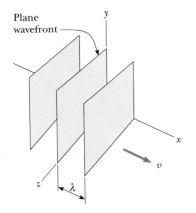

Figure 14.7 A representation of a plane wave moving in the positive x direction with a speed of v. The wavefronts are planes parallel to the yz plane.

EXAMPLE 14.5   Intensity Variations of a Point Source

A small source emits sound waves with a power output of 80 W.
(a) Find the intensity 3.0 m from the source.

Solution   A small source emits energy in the form of spherical waves (see Fig. 14.4). Let P_{av} be the average power output of the source. At a distance of r from the source, the power is distributed over the surface area of a sphere, $4\pi r^2$. Therefore, the intensity at distance r from the source is given by Equation 14.8. Because $P_{av} = 80$ W and $r = 3.0$ m, we find that

$$I = \frac{P_{av}}{4\pi r^2} = \frac{80 \text{ W}}{4\pi(3.0 \text{ m})^2} = \boxed{0.71 \text{ W/m}^2}$$

which is close to the threshold of pain.

(b) Find the distance at which the sound level is 40 dB.

Solution   We can find the intensity at the 40-dB intensity level by using Equation 14.7 with $I_0 = 1.0 \times 10^{-12}$ W/m^2:

$$40 = 10 \log(I/I_0)$$
$$4 = \log(I/I_0)$$
$$10^4 = I/I_0$$
$$I = 10^4 I_0 = 1.0 \times 10^{-8} \text{ W/m}^2$$

When this value for I is used in Equation 14.8, solving for r gives

$$r = \left(\frac{P_{av}}{4\pi I}\right)^{1/2} = \left(\frac{80 \text{ W}}{4\pi \times 10^{-8} \text{ W/m}^2}\right)^{1/2} = \boxed{2.5 \times 10^4 \text{ m}}$$

which is approximately 15 mi!

14.6   THE DOPPLER EFFECT

If a car or truck is moving while its horn is blowing, the frequency of the sound you hear is higher as the vehicle approaches you and lower as it moves away from you. This is one example of the Doppler effect, named for the Austrian physicist Christian Doppler (1803–1853), who discovered it.

In general, a **Doppler effect** is experienced whenever there is relative motion between a source of waves and an observer. When the source and observer are moving toward each other, the observer hears a frequency higher than the frequency of the source in the absence of relative motion. When the source and observer are moving away from each other, the observer hears a frequency lower than the source frequency.

Although the Doppler effect is most commonly experienced with sound waves, it is a phenomenon common to all waves. For example, the frequencies of light waves are also shifted by the relative motion of source and observer.

First let us consider the case in which the observer is moving and the sound source is stationary. For simplicity, we assume that the air is also stationary and that all velocity measurements are made relative to this stationary medium. Once a sound leaves its source, the speed of the wave is independent of the speed of the source. Figure 14.8 describes the situation in which the observer is moving with a speed of v_0 toward the source (considered a point source), which is at rest ($v_s = 0$).

We shall take the frequency of the source to be f, the wavelength to be λ, and the speed of sound in air to be v. Clearly, if both observer and source were stationary, the observer would detect f wavefronts per second. (That is, when $v_0 = 0$ and $v_s = 0$, the observed frequency equals the source frequency.) When the observer is moving toward the source, he or she moves a distance of $v_0 t$ in t seconds. During this interval, **the observer detects an additional number of wavefronts.** The number of extra wavefronts detected is equal to the distance traveled, $v_0 t$, divided by the wavelength, λ. Thus,

$$\text{Additional wavefronts detected} = \frac{v_0 t}{\lambda}$$

The number of additional wavefronts detected *per second* is v_0/λ. Hence, the frequency f' heard by the observer is *increased* to

$$f' = f + \frac{v_0}{\lambda}$$

Using the fact that $\lambda = v/f$, we see that $v_0/\lambda = (v_0/v)f$. Hence, f' can be expressed as

$$f' = f\left(\frac{v + v_0}{v}\right) \qquad [14.9]$$

An observer traveling *away* from the source, as in Figure 14.9, **detects fewer wavefronts per second.** Thus, it follows that the frequency heard by the observer in this case is *lowered* to

$$f' = f\left(\frac{v - v_0}{v}\right) \qquad [14.10]$$

We can incorporate these two equations into one:

$$f' = f\left(\frac{v \pm v_0}{v}\right) \qquad [14.11]$$

This general equation applies when an observer is moving with a speed of v_0 relative to a stationary source. **The positive sign is used when the observer is moving toward the source, and the negative sign is used when the observer is moving away from the source.**

Now consider the situation in which the source is in motion and the observer is at rest. If the source is moving directly toward observer A in Figure 14.10a, the

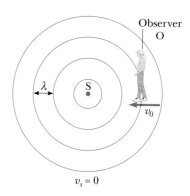

Figure 14.8 An observer moving with a speed of v_0 *toward* a stationary point source (S) hears a frequency, f', that is *greater* than the source frequency, f.

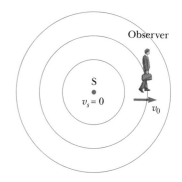

Figure 14.9 An observer moving with a speed of v_0 *away* from a stationary source hears a frequency, f', that is *lower* than the source frequency.

◀ Observed frequency – observer in motion

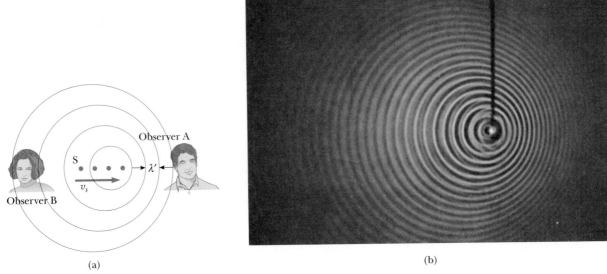

Figure 14.10   (a) A source, S, moving with speed v_s toward stationary observer A and away from stationary observer B. Observer A hears an *increased* frequency, and observer B hears a *decreased* frequency. (b) The Doppler effect in water, observed in a ripple tank. The source producing the water waves is moving to the right. *(Courtesy Educational Development Center, Newton, Mass.)*

wavefronts heard by A are closer together because the source is moving in the direction of the outgoing wave. As a result, the wavelength λ' measured by observer A is shorter than the true wavelength, λ, of the source. During each vibration, which lasts for an interval of T (the period), the source moves a distance of $v_sT = v_s/f$ and **the wavelength is shortened by this amount.** Therefore, the observed wavelength, λ', is given by

$$\lambda' = \lambda - \frac{v_s}{f}$$

Because $\lambda = v/f$, the frequency heard by observer A is

$$f' = \frac{v}{\lambda'} = \frac{v}{\lambda - \dfrac{v_s}{f}} = \frac{v}{\dfrac{v}{f} - \dfrac{v_s}{f}} = f\left(\frac{v}{v - v_s}\right) \qquad [14.12]$$

That is, **the observed frequency increases when the source is moving toward the observer.**

In Figure 14.10a the source is moving away from observer B, who is at rest to the left of the source. Thus, observer B measures a wavelength that is *greater* than λ and hears a *decreased* frequency of

$$f' = f\left(\frac{v}{v + v_s}\right) \qquad [14.13]$$

Combining Equations 14.12 and 14.13, we can express the general relationship for the observed frequency when the source is moving and the observer is at rest:

$$f' = f\left(\frac{v}{v \mp v_s}\right) \qquad [14.14]$$

◀ Observed frequency – source in motion

The *negative* sign is used when the source is moving *toward* the observer, and the *positive* sign is used when the source is moving *away from* the observer.

Finally, if both the source and the observer are in motion, one finds the following general relationship for the observed frequency:

$$f' = f\left(\frac{v \pm v_0}{v \mp v_s}\right) \qquad [14.15]$$

◀ Observed frequency – observer and source in motion

In this expression the *upper* signs ($+ v_0$ and $- v_s$) refer to motion of one *toward* the other, and the lower signs ($- v_0$ and $+ v_s$) refer to motion of one *away* from the other.

Thinking Physics 2

Suppose you place your stereo speakers far apart and run past them from right to left or left to right. If you run rapidly enough and have excellent pitch discrimination, you may notice that the music that is playing seems to be out of tune when you are between the speakers. Why?

Explanation   When you are between the speakers, you are running away from one of them and toward the other. Thus, there is a Doppler shift downward for the sound from the speaker behind you and a Doppler shift upward for the sound from the speaker ahead of you. As a result, the sound from the two speakers will not be in tune. A calculation shows that a world-class sprint runner could run fast enough to generate about a semitone difference in the sound from the two speakers.

EXAMPLE 14.6   Listen, But Don't Stand on the Track

A train moving at a speed of 40.0 m/s sounds its whistle, which has a frequency of 500 Hz. Determine the frequency heard by a stationary observer as the train approaches the observer.

Solution   We can use Equation 14.12 to get the apparent frequency as the train approaches the observer. Taking $v = 345$ m/s as the speed of sound in air, we have

$$f' = f\left(\frac{v}{v - v_s}\right) = (500 \text{ Hz})\left(\frac{345 \text{ m/s}}{345 \text{ m/s} - 40.0 \text{ m/s}}\right) = 566 \text{ Hz}$$

Exercise   Determine the frequency heard by the stationary observer as the train recedes from the observer.

Answer   448 Hz

EXAMPLE 14.7   The Noisy Siren

An ambulance travels down a highway at a speed of 75.0 mi/h, its siren emitting sound at a frequency of 400 Hz. What frequency is heard by a passenger in a car traveling at 55.0 mi/h in the opposite direction as the car (a) approaches? (b) moves away from the ambulance?

Solution   Let us take the velocity of sound in air to be $v = 345$ m/s and use the conversion 1.00 mi/h $= 0.447$ m/s. Therefore, $v_s = 75.0$ mi/h $= 33.5$ m/s and $v_0 = 55.0$ mi/h $= 24.6$ m/s. We can use Equation 14.15 in both cases.
(a) As the ambulance and car approach each other, the observed frequency is

$$f' = f\left(\frac{v + v_0}{v - v_s}\right) = (400 \text{ Hz})\left(\frac{345 \text{ m/s} + 24.6 \text{ m/s}}{345 \text{ m/s} - 33.5 \text{ m/s}}\right) = \boxed{475 \text{ Hz}}$$

(b) As the two vehicles recede from each other, the passenger in the car hears a frequency of

$$f' = f\left(\frac{v - v_0}{v + v_s}\right) = (400 \text{ Hz})\left(\frac{345 \text{ m/s} - 24.6 \text{ m/s}}{345 \text{ m/s} + 33.5 \text{ m/s}}\right) = \boxed{339 \text{ Hz}}$$

Shock Waves

Now let us consider what happens when the source speed, v_s, *exceeds* the wave velocity, v. Figure 14.11a describes this situation graphically. The circles represent spherical wavefronts emitted by the source at various times during its motion. At

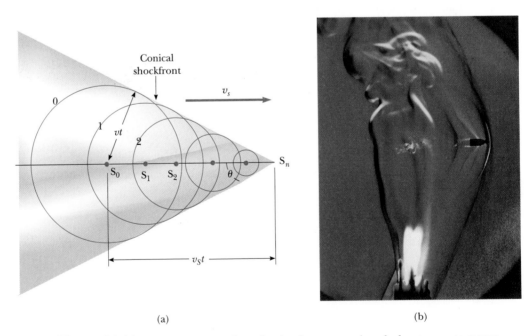

(a)   (b)

Figure 14.11   (a) A representation of a shock wave, produced when a source moves from S_0 to S_n with a speed, v_s, that is *greater* than the wave speed, v, in that medium. The envelope of the wavefronts forms a cone whose half-angle is $\sin\theta = v/v_s$. (b) A strobo-scopic photograph of a bullet moving at supersonic speed through the hot air above a candle. *(Harold Edgerton, Courtesy of Palm Press, Inc.)*

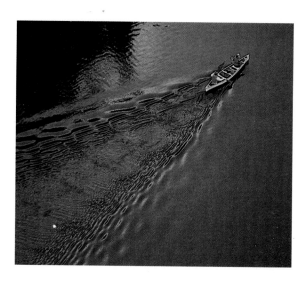

Figure 14.12 The V-shaped bow wave of a boat is formed because the boat travels at a speed greater than the speed of water waves. A bow wave is analogous to a shock wave formed by an airplane traveling faster than sound. *(© 1994, Comstock)*

$t = 0$, the source is at point S_0, and at some later time t, the source is at point S_n. In the interval t, the wavefront centered at S_0 reaches a radius of vt. In this same interval, the source travels to S_n, a distance of $v_s t$. At the instant the source is at S_n, the waves just beginning to be generated at this point have wavefronts of zero radius. The line drawn from S_n to the wavefront centered on S_0 is tangent to all other wavefronts generated at intermediate times. All such tangent lines lie on the surface of a cone. The angle, θ, between one of these tangent lines and the direction of travel is given by

$$\sin \theta = \frac{v}{v_s}$$

The ratio v_s/v is referred to as the **Mach number.** The conical wavefront produced when $v_s > v$ (supersonic speeds) is known as a **shock wave.** Figure 14.11b is a photograph of a bullet traveling at supersonic speed through the hot air rising above a candle. Note the shock waves in the vicinity of the bullet. Another interesting example of shock waves is the V-shaped wavefront produced by a boat (the bow wave) when the boat's speed exceeds the speed of the water waves (Fig. 14.12).

Jet airplanes traveling at supersonic speeds produce shock waves, which are responsible for the loud explosion, or sonic boom, heard on the ground. A shock wave carries a great deal of energy concentrated on the surface of the cone, with correspondingly great pressure variations. Shock waves are unpleasant to hear and can damage buildings when aircraft fly supersonically at low altitudes. In fact, an airplane flying at supersonic speeds produces a double boom because two shock waves are formed, one from the nose of the plane and one from the tail (Fig. 14.13).

Pressure

Atmospheric pressure

Figure 14.13 The two shock waves produced by the nose and tail of a jet airplane traveling at supersonic speed.

APPLICATION

Sonic Booms.

14.7 INTERFERENCE OF SOUND WAVES

Sound waves can be made to interfere with each other. This can be demonstrated with the device shown in Figure 14.14. Sound from a loudspeaker at S is sent into a tube at *P*, where there is a T-shaped junction. The sound splits and follows two

Figure 14.14 An acoustical system for demonstrating interference of sound waves. Sound from the speaker enters the tube and splits into two parts at P. The two waves combine at the opposite side and are detected at R. The upper path length is varied by the sliding section.

separate pathways indicated by the red arrows. Half of the sound travels upward, and half downward. Finally, the two sounds merge at an opening where a listener places her ear. If the two paths r_1 and r_2 are of the same length, a crest of the wave that enters the junction will separate into two halves, travel the two paths, and then combine again at the ear. This reuniting of the two waves produces constructive interference, and thus a loud sound is heard by the listener.

Suppose, however, that one of the path lengths (r_2) is adjusted by sliding the upper U-shaped tube upward so that the upper path is half a wavelength *longer* than the lower path r_1. In this case, an entering sound wave splits and travels the two paths as before, but now the wave along the upper path must travel a distance equivalent to half a wavelength farther than the wave traveling along the lower path. As a result, the crest of one wave meets the trough of the other when they merge at the receiver. Because this is the condition for destructive interference, no sound is detected at the receiver.

You should be able to predict what will be heard if the upper path is adjusted to be one full wavelength longer than the lower path. In this case, constructive interference of the two waves occurs, and a loud sound is detected at the receiver.

Nature provides many other examples of interference phenomena. In a later chapter we shall describe several interesting interference effects involving light waves.

When connecting the wires between your stereo system and your loudspeakers, you may notice that the wires are color-coded and that the speakers have positive and negative signs on the connections. The reason for this is that the speakers need to be connected with the same "polarity." If they are not, then a sound that is fed to both speakers will result in one speaker cone moving outward at the same time that the other speaker cone is moving inward. In this case, the sounds leaving the two speakers will be 180° out of phase with each other. If you are sitting midway between the speakers, the sounds from both speakers travel the same distance and preserve the phase difference they had when they left. In an ideal situation, for a 180° phase difference you would get complete destructive interference and no sound! In reality, the cancellation is not complete and is much more significant for bass notes (which have a long wavelength) than for the shorter wavelength treble notes. Nevertheless, to avoid a significant reduction in the intensity of bass notes, the color-coded wires and the signs on the speaker connections should be noted carefully.

APPLICATION

Connecting Your Stereo Speakers.

EXAMPLE 14.8 Interference from Two Loudspeakers

Two loudspeakers are placed as in Figure 14.15 and driven by the same source at a frequency of 2000 Hz. The top speaker is then moved left to position A. At this location, an observer at a great distance from the speakers and directly in front of them notices that the intensity of the sound from the two sources has decreased to a minimum. What is the minimum distance the top speaker has been moved? Assume that the speed of sound in air is 345 m/s.

Figure 14.15 (Example 14.8) Two loudspeakers driven by the same source can produce interference.

Solution Initially, both speakers are at the same distance from the observer. Hence, the sound from each speaker must travel the same distance, and the observer hears a loud sound corresponding to constructive interference. When the top speaker is at position A, its sound must travel farther to reach the observer. Because the observer notices a minimum in the sound level when the top speaker is at A, destructive interference is taking place between the two separate sound signals. This means that the top speaker has been moved half a wavelength. With the speed of sound in air equal to 345 m/s, we can calculate the wavelength:

$$\lambda = \frac{v}{f} = \frac{345 \text{ m/s}}{2000 \text{ s}^{-1}} = 0.173 \text{ m}$$

Therefore, the distance moved is half this value, or 0.0865 m.

14.8 STANDING WAVES

Standing waves can be set up in a stretched string by connecting one end of the string to a stationary clamp and connecting the other end to a vibrating object, such as the end of a tuning fork, or by shaking your hand up and down at a steady rate (Fig. 14.16). In this situation, traveling waves reflect from the ends, creating waves traveling in both directions on the string. The incident and reflected waves combine according to the superposition principle. If the string is vibrated at exactly the right frequency, the wave appears to stand still—hence its name, **standing wave.** A **node** occurs where the two traveling waves always have the same magnitude of displacement but of opposite sign, so that the net displacement is zero at that point. There is no motion in the string at the nodes but midway between two adjacent nodes, at an **antinode,** the string vibrates with the largest amplitude.

Figure 14.17 shows the oscillation of a standing wave during half of a cycle. Notice that **all points on the string oscillate vertically with the same frequency except for the node, which is stationary.** (The points at which the string is attached to the wall are also nodes, labeled N in Figure 14.17a.) Furthermore, different points have different amplitudes of motion. From the figure observe that the distance between adjacent nodes is one-half the wavelength of the wave:

$$d_{NN} = \tfrac{1}{2}\lambda$$

Vibrating blade

Figure 14.16 Standing waves can be set up in a stretched string by connecting one end of the string to a vibrating blade. When the blade vibrates at one of the natural frequencies of the string, large-amplitude standing waves are created.

N   t = 0   N   N
(a)

(b)
     t = T/8

(c)   t = T/4

(d)   t = 3T/8

(e)   t = T/2

Figure 14.17   A standing wave pattern in a stretched string, shown by snapshots of the string during one half of a cycle.

Consider a string of length L that is fixed at both ends, as in Figure 14.18. The string has a number of natural patterns of vibration, called normal modes. Three of these are pictured in Figures 14.18b, 14.18c, and 14.18d. Each has a characteristic frequency, which we shall now calculate.

First, note that **the ends of the string must be nodes because these points are fixed.** If the string is displaced at its midpoint and released, the vibration shown in Figure 14.18b can be produced, in which case the center of the string is an antinode. For this normal mode, the length of the string equals $\lambda/2$ (the distance between nodes). Thus,

$$L = \frac{\lambda_1}{2} \quad \text{or} \quad \lambda_1 = 2L$$

and the frequency of this vibration is

$$f_1 = \frac{v}{\lambda_1} = \frac{v}{2L} \qquad \text{[14.16]}$$

In Chapter 13 (Equation 13.17), the speed of a wave on a string was given as $v = \sqrt{F/\mu}$, where F is the tension in the string and μ is its mass per unit length. Thus, we can express Equation 14.16 as

$$f_1 = \frac{1}{2L}\sqrt{\frac{F}{\mu}} \qquad \text{[14.17]}$$

This lowest frequency of vibration is called the **fundamental frequency** of the vibrating string.

The next normal mode, of wavelength λ_2 (Fig. 14.18c), occurs when the length of the string equals one wavelength — that is, when $\lambda_2 = L$. Hence,

$$f_2 = \frac{v}{L} = \frac{2v}{2L} = 2f_1 \qquad \text{[14.18]}$$

Note that this frequency is equal to *twice* the fundamental frequency. You should convince yourself that the next highest frequency of vibration, shown in Figure 14.18d, is

Figure 14.18   (a) Standing waves in a stretched string of length L, fixed at both ends. The normal frequencies of vibration form a harmonic series: (b) the fundamental frequency, or first harmonic; (c) the second harmonic; and (d) the third harmonic.

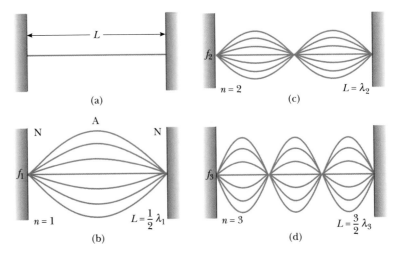

(a)

f_2 n = 2   (c) $L = \lambda_2$

A
N   N
f_1
n = 1   (b) $L = \frac{1}{2}\lambda_1$

f_3 n = 3   (d) $L = \frac{3}{2}\lambda_3$

Multiflash photographs of standing wave patterns in a cord driven by a vibrator at the left end. The single loop pattern at the top left represents the fundamental ($n = 1$), the two-loop pattern at the right represents the second harmonic ($n = 2$), and the three-loop pattern at the lower left represents the third harmonic ($n = 3$). *(Richard Megna, Fundamental Photographs, NYC)*

$$f_3 = \frac{3v}{2L} = 3f_1 \qquad\qquad \textbf{[14.19]}$$

In general, the characteristic frequencies are given by

$$f_n = nf_1 = \frac{n}{2L}\sqrt{\frac{F}{\mu}} \qquad\qquad \textbf{[14.20]}$$

where $n = 1, 2, 3, \ldots$. In other words, the frequencies are integral multiples of the fundamental frequency. The frequencies f_1, $2f_1$, $3f_1$, and so on form a **harmonic series.** The fundamental f_1 corresponds to the **first harmonic;** the frequency $f_2 = 2f_1$, to the **second harmonic;** and so on.

When a stretched string is distorted to a shape that corresponds to any one of its harmonics, after being released it vibrates only at the frequency of that harmonic. If the string is struck or bowed, however, the resulting vibration includes frequencies of various harmonics, including the fundamental. Waves not in the harmonic series are quickly damped out on a string fixed at both ends. In effect, when disturbed, the string "selects" the normal-mode frequencies. As we shall see later, the presence of several harmonics on a string gives stringed instruments their characteristic sound, which enables us to distinguish one from another even when they are producing identical fundamental frequencies.

The frequency of a string on a musical instrument can be changed either by varying the tension or by changing the length. For example, the tension in guitar and violin strings is varied by turning pegs on the neck of the instrument. As the tension is increased, the frequency of the normal modes increases according to Equation 14.20. Once the instrument is tuned, the musician varies the frequency by pressing the strings against the neck at a variety of positions, thereby changing the effective lengths of the vibrating portions of the strings. As the length is reduced, the frequency increases, as Equation 14.20 indicates.

Finally, Equation 14.20 shows that a string of fixed length can be made to vibrate at a lower fundamental frequency by increasing its mass per unit length.

A concert-style harp. *(Lyon & Healy Harps, Chicago)*

APPLICATION

Tuning a Musical Instrument.

This is achieved in the bass strings of guitars and pianos by wrapping them with metal windings.

EXAMPLE 14.9 Harmonics of a Stretched String

Find the first four harmonics of a 1.0-m-long string if the string has a mass per unit length of 2.0×10^{-3} kg/m and is under a tension of 80 N.

Solution The speed of the wave on the string is

$$v = \sqrt{\frac{F}{\mu}} = \sqrt{\frac{80 \text{ N}}{2.0 \times 10^{-3} \text{ kg/m}}} = 200 \text{ m/s}$$

The fundamental frequency can be found using Equation 14.16:

$$f_1 = \frac{v}{2L} = \frac{200 \text{ m/s}}{2(1.0 \text{ m})} = \boxed{100 \text{ Hz}}$$

The frequencies of the next three modes are $f_2 = 2f_1$, $f_3 = 3f_1$, and $f_4 = 4f_1$. Thus,

$$f_2 = 200 \text{ Hz}, f_3 = 300 \text{ Hz, and } f_4 = 400 \text{ Hz.}$$

Exercise Find the tension in the string if the fundamental frequency is increased to 120 Hz.

Answer 120 N

14.9 FORCED VIBRATIONS AND RESONANCE

In Chapter 13 we learned that the energy of a damped oscillator decreases in time because of friction. It is possible to compensate for this energy loss by applying an external force that does positive work on the system.

For example, suppose a mass-spring system having some natural frequency of vibration, f_0, is pushed back and forth with a periodic force whose frequency is f. The system vibrates at the frequency, f, of the driving force. This type of motion is referred to as a **forced vibration.** Its amplitude reaches a maximum when the frequency of the driving force equals the natural frequency of the system, f_0, called the **resonant frequency** of the system. Under this condition, the system is said to be in **resonance.**

In Section 14.8 we learned that a stretched string can vibrate in one or more of its natural modes. Here again, if a periodic force is applied to the string, the amplitude of vibration increases as the frequency of the applied force approaches one of the natural frequencies of vibration.

Resonance vibrations occur in a wide variety of circumstances. Figure 14.19 illustrates one experiment that demonstrates a resonance condition. Several pendulums of different lengths are suspended from a flexible beam. If one of them, such as A, is set in motion, the others begin to oscillate because of vibrations in the flexible beam. Pendulum C, the same length as A, oscillates with the greatest amplitude because its natural frequency matches that of pendulum A (the driving force).

Another simple example of resonance is a child being pushed on a swing, which is essentially a pendulum with a natural frequency that depends on the length. The swing is kept in motion by a series of appropriately timed pushes. For its amplitude

Figure 14.19 Resonance. If pendulum A is set in oscillation, only pendulum C, whose length matches that of A, will eventually oscillate with a large amplitude, or resonate. The arrows indicate motion perpendicular to the page.

Figure 14.20 (*Left*) Standing wave pattern in a vibrating wine glass. The wine glass will shatter if the amplitude of vibration becomes too large. (*Courtesy of Professor Thomas D. Rossing, Northern Illinois University*) (*Right*) A wine glass shattered by the amplified sound of a human voice. (*Ben Rose/The IMAGE Bank*)

to be increased, the swing must be pushed each time it returns to the person's hands. This corresponds to a frequency equal to the natural frequency of the swing. If the energy put into the system per cycle of motion exactly equals the energy lost due to friction, the amplitude remains constant.

Opera singers have been known to set crystal goblets in audible vibration with their powerful voices, as shown in Figure 14.20. This is yet another example of resonance: The sound waves emitted by the singer can set up large-amplitude vibrations in the glass. If a highly amplified sound wave has the right frequency, the amplitude of forced vibrations in the glass increases to the point where the glass becomes heavily strained and shatters.

A striking example of structural resonance occurred in 1940, when the Tacoma Narrows bridge in the state of Washington was set in oscillation by the wind (Fig. 14.21). The amplitude of the oscillations increased steadily until the bridge col-

APPLICATION

Shattering Goblets with the Voice.

APPLICATION

Structural Resonance in Bridges and Buildings.

Figure 14.21 The collapse of the Tacoma Narrows suspension bridge in 1940 was a vivid demonstration of mechanical resonance. High winds set up standing waves in the bridge, causing it to oscillate at one of its natural frequencies. Once established, this resonance condition led to the bridge's collapse. (*United Press International Photo*)

lapsed. A more recent example of destruction by structural resonance occurred during the Loma Prieta earthquake near Oakland, California, in 1989. In one section—almost a mile long—of the double-decker Nimitz Freeway, the upper deck collapsed onto the lower deck, killing several people. The collapse of this particular section of roadway, while other sections escaped serious damage, has been traced to the fact that the earthquake waves had a frequency of approximately 1.5 Hz—very close to a natural resonant frequency of the section of roadway that gave way.

Thinking Physics 3

Passing ocean waves sometimes cause the water in a harbor to undergo very large oscillations, called *seiches*. Why would this happen?

Explanation   Water in a harbor is enclosed and possesses a natural frequency based on the size of the harbor. This is similar to the natural frequency of the enclosed air in a bottle, which can be excited by blowing across the edge of the opening. Ocean waves pass by the opening of the harbor at a certain frequency. If this frequency matches that of the enclosed harbor, then a large standing wave can be set up in the water by resonance. This can be simulated by carrying a fish tank with water. If your walking frequency matches the natural frequency of the water as it sloshes back and forth, a large standing wave in the fish tank can be established.

14.10   STANDING WAVES IN AIR COLUMNS

Standing longitudinal waves can be set up in a tube of air, such as an organ pipe, as the result of interference between sound waves traveling in opposite directions. The relationship between the incident wave and the reflected wave depends on whether the reflecting end of the tube is open or closed. A portion of the sound wave is reflected back into the tube at the open end. **If the reflecting end is closed, a node must exist at this end because the movement of air molecules is restricted. If the end is open, the air molecules have complete freedom of motion, and an antinode exists.**

Figure 14.22a shows the first three modes of vibration of a pipe open at both ends. When air is directed against an edge at the left, longitudinal standing waves are formed and the pipe vibrates at its natural frequencies. Note that for the fundamental frequency the wavelength is twice the length of the pipe and hence $f_1 = v/2L$. Similarly, one finds that the frequencies of the second and third harmonics are $2f_1$, $3f_1$, ... Thus, **in a pipe open at both ends, the natural frequencies of vibration form a series in which all harmonics are present. These harmonics are equal to integral multiples of the fundamental frequency.** We can express this harmonic series as

Pipe open at both ends; all ▷
harmonics present

$$f_n = n \frac{v}{2L} \qquad n = 1, 2, 3, \ldots \qquad [14.21]$$

where v is the speed of sound in air.

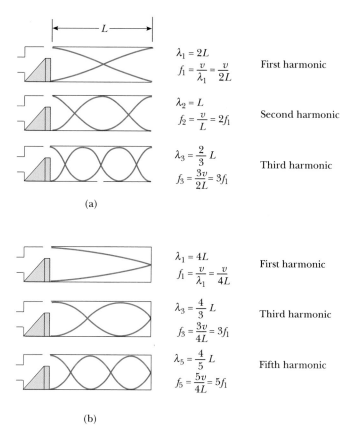

First harmonic
$$\lambda_1 = 2L$$
$$f_1 = \frac{v}{\lambda_1} = \frac{v}{2L}$$

Second harmonic
$$\lambda_2 = L$$
$$f_2 = \frac{v}{L} = 2f_1$$

Third harmonic
$$\lambda_3 = \frac{2}{3}L$$
$$f_3 = \frac{3v}{2L} = 3f_1$$

(a)

First harmonic
$$\lambda_1 = 4L$$
$$f_1 = \frac{v}{\lambda_1} = \frac{v}{4L}$$

Third harmonic
$$\lambda_3 = \frac{4}{3}L$$
$$f_3 = \frac{3v}{4L} = 3f_1$$

Fifth harmonic
$$\lambda_5 = \frac{4}{5}L$$
$$f_5 = \frac{5v}{4L} = 5f_1$$

(b)

Figure 14.22 (a) Standing longitudinal waves in an organ pipe open at both ends. The natural frequencies f_1, $2f_1$, $3f_1$ . . . form a harmonic series. (b) Standing longitudinal waves in an organ pipe closed at one end. Only *odd* harmonics are present, and the natural frequencies are f_1, $3f_1$, $5f_1$, and so on.

If a pipe is closed at one end and open at the other, the closed end is a node (Fig. 14.22b). In this case, the wavelength of the fundamental mode is four times the length of the tube. Hence, $f_1 = v/4L$ and the frequencies of the third and fifth harmonics are $3f_1$, $5f_1$, ... That is, **in a pipe closed at one end and open at the other, only odd harmonics are present.** These are given by

$$f_n = n\frac{v}{4L} \qquad n = 1, 3, 5, \ldots \qquad [14.22]$$

◀ Pipe closed at one end; odd harmonics present

Applying Physics 1

If an orchestra doesn't warm up before a performance, the strings go flat and the wind instruments go sharp during the performance. Why?

Explanation Without warming up, all the instruments will be at room temperature at the beginning of the concert. As the wind instruments are played, they fill with warm air from the player's exhalation. The increase in temperature of the air in the instrument causes an increase in the speed of sound, which raises the resonance frequencies of the air columns. As a result, the instruments go sharp. The strings on the stringed instruments also increase in temperature due to the friction of rubbing with the bow. This results in thermal expansion, which causes a decrease in tension in the strings. With a decrease in tension, the wave speed on the strings drops, and the fundamental frequencies decrease. Thus, the stringed instruments go flat.

Applying Physics 2

A bugle has no valves, keys, slides, or finger holes. How can it play a song?

Explanation   Songs for the bugle are limited to harmonics of the fundamental frequency, because there is no control over frequencies without valves, keys, slides, or finger holes. The player obtains different notes by changing the tension in the lips as the bugle is played, in order to excite different harmonics. The normal playing range of a bugle is among the third, fourth, fifth, and sixth harmonics of the fundamental. For example "Reveille" is played with just the three notes G, C, and F. And "Taps" is played with these three notes and the G one octave above the lower G.

EXAMPLE 14.10   Harmonics of a Pipe

A pipe is 2.46 m long.
(a) Determine the frequencies of the first three harmonics if the pipe is open at both ends. Take 345 m/s as the speed of sound in air.

Solution   The fundamental frequency of a pipe open at both ends can be found from Equation 14.21, with $n = 1$.

$$f_1 = \frac{v}{2L} = \frac{345 \text{ m/s}}{2(2.46 \text{ m})} = \boxed{70.0 \text{ Hz}}$$

Because all harmonics are present in a pipe open at both ends, the second and third harmonics have frequencies of $f_2 = 2f_1 = 140$ Hz and $f_3 = 3f_1 = 210$ Hz.

(b) What are the three lowest possible frequencies if the pipe is closed at one end and open at the other?

Solution   The fundamental frequency of a pipe closed at one end can be found from Equation 14.22, with $n = 1$.

$$f_1 = \frac{v}{4L} = \frac{345 \text{ m/s}}{4(2.46 \text{ m})} = \boxed{35.0 \text{ Hz}}$$

In this case, only odd harmonics are present, and so the third and fifth harmonics have frequencies of $f_3 = 3f_1 = 105$ Hz and $f_5 = 5f_1 = 175$ Hz.

Exercise   If the pipe is open at one end, how many harmonics are possible in the normal hearing range, 20 to 20 000 Hz?

Answer   286

QUICKLAB

Using an empty 1-liter soft drink container, blow over the open end and listen to the sound that is produced. Add some water to the container to change the height of the air column and repeat the experiment at various heights of the column. How does the frequency that you hear change with the height of the air column?

EXAMPLE 14.11 Resonance in a Tube of Variable Length

Figure 14.23 shows a simple apparatus for demonstrating resonance in a tube. A long tube open at both ends is partially submerged in a beaker of water, and a vibrating tuning fork of unknown frequency is placed near the top of the tube. The length of the air column, L, is adjusted by moving the tube vertically. The sound waves generated by the fork are reinforced when the length of the air column corresponds to one of the resonant frequencies of the tube. The smallest value of L for which a peak occurs in the sound intensity is 9.00 cm. For this measurement, determine the frequency of the tuning fork and the value of L for the next two resonant vibrations.

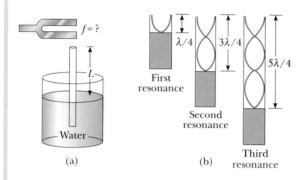

Figure 14.23 (Example 14.11) (a) Apparatus for demonstrating the resonance of sound waves in a tube closed at one end. The length, L, of the air column is varied by moving the tube vertically while it is partially submerged in water. (b) The first three normal frequencies of vibration for the system.

Reasoning and Solution Once the tube is in the water, this setup represents a pipe closed at one end, and the fundamental has a frequency of $v/4L$ (Fig. 14.23b). If we take $v = 345$ m/s for the speed of sound in air, and $L = 0.0900$ m, we get

$$f_1 = \frac{v}{4L} = \frac{345 \text{ m/s}}{4(0.0900 \text{ m})} = \boxed{958 \text{ Hz}}$$

The fundamental wavelength of the pipe is given by $\lambda = 4L = 0.360$ m. Because the frequency of the source is constant, we see that the next resonance positions (Fig. 14.23b) correspond to lengths of $3\lambda/4 = 0.270$ m and $5\lambda/4 = 0.450$ m.

This arrangement is often used to measure the speed of sound, in which case the frequency of the tuning fork and the lengths at which resonance occurs must be known.

14.11 BEATS

The interference phenomena we have been discussing so far have involved the superposition of two or more waves with the same frequency, traveling in opposite directions. Let us now consider another type of interference effect that results from the superposition of two waves with slightly different frequencies. In this situation, the waves at some fixed point are periodically in and out of phase, corresponding to an alternation in time between constructive and destructive interference. In order to understand this phenomenon, consider Figure 14.24. The two waves in Figure 14.24a were emitted by two tuning forks having slightly different frequencies; Figure 14.24b shows the superposition of these two waves. At some time t_a, the two

Figure 14.24 Beats are formed by the combination of two waves of slightly different frequencies traveling in the same direction. (a) The individual waves. (b) The combined wave has an amplitude (dashed line) that oscillates in time.

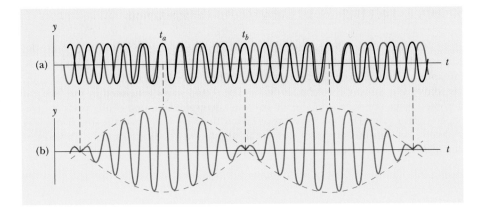

waves are in phase, and constructive interference occurs, as demonstrated by the resultant curve in Figure 14.24b. At some later time, however, the vibrations of the two forks move out of step with each other. At time t_b, one fork emits a compression while the other emits a rarefaction, and destructive interference occurs, as demonstrated by the curve in Figure 14.24b. As time passes, the vibrations of the two forks move out of phase, then into phase, and so on. As a consequence, a listener at some fixed point hears an alternation in loudness, known as **beats.** The number of beats per second, or beat frequency, equals the difference in frequency between the two sources. One can tune a stringed instrument, such as a piano, by beating a note on the instrument against a note of known frequency. The string can then be tuned to the desired frequency by adjusting the tension until no beats are heard.

APPLICATION

Another Way to Tune a Musical Instrument.

EXAMPLE 14.12 Sour Notes

A particular piano string is supposed to vibrate at a frequency of 440 Hz. In order to check its frequency, a tuning fork known to vibrate at a frequency of 440 Hz is sounded at the same time the piano key is struck, and a beat frequency of 4 beats per second is heard. Find the possible frequencies at which the string could be vibrating.

Solution The number of beats per second is equal to the difference in frequency between the two sound sources. In this case, because one of the source frequencies is 440 Hz, 4 beats per second would be heard if the frequency of the string (the second source) were either 444 Hz or 436 Hz.

Optional Section

14.12   **QUALITY OF SOUND**

The sound-wave patterns produced by most musical instruments are very complex. Figure 14.25 shows characteristic waveforms produced by a tuning fork, a flute, and a clarinet, each playing the same steady tone. Although each instrument has its own

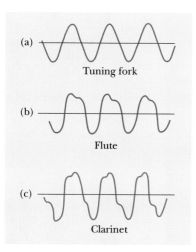

Figure 14.25 Waveforms produced by (a) a tuning fork, (b) a flute, and (c) a clarinet, all at approximately the same frequency. *(Adapted from C. A. Culver, Musical Acoustics, 4th ed., New York, McGraw-Hill, 1956)*

characteristic pattern, the figure reveals that each of the waveforms is periodic. Note that the tuning fork produces only one harmonic (the fundamental frequency), but the two instruments emit mixtures of harmonics. Figure 14.26 graphs the harmonics of the waveforms of Figure 14.25. When this note is played on the flute (Fig. 14.26b), part of the sound consists of a vibration at the fundamental frequency, an even higher intensity is contributed by the second harmonic, the fourth harmonic produces about the same intensity as the fundamental, and so on. These sounds add together according to the principle of superposition to give the complex waveform shown. The clarinet emits a certain intensity at a frequency of the first harmonic, about half as much intensity at the frequency of the second harmonic, and so forth. The resultant superposition of these frequencies produces the pattern shown in Figure 14.25c. The tuning fork (Figs. 14.25a and 14.26a) emits sound only at the frequency of the first harmonic.

In music, the mixture of harmonics that produces the characteristic sound of any instrument is referred to as the *quality*, or *timbre*, of the sound. We say that the

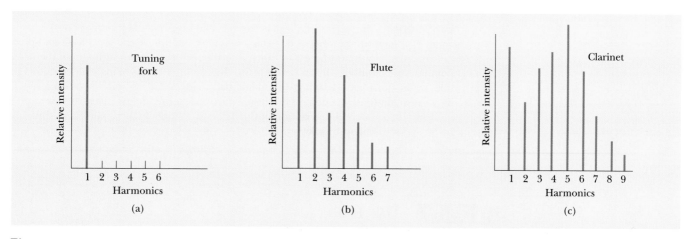

Figure 14.26 Harmonics of the waveforms in Figure 14.25. Note their variation in intensity. *(Adapted from C. A. Culver, Musical Acoustics, 4th ed., New York, McGraw-Hill, 1956)*

Each musical instrument has its own characteristic sound and mixture of harmonics. Instruments shown are (a) the violin, (b) the saxophone, and (c) the trumpet. *(Photographs courtesy of (a) © 1989 Gary Buss/FPG; (b) and (c) © 1989 Richard Laird/FPG)*

(a)   (b)   (c)

note C on a flute differs in quality from the same C on a clarinet. Instruments such as the bugle, trumpet, violin, and tuba are rich in harmonics. A musician playing a wind instrument can emphasize one or another of these harmonics by changing the configuration of his or her lips, and can thus play different musical notes with the same valve openings.

Thinking Physics 4

A professor performs a demonstration in which he breathes helium and then speaks with a comical voice. One student explains, "The velocity of sound in helium is higher than in air, so the fundamental frequency of the standing waves in the mouth is increased." Another student says, "No, the fundamental frequency is determined by the vocal folds and cannot be changed. Only the quality of the voice has changed." Which student is correct?

Explanation   The second student is correct. The fundamental frequency of the complex tone from the voice is determined by the vibration of the vocal folds and is not changed by substituting a different gas in the mouth. The introduction of the helium into the mouth results in harmonics of higher frequencies being excited more than in the normal voice, but the fundamental frequency of the voice is the same — only the quality has changed. The unusual inclusion of the higher frequency harmonics results in a common description of this effect as a "high-pitched" voice, but this is incorrect (it is really a quacky timbre).

14.13   THE EAR

The human ear is divided into three regions: the outer ear, the middle ear, and the inner ear (Fig. 14.27). The *outer ear* consists of the ear canal (open to the atmosphere), which terminates at the eardrum (tympanum). Sound waves travel

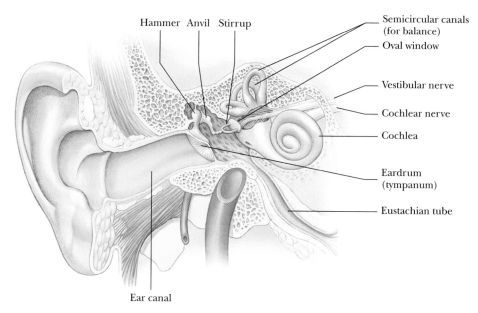

Hammer Anvil Stirrup

Semicircular canals
(for balance)

Oval window

Vestibular nerve

Cochlear nerve

Cochlea

Eardrum
(tympanum)

Eustachian tube

Ear canal

Figure 14.27 The structure of the human ear. The three tiny bones (ossicles) that con-
nect the eardrum to the window of the cochlea act as a double-lever system to decrease the
amplitude of vibration and hence increase the pressure on the fluid in the cochlea.

down the ear canal to the eardrum, which vibrates in and out in phase with the
pushes and pulls caused by the alternating high and low pressures of the sound
wave. Behind the eardrum are three small bones of the *middle ear*, called the *hammer*,
the *anvil*, and the *stirrup* because of their shapes. These bones transmit the vibration
to the *inner ear*, which contains the *cochlea*, a snail-shaped tube about 2 cm long.
The cochlea makes contact with the stirrup at the oval window and is divided along
its length by the basilar membrane, which consists of small hairs and nerve fibers.
This membrane varies in mass per unit length and in tension along its length, and
different portions of it resonate at different frequencies. (Recall that the natural
frequency of a string depends on its mass per unit length and on the tension on
it.) Along the basilar membrane are numerous nerve endings, which sense the
vibration of the membrane and in turn transmit impulses to the brain. The brain
interprets the impulses as sounds of varying frequency, depending on the locations
along the basilar membrane of the impulse-transmitting nerves and on the rates at
which the impulses are transmitted. This is a simplified description of the complex
functioning of the ear.

Figure 14.28 shows the frequency response curves of an average human ear for
sounds of equal loudness, ranging from 0 to 120 dB. To interpret this series of
graphs, take the bottom curve as the threshold of hearing. Compare the intensity
level on the vertical axis for the two frequencies 100 Hz and 1000 Hz. The vertical
axis shows that the 100-Hz sound must be about 38 dB greater than the 1000-Hz
sound to be at the threshold of hearing. Thus, we see that the threshold of hearing
is very strongly dependent on frequency. The easiest frequencies to hear are around
3300 Hz, whereas frequencies above 12 000 Hz or below about 50 Hz must be relatively intense to be heard.

Figure 14.28 Curves of intensity level versus frequency for sounds that are perceived to be of equal loudness. Note that the ear is most sensitive at a frequency of about 3300 Hz. The lowest curve corresponds to the threshold of hearing for only about 1% of the population.

Now consider the curve labeled 80. This curve uses as its reference a 1000-Hz tone at an intensity level of 80 dB. The curve shows that a tone of frequency 100 Hz would have to be about 4 dB louder than the 80-dB, 1000-Hz tone in order to sound as loud. Notice that the curves flatten out as the intensity levels of the sounds increase. That is, when sounds are loud, all frequencies can be heard equally well.

The small bones in the middle ear represent an intricate lever system that increases the force on the oval window. The pressure is greatly magnified because the surface area of the eardrum is about 20 times that of the oval window (in analogy with a hydraulic press). The middle ear, together with the eardrum and oval window, in effect acts as a matching network between the air in the outer ear and the liquid in the inner ear. The overall energy transfer between the outer ear and inner ear is highly efficient, with pressure amplification factors of several thousand. In other words, pressure variations in the inner ear are much greater than those in the outer ear.

The ear has its own built-in protection against loud sounds. The muscles connecting the three middle-ear bones to the walls control the volume of the sound by changing the tension on the bones as sound builds up, thus hindering their ability to transmit vibrations. In addition, the eardrum becomes stiffer as the sound intensity increases. These two occurrences make the ear less sensitive to loud incoming sounds. There is a time delay between the onset of loud sound and the ear's protective reaction, however, so a very sudden loud sound can still damage the ear.

The complex structure of the human ear is believed to be related to the fact that mammals have evolved from sea-going creatures. In comparison, insect ears are considerably simpler in design because they have always been land residents. A typical insect ear consists of an eardrum exposed directly to the air on one side and to an air filled cavity on the other side. Nerve cells communicate directly with the cavity and the brain without the need for the complex intermediary of an inner and middle ear. This simple design allows the ear to be placed virtually anywhere on the body. For example, a grasshopper has its ears on its legs. One advantage of this location is that the distance and orientation of the ears can be varied so that sound location is easier to accomplish.

SUMMARY

Sound waves are longitudinal waves. **Audible waves** are sound waves with frequencies between 20 and 20 000 Hz. **Infrasonic waves** have frequencies below the audible range, and **ultrasonic waves** have frequencies above the audible range.

The speed of sound in a medium of bulk modulus B and density ρ is

$$v = \sqrt{\frac{B}{\rho}} \qquad [14.1]$$

The speed of sound also depends on the temperature of the medium. The relationship between temperature and speed for sound in air is

$$v = (331 \text{ m/s}) \sqrt{1 + \frac{T}{273}} \qquad [14.4]$$

where T is the temperature in degrees Celsius and 331 m/s is the speed of sound in air at 0°C.

The **intensity level** of a sound wave, in decibels, is given by

$$\beta \equiv 10 \log \left(\frac{I}{I_0}\right) \qquad [14.7]$$

The constant I_0 is a reference intensity, usually taken to be at the threshold of hearing ($I_0 = 1.0 \times 10^{-12}$ W/m^2), and I is the intensity with level β, where β is measured in **decibels (dB).**

The **intensity** of a *spherical wave* produced by a point source is proportional to the average power emitted and inversely proportional to the square of the distance from the source:

$$I = \frac{P_{av}}{4\pi r^2} \qquad [14.8]$$

The change in frequency heard by an observer whenever there is relative motion between the source and observer is called the **Doppler effect.** If the observer is moving with a speed v_0 and the source is at rest, the observed frequency, f', is

$$f' = f\left(\frac{v \pm v_0}{v}\right) \qquad [14.11]$$

The positive sign is used when the observer is moving toward the source, and the negative sign when the observer is moving away from the source.

If the source is moving with a speed v_s and the observer is at rest, the observed frequency is

$$f' = f\left(\frac{v}{v \mp v_s}\right) \qquad [14.14]$$

where $-v_s$ refers to motion toward the observer and $+v_s$ refers to motion away from the observer.

When the observer and source are both moving, the observed frequency is

$$f' = f\left(\frac{v \pm v_0}{v \mp v_s}\right)$$ [14.15]

When waves interfere, the resultant wave is found by adding the individual waves together point by point. When crest meets crest and trough meets trough, the waves undergo **constructive interference.** When crest meets trough, **destructive interference** occurs.

Standing waves are formed when two waves having the same frequency, amplitude, and wavelength travel in opposite directions through a medium. One can set up standing waves of specific frequencies in a stretched string. The natural frequencies of vibration of a stretched string of length L, fixed at both ends, are

$$f_n = \frac{n}{2L}\sqrt{\frac{F}{\mu}}   \qquad n = 1, 2, 3, \ldots$$ [14.20]

where F is the tension in the string and μ is its mass per unit length. The natural frequencies of vibration form a **harmonic series;** that is, the frequencies are integral multiples of the fundamental (lowest) frequency.

A system capable of oscillating is said to be in **resonance** with some driving force whenever the frequency of the driving force matches one of the natural frequencies of the system. When the system is resonating, it oscillates with maximum amplitude.

Standing waves can be produced in a tube of air. If the reflecting end of the tube is *open*, all harmonics are present and the natural frequencies of vibration are

$$f_n = n\frac{v}{2L}   \qquad n = 1, 2, 3, \ldots$$ [14.21]

If the tube is *closed* at the reflecting end, only the odd harmonics are present, and the natural frequencies of vibration are

$$f_n = n\frac{v}{4L}   \qquad n = 1, 3, 5, \ldots$$ [14.22]

The phenomenon of **beats** is an interference effect that occurs when two waves of slightly different frequencies travel in the same direction. For sound waves, the loudness of the resultant sound changes periodically with time.

MULTIPLE-CHOICE QUESTIONS

1. Two sirens are sounding so that the frequency from A is twice the frequency from B. Compared to the speed of sound from A, the speed of sound from B is
   (a) twice as fast   (b) half as fast   (c) four times as fast
   (d) one fourth as fast   (e) the same

2. A hollow pipe (such as an organ pipe open at both ends) is made to go into resonance at frequency f_{open}.

One end of the pipe is now covered and the pipe is again made to go into resonance, this time at frequency f_{closed}. Both resonances are first harmonics. How do these two resonances compare?
   (a) They are the same.   (b) $f_{open} = 2f_{closed}$
   (c) $f_{closed} = 2f_{open}$ (d) $f_{open} = f_{closed}$
   (e) $f_{closed} = \frac{3}{2}f_{open}$

3. When two tuning forks are sounded at the same time, a beat frequency of 5 Hz occurs. If one of the tuning forks has a frequency of 245 Hz, what is the frequency of the other tuning fork?
   (a) 240 Hz   (b) 242.5 Hz
   (c) 247.5 Hz   (d) 250 Hz
   (e) More than one answer could be correct.
4. Doubling the power output from a sound source emit-ting a single frequency will result in what increase in level?
   (a) 0.5 dB   (b) 2.0 dB   (c) 3.0 dB   (d) 20 dB
5. If a 1000-Hz sound source moves at a speed of 50.0 m/s toward a listener who moves at a speed of 30.0 m/s in a direction away from the source, what is the apparent frequency heard by the listener? (The velocity of sound is 340 m/s.)
   (a) 937 Hz   (b) 947 Hz   (c) 1060 Hz   (d) 1070 Hz

CONCEPTUAL QUESTIONS

1. An autofocus camera sends out a pulse of sound and measures the time for the pulse to reach an object, re-flect off of it and return to be detected. Can the temper-ature affect your focus for such a camera?
2. To keep animals away from their cars, some people mount short, thin pipes on the fenders. The pipes give out a high-pitched wail when the cars are moving. How do they create the sound?
3. Secret agents in the movies always want to get to a se-cure phone where a voice scrambler is in use. How do these devices work?
4. When a bell is rung, standing waves are set up around the bell's circumference. What boundary conditions must be satisfied by the resonant wavelengths? How does a crack in the bell, such as in the Liberty Bell, affect the satisfying of the boundary conditions and the sound em-anating from the bell?
5. Suppose the wind blows. Does this cause a Doppler ef-fect as you hear it for a sound moving through the air? Is it like a moving source or a moving observer?
6. Explain why your voice seems to sound richer than usual when you sing in the shower.
7. You are driving toward a cliff and you honk your horn. Is there a Doppler shift of the sound when you hear the echo? Is it like a moving source or moving observer? What if the reflection occurs not from a cliff but from the forward edge of a huge alien spacecraft that is mov-ing toward you as you drive?
8. Of the following sounds, state which is most likely to have an intensity level of 60 dB: a rock concert, the turning of a page in this text, a normal conversation, a cheering crowd at a football game, or background noise at a church?
9. Guitarists sometimes play a "harmonic" by lightly touch-ing a string at the exact center and plucking the string. The result is a clear note one octave higher than the fundamental of the string, even though the string is not pressed to the fingerboard. Why does this happen?
10. By listening to a band or orchestra, how can you deter-mine that the speed of sound is the same for all fre-quencies?
11. An archer shoots an arrow from a bow. Does the string of the bow exhibit standing waves after the arrow leaves? If so, and if the bow is perfectly symmetric so that the arrow leaves from the center of the string, what harmon-ics are excited?
12. The radar systems used by police to detect speeders are sensitive to the Doppler shift of a pulse of radio waves. Discuss how this sensitivity can be used to measure the speed of a car.
13. As oppositely moving pulses of the same shape (one up-ward, one downward) on a string pass through each other, there is one instant at which the string shows no displacement from the equilibrium position at any point. Has the energy carried by the pulses disappeared at this instant of time? If not, where is it?
14. A soft-drink bottle resonates as air is blown across its top. What happens to the resonant frequency as the level of fluid in the bottle decreases? (See QuickLab, p. 466.)
15. In Balboa Park in San Diego, California, there is a huge outdoor organ. Does the fundamental frequency of a particular pipe on this organ change on hot and cold days? How about on days with high and low atmospheric pressure?
16. Despite a reasonably steady hand, a person often spills his coffee when carrying it to his seat. Discuss resonance as a possible cause of this difficulty, and devise a means for solving the problem.
17. If you have a series of identical glass bottles, with varying amounts of water in them, you can play musical notes by either striking the bottles with a spoon or blowing across the open tops of the bottles. When hitting the bottles, the frequency of the note decreases as the water level rises. When blowing on the bottles, the frequency of the note increases as the water level rises. Why is the behav-ior of the frequency different in these two cases?

18. An airplane mechanic notices that the sound from a twin-engine aircraft rapidly varies in loudness when both engines are running. What could be causing this variation from loud to soft?

19. Why does a vibrating guitar string sound louder when placed on the instrument than it would if allowed to vibrate in the air while off the instrument?

PROBLEMS

1, 2, 3 = straightforward, intermediate, challenging   ☐ = full solution available in Study Guide/Student Solutions Manual
WEB = solution posted at **http://www.saunderscollege.com/physics/college**   🐟 = biomedical application   💻 = Interactive Physics

Section 14.2 Characteristics of Sound Waves

Section 14.3 The Speed of Sound

Note: Unless otherwise stated, use 345 m/s as the speed of sound in air.

🐟 1. A dolphin located in seawater at a temperature of 25°C emits a sound directed toward the bottom of the ocean 150 m below. How much time passes before it hears an echo?

2. A sound wave has a frequency of 700 Hz in air and a wavelength of 0.50 m. What is the temperature of the air?

🐟 ☐ 3. The range of human hearing extends from approximately 20 Hz to 20 000 Hz. Find the wavelengths of these extremes at a temperature of 27°C.

4. A group of hikers hear an echo 3.00 s after they shout. If the temperature is 22.0°C, how far away is the mountain that reflected the sound wave?

5. The greatest value ever achieved for the speed of sound in air is about 1.0×10^4 m/s, and the highest frequency ever produced is about 2.0×10^{10} Hz. Find the wavelength of this wave.

6. A stone is dropped from rest into a well. The sound of the splash is heard exactly 2.00 s later. Find the depth of the well if the air temperature is 10.0°C.

Section 14.4 Energy and Intensity of Sound Waves

🐟 7. The area of a typical eardrum is about 5.0×10^{-5} m². Calculate the sound power (the energy per second) incident on an eardrum at **(a)** the threshold of hearing and **(b)** the threshold of pain.

8. A microphone in the ocean is sensitive to sounds emitted by porpoises. To produce a usable signal, sound waves striking the microphone must have an intensity of 10 dB. If porpoises emit sound waves with a power of 0.050 W, how far can a porpoise be from the microphone and still be heard? Disregard absorption of sound waves by the water.

WEB ☐ 9. The intensity level of an orchestra is 85 dB. A single violin achieves a level of 70 dB. How does the intensity of the sound of the full orchestra compare with that of the violin's sound?

10. A noisy machine in a factory produces a decibel rating of 80 dB. How many identical machines could you add to the factory without exceeding the 90-dB limit?

11. Two sounds have measured intensities of $I_1 = 100$ W/m² and $I_2 = 200$ W/m². By how many decibels is the level of sound 1 lower than that of sound 2?

Section 14.5 Spherical and Plane Waves

12. A stereo speaker (considered a small source) emits sound waves with a power output of 100 W. **(a)** Find the intensity 10.0 m from the source. **(b)** Find the intensity level, in decibels, at this distance. **(c)** At what distance would you experience the sound at the threshold of pain, 120 dB?

☐ 13. A train sounds its horn as it approaches an intersection. The horn can just be heard at a level of 50 dB by an observer 10 km away. **(a)** What is the average power generated by the horn? **(b)** What intensity level of the horn's sound is observed by someone waiting at an intersection 50 m from the train? Treat the horn as a point source and neglect any absorption of sound by the air. Suppose the ground absorbs all the sound that is radiated downward.

14. A man shouting loudly produces a 70-dB sound at a distance of 5.0 m. How many watts of power does the man emit? (Treat the man as a point source.)

15. A skyrocket explodes 100 m above the ground (Fig. P14.15). Three observers are spaced 100 m apart, with the first (A) is directly under the point of the explosion.

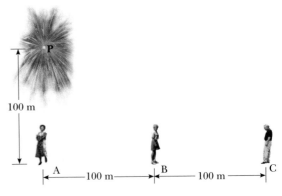

Figure P14.15

(a) What is the ratio of sound intensities heard by observers A and B? (b) What is the ratio of intensities heard by observers A and C?

Section 14.6 The Doppler Effect

16. A train at rest emits a sound at a frequency of 1000 Hz. An observer in a car travels away from the sound source at a speed of 30 m/s. What is the frequency heard by the observer?

17. An airplane traveling with half the speed of sound ($v = 172$ m/s) emits a sound of frequency 5.00 kHz. At what frequency does a stationary listener hear the sound (a) as the plane approaches? (b) after it passes?

18. At rest, a car's horn sounds the note A (440 Hz). The horn is sounded while the car is moving down the street. A bicyclist moving in the same direction with one third the car's speed hears a frequency of 415 Hz. What is the speed of the car? Is the cyclist ahead of or behind the car?

19. Two trains on separate tracks move toward one another. Train 1 has a speed of 130 km/h and train 2 a speed of 90.0 km/h. Train 2 blows its horn, emitting a frequency of 500 Hz. What is the frequency heard by the engineer on train 1?

20. A bat flying at 5.0 m/s emits a chirp at 40 kHz. If this sound pulse is reflected by a wall, what is the frequency of the echo received by the bat?

WEB 21. An alert physics student stands beside the tracks as a train rolls slowly past. He notes that the frequency of the train whistle is 442 Hz when the train is approaching him and 441 Hz when the train is receding from him. From this he can find the speed of the train. What value does he find?

22. A supersonic jet traveling at Mach 3 (which means the speed of the jet is 3 times the speed of sound in air) at an altitude of 20 000 m is directly overhead at time $t = 0$, as in Figure P14.22. (a) How long will it be before the ground observer encounters the shock wave? (b) Where will the plane be when it is finally heard?

(Assume that the speed of sound in air is uniform at 345 m/s.)

23. The Concorde flies at Mach 1.5. What is the angle between the direction of propagation of the shock wave and the direction of the plane's velocity?

Section 14.7 Interference of Sound Waves

24. The sound interferometer shown in Figure 14.14 is driven by a speaker emitting a 400-Hz note. If *destructive* interference occurs at a particular instant, how much must the path length in the U-shaped tube be increased in order to hear (a) constructive interference and (b) destructive interference once again?

25. The ship in Figure P14.25 travels along a straight line parallel to the shore and 600 m from the shore. The ship's radio receives simultaneous signals of the same frequency from antennas A and B. The signals interfere constructively at point C, which is equidistant from A and B. The signal goes through the first minimum at point D. Determine the wavelength of the radio waves.

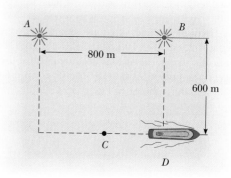

A   B
← 800 m →
600 m
C
D

Figure P14.25

26. Two loudspeakers are placed above and below one another, as in Figure 14.15, and driven by the same source at a frequency of 500 Hz. (a) What minimum distance should the top speaker be moved back in order to create destructive interference between the two speakers? (b) If the top speaker is moved back twice the distance calculated in part (a), will constructive or destructive interference occur?

27. A pair of speakers separated by 0.700 m are driven by the same oscillator at a frequency of 690 Hz. An observer, originally positioned at one of the speakers, begins to walk along a line perpendicular to the line joining the two speakers. (a) How far must the observer walk before reaching a relative maximum in intensity? (b) How far will the observer be from the speaker when the first relative minimum is detected in the intensity?

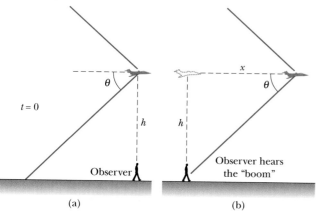

θ

$t = 0$

x

θ

h h

Observer

Observer hears the "boom"

(a)   (b)

Figure P14.22

Section 14.8   Standing Waves

28. A steel wire in a piano has a length of 0.7000 m and a mass of 4.300×10^{-3} kg. To what tension must this wire be stretched in order that the fundamental vibration correspond to middle C ($f_C = 261.6$ Hz on the chromatic musical scale)?

29. A stretched string fixed at each end has a mass of 40 g and a length of 8.0 m. The tension in the string is 49 N. **(a)** Determine the positions of the nodes and antinodes for the third harmonic. **(b)** What is the vibration frequency for this harmonic?

30. Resonance of sound waves can be produced within an aluminum rod by holding the rod at its midpoint and stroking it with an alcohol-saturated paper towel. In this resonance mode, the middle of the rod is a node and the ends are antinodes, there being no other nodes or antinodes present. What is the frequency of the resonance if the rod is 1 m long?

31. A 0.300-g wire is stretched between two points 70.0 cm apart. If the tension in the wire is 600 N, find the wire's first, second, and third harmonics.

32. A 12-kg mass hangs in equilibrium from a string of total length $L = 5.0$ m and linear mass density $\mu = 0.0010$ kg/m. The string is wrapped around two light, frictionless pulleys that are separated by the distance $d = 2.0$ m (Fig. P14.32a). **(a)** Determine the tension in the string. **(b)** At what frequency must the string between the pulleys vibrate in order to form the standing wave pattern shown in Figure P14.32b?

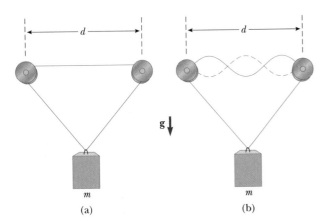

Figure P14.32

33. In the arrangement shown in Figure P14.33, a mass, $m = 5.0$ kg, hangs from a cord around a light pulley. The length of the cord between point P and the pulley is $L = 2.0$ m. **(a)** When the vibrator is set to a frequency of 150 Hz, a standing wave with six loops is formed. What must be the linear mass density of the cord? **(b)** How many loops (if any) will result if m is changed

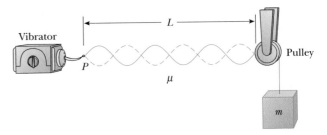

Figure P14.33

to 45 kg? **(c)** How many loops (if any) will result if m is changed to 10 kg?

34. Two pieces of steel wire with identical cross-sections have lengths of L and $2L$. Each of the wires is fixed at both ends and stretched so that the tension in the longer wire is four times greater than that in the shorter wire. If the fundamental frequency in the shorter wire is 60 Hz, what is the frequency of the second harmonic in the longer wire?

Section 14.9   Forced Vibrations and Resonance

35. A 5.0-kg mass connected to a spring is found to resonate when it is pushed at the frequency 2.4 Hz. Determine the spring constant for the spring.

36. The chains suspending a child's swing are 2.00 m long. At what frequency should a big brother push to make the child swing with largest amplitude?

Section 14.10   Standing Waves in Air Columns

37. A tuning fork is sounded above a resonating tube, as in Figure 14.23. The first resonant point is 0.080 m from the top of the tube, and the second is at 0.24 m. **(a)** Where is the third resonant point? **(b)** What is the frequency of the tuning fork?

38. The fundamental frequency of an open organ pipe corresponds to middle C (261.6 Hz on the chromatic musical scale). The third resonance of a closed organ pipe has the same frequency. What are the lengths of the two pipes?

39. The human ear canal is about 2.8 cm long. If it is regarded as a tube open at one end and closed at the eardrum, what is the fundamental frequency around which we would expect hearing to be best? Take the speed of sound to be 340 m/s.

40. A pipe open at each end has a fundamental frequency of 300 Hz when the speed of sound in air is 333 m/s. **(a)** How long is the pipe? **(b)** What is the frequency of the second harmonic when the temperature of the air is increased so that the speed of sound in the pipe is 344 m/s?

WEB **41.** A pipe open at both ends has a fundamental frequency of 300 Hz when the temperature is 0°C. (a) What is the length of the pipe? (b) What is the fundamental frequency at a temperature of 30°C?

42. A 2.00-m-long air column is open at both ends. The frequency of a certain harmonic is 410 Hz, and the frequency of the next higher harmonic is 492 Hz. Determine the speed of sound in the air column.

Section 14.11 Beats

43. Two identical mandolin strings under 200 N of tension are sounding tones with frequencies of 523 Hz. The peg of one string slips slightly, and the tension in it drops to 196 N. How often are beats heard?

44. The G string on a violin has a fundamental frequency of 196 Hz. It is 30.0 cm long and has a mass of 0.500 g. While this string is sounding, a nearby violinist effectively shortens (by sliding her finger down the string) the G string on her identical violin until a beat frequency of 2.00 Hz is heard between the two strings. When this occurs, what is the effective length of her string?

45. Two train whistles have identical frequencies of 180 Hz. When one train is at rest in the station, sounding its whistle, a beat frequency of 2 Hz is heard from a moving train. What two possible speeds and directions can the moving train have?

46. Two pipes, equal in length, are each open at one end. Each has a fundamental frequency of 480 Hz at 300 K. In one pipe the air temperature is increased to 305 K. If the two pipes are sounded together, what beat frequency results?

Section 14.13 The Ear

47. If a human ear canal can be thought of as resembling an organ pipe, closed at one end, that resonates at a fundamental frequency of 3000 Hz, what is the length of the canal? Use normal body temperature for your determination of the speed of sound in the canal.

48. Some studies indicate that the upper frequency limit of hearing is determined by the diameter of the eardrum. The wavelength of the sound wave and the diameter of the eardrum are approximately equal at this upper limit. If this is so, what is the diameter of the eardrum of a person capable of hearing 20 000 Hz? (Assume a body temperature of 37° C.)

ADDITIONAL PROBLEMS

49. Two cars are traveling in the same direction, both at a speed of 55 mph (80.7 ft/s). The driver of the trailing car sounds his horn, which has a frequency of 300 Hz. If the speed of sound is 1100 ft/s, what sound frequency is heard by the driver of the leading car? (*Hint:* Consider the relative motion between the source and observer in this case.)

50. A commuter train blows its horn as it passes a passenger platform at a constant speed of 40.0 m/s. The train horn sounds at a frequency of 320 Hz when the train is at rest. (a) What is the frequency observed by a person on the platform as the train approaches and (b) as the train recedes from him? (c) What wavelength does the observer find in each case?

51. On a workday the average decibel level of a busy street is 70 dB, with 100 cars passing a given point every minute. If the number of cars is reduced to 25 every minute on a weekend, what is the decibel level of the street?

52. A variable-length air column is placed just below a vibrating wire that is fixed at both ends. The length of the air column is gradually increased from zero until the first position of resonance is observed at $L = 34$ cm. The wire is 120 cm long and is vibrating in its third harmonic. If the speed of sound in air is 340 m/s, what is the speed of transverse waves in the wire?

53. A flute is designed so that it plays a frequency of 261.6 Hz, middle C, when all the holes are covered and the temperature is 20.0°C. (a) Consider the flute to be a pipe open at both ends, and find its length, assuming that the middle C frequency is the fundamental. (b) A second player, nearby in a colder room, also attempts to play middle C on an identical flute. A beat frequency of 3.00 beats/s is heard. What is the temperature of the room?

54. When at rest, two trains have sirens that emit a frequency of 300 Hz. The two trains travel toward one another and toward an observer stationed between them. One of the trains moves at 30 m/s, and the observer hears a beat frequency of 3 beats per second. What is the velocity of the second train, which travels faster than the first?

55. A speaker at the front of a room and an identical speaker at the rear of the room are being driven at 456 Hz by the same sound source. A student walks at a uniform rate of 1.50 m/s away from one speaker and toward the other. How many beats does the student hear per second?

56. A tree branch touches the surface of a smoothly gliding river. The wind makes the branch vibrate steadily to produce ripples that move outward with speed 0.500 m/s relative to the water. If the wavelength of the waves downstream is 50.0% longer than the wavelength of the ripples upstream from the branch, how fast is the river flowing?

57. A flower pot is knocked off a balcony 20.0 m above the sidewalk heading for a 1.75-m-tall man standing below. How high from the ground can the flower pot be after which it would be too late for a shouted warning to reach the man in time? Assume that the man below requires 0.300 s to respond to the warning.

58. A rescue plane flies horizontally at a constant speed searching for a disabled boat. When the plane is directly above the boat, the boat's crew blows a loud horn. By the time the plane's sound detector perceives the horn's sound, the plane has traveled a distance equal to one half its altitude above the ocean. If it takes the sound 2.00 s to reach the plane, determine (a) the speed of the plane and (b) its altitude. Take the speed of sound to be 343 m/s.

59. A tuning fork vibrating at 512 Hz falls from rest and accelerates at 9.80 m/s². How far below the point of release is the tuning fork when waves of frequency 485 Hz reach the release point? Take the speed of sound in air to be 340 m/s.

60. A block with a speaker bolted to it is connected to a spring having spring constant $k = 20.0$ N/m, as in Figure P14.60. The total mass of the block and speaker is

Figure P14.60

5.00 kg, and the amplitude of this unit's motion is 0.500 m. If the speaker emits sound waves of frequency 440 Hz, determine the range in frequencies heard by the person to the right of the speaker. Assume that the speed of sound is 343 m/s.

61. In order to be able to determine her speed, a skydiver carries a tone generator. A friend on the ground at the landing site has equipment for receiving and analyzing sound waves. While the skydiver is falling at terminal speed, her tone generator emits a steady tone of 1800 Hz. (Assume that the air is calm and that the sound speed is 343 m/s, independent of altitude.) (a) If her friend on the ground (directly beneath the skydiver) receives waves of frequency 2150 Hz, what is the skydiver's speed of descent? (b) If the skydiver were also carrying sound-receiving equipment sensitive enough to detect waves reflected from the ground, what frequency would she receive?

62. Two identical speakers separated by 10.0 m are driven by the same oscillator with a frequency of $f = 21.5$ Hz (Fig. P14.62). (a) Explain why a receiver at A records a

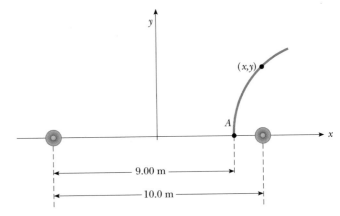

Figure P14.62

minimum in sound intensity from the two speakers. (b) If the receiver is moved in the plane of the speakers, what path should it take so that the intensity remains at a minimum? That is, determine the relationship between x and y (the coordinates of the receiver) that causes the receiver to record a minimum in sound intensity. Take the speed of sound to be 344 m/s.

WEB 63. By proper excitation, it is possible to produce both longitudinal and transverse waves in a long metal rod. In a particular case, the rod is 150 cm long and 0.200 cm in radius and has a mass of 50.9 g. Young's modulus for the material is 6.80×10^{10} Pa. Determine the required tension in the rod so that the ratio of the speed of longitudinal waves to the speed of transverse waves is 8.

64. A student stands several meters in front of a smooth reflecting wall, holding a board on which a wire is fixed at each end. The wire, vibrating in its third harmonic, is 75.0 cm long, has a mass of 2.25 g, and is under a tension of 400 N. A second student, moving toward the wall, hears 8.30 beats per second. What is the speed of the student approaching the wall? Use 340 m/s as the speed of sound in air.

65. Two ships are moving along a line due east. The trailing vessel has a speed of 64.0 km/h relative to a land-based observation point, and the leading ship has a speed of 45.0 km/h relative to the same station. The two ships are in a region of the ocean where the current is moving uniformly due west at 10.0 km/h. The trailing ship transmits a sonar signal at a frequency of 1200 Hz. What frequency is monitored by the leading ship? (Use 1520 m/s as the speed of sound in ocean water.)

66. The Doppler equation presented in the text is valid when the motion between the observer and the source occurs on a straight line, so that the source and observer are moving either directly toward or directly away from each other. If this restriction is relaxed, one must use the more general Doppler equation

$$f' = \left[\frac{v + v_0 \cos(\theta_0)}{v - v_s \cos(\theta_s)} \right] f$$

where θ_0 and θ_s are defined in Figure P14.66a. **(a)** If both observer and source are moving away from each other, show that the preceding equation reduces to the Doppler equations used in the text, with lower signs. **(b)** Use the preceding equation to solve the following problem. A train moves at a constant speed of 25.0 m/s toward the intersection shown in Figure P14.66b. A car is stopped near the intersection, 30.0 m from the tracks. If the train's horn emits a frequency of 500 Hz, what is the frequency heard by the passengers in the car when the train is 40 m from the intersection? Take the speed of sound to be 343 m/s.

(a) (b)

Figure P14.66

"I love hearing that lonesome wail of the train whistle as the magnitude of the frequency of the wave changes due to the Doppler effect."

Mathematical Review

A.1 SCIENTIFIC NOTATION

Many quantities that scientists deal with often have very large or very small values. For example, the speed of light is about 300 000 000 m/s and the ink required to make the dot over an i in this textbook has a mass of about 0.000 000 001 kg. Obviously, it is very cumbersome to read, write, and keep track of numbers such as these. We avoid this problem by using a method dealing with powers of the number 10:

$$10^0 = 1$$

$$10^1 = 10$$

$$10^2 = 10 \times 10 = 100$$

$$10^3 = 10 \times 10 \times 10 = 1000$$

$$10^4 = 10 \times 10 \times 10 \times 10 = 10\ 000$$

$$10^5 = 10 \times 10 \times 10 \times 10 \times 10 = 100\ 000$$

and so on. The number of zeros corresponds to the power to which 10 is raised, called the **exponent** of 10. For example, the speed of light, 300 000 000 m/s, can be expressed as 3×10^8 m/s.

For numbers less than one, we note the following:

$$10^{-1} = \frac{1}{10} = 0.1$$

$$10^{-2} = \frac{1}{10 \times 10} = 0.01$$

$$10^{-3} = \frac{1}{10 \times 10 \times 10} = 0.001$$

$$10^{-4} = \frac{1}{10 \times 10 \times 10 \times 10} = 0.0001$$

$$10^{-5} = \frac{1}{10 \times 10 \times 10 \times 10 \times 10} = 0.000\ 01$$

In these cases, the number of places the decimal point is to the left of the digit 1 equals the value of the (negative) exponent. Numbers that are expressed as some power of 10 multiplied by another number between 1 and 10 are said to be in

scientific notation. For example, the scientific notation for 5 943 000 000 is 5.943×10^9 and that for 0.000 083 2 is 8.32×10^{-5}.

When numbers expressed in scientific notation are being multiplied, the following general rule is very useful:

$$10^n \times 10^m = 10^{n+m} \qquad \text{[A.1]}$$

where n and m can be *any* numbers (not necessarily integers). For example, $10^2 \times 10^5 = 10^7$. The rule also applies if one of the exponents is negative. For example, $10^3 \times 10^{-8} = 10^{-5}$.

When dividing numbers expressed in scientific notation, note that

$$\frac{10^n}{10^m} = 10^n \times 10^{-m} = 10^{n-m} \qquad \text{[A.2]}$$

Exercises

With help from the above rules, verify the answers to the following:

1. $86\ 400 = 8.64 \times 10^4$
2. $9\ 816\ 762.5 = 9.816\ 762\ 5 \times 10^6$
3. $0.000\ 000\ 039\ 8 = 3.98 \times 10^{-8}$
4. $(4.0 \times 10^8)(9.0 \times 10^9) = 3.6 \times 10^{18}$
5. $(3.0 \times 10^7)(6.0 \times 10^{-12}) = 1.8 \times 10^{-4}$
6. $\dfrac{75 \times 10^{-11}}{5.0 \times 10^{-3}} = 1.5 \times 10^{-7}$
7. $\dfrac{(3 \times 10^6)(8 \times 10^{-2})}{(2 \times 10^{17})(6 \times 10^5)} = 2 \times 10^{-18}$

A.2   ALGEBRA

A. Some Basic Rules

When algebraic operations are performed, the laws of arithmetic apply. Symbols such as x, y, and z are usually used to represent quantities that are not specified, what are called the **unknowns.**

First, consider the equation

$$8x = 32$$

If we wish to solve for x, we can divide (or multiply) each side of the equation by the same factor without destroying the equality. In this case, if we divide both sides by 8, we have

$$\frac{8x}{8} = \frac{32}{8}$$

$$x = 4$$

Next consider the equation

$$x + 2 = 8$$

In this type of expression, we can add or subtract the same quantity from each side. If we subtract 2 from each side, we get

$$x + 2 - 2 = 8 - 2$$
$$x = 6$$

In general, if $x + a = b$, then $x = b - a$.

Now consider the equation

$$\frac{x}{5} = 9$$

If we multiply each side by 5, we are left with x on the left by itself and 45 on the right:

$$\left(\frac{x}{5}\right)(5) = 9 \times 5$$
$$x = 45$$

In all cases, **whatever operation is performed on the left side of the equality must also be performed on the right side.**

The following rules for multiplying, dividing, adding, and subtracting fractions should be recalled, where a, b, and c are three numbers:

	Rule	Example
Multiplying	$\left(\dfrac{a}{b}\right)\left(\dfrac{c}{d}\right) = \dfrac{ac}{bd}$	$\left(\dfrac{2}{3}\right)\left(\dfrac{4}{5}\right) = \dfrac{8}{15}$
Dividing	$\dfrac{(a/b)}{(c/d)} = \dfrac{ad}{bc}$	$\dfrac{2/3}{4/5} = \dfrac{(2)(5)}{(4)(3)} = \dfrac{10}{12}$
Adding	$\dfrac{a}{b} \pm \dfrac{c}{d} = \dfrac{ad \pm bc}{bd}$	$\dfrac{2}{3} - \dfrac{4}{5} = \dfrac{(2)(5) - (4)(3)}{(3)(5)} = -\dfrac{2}{15}$

Exercises

In the following exercises, solve for x:

Answers

1. $a = \dfrac{1}{1 + x}$ $x = \dfrac{1 - a}{a}$

2. $3x - 5 = 13$ $x = 6$

3. $ax - 5 = bx + 2$ $x = \dfrac{7}{a - b}$

4. $\dfrac{5}{2x + 6} = \dfrac{3}{4x + 8}$ $x = -\dfrac{11}{7}$

B. Powers

When powers of a given quantity x are multiplied, the following rule applies:

$$x^n x^m = x^{n+m} \qquad \text{[A.3]}$$

For example, $x^2 x^4 = x^{2+4} = x^6$.

When dividing the powers of a given quantity, note that

$$\frac{x^n}{x^m} = x^{n-m} \qquad \text{[A.4]}$$

For example, $x^8 / x^2 = x^{8-2} = x^6$.

A power that is a fraction, such as $\frac{1}{3}$, corresponds to a root as follows:

$$x^{1/n} = \sqrt[n]{x} \qquad \text{[A.5]}$$

For example, $4^{1/3} = \sqrt[3]{4} = 1.5874$. (A scientific calculator is useful for such calculations.)

Finally, any quantity x^n that is raised to the mth power is

$$(x^n)^m = x^{nm} \qquad \text{[A.6]}$$

Table A.1 summarizes the rules of exponents.

TABLE A.1
Rules of Exponents

$$x^0 = 1$$
$$x^1 = x$$
$$x^n x^m = x^{n+m}$$
$$x^n / x^m = x^{n-m}$$
$$x^{1/n} = \sqrt[n]{x}$$
$$(x^n)^m = x^{nm}$$

Exercises

Verify the following:

1. $3^2 \times 3^3 = 243$
2. $x^5 x^{-8} = x^{-3}$
3. $x^{10} / x^{-5} = x^{15}$
4. $5^{1/3} = 1.709\ 975$ (Use your calculator.)
5. $60^{1/4} = 2.783\ 158$ (Use your calculator.)
6. $(x^4)^3 = x^{12}$

C. Factoring

Some useful formulas for factoring an equation are

$$ax + ay + az = a(x + y + z) \qquad \text{common factor}$$
$$a^2 + 2ab + b^2 = (a + b)^2 \qquad \text{perfect square}$$
$$a^2 - b^2 = (a + b)(a - b) \qquad \text{differences of squares}$$

D. Quadratic Equations

The general form of a quadratic equation is

$$ax^2 + bx + c = 0 \qquad \text{[A.7]}$$

where x is the unknown quantity and a, b, and c are numerical factors referred to as **coefficients** of the equation. This equation has two roots, given by

$$x = \frac{-b \pm \sqrt{b^2 - 4ac}}{2a} \qquad \text{[A.8]}$$

If $b^2 \geq 4ac$, the roots will be real.

EXAMPLE

The equation $x^2 + 5x + 4 = 0$ has the following roots corresponding to the two signs of the square-root term:

$$x = \frac{-5 \pm \sqrt{5^2 - (4)(1)(4)}}{2(1)} = \frac{-5 \pm \sqrt{9}}{2} = \frac{-5 \pm 3}{2}$$

that is,

$$x_+ = \frac{-5 + 3}{2} = -1 \qquad x_- = \frac{-5 - 3}{2} = -4$$

where x_+ refers to the root corresponding to the positive sign and x_- refers to the root corresponding to the negative sign.

Exercises

Solve the following quadratic equations:

Answers

1. $x^2 + 2x - 3 = 0$ $x_+ = 1$ $x_- = -3$
2. $2x^2 - 5x + 2 = 0$ $x_+ = 2$ $x_- = 1/2$
3. $2x^2 - 4x - 9 = 0$ $x_+ = 1 + \sqrt{22}/2$ $x_- = 1 - \sqrt{22}/2$

E. Linear Equations

A linear equation has the general form

$$y = ax + b \qquad \text{[A.9]}$$

where a and b are constants. This equation is referred to as being linear because the graph of y versus x is a straight line, as shown in Figure A.1. The constant b, called the **intercept,** represents the value of y at which the straight line intersects the y axis. The constant a is equal to the **slope** of the straight line and is also equal to the tangent of the angle that the line makes with the x axis. If any two points on the straight line are specified by the coordinates (x_1, y_1) and (x_2, y_2), as in Figure A.1, then the slope of the straight line can be expressed

$$\text{Slope} = \frac{y_2 - y_1}{x_2 - x_1} = \frac{\Delta y}{\Delta x} \qquad \text{[A.10]}$$

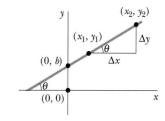

Figure A.1

Note that a and b can have either positive or negative values. If $a > 0$, the straight line has a *positive* slope, as in Figure A.1. If $a < 0$, the straight line has a *negative* slope. In Figure A.1, both a and b are positive. Three other possible situations are shown in Figure A.2: $a > 0, b < 0$; $a < 0, b > 0$; and $a < 0, b < 0$.

Exercises

1. Draw graphs of the following straight lines:
   (a) $y = 5x + 3$ (b) $y = -2x + 4$ (c) $y = -3x - 6$
2. Find the slopes of the straight lines described in Exercise 1.
   Answers: (a) 5 (b) -2 (c) -3
3. Find the slopes of the straight lines that pass through the following sets of points:
   (a) $(0, -4)$ and $(4, 2)$, (b) $(0, 0)$ and $(2, -5)$, and (c) $(-5, 2)$ and $(4, -2)$
   Answers: (a) $3/2$ (b) $-5/2$ (c) $-4/9$

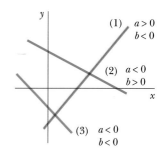

Figure A.2

F. Solving Simultaneous Linear Equations

Consider an equation such as $3x + 5y = 15$, which has two unknowns, x and y. Such an equation does not have a unique solution. That is, $(x = 0, y = 3)$, $(x = 5, y = 0)$, and $(x = 2, y = 9/5)$ are all solutions to this equation.

If a problem has two unknowns, a unique solution is possible only if we have *two* independent equations. In general, if a problem has n unknowns, its solution requires n independent equations. In order to solve two simultaneous equations involving two unknowns, x and y, we solve one of the equations for x in terms of y and substitute this expression into the other equation.

EXAMPLE

Solve the following two simultaneous equations:

$$(1) \quad 5x + y = -8 \qquad (2) \quad 2x - 2y = 4$$

Solution　From (2), we find that $x = y + 2$. Substitution of this into (1) gives

$$5(y + 2) + y = -8$$
$$6y = -18$$
$$y = -3$$
$$x = y + 2 = \boxed{-1}$$

Alternate solution: Multiply each term in (1) by the factor 2 and add the result to (2):

$$10x + 2y = -16$$
$$\underline{2x - 2y = 4}$$
$$12x = -12$$
$$x = \boxed{-1}$$
$$y = x - 2 = \boxed{-3}$$

Two linear equations with two unknowns can also be solved by a graphical method. If the straight lines corresponding to the two equations are plotted in a conventional coordinate system, the intersection of the two lines represents the solution. For example, consider the two equations

$$x - y = 2$$
$$x - 2y = -1$$

These are plotted in Figure A.3. The intersection of the two lines has the coordinates $x = 5$, $y = 3$. This represents the solution to the equations. You should check this solution by the analytical technique discussed above.

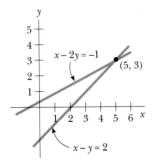

Figure A.3

Exercises

Solve the following pairs of simultaneous equations involving two unknowns:

	Answers
1. $x + y = 8$	$x = 5, y = 3$
$x - y = 2$	
2. $98 - T = 10a$	$T = 65, a = 3.27$
$T - 49 = 5a$	
3. $6x + 2y = 6$	$x = 2, y = -3$
$8x - 4y = 28$	

G. Logarithms

Suppose that a quantity x is expressed as a power of some quantity a:

$$x = a^y \qquad \text{[A.11]}$$

The number a is called the **base** number. The **logarithm** of x with respect to the base a is equal to the exponent to which the base must be raised in order to satisfy the expression $x = a^y$:

$$y = \log_a x \qquad \text{[A.12]}$$

Conversely, the **antilogarithm** of y is the number x:

$$x = \text{antilog}_a y \qquad \text{[A.13]}$$

In practice, the two bases most often used are base 10, called the *common* logarithm base, and base $e = 2.718\ldots$, called the *natural* logarithm base. When common logarithms are used,

$$y = \log_{10} x \qquad (\text{or } x = 10^y) \qquad \text{[A.14]}$$

When natural logarithms are used,

$$y = \ln_e x \qquad (\text{or } x = e^y) \qquad \text{[A.15]}$$

For example, $\log_{10} 52 = 1.716$, so that $\text{antilog}_{10} 1.716 = 10^{1.716} = 52$. Likewise, $\ln_e 52 = 3.951$, so $\text{antiln}_e 3.951 = e^{3.951} = 52$.

In general, note that you can convert between base 10 and base e with the equality

$$\ln_e x = (2.302\ 585)\log_{10} x \qquad \text{[A.16]}$$

Finally, some useful properties of logarithms are

$$\log (ab) = \log a + \log b \qquad \ln e = 1$$

$$\log (a/b) = \log a - \log b \qquad \ln e^a = a$$

$$\log (a^n) = n \log a \qquad \ln\left(\frac{1}{a}\right) = -\ln a$$

A.3   GEOMETRY

Table A.2 gives the areas and volumes for several geometric shapes used throughout this text:

TABLE A.2   Useful Information for Geometry

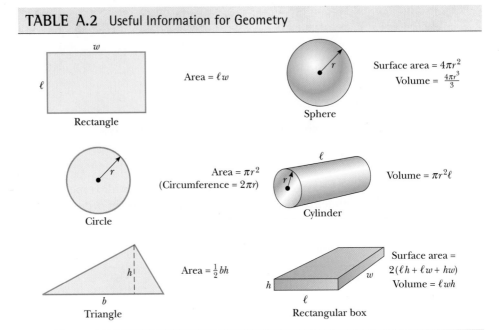

Rectangle — Area $= \ell w$

Sphere — Surface area $= 4\pi r^2$, Volume $= \frac{4\pi r^3}{3}$

Circle — Area $= \pi r^2$ (Circumference $= 2\pi r$)

Cylinder — Volume $= \pi r^2 \ell$

Triangle — Area $= \frac{1}{2}bh$

Rectangular box — Surface area $= 2(\ell h + \ell w + hw)$, Volume $= \ell wh$

A.4   TRIGONOMETRY

Some of the most basic facts concerning trigonometry are presented in Chapter 1, and we encourage you to study the material presented there if you are having trouble with this branch of mathematics. In addition to the discussion of Chapter 1, certain useful trig identities that can be of value to you follow.

$$\sin^2 \theta + \cos^2 \theta = 1$$

$$\sin \theta = \cos(90° - \theta)$$

$$\cos \theta = \sin(90° - \theta)$$

$$\sin 2\theta = 2 \sin \theta \cos \theta$$

$$\cos 2\theta = \cos^2 \theta - \sin^2 \theta$$

$$\sin(\theta \pm \phi) = \sin \theta \cos \phi \pm \cos \theta \sin \phi$$

$$\cos(\theta \pm \phi) = \cos \theta \cos \phi \mp \sin \theta \sin \phi$$

An Abbreviated Table of Isotopes

Atomic Number, Z	Element	Symbol	Chemical Atomic Mass (u)	Mass Number (* Indicates Radioactive) A	Atomic Mass (u) [a]	Percentage Abundance	Half-Life (if Radioactive) $T_{1/2}$
0	(Neutron)	n		1*	1.008 665		10.4 min
1	Hydrogen	H	1.0079	1	1.007 825	99.985	
	Deuterium	D		2	2.014 102	0.015	
	Tritium	T		3*	3.016 049		12.33 y
2	Helium	He	4.00260	3	3.016 029	0.00014	
				4	4.002 602	99.99986	
3	Lithium	Li	6.941	6	6.015 121	7.5	
				7	7.016 003	92.5	
4	Beryllium	Be	9.0122	7*	7.016 928		53.3 days
				9	9.012 174	100	
5	Boron	B	10.81	10	10.012 936	19.9	
				11	11.009 305	80.1	
6	Carbon	C	12.011	11*	11.011 433		20.4 min
				12	12.000 000	98.90	
				13	13.003 355	1.10	
				14*	14.003 242		5730 y
7	Nitrogen	N	14.0067	13*	13.005 738		9.96 min
				14	14.003 074	99.63	
				15	15.000 108	0.37	
8	Oxygen	O	15.9994	15*	15.003 065		122 s
				16	15.994 915	99.761	
				18	17.999 160	0.20	
9	Fluorine	F	18.99840	19	18.998 404	100	
10	Neon	Ne	20.180	20	19.992 435	90.48	
				22	21.991 383	9.25	
11	Sodium	Na	22.98987	22*	21.994 434		2.61 y
				23	22.989 770	100	
				24*	23.990 961		14.96 h
12	Magnesium	Mg	24.305	24	23.985 042	78.99	
				25	24.985 838	10.00	
				26	25.982 594	11.01	
13	Aluminum	Al	26.98154	27	26.981 538	100	
14	Silicon	Si	28.086	28	27.976 927	92.23	
15	Phosphorus	P	30.97376	31	30.973 762	100	
				32*	31.973 908		14.26 days

(Table continues)

[a] The masses in the sixth column are atomic masses, which include the mass of Z electrons. Data are from the National Nuclear Data Center, Brookhaven National Laboratory, prepared by Jagdish K. Tuli, July 1990. The data are based on experimental results reported in *Nuclear Data Sheets* and *Nuclear Physics* and also from *Chart of the Nuclides*, 14th ed. Atomic masses are based on those by A. H. Wapstra, G. Audi, and R. Hoekstra. Isotopic abundances are based on those by N. E. Holden.

Appendix B An Abbreviated Table of Isotopes

Atomic Number, Z	Element	Symbol	Chemical Atomic Mass (u)	Mass Number (* Indicates Radioactive) A	Atomic Mass (u) [a]	Percentage Abundance	Half-Life (if Radioactive) $T_{1/2}$
16	Sulfur	S	32.066	32	31.972 071	95.02	
				35*	34.969 033		87.5 days
17	Chlorine	Cl	35.453	35	34.968 853	75.77	
				37	36.965 903	24.23	
18	Argon	Ar	39.948	40	39.962 384	99.600	
19	Potassium	K	39.0983	39	38.963 708	93.2581	
				40*	39.964 000	0.0117	1.28×10^9 y
20	Calcium	Ca	40.08	40	39.962 591	96.941	
21	Scandium	Sc	44.9559	45	44.955 911	100	
22	Titanium	Ti	47.88	48	47.947 947	73.8	
23	Vanadium	V	50.9415	51	50.943 962	99.75	
24	Chromium	Cr	51.996	52	51.940 511	83.79	
25	Manganese	Mn	54.93805	55	54.938 048	100	
26	Iron	Fe	55.847	56	55.934 940	91.72	
27	Cobalt	Co	58.93320	59	58.933 198	100	
				60*	59.933 820		5.27 y
28	Nickel	Ni	58.693	58	57.935 346	68.077	
				60	59.930 789	26.223	
29	Copper	Cu	63.54	63	62.929 599	69.17	
				65	64.927 791	30.83	
30	Zinc	Zn	65.39	64	63.929 144	48.6	
				66	65.926 035	27.9	
				68	67.924 845	18.8	
31	Gallium	Ga	69.723	69	68.925 580	60.108	
				71	70.924 703	39.892	
32	Germanium	Ge	72.61	70	69.924 250	21.23	
				72	71.922 079	27.66	
				74	73.921 177	35.94	
33	Arsenic	As	74.9216	75	74.921 594	100	
34	Selenium	Se	78.96	78	77.917 307	23.78	
				80	79.916 519	49.61	
35	Bromine	Br	79.904	79	78.918 336	50.69	
				81	80.916 287	49.31	
36	Krypton	Kr	83.80	82	81.913 481	11.6	
				83	82.914 136	11.5	
				84	83.911 508	57.0	
				86	85.910 615	17.3	
37	Rubidium	Rb	85.468	85	84.911 793	72.17	
				87*	86.909 186	27.83	4.75×10^{10} y
38	Strontium	Sr	87.62	86	85.909 266	9.86	
				88	87.905 618	82.58	
				90*	89.907 737		29.1 y
39	Yttrium	Y	88.9058	89	88.905 847	100	
40	Zirconium	Zr	91.224	90	89.904 702	51.45	
				91	90.905 643	11.22	
				92	91.905 038	17.15	
				94	93.906 314	17.38	

(Table continues)

Atomic Number, Z	Element	Symbol	Chemical Atomic Mass (u)	Mass Number (* Indicates Radioactive) A	Atomic Mass (u)^a	Percentage Abundance	Half-Life (if Radioactive) $T_{1/2}$
41	Niobium	Nb	92.9064	93	92.906 376	100	
42	Molybdenum	Mo	95.94	92	91.906 807	14.84	
				95	94.905 841	15.92	
				96	95.904 678	16.68	
				98	97.905 407	24.13	
43	Technetium	Tc		98*	97.907 215		4.2×10^6 y
				99*	98.906 254		2.1×10^5 y
44	Ruthenium	Ru	101.07	99	98.905 939	12.7	
				100	99.904 219	12.6	
				101	100.905 558	17.1	
				102	101.904 348	31.6	
				104	103.905 428	18.6	
45	Rhodium	Rh	102.9055	103	102.905 502	100	
46	Palladium	Pd	106.42	104	103.904 033	11.14	
				105	104.905 082	22.33	
				106	105.903 481	27.33	
				108	107.903 893	26.46	
				110	109.905 158	11.72	
47	Silver	Ag	107.868	107	106.905 091	51.84	
				109	108.904 754	48.16	
48	Cadmium	Cd	112.41	110	109.903 004	12.49	
				111	110.904 182	12.80	
				112	111.902 760	24.13	
				113*	112.904 401	12.22	9.3×10^{15} y
				114	113.903 359	28.73	
49	Indium	In	114.82	115*	114.903 876	95.7	4.4×10^{14} y
50	Tin	Sn	118.71	116	115.901 743	14.53	
				118	117.901 605	24.22	
				120	119.902 197	32.59	
51	Antimony	Sb	121.76	121	120.903 820	57.36	
				123	122.904 215	42.64	
52	Tellurium	Te	127.60	126	125.903 309	18.93	
				128*	127.904 463	31.70	$>8 \times 10^{24}$ y
				130*	129.906 228	33.87	$\leqslant 1.25 \times 10^{21}$ y
53	Iodine	I	126.9045	127	126.904 474	100	
				129*	128.904 984		1.6×10^7 y
				131*	130.906 118		8.04 days
54	Xenon	Xe	131.29	129	128.904 779	26.4	
				131	130.905 069	21.2	
				132	131.904 141	26.9	
				134	133.905 394	10.4	
				136*	135.907 215	8.9	$\geqslant 2.36 \times 10^{21}$ y
55	Cesium	Cs	132.9054	133	132.905 436	100	
56	Barium	Ba	137.33	137	136.905 816	11.23	
				138	137.905 236	71.70	
				144*	143.922 673		11.9 s

(Table continues)

Atomic Number, Z	Element	Symbol	Chemical Atomic Mass (u)	Mass Number (* Indicates Radioactive) A	Atomic Mass (u) [a]	Percentage Abundance	Half–Life (if Radioactive) $T_{1/2}$
57	Lanthanum	La	138.905	139	138.906 346	99.9098	
58	Cerium	Ce	140.12	140	139.905 434	88.43	
				142*	141.909 241	11.13	$>5 \times 10^{16}$ y
59	Praseodymium	Pr	140.9076	141	140.907 647	100	
60	Neodymium	Nd	144.24	142	141.907 718	27.13	
				144*	143.910 082	23.80	2.3×10^{15} y
				146	145.913 113	17.19	
61	Promethium	Pm		145*	144.912 745		17.7 y
62	Samarium	Sm	150.36	147*	146.914 894	15.0	1.06×10^{11} y
				149*	148.917 180	13.8	$>2 \times 10^{15}$ y
				152	151.919 728	26.7	
				154	153.922 206	22.7	
63	Europium	Eu	151.96	151	150.919 846	47.8	
				153	152.921 226	52.2	
64	Gadolinium	Gd	157.25	156	155.922 119	20.47	
				158	157.924 099	24.84	
				160	159.927 050	21.86	
65	Terbium	Tb	158.9253	159	158.925 345	100	
66	Dysprosium	Dy	162.50	162	161.926 796	25.5	
				163	162.928 729	24.9	
				164	163.929 172	28.2	
67	Holmium	Ho	164.9303	165	164.930 316	100	
68	Erbium	Er	167.26	166	165.930 292	33.6	
				167	166.932 047	22.95	
				168	167.932 369	27.8	
69	Thulium	Tm	168.9342	169	168.934 213	100	
70	Ytterbium	Yb	173.04	172	171.936 380	21.9	
				173	172.938 209	16.12	
				174	173.938 861	31.8	
71	Lutecium	Lu	174.967	175	174.940 772	97.41	
72	Hafnium	Hf	178.49	177	176.943 218	18.606	
				178	177.943 697	27.297	
				179	178.945 813	13.629	
				180	179.946 547	35.100	
73	Tantalum	Ta	180.9479	181	180.947 993	99.988	
74	Tungsten	W	183.85	182	181.948 202	26.3	
				183	182.950 221	14.28	
				184	183.950 929	30.7	
				186	185.954 358	28.6	
75	Rhenium	Re	186.207	185	184.952 951	37.40	
				187*	186.955 746	62.60	4.4×10^{10} y
76	Osmium	Os	190.2	188	187.955 832	13.3	
				189	188.958 139	16.1	
				190	189.958 439	26.4	
				191*	190.960 94		15.4 days
				192	191.961 468	41.0	

(Table continues)

Atomic Number, Z	Element	Symbol	Chemical Atomic Mass (u)	Mass Number (* Indicates Radioactive) A	Atomic Mass (u) [a]	Percentage Abundance	Half-Life (if Radioactive) $T_{1/2}$
77	Iridium	Ir	192.2	191	190.960 585	37.3	
				193	192.962 916	62.7	
78	Platinum	Pt	195.08	194	193.962 655	32.9	
				195	194.964 765	33.8	
				196	195.964 926	25.3	
79	Gold	Au	196.9665	197	196.966 543	100	
80	Mercury	Hg	200.59	199	198.968 253	16.87	
				200	199.968 299	23.10	
				201	200.970 276	13.10	
				202	201.970 617	29.86	
81	Thallium	Tl	204.383	203	202.972 320	29.524	
				205	204.974 400	70.476	
		(Th C″)		208*	207.981 992		3.053 min
				210*	209.990 069		1.3 min
82	Lead	Pb	207.2	204*	203.973 020	1.4	$\geq 1.4 \times 10^{17}$ y
				206	205.974 440	24.1	
				207	206.975 871	22.1	
				208	207.976 627	52.4	
		(Ra D)		210*	209.984 163		22.3 y
		(Ac B)		211*	210.988 734		36.1 min
		(Th B)		212*	211.991 872		10.64 h
		(Ra B)		214*	213.999 798		26.8 min
83	Bismuth	Bi	208.9803	209	208.980 374	100	
		(Th C)		211*	210.987 254		2.14 min
84	Polonium	Po					
		(Ra F)		210*	209.982 848		138.38 days
		(Ra C′)		214*	213.995 177		164 μs
85	Astatine	At		218*	218.008 685		1.6 s
86	Radon	Rn		222*	222.017 571		3.823 days
87	Francium	Fr					
		(Ac K)		223*	223.019 733		22 min
88	Radium	Ra		226*	226.025 402		1600 y
		(Ms Th₁)		228*	228.031 064		5.75 y
89	Actinium	Ac		227*	227.027 749		21.77 y
90	Thorium	Th	232.0381				
		(Rd Th)		228*	228.028 716		1.913 y
				232*	232.038 051	100	1.40×10^{10} y
91	Protactinium	Pa		231*	231.035 880		32.760 y
92	Uranium	U	238.0289	232*	232.037 131		69 y
				233*	233.039 630		1.59×10^5 y
		(Ac U)		235*	235.043 924	0.720	7.04×10^8 y
				236*	236.045 562		2.34×10^7 y
		(UI)		238*	238.050 784	99.2745	4.47×10^9 y
93	Neptunium	Np		237*	237.048 168		2.14×10^6 y
94	Plutonium	Pu		239*	239.052 157		2.412×10^3 y
				242*	242.058 737		3.73×10^5 y
				244*	244.064 200		8.1×10^7 y

Appendix C

Some Useful Tables

TABLE C.1 Mathematical Symbols Used in the Text and Their Meaning

Symbol	Meaning
$=$	is equal to
$\neq$	is not equal to
$\equiv$	is defined as
$\propto$	is proportional to
$>$	is greater than
$<$	is less than
$\gg$	is much greater than
$\ll$	is much less than
$\approx$	is approximately equal to
Δx	change in x or uncertainty in x
Σx_i	sum of all quantities x_i
$\lvert x \rvert$	absolute value of x (always a positive quantity)

TABLE C.2 Standard Abbreviations and Symbols for Units

Symbol	Unit	Symbol	Unit
A	ampere	J	joule
Å	angstrom	K	kelvin
atm	atmosphere	kcal	kilocalorie
Btu	British thermal unit	kg	kilogram
C	coulomb	km	kilometer
°C	degree Celsius	kmol	kilomole
cal	calorie	lb	pound
cm	centimeter	m	meter
deg	degree (angle)	min	minute
eV	electron volt	N	newton
°F	degree Fahrenheit	nm	nanometer
F	farad	rev	revolution
ft	foot	s	second
G	Gauss	T	tesla
g	gram	u	atomic mass unit
H	henry	V	volt
h	hour	W	watt
hp	horsepower	Wb	weber
Hz	hertz	μm	micrometer
in.	inch	Ω	ohm

TABLE C.3 The Greek Alphabet

Alpha	A	α	Iota	I	ι	Rho	P	ρ
Beta	B	β	Kappa	K	κ	Sigma	Σ	σ
Gamma	Γ	γ	Lambda	Λ	λ	Tau	T	τ
Delta	Δ	δ	Mu	M	μ	Upsilon	Y	υ
Epsilon	E	ϵ	Nu	N	ν	Phi	Φ	ϕ
Zeta	Z	ζ	Xi	Ξ	ξ	Chi	X	χ
Eta	H	η	Omicron	O	o	Psi	Ψ	ψ
Theta	Θ	θ	Pi	Π	π	Omega	Ω	ω

TABLE C.4 Physical Data Often Used [a]

Average Earth-Moon distance	3.84×10^8 m
Average Earth-Sun distance	1.496×10^{11} m
Average radius of the Earth	6.37×10^6 m
Density of air (20°C and 1 atm)	1.20 kg/m^3
Density of water (20°C and 1 atm)	1.00×10^3 kg/m^3
Free-fall acceleration	9.80 m/s^2
Mass of the Earth	5.98×10^{24} kg
Mass of the Moon	7.36×10^{22} kg
Mass of the Sun	1.99×10^{30} kg
Standard atmospheric pressure	1.013×10^5 Pa

[a] These are the values of the constants as used in the text.

Appendix D

SI Units

TABLE D.1 SI Base Units

Base Quantity	SI Base Unit	
	Name	Symbol
Length	Meter	m
Mass	Kilogram	kg
Time	Second	s
Electric current	Ampere	A
Temperature	Kelvin	K
Amount of substance	Mole	mol
Luminous intensity	Candela	cd

TABLE D.2 Derived SI Units

Quantity	Name	Symbol	Expression in Terms of Base Units	Expression in Terms of Other SI Units
Plane angle	Radian	rad	m/m	
Frequency	Hertz	Hz	s^{-1}	
Force	Newton	N	$kg \cdot m/s^2$	J/m
Pressure	Pascal	Pa	$kg/m \cdot s^2$	N/m^2
Energy: work	Joule	J	$kg \cdot m^2/s^2$	$N \cdot m$
Power	Watt	W	$kg \cdot m^2/s^3$	J/s
Electric charge	Coulomb	C	$A \cdot s$	
Electric potential (emf)	Volt	V	$kg \cdot m^2/A \cdot s^3$	$W/A, J/C$
Capacitance	Farad	F	$A^2 \cdot s^4/kg \cdot m^2$	C/V
Electric resistance	Ohm	Ω	$kg \cdot m^2/A^2 \cdot s^3$	V/A
Magnetic flux	Weber	Wb	$kg \cdot m^2/A \cdot s^2$	$V \cdot s, T \cdot m^2$
Magnetic field intensity	Tesla	T	$kg/A \cdot s^2$	Wb/m^2
Inductance	Henry	H	$kg \cdot m^2/A^2 \cdot s^2$	Wb/A

Answers to Selected Questions and Problems

CHAPTER 1

Multiple-Choice Questions
1. b   2. d   3. b   4. d   5. d
6. d   7. d

Conceptual Questions
1. (a) ≈ 0.1 m   (b) ≈ 1 m   (c) ≈ 30 m   (d) ≈ 10 m
   (e) ≈ 100 m
3. $\approx 10^9$ s
5. The length of a hand varies from person to person, so it is not a useful standard of length.
7. The ark has an approximate volume of 4×10^4 m^3, whereas a home has an approximate volume of 10^3 m^3.
9. $\approx 10^8$ revolutions

Problems
1. (b) $A_{\text{cylinder}} = \pi R^2$, $A_{\text{rectangular solid}} = lw$.
5. (a) $\dfrac{L}{T^2}$ (b) L
7. (a) 3 significant figures   (b) 4 significant figures
   (c) 3 significant figures   (d) 2 significant figures.
9. (a) 2.96×10^9 (b) 6.876×10^{-2} (If you disagree with either the significant figures or the power of ten of either of these answers, please check your calculator manual.)
11. 115.9 m
13. 10^{17} ft
15. 6.71×10^8 mi/h
17. 36 m^2
19. 10^{10} lb, 4.45×10^{10} N
21. 3.08×10^4 m^3
23. 9.82 cm
25. (a) 1000 kg
    (b) $m_{\text{cell}} = 5.2 \times 10^{-16}$ kg
    $m_{\text{kidney}} = 0.27$ kg
    $m_{\text{fly}} = 1.3 \times 10^{-5}$ kg
27. 1800 balls (assumes 81 games per season, 9 innings per game, an average of 10 hitters per inning, and 1 ball lost for every four hitters)
29. (2.1, 1.4)
31. $r = 2.24$ m, $\theta = 26.6°$
33. (a) 6.71 m   (b) 0.894   (c) 0.746
35. (a) 3   (b) 3   (c) 4/5   (d) 4/5   (e) 4/3
37. $\frac{5}{7}$; the angle itself is 35.5°
39. (a) 0.677 g/cm^3 (b) 4.63×10^{17} ft^2

41. The value of k, a dimensionless constant, cannot be found by dimensional analysis.
43. (a) on the order of 100 kg   (b) on the order of 1000 kg
45. Both sides have the units of length, L.
47. (a) 3.16×10^7 s   (b) 6.0×10^{10} y
49. (a) 127 y   (b) 15 500 times

CHAPTER 2

Multiple-Choice Questions
1. c   2. a   3. b   4. c   5. d

Conceptual Questions
1. Yes. This occurs when a car is slowing down, so that the direction of its acceleration is opposite to its direction of motion.
3. No. For an accelerating particle to stop at all, the velocity and acceleration must have opposite signs, so that the speed is decreasing. Given that this is the case, the particle will eventually come to rest. If the acceleration remains constant, however, the particle must begin to move again, opposite to the direction of its original motion. If the particle comes to rest and then stays at rest, the acceleration has disappeared at the moment the motion stops. This is the case for a braking car—the acceleration is negative and goes to zero as the car comes to rest.
5. Once the objects leave the hand, both are freely falling objects, and both experience the same downward acceleration equal to the free-fall acceleration, $-g$.
7. Yes. Yes.
9. (a) The velocity of the ball changes continuously but its acceleration is always the free-fall acceleration, $g = 9.8$ m/s^2 downward. As the ball travels upward, its speed decreases by 9.8 m/s during each second of its motion, and its velocity vector is upward. When it reaches the peak of its motion, its speed becomes zero. (If the acceleration were zero at the peak when the velocity is zero, there would be no change in velocity thereafter, so the ball would stop at the peak and remain there, which is not what happens.) (b) The acceleration of the ball remains constant in magnitude and direction throughout its free flight, from the instant it leaves the hand until the

instant before it strikes the ground. Its magnitude is the free-fall acceleration, $g = 9.8$ m/s^2.

11. No. Car A might have greater acceleration than B, but they might both have zero acceleration, or otherwise equal accelerations: or the driver of B might have stepped very hard on the gas pedal in the recent past.

13. They are the same! You can see this for yourself by solving the kinematic equations for the two cases.

15. Ignoring air resistance, in 16 s the pebble would fall a distance of $\frac{1}{2}gt^2 = \frac{1}{2}(9.80$ m/s$^2)(16.0$ s$) = 1250$ m $= 1.25$ km. Air resistance is an important force after the first few seconds, when the pebble has attained high speed. Also, part of the 16-s time interval must be occupied by the sound returning up the well. Thus the depth of the well is less than 1.25 km.

Problems

1. (a) 52.9 km/h   (b) 90.0 km
3. (a) Boat A by 60 km   (b) 0
5. 12.2 mph
7. (a) 2.34 min   (b) 64.2 miles
9. 2.81 h, 218 km
11. 76.1 m/s (274 km/h)
13. (a) 1.17 m/s   (b) 1.40 m/s
15. (a) 4.00 m/s   (b) -4.00 m/s   (c) 0
    (d) 2.00 m/s
17. 0.75 m/s^2
19. 3.7 s
21. (a) 8.00 m/s^2 (b) 11.0 m/s^2
23. (a) -8.0 m/s^2 (b) 100 m
25. (a) 35 s   (b) 16 m/s
27. (a) 3.0×10^{-10} s   (b) 1.26×10^{-4} m
29. (a) 1.3 m/s^2 (b) 8.0 s
31. (a) 5.51 km
    (b) 20.8 m/s, 41.6 m/s, 20.8 m/s, 38.7 m/s
33. (a) 110 m   (b) 1.49 m/s^2
35. 29.1 s
37. (a) 31.9 m   (b) 2.55 s   (c) 2.55 s
    (d) -25.0 m/s
39. (a) -21.1 m/s   (b) 19.6 m   (c) -18.1 m/s, 19.6 m
41. (a) 9.80 m/s   (b) 4.90 m
43. (a) 2.33 s   (b) -32.9 m/s
45. (a) -3.5×10^5 m/s^2 (b) 2.86×10^{-4} s
47. 1.03 s
49. 3.10 m/s
51. (a) 3.0 s   (b) -15.2 m/s
    (c) -31.4 m/s, -34.8 m/s
53. (a) 2.17 s   (b) 21.3 m/s   (c) 2.24 s
55. (a) 5500 ft   (b) 367 ft/s   (c) The plane would travel 0.002 ft in the time it takes light from the bolt to reach the eye.
57. $a = 1500$ m/s^2
59. (a) 2.64 s   (b) -20.9 m/s
    (c) 1.62 s, and 20.9 m/s down

CHAPTER 3

Multiple-Choice Questions
1. a   2. c   3. b   4. c   5. d

Conceptual Questions

1. (a) Yes. Although its speed may be constant, the direction of the motion may change, resulting in an acceleration. For example, an object moving in a circular path with constant speed has an acceleration because the direction of its velocity vector is changing. (b) No. An object that moves with constant velocity has zero acceleration. Note that constant velocity means that both the direction and magnitude of the velocity remain constant.

3. A constant velocity is interpreted as constant in both magnitude and direction. Thus, a constant velocity clearly implies a speed that does not change. On the other hand, a constant speed indicates that only the magnitude of the velocity vector is constant—the direction can change. Thus, as long as an object covers the same distance in equal time intervals, the speed is constant, and its direction can change continuously, giving a varying velocity. A familiar example is an object moving in a circular path at constant speed, such that the velocity is constantly changing direction.

5. You should throw it straight up in the air. Because the ball was moving along with you, it will follow the path of a projectile moving with a constant horizontal velocity that is the same as yours.

7. The components of **A** will both be negative when **A** lies in the third quadrant. The components of **A** will have opposite signs when **A** lies in either the second quadrant or the fourth quadrant.

9. (a) At the top of its flight, the velocity is horizontal and the acceleration is downward. This is the only point at which the velocity and acceleration vectors are perpendicular. (b) If the object is thrown straight up or down, then the velocity and acceleration will be parallel throughout the motion. For any other kind of projectile motion, the velocity and acceleration vectors are never parallel.

11. (a) The acceleration is zero, because |**v**| and its direction remain constant. (b) The particle has an acceleration because the direction of **v** changes.

13. The spacecraft will follow a parabolic path. This is equivalent to a projectile thrown off a cliff with a horizontal velocity. For the projectile, gravity provides an acceleration that is always perpendicular to the initial velocity, resulting in a parabolic path. For the spacecraft, the initial velocity plays the role of the horizontal velocity of the projectile. The acceleration provided by the leaking gas provides an acceleration that plays the role of gravity for the projectile. If the orientation of the spacecraft were to change in response to the gas leak (which is by far the more likely result), then the acceleration would change

direction and the motion could become quite complicated.

15. For angles $\theta < 45°$, the projectile thrown at angle θ will be in the air for a shorter time interval. For a small angle, the vertical component of the initial velocity is smaller than for the larger angle. Thus, the projectile thrown at the smaller angle will not go as high into the air and will spend less time in the air before landing.

17. Yes. The projectile is a freely falling body, because nothing counteracts the force of gravity. The vertical acceleration will be the local gravitational acceleration, g; the horizontal acceleration will be zero.

19. (a) The driver must drive in a straight line. (b) The driver must hold the steering wheel at a steady angle, so that the car will turn at a constant radius from the center of the turn.

Review Problem

(a) 246.2 ft at 35.7° west of south (from C to A)
(b) 0.714 acres

Problems

1. 7.92 m at 4.34° north of west
3. (a) 5.2 m at 60° above x axis
   (b) 3.0 m at 30° below x axis
   (c) 3.0 m at 150° with the x axis
   (d) 5.2 m at 60° below x axis
5. 15.3 m at 58° south of east
7. 8.07 m at 42° south of east
9. (a) 5 blocks at 53.1° north of east   (b) 13 blocks
11. 157 km
13. 245 km at 21.4° west of north
15. 196 cm at 14.7° below x axis
17. 2.8 m from base of table, $v_x = 5.0$ m/s, $v_y = -5.4$ m/s
19. 25 m
21. (a) 32.5 m horizontally from base of cliff   (b) 1.78 s
23. (a) 52 m/s directed horizontally   (b) 210 m
25. 61 s
27. 249 ft upstream
29. 18 s
31. (a) 10.0 m   (b) 15.7 m   (c) 0
33. (a) approximately 2.3 m/s horizontally
35. (a) 57.7 km/h at 60° west of vertical
    (b) 28.9 km/h downward
37. (a) 1.52×10^3 m   (b) 36.1 s   (c) 4.05×10^3 m
39. 18 m on the Moon, 7.9 m on Mars
41. (a) 23 m/s   (b) 360 m horizontally from base of cliff
43. (a) 42 m/s   (b) 3.8 s
    (c) $v_x = 34$ m/s, $v_y = -13$ m/s, $v = 37$ m/s
45. 7.5 m/s in the direction the ball was thrown
49. $R/2$
51. 7.5 min
53. 10.8 m above cannon

55. (a) and (b) $y = -\dfrac{1}{2} gt^2 = -\dfrac{g}{2}\left(\dfrac{x}{v_0}\right)^2 = Ax^2$
    where $A = -\dfrac{g}{2v_0{}^2}$ (c) 14.6 m/s
57. 227 paces at 165°
59. (a) 20.0°   (b) 3.05 s

CHAPTER 4

Multiple-Choice Questions
1. e   2. c   3. b   4. b   5. b   6. a

Conceptual Questions
1. (a) If a single force acts on it, the object must accelerate. If an object accelerates, at least one force must act on it. (b) If an object has no acceleration, you cannot conclude that no forces act on it. In this case, you can only say that the net force on the object is zero.
3. Motion can occur in the absence of a net force. Newton's first law holds that an object will continue to move with a constant speed and in a straight line if there is no net force acting on it.
5. The inertia of the suitcase would keep it moving forward as the bus stops. There would be no tendency for the suitcase to be thrown backward toward the passenger. The case should be dismissed.
7. The force causing an automobile to move is the force of friction between the tires and the roadway as the automobile attempts to push the roadway backward. The force driving a propeller airplane forward is the reaction force of the air on the propeller as the rotating propeller pushes the air backward (the action). In a rowboat, the rower pushes the water backward with the oars (the action). The water pushes forward on the oars and hence the boat (the reaction).
9. When the bus starts moving, the mass of Claudette is accelerated by the force of the back of the seat on her body. Clark is standing, however, and the only force on him is the friction between his shoes and the floor of the bus. Thus, when the bus starts moving, his feet start accelerating forward, but the rest of his body experiences almost no accelerating force (except that due to his being attached to his accelerating feet!). As a consequence, his body tends to stay almost at rest, according to Newton's first law, relative to the ground. Relative to Claudette, however, he is moving toward her and falls into her lap.
11. The tension in the rope is the maximum force that occurs in *both* directions. In this case, then, because both are pulling with a force of 200 N, the tension is 200 N. If the rope does not move, then the force on each athlete must equal zero. Therefore, each athlete exerts 200 N against the ground.

13. As a man takes a step, the action is the force his foot exerts on the Earth; the reaction is the force of the Earth on his foot. In the second case, the action is the force exerted on the girl's back by the snowball; the reaction is the force exerted on the snowball by the girl's back. The third action is the force of the glove on the ball; the reaction is the force of the ball on the glove. The fourth action is the force exerted on the window by the air molecules; the reaction is the force on the air molecules exerted by the window.

15. The brakes may lock and the car will slide farther than it would if the wheels continued to roll because the coefficient of kinetic friction is less than the coefficient of static friction. Hence, the force of kinetic friction is less than the maximum force of static friction.

17. (a) The crate accelerates because of a friction force exerted on it by the floor of the truck. (b) If the driver slams on the brakes, the inertia of the crate tends to keep it moving forward.

Problems

1. (a) 12 N   (b) 3.0 m/s^2
3. 3.71 N, 58.7 N, 2.27 kg
5. 9.6 N
7. (a) 0.20 m/s^2 (b) 10.0 m   (c) 2.00 m/s
9. 1.59 m/s^2 at 65.2° north of east
11. 796 N in the horizontal cable, 997 N in the inclined cable, 600 N in the vertical cable.
13. 75 N (right side cable), 130 N (left side cable), 150 N (vertical cable)
15. (a) 34 N   (b) 39 N
17. 64 N
19. 7500 N, 50.0 m
21. (a) 7.0 m/s^2 (b) 21 N   (c) 14 N toward the right
23. 13 N down the incline
25. 6.53 m/s^2, 32.7 N
27. (a) 2.15×10^3 N   (b) 645 N   (c) 645 N to rear
    (d) 1.02×10^4 N at 15.9° to rear of vertically downward
29. $\mu_S = 0.38$, $\mu_k = 0.31$
31. (a) 0.256   (b) 0.509 m/s^2
33. 32.1 N
35. (a) 32.7 m/s   (b) No. The object can speed up to 32.7 m/s from any higher speed or slow down to 32.7 m/s from any higher speed.
37. 0.288
39. 0.436
41. 3.32 m/s^2
43. (a) 0.404   (b) 45.8 lb
45. (a) 84.9 N vertically   (b) 84.9 N
47. 50 m
49. (a) friction between the box and truck   (b) 2.94 m/s^2
51. (a) 0.232 m/s^2 (b) 9.68 N
53. (a) 1.78 m/s^2 (b) 0.368   (c) 9.37 N
    (d) 2.67 m/s

55. (a) 1.7 m/s^2, 17 N   (b) 0.69 m/s^2, 17 N
57. 100 N, 204 N
59. (a) $T_1 = 78$ N, $T_2 = 35.9$ N   (b) $\mu_k = 0.66$
61. 23 m/s
63. (a) 7250 N   (b) 4.57 m/s^2
65. 72.0 N
67. 9.95 s
69. (a) 0.408 m/s^2 (b) 83.3 N

CHAPTER 5

Multiple-Choice Questions
1. c   2. b   3. b   4. d   5. c

Conceptual Questions
1. Because there is no motion taking place, the rope experiences no displacement. Thus, no work is done on it. For the same reason, no work is being done on the pullers or the ground. Work is only being done within the bodies of the pullers. For example, the heart of each puller is applying forces on the blood to move blood through the body.

3. Less force will be necessary with a longer ramp, but the force must act over a longer distance to do the same amount of work. Suppose a refrigerator is rolled up the ramp at constant speed. The normal force does no work because it acts at 90° to the motion. The work by gravity is just the weight of the refrigerator times the vertical height through which it is displaced ($W_g = -mgh$). Therefore, the movers must do work mgh on the refrigerator, however long the ramp.

5. If we ignore any effects due to rolling friction on the tires of the car, the same amount of work would be done in driving up the switchback and driving straight up the mountain, because the weight of the car is moved upward against gravity by the same vertical distance in each case. If we include friction, there is more work done in driving the switchback, because the distance over which the friction force acts is much longer. So why do we use switchbacks? The answer lies in the force required, not the work. The force from the engine required to follow a gentle rise is much smaller than that required to drive straight up the hill. Roadways running straight uphill would require redesigning engines in order to be able to apply much larger forces. This is similar to the relative ease with which heavy objects can be moved up ramps into trucks as in Conceptual Question 3.

7. (a) The tension in the supporting cord does no work, because the motion of the pendulum is always perpendicular to the cord, and therefore to the tension force. (b) The air resistance does negative work at all times, because the air resistance is always acting in a direction opposite to the motion. (c) The force of gravity always acts downward; therefore, the work done by gravity is positive on the downswing and negative on the upswing.

9. Because the time periods are the same for both cars, we need only to compare the work done. Because the sports car is moving twice as fast as the older car at the end of the time interval, it has four times the kinetic energy. Thus, according to the work-kinetic energy theorem, four times as much work was done, and the engine must have expended four times the power.

11. If a crate is located on the bed of a truck, and the truck accelerates, the friction force acting on the crate causes it to undergo the same acceleration as the truck, assuming the crate doesn't slip. Another example is a car that accelerates because of the frictional forces between the road surface and its tires. This force is in the direction of motion of the car and produces an increase in the car's kinetic energy.

13. Yes, the total mechanical energy of the system is conserved because the only forces acting are conservative—the force of gravity and the spring force. There are two forms of potential energy in this case, gravitational potential energy and elastic potential energy stored in the spring.

15. Let us assume you lift the book slowly. In this case there are two forces on the book that are almost equal in magnitude. They are the lifting force and the force of gravity acting on the book. Thus, the positive work done by you and the negative work done by gravity cancel. There is no net work performed, and no net change in the kinetic energy—the work-kinetic energy theorem is satisfied.

17. As the satellite moves in a circular orbit about the Earth, its displacement along the circular path during any small time interval is always perpendicular to the gravitational force, which always acts toward the center of the Earth. Therefore, the work done by the gravitational force during any displacement is zero. (Recall that the work done by a force is defined to be $Fs \cos \theta$, where θ is the angle between the force and the displacement. In this case, the angle is 90°, so the work done is zero.) Because the work-kinetic energy theorem says that the net work done on an object during any displacement is equal to the change in its kinetic energy, and the work done in this case is zero, the change in the satellite's kinetic energy is zero; hence its speed remains constant.

Review Problem
(a) 9.3 m/s   (b) 9.3 m/s

Problems
1. 700 J
3. 15.0 MJ
5. (a) 61 J   (b) −46 J   (c) 0
7. (a) 2.00 m/s   (b) 200 N
9. (a) 560 J   (b) 1.2 m
11. 0.265 m/s
13. (a) 90 J   (b) 180 N
15. 1.0 m/s

17. (a) 3.15×10^5 J, 0, 3.15×10^5 J
    (b) 0, -3.15×10^5 J, 3.15×10^5 J
19. 1.9 m/s
21. (a) 80 J   (b) 11 J   (c) 0
23. 0.459 m
25. (a) 0.588 J   (b) 0.588 J   (c) 2.42 m/s
    (d) $PE_C = 0.392$ J, $KE_C = 0.196$ J
27. 5.1 m/s
29. 61 m
31. (a) 9.90 m/s   (b) 11.8 J
33. 2.1×10^3 N
35. 3.8 m/s
37. (a) 5.42 m/s   (b) 0.30   (c) 147 J
39. 289 m
41. (a) 2.1×10^4 J   (b) 0.92 hp
43. 590 MW
45. (a) 2.39×10^4 W   (b) 4.77×10^4 W
47. 2.9 m/s
49. 9.80 m/s
51. 1.4 m/s
53. 3.9×10^3 J
55. (a) 2.3×10^3 N   (b) 4.7×10^2 N
57. (a) 28.0 m/s   (b) 30.0 m   (c) 89.0 m
59. (a) 310 J   (b) −150 J   (c) 0   (d) 150 J
61. (a) javelin requires 349 J, discus requires 676 J, and shot requires 741 J
    (b) 175 N on javelin, 338 N on discus, 371 N on shot
    (c) Yes
63. (a) 6.15 m/s   (b) 9.87 m/s
67. (a) 100 J   (b) 0.410 m   (c) 2.84 m/s
    (d) −9.80 mm   (e) 2.85 m/s
69. 914 N/m

CHAPTER 6

Multiple-Choice Questions
1. a   2. b   3. d   4. c   5. b

Conceptual Questions
1. If all the kinetic energy disappears, there must be no motion of either of the objects after the collision. If nothing is moving there is no momentum. Thus, if the final momentum of the system is zero, then the initial momentum of the system must also have been zero. A situation in which this could be true would be the head-on collision of two objects, each having equal but opposite momenta.

3. Initially the clay has momentum directed toward the wall. When it collides and sticks to the wall, nothing appears to have any momentum. Thus, it is tempting to conclude (wrongly) that momentum is not conserved. However, the "lost" momentum is actually imparted to the wall and Earth, causing both to move. Because of the enormous mass of the Earth, its recoil speed is too small to detect.

5. Before the step the momentum was zero, so afterward the net momentum must also be zero. Obviously, you have some momentum, so something must have momentum in the opposite direction. That something is the Earth. As noted in Conceptual Question 3, the enormous mass of the Earth ensures that its recoil speed will be too small to detect, but if you want to make the Earth move, it is as simple as taking a step.

7. The momentum of the system remains constant, so if the boy gains momentum away from the shore, something must have an equal and opposite momentum toward the shore so that the total momentum of the system remains zero. This something is the raft.

9. There are no external horizontal forces acting on the box, so its momentum cannot change as it moves along the horizontal surface. As the box slowly fills with water, its mass increases with time. Because the product mv must be a constant, and m is increasing, the speed of the box must decrease.

11. It will be easiest to catch the medicine ball when its speed (and kinetic energy) is lowest. The first option — throwing the medicine ball at the same velocity — will be the most difficult, because the speed will not be reduced at all. The second option, throwing the medicine ball with the same momentum, will reduce the velocity by the ratio of the masses. Because $m_t v_t = m_m v_m$,

$$v_m = v_t \left(\frac{m_t}{m_m} \right)$$

The third option, throwing the medicine ball with the same kinetic energy, will also reduce the velocity, but only by the square root of the ratio of the masses. Because $\frac{1}{2} m_t v_t^2 = \frac{1}{2} m_m v_m^2$,

$$v_m = v_t \sqrt{\frac{m_t}{m_m}}$$

Thus, the slowest — and easiest — throw will be made when the momentum is held constant. If you wish to check this, try substituting in values of $v_t = 1$ m/s, $m_t = 1$ kg, and $m_m = 100$ kg. The same-momentum throw will be caught at 1 cm/s, and the same-energy throw will be caught at 10 cm/s.

13. The follow-through keeps the club in contact with the ball as long as possible, maximizing the impulse. Thus, the ball experiences a larger change in momentum than without the follow-through, and it leaves the club with a higher velocity and it travels farther. With a short shot to the green, the primary factor is control, not distance. Thus, there is little or no follow-through, allowing for the golfer to have a better feel for how hard he or she is striking the ball.

15. It is the product mv that is the same for both the bullet and the gun. The bullet has a large velocity and a small mass, and the gun has a small velocity and a large mass. Furthermore, the bullet carries much more kinetic energy than the gun.

Problems

3. (a) 0   (b) 1.1 kg m/s
5. 7.5×10^4 N
7. (a) -7.50 kg m/s   (b) 375 N
9. (a) 12.0 N s   (b) 6.0 m/s   (c) 4.0 m/s
11. (a) 0.096 s   (b) 3.65×10^5 N   (c) 26.5 g
13. (a) 6.3 kg m/s
     (b) 3.2×10^3 N (both directed toward the pitcher)
15. 62 s
17. (a) 0.49 m/s   (b) 2.0×10^{-2} m/s
19. $v_{\text{thrower}} = 2.48$ m/s, $v_{\text{catcher}} = 2.25 \times 10^{-2}$ m/s
21. 2.66 m/s
23. (a) 1.80 m/s   (b) 2.16×10^4 J
25. 57 m
27. 273 m/s
29. (a) -6.67 cm/s, 13.3 cm/s   (b) 0.88
31. 17.1 cm/s (25-g object), 22.1 cm/s(10-g object)
33. (a) 2.9 m/s at 32° with respect to initial direction of travel of fullback   (b) 780 J
35. 5.59 m/s
37. (a) 2.5 m/s at $-60°$ with respect to the original line of motion of incident ball   (b) elastic
39. 14.8 kg·m/s in the direction of the final velocity of the ball
41. (a) $\frac{8}{3}$ m/s (incident particle), $\frac{32}{3}$ m/s (target)
     (b) $\frac{-16}{3}$ m/s (incident particle), $\frac{8}{3}$ m/s (target)
     (c) 7.1×10^{-3} J, in case (a), 2.8×10^{-3} J in case (b). The incident particle loses more energy in case (a).
43. 0.96 m above the level of point B
45. 91.2 m/s
47. (a) -2.33 m/s, 4.67 m/s   (b) 0.277 m   (c) 2.98 m
     (d) 1.49 m
49. (a) -0.67 m/s   (b) 0.953 m
51. (a) 300 m/s   (b) 3.75 m/s   (c) 1.20 m
53. 2.36 cm
55. $v = \frac{4M}{m} \sqrt{g\ell}$
57. about 10^{-23} m/s
59. (a) 90°   (b) $v_1 = 3.46$ m/s and $v_2 = 2.00$ m/s

CHAPTER 7

Multiple-Choice Questions

1. e   2. a   3. b   4. e   5. a

Conceptual Questions

1. An object can move in a circle even if the total force on it is not perpendicular to its velocity, but then its speed will change. Resolve the total force into an inward radial component and a perpendicular tangential component. If the tangential force is forward, the object will speed up, and if the tangential force acts backward, it will slow down.

3. Yes, all points on the wheel have the same angular speed. This is why we use angular quantities to describe rotational motion. Not all points on the wheel have the same linear speed. The point at $r = 0$ has zero linear speed and zero linear acceleration; a point at $r = R/2$ has a linear speed $v = R\omega/2$ and a linear acceleration equal to the centripetal acceleration $v^2/(R/2) = R\omega^2/2$. (The tangential acceleration is zero at all points because ω is a constant.) A point on the rim at $r = R$ has a linear speed $v = R\omega$ and a linear acceleration $R\omega^2$.

5. To a good first approximation, your bathroom scale reading is unaffected because you, Earth, and the scale are all in free fall in the Sun's gravitational field, in orbit around the Sun. To a precise second approximation, you weigh slightly less at noon and at midnight than you do at sunrise or sunset. The Sun's gravitational field is a little weaker at the center of the Earth than at the surface subsolar point, and a little weaker still on the far side of the planet. When the Sun is high in your sky, its gravity pulls up on you a little more strongly than on the Earth as a whole. At midnight the Sun pulls down on you a little less strongly than it does on the Earth below you. So you can have another doughnut with lunch, and your bedsprings will still last a little longer.

7. Air resistance causes a decrease in the energy of the satellite-Earth system. This reduces the diameter of the orbit, bringing the satellite closer to the surface of the Earth. A satellite in a smaller orbit, however, must travel faster. Thus, the effect of air resistance is to speed up the satellite!

9. Kepler's third law, which applies to all planets, tells us that the period of a planet is proportional to $r^{3/2}$. Because Saturn and Jupiter are farther from the Sun than Earth, they have longer periods. The Sun's gravitational field is much weaker at a distant Jovian planet. Thus, an outer planet experiences much smaller centripetal acceleration than Earth and a correspondingly longer period.

11. Consider an individual standing against the inside wall of the cylinder with her head pointed toward the axis of the cylinder. As the cylinder rotates, the person tends to move in a straight-line path tangent to the circular path followed by the cylinder wall. As a result, the person is forced against the wall, and the normal force exerted on her provides the radial force required to keep her moving in a circular path. If the rotational speed is adjusted such that this normal force is equal in magnitude to her weight on Earth, she would not be able to distinguish between the artificial gravity of the colony and ordinary gravity.

13. (a) As the object moves in its circular path with constant speed, the *direction* of the velocity vector changes. Thus, the velocity of the object is not constant. (b) The magnitude of its acceleration remains constant, and is equal to v_t^2/r. The acceleration vector is always directed toward the center of the circular path.

15. Any object that moves such that the *direction* of its velocity changes has an acceleration. A car moving in a circular path will always have a centripetal acceleration.

17. When a pilot is pulling out of a dive, blood leaves the head because there is not a great enough radial force to cause it to follow the circular path of the airplane. The loss of blood from the brain can cause the pilot to black out.

19. The tendency of the water is to move in a straight-line path tangent to the circular path followed by the container. As a result, at the top of the circular path, the water is forced against the bottom of the pail, and the normal force exerted on the water by the pail provides the radial force required to keep the water moving in its circular path.

Review Problem
(a) 3.77 m/s^2 (b) 3.26 s

Problems
1. (a) $3.2 \times 10^8 \text{ rad}$ (b) 5.0×10^7 revolutions
3. (a) 3.5 rad/s (b) 5.2 rad
5. $4.2 \times 10^{-2} \text{ rad/s}^2$
7. 3.2 rad
9. (a) 5.24 s (b) 27.4 rad
11. 13.7 rad/s^2
13. 1.02 m
15. (a) -0.17 rad/s^2 (b) 5.5 rev (c) 0.48 m/s
17. $5.0 \times 10^{-2} \text{ rad/s}$
19. (a) $3.5 \times 10^{-1} \text{ m/s}^2$ (b) 1.0 m/s
    (c) $3.5 \times 10^{-1} \text{ m/s}^2$, 0.94 m/s^2, 1.0 m/s^2 at $20°$ with respect to the direction of $\mathbf{a}_r$
21. 150 rev/s
23. (a) 18.0 m/s^2 (b) 900 N (c) 1.84; a coefficient of friction greater than 1 is unreasonable. She will not be able to stay on the merry-go-round.
25. (b) $20.1°$
27. The required tension in the vine is $1.4 \times 10^3 \text{ N}$. He does not make it.
29. (a) $2.49 \times 10^4 \text{ N}$ (b) 12.1 m/s
31. 321 N toward Earth
33. $1.1 \times 10^{-10} \text{ N}$ at $72°$ from $+x$ axis
35. $6.01 \times 10^{24} \text{ kg}$. The estimate is high because the Moon actually orbits about the center of mass of the Earth–Moon system not about the center of the Earth.
37. (a) $5.58 \times 10^3 \text{ m/s}$ (b) 239 min (c) $1.47 \times 10^3 \text{ N}$
39. (a) $-4.76 \times 10^9 \text{ J}$ (b) $5.68 \times 10^2 \text{ N}$
41. (a) 88.5 min (b) 7.79 km/s (c) $6.43 \times 10^9 \text{ J}$
43. (a) $2.3 \times 10^{-10} \text{ N}$ (in $-x$ direction)
    (b) $1.0 \times 10^{-10} \text{ N}$ (in $+x$ direction)
45. $1.90 \times 10^{27} \text{ kg}$
47. (a) $7.8 \times 10^3 \text{ m/s}$ (b) $5.4 \times 10^3 \text{ s}$ (89 min)
49. (a) $n = mg - \left(\dfrac{mv^2}{r}\right)$ (b) 17.1 m/s
51. (a) $2.0 \times 10^{12} \text{ m/s}^2$ (b) $2.4 \times 10^{11} \text{ N}$
    (c) $1.4 \times 10^{12} \text{ J}$
53. 0.131
55. (a) 2.1 m/s (b) $54°$ (c) 4.7 m/s
57. 35.2 A.U.

59. (a) 1.31×10^{17} N   (b) 2.62×10^{12} N/kg
63. (a) 15.1 N at A, 29.8 N at B (b) 0.166
65. (a) $v_{min} = [Rg(\tan \theta - \mu)/(1 + \mu \tan \theta)]^{1/2}$,
    $v_{max} = [Rg(\tan \theta + \mu)/(1 - \mu \tan \theta)]^{1/2}$
    (b) 8.57 m/s to 16.6 m/s

CHAPTER 8

Multiple-Choice Questions
1. b   2. c   3. c   4. d   5. a

Conceptual Questions
1. There are two major differences between torque and work. The primary difference is that the displacement in the expression for work is directed along the force, whereas the important distance in the torque expression is perpendicular to the force. The second difference involves whether there is motion or not—in the case of work, there is work done only if the force succeeds in causing a displacement of the point of application of the force. However, a force applied at a perpendicular distance from a rotation axes results in a torque whether or not there is motion.

   As far as units are concerned, the mathematical expressions for both work and torque result in the product of newtons and meters, but this product is called a joule in the case of work and remains as a newton-meter in the case of torque.

3. On a frictionless table, the only forces on the pencil are the gravitational force, at the center of the mass, and the normal force on the pencil. These forces result in a torque on the pencil causing it to rotate. Both of these forces, however, are vertical—there are no horizontal forces on the pencil in the absence of friction. As a result, the center of mass of the pencil must fall straight downward. As the pencil falls, the tip slides to the side, so that the center of mass of the pencil follows a straight line downward to the floor.

5. As the motorcycle leaves the ground, the friction between the tire and the ground suddenly disappears. If the motorcycle driver keeps the throttle open while leaving the ground, the rear tire will increase its angular velocity and, hence, its angular momentum. The airborne motorcycle is now an isolated system, and its angular momentum must be conserved. The increase in angular momentum of the tire, say, directed clockwise must be compensated by an increase in angular momentum of the entire motorcycle counterclockwise. This rotation results in the nose of the motorcycle rising and the tail dropping.

7. In general, you want the rotational kinetic energy of the system to be as small a fraction of the total energy as possible—you want translation, not rotation. You want the wheels to have as little moment of inertia as possible,

so that they represent the lowest resistance to changes in rotational motion. Disk-like wheels would have lower moments of inertia than hoop-like wheels, so disks are preferable. The lower the mass of the wheels, the less is the moment of inertia, so light wheels are preferable. The smaller the radius of the wheels, the less is the moment of inertia, so smaller wheels are preferable, within limits— you want the wheels to be large enough to be able to travel relatively smoothly over irregularities in the road.

9. The angular momentum of the gas cloud is conserved. Thus the product $I\omega$ remains constant. Hence, as the cloud shrinks in size, its moment of inertia decreases, so its angular speed ω must increase.

11. The Earth already bulges slightly at the Equator and is slightly flat at the poles. If more mass moved toward the Equator, it would essentially move the mass to a greater distance from the axis of rotation, and increase the moment of inertia. Because conservation of angular momentum requires that $I_z\omega_z = $ constant, an increase in the moment of inertia would decrease the angular velocity and slow down the spinning of the Earth. Thus, the length of each day would increase.

13. We can assume fairly accurately that the driving motor will run at a constant angular speed and at a constant torque. Therefore, as the radius of the takeup reel increases, the tension in the tape will decrease.

   (1) $T = \tau_{const}/R_{takeup}$

   (2) $\tau_{source} = FR_{source} = \tau_{const}R_{source}/R_{takeup}$

   As the radius of the source reel decreases, given a decreasing tension, the torque in the source reel will **decrease** even faster.

   This torque will be partly absorbed by friction in the feed heads (which we assume to be small); some will be absorbed by friction in the source reel. Another small amount of the torque will be absorbed by the increasing angular velocity of the source reel. However, in the case of a sudden jerk on the tape, the changing angular velocity of the source reel becomes important. If the source reel is full, then the moment of inertia will be large, and the tension in the tape will be large. If it is nearly empty, then the angular acceleration will be large instead. Thus, the tape will be more likely to break when the source reel is nearly **full.** One sees the same effect in the case of paper towels; it is easier to snap a towel free when the roll is new than when it is nearly empty.

15. The initial angular momentum of the system (mouse plus turntable) is zero. As the mouse begins to walk clockwise, its angular momentum increases, so the turntable must rotate in the counterclockwise direction with an angular momentum whose magnitude equals that of the mouse. This follows from the fact that the final angular momen-

tum of the system must equal the initial angular momentum, which is zero in this case.

17. When a ladder leans against a wall, both the wall and floor exert forces of friction on the ladder. If the floor is perfectly smooth, it can exert no frictional force in the horizontal direction to counterbalance the wall's normal force. Therefore a ladder on a smooth floor cannot stand in equilibrium. However, a smooth wall can still exert a normal force to hold the ladder in equilibrium against horizontal motion. The counterclockwise torque of this force prevents rotation about the foot of the ladder. So you should choose a rough floor.

Review Problem

(a) 3.46×10^5 km from center of Earth
(b) 4.68×10^3 km from center of Earth. No, the two points do not coincide.

Problems

1. 133 N
3. 5.1 N·m
5. (a) 29.6 N·m (counterclockwise)
   (b) 35.6 N·m (counterclockwise)
7. $x_{cg} = 6.69 \times 10^{-3}$ nm, $y_{cg} = 0$
9. $x_{cg} = 3.33$ ft, $y_{cg} = 1.67$ ft
11. 312 N
13. $T = 157$ N, $R = 107$ N
15. T(left wire) $= \frac{1}{3}W$, T(right wire) $= \frac{2}{3}W$
17. (a) 443 N (b) 222 N (to right) 216 N (upward)
19. $T_1 = 501$ N, $T_2 = 672$ N, $T_3 = 384$ N
21. 6.2 m
23. 209 N
25. (a) 99.0 kg·m² (b) 44.0 kg·m² (c) 140 kg·m²
27. $\mu_k = 0.31$
29. 177 N
31. (a) 5.35 m/s² downward (b) 42.8 m
    (c) 8.91 rad/s²
33. 17.3 rad/s
35. (a) 500 J (b) 250 J (c) 750 J
37. 276 J
39. 149 rad/s
41. 17.3 rad/s
43. 17.5 J·s counterclockwise
45. 8.0 rev/s
47. (a) 1.9 rad/s (b) $KE_i = 2.5$ J, $KE_f = 6.4$ J
49. (a) 3.6 rad/s (b) 540 J (difference results from work done by the man as he walks inward)
51. (a) 1.2 kg (b) at the 59.6 cm mark
53. $T = 2.70 \times 10^3$ N, $R_x = 2.65 \times 10^3$ N
55. 36.9 °
57. $a_{\text{sphere}} = \dfrac{g \sin \theta}{1.4}$, $a_{\text{cylinder}} = \dfrac{g \sin \theta}{1.5}$, $a_{\text{ring}} = \dfrac{g \sin \theta}{2.0}$
   Thus, sphere wins and ring comes in last.
59. (a) 1.1 m/s² (b) $T_1 = 22$ N $T_2 = 44$ N
61. (a) 24.0 N·m (b) 0.0356 rad/s² (c) 1.07 m/s²
63. (a) 3750 kg·m²/s (b) 1.88 kJ (c) 3750 kg·m²/s
    (d) 10.0 m/s (e) 7.5 kJ (f) 5.62 kJ
65. $T_R = 1589$ N, $T_L = 1006$ N
67. (a) 0.36 rad/s, counterclockwise (b) 99.9 J
69. 24 m
71. $\frac{3}{8} w$

CHAPTER 9

Multiple-Choice Questions

1. b   2. d   3. c   4. c   5. e

Conceptual Questions

1. She can exert enough pressure on the floor to dent or puncture the floor covering. The large pressure is caused by the fact that her weight is distributed over the very small cross-sectional area of her high heels. If you are the homeowner, you might want to suggest that she remove her high heels and put on some slippers.

3. If you think of the grain stored in the silo as a fluid, the pressure the grain exerts on the walls of the silo increases with increasing depth just as water pressure in a lake increases with increasing depth. Thus, the spacing between bands is made smaller at the lower portions to overcome the larger outward forces on the walls in these regions.

5. The level of floating of a ship is unaffected by atmospheric pressure. The buoyant force results from the pressure differential in the fluid. On a high-pressure day, the pressure at all points in the water is higher than on a low-pressure day. Because water is almost incompressible, however, the rate of change of pressure with depth is the same, resulting in no change in the buoyant force.

7. In the ocean, the ship floats due to the buoyant force from *salt water*. Salt water is denser than fresh water. As the ship is pulled up the river, the buoyant force from the fresh water in the river is not sufficient to support the weight of the ship, and it sinks.

9. The balance will not be in equilibrium—the lead side will be lower. Despite the fact that the weights on both sides of the balance are the same, the styrofoam, due to its larger volume, will experience a larger buoyant force from the surrounding air. Thus, the net force of the weight and the buoyant force is larger, in the downward direction, for the lead than for the styrofoam.

11. The level of the pond falls. This is because the anchor displaces more water while in the boat. A floating object displaces a volume of water whose weight is equal to the weight of the object. A submerged object displaces a volume of water equal to the volume of the object. Because

the density of the anchor is greater than that of water, a volume of water that weighs the same as the anchor will be greater than the volume of the anchor.

13. According to Archimedes's principle, the magnitude of the buoyant force on the ship is equal to the weight of the water displaced by the ship. Because the density of salty ocean water is greater than fresh lake water, less ocean water needs to be displaced to enable the ship to float. Thus, the boat floats higher in the ocean than in the inland lake.

15. As the truck passes, the air between your car and the truck is compressed into the channel between you and the truck and moves at a higher speed than when your car is in the open. According to Bernoulli's principle, this high-speed air has a lower pressure than the air on the outer side of your car. The difference in pressure provides a net force on your car toward the truck.

Problems

1. 1.8×10^6 Pa
3. 3.5×10^8 Pa
5. 5.5 mm
7. 2.4×10^{-2} mm
9. (a) 1.88×10^5 Pa　　(b) 2.65×10^5 Pa
11. 1.2×10^6 Pa
13. 3.58×10^6 Pa
15. 1.4 atm
17. 1.05×10^5 Pa
19. 2.0 N·m
21. 9.41×10^3 N
23. 5.57 N
25. (a) 1.46×10^{-2} m^3　　(b) 2.10×10^3 kg/m^3
27. 1.07 m/s^2
29. 78 kg
31. 154 in/s
33. (a) 80 g/s　　(b) 2.7×10^{-2} cm/s
35. (a) 11.0 m/s　　(b) 2.64×10^4 Pa
37. 4.4×10^{-2} Pa
39. (a) 17.7 m/s　　(b) 1.73 mm
41. (a) 2.7 m/s　　(b) 2.3×10^4 Pa
43. 7.32×10^{-2} N/m
45. 5.6 m
47. 0.694 mm
49. 0.12 N
51. 1.5 m/s
53. 0.41 mm
55. 8.0 cm/s
57. 9.5×10^{-10} m^2/s
59. 1.02×10^3 kg/m^3
61. 2.5×10^7 capillaries
63. 833 kg/m^3
65. 1.9 m
67. 15 m
69. (a) 18.34 mm　　(b) 14.31 mm　　(c) 8.56 mm
71. (a) 1.25 cm　　(b) 13.8 m/s
73. 0.72 mm
75. (b) 2.6×10^4 N
77. 1.71 cm

CHAPTER 10

Multiple-Choice Questions

1. b　　2. d　　3. b　　4. a　　5. b

Conceptual Questions

1. The accurate answer is that it doesn't matter! Temperatures on the Kelvin and Celsius scales differ by only 273 degrees. This difference is insignificant for temperatures on the order of 10^7 degrees. If we imagine that the temperature is given in kelvin, and ignore any problems with significant figures, then the Celsius temperature is 1.4999727×10^{7}°C.

3. Mercury must have the larger coefficient of expansion. As the temperature of a thermometer rises, both the mercury and the glass expand. If they both had the same coefficient of linear expansion, the mercury and the cavity in the glass would both expand by the same amount, and there would be no apparent movement of the end of the mercury column relative to the calibration scale on the glass. If the glass expanded more than the mercury, the reading would go down as the temperature went up! Now that we have argued this conceptually, we can look in a table and find that the coefficient for mercury is about 20 times as large as for glass, so that the expansion of the glass can sometimes be ignored.

5. This occurs because the water in the center of the glass is being heated, but the glass itself is not rising in temperature as rapidly as the water. The bubbles associated with boiling usually first appear at irregularities on the surface of the glass, but if the glass is not yet at 100°C, boiling does not take place. The superheated liquid is metastable and lacks nuclei for bubble formation.

7. A cavity in a material expands in exactly the same way as if the cavity were filled with material. Thus, both spheres will expand by the same amount.

9. The chip bags contain a sealed sample of air. When the bags are taken up the mountain, the external atmospheric pressure on the bags is reduced. As a result, the difference between the pressure of the air inside the bags and the reduced pressure outside results in a net force pushing the plastic of the bag outward.

11. On a cold day, the trapped air in the bubbles would be reduced in pressure, according to the ideal gas law. Thus, the volume of the bubbles may be smaller than on a hot day, and the material would not be as effective in cushioning the package contents.

13. One can think of each bacterium as being a small bag of liquid containing bubbles of gas at a very high pressure. If the bacterium is raised rapidly to the surface, the ideal gas law indicates that the volume must increase dramatically. In fact, this increase in volume is sufficient to rupture the cell.

15. The existence of an atmosphere on a planet is due to the gravitational force holding the gas of the atmosphere to the planet. On a small planet, the gravitational force is very small, and the escape speed is correspondingly small. If a small planet starts its existence with an atmosphere, the molecules of the gas will have a distribution of speeds, according to kinetic theory. Some of these molecules will have speeds higher than the escape speed of the planet and will leave the atmosphere. As the remaining atmosphere is warmed by radiation from the Sun, more molecules will attain speeds high enough to escape. As a result, the atmosphere bleeds off into space.

17. The measurements are too short. At 22°C the tape would read the width of the object accurately, but an increase in temperature causes the divisions ruled on the tape to be farther apart than they should be. This "too long" ruler will, then, measure objects to be shorter than they really are.

19. The glass surrounding the mercury expands before the mercury does, causing the level of the mercury to drop slightly. The mercury rises after it begins to heat up and approach the temperature of the hot water, because its temperature coefficient of expansion is greater than that for glass.

21. Imagine the balloon rising into air at uniform temperature. The air cannot be uniform in pressure because the lower layers support the weight of all the air above them. The rubber in a typical balloon stretches or contracts until interior and exterior pressures are nearly equal. So as the balloon rises, it expands; this can be considered as a constant temperature expansion with P decreasing as V increases by the same factor in $PV = nRT$. If the rubber wall is strong enough, it will eventually contain the helium at a pressure higher than the air outside but at the same density, so that the balloon will stop rising. It is more likely that the rubber will stretch and rupture, releasing the helium, which in turn will "boil out" of the Earth's atmosphere.

Problems
1. (a) $-459.67°F$ (b) $37.0°C$ (c) $-280°F$
3. (a) $-423°F$, 20.28 K (b) $68°F$, 293 K
5. (a) $-273.5°C$ (b) $P_f = 1.272$ atm, $P_b = 1.737$ atm
9. 2.171 cm
11. 55.0°C
13. (a) 263.5°C (b) $-262.2°C$
15. (a) 437°C (b) 2099°C; aluminum melts at 660°C
17. 270 N

19. 1020 gallons
21. (a) 627°C (b) 927°C
23. (a) 2.5×10^{19} molecules (b) 4.1×10^{-21} mol
25. 290°C
27. 4.8 g
29. 3.84 m
31. 0.131 kg/m³
33. 6.21×10^{-21} J
35. 6.64×10^{-27} kg
37. (a) $v_{H_2} = 1.73$ km/s (b) $v_{CO_2} = 0.369$ km/s (c) Hydrogen escapes; carbon dioxide does not.
39. 16 N
41. 5.1×10^{-21} J/molecule
43. 28 m
45. 2.4 m
47. 800°C
49. (a) 343 K (b) 12.5% of original mass
51. $L_C = 9.17$ cm, $L_S = 14.17$ cm
53. (a) 0.169 m (b) 1.35×10^5 Pa
55. 0.417 liters
57. 3.3 cm

CHAPTER 11

Multiple-Choice Questions
1. d 2. c 3. e 4. c 5. b

Conceptual Questions
1. When you rub the surface, you increase the temperature of the rubbed region. With the metal surface, some of this energy is transferred away from the rubbing site by conduction. Thus, the temperature in the rubbed area is not as high for the metal as it is for the wood, and it feels relatively cooler than the wood.

3. If the fruit were open to the cold air, it is possible that the temperature of the fruit could drop to a point several degrees below freezing, causing damage. By spraying with water, a thin film of ice can form on the fruit. This provides some insulation to the fruit so that its temperature can remain somewhat higher than that of the air.

5. The operation of the immersion coil depends on the convection of water to maintain a safe temperature. As the water near the coils warms up, it floats to the top due to Archimedes's principle. The temperature of the coils cannot go higher than the boiling temperature of water, 100°C. If the coils are operated in air, the convection process is reduced, and the upper limit of 100°C is removed. As a result, the coils can become hot enough to be damaged. If the coils are used in an attempt to warm a thick liquid like stew, the convection process cannot occur fast enough to carry energy away from the coils, so that they again become hot enough to be damaged.

7. One of the ways that objects transfer energy is by radiation. If we consider a mailbox, the top of the box is oriented toward the clear sky. Radiation emitted by the top of the mailbox goes upward and into space. There is little radiation coming down from space to the top of the mailbox. Radiation leaving the sides of the mailbox is absorbed by the environment. Radiation from the environment (tree, houses, cars, etc.), however, can enter the sides of the mailbox, keeping them warmer than the top. As a result, the top is the coldest portion and frost forms there first.

9. The tile is a better conductor of heat than carpet. Thus, energy is conducted away from your feet more rapidly by the tile than by the carpeted floor.

11. The large amount of energy stored in the concrete during the day as the sun falls on it is released at night resulting in an overall higher average temperature than the countryside. The heated air in a city rises to be replaced by cooler air drawn in from the countryside. Thus, evening breezes tend to blow from country to city.

13. The fingers are wetted to create a layer of steam between the fingers and the molten lead. The steam acts as an insulator and prevents serious burns. The molten lead demonstration is dangerous, and we do not recommend it.

15. The increase in temperature of the ethyl alcohol will be about twice that of the water.

Review Problem

20.02 cm

Problems

1. 1.17 Cal
3. 1.03×10^3 J
5. 176°C
7. 10.1°C
9. 88 W
11. 467 pellets
13. 185 g
15. 35°C
17. 1.7 kg
19. 0.26 kg
21. 66°C
23. 21 g
25. 2.3 km
27. 16°C
29. (a) all ice melts, $T_f = 40°C$ (b) 8.0 g, 0°C
31. (a) 470 J/s into house (b) 1.7×10^3 J/s to outdoors
33. 52 J/s (Thermopane), 1.8×10^3 J/s (single pane)
35. 9.0 cm
37. 110 W
39. 2.6×10^3°C
41. 1.3×10^4 J/s
43. 89.8°C (silver), 89.7°C (copper); copper wins
45. 2.4 kg
47. 1.4 kg
49. 12 h
51. (a) 75°C (b) 3.6×10^4 J
53. 38°C
55. 28°C
57. (a) 25.8°C (b) no
59. 51.2°C
61. 400 cm³/h
63. (a) 74 stops (b) Answer assumes no heat loss to surroundings and that all internal energy generated remains with the brakes.

CHAPTER 12

Multiple-Choice Questions

1. b   2. c   3. d   4. d   5. c

Conceptual Questions

1. If there is no change in internal energy, then, according to the first law, the heat is equal to the work done by the gas, $Q = W$.

3. The energy that is leaving the body by work and heat is replaced by means of biological processes that transform chemical energy in the food that the individual ate into internal energy. Thus, the temperature of the body can be maintained.

5. The statement shows a misunderstanding of the concept of heat. Heat is energy in the process of being transferred, not a form of energy that is held or contained. If you wish to speak of energy that is "contained," you speak of *internal energy*, not heat. Correct statements would be, (1) "Given any two objects in thermal contact, the one with the higher temperature will transfer heat to the other." (2) "Given any two objects of equal mass, the one with the higher product of absolute temperature and specific heat contains more internal energy."

7. Although no heat is transferred into or out of the system, work is done on the system as the result of the agitation. As a consequence, both the temperature and the internal energy of the coffee increase.

9. The most basic limit of efficiency for automobile engines is the Carnot cycle; the efficiency can **never** exceed this ideal model. The efficiency is therefore affected by the maximum temperature that the engine block can stand and by the intake and exhaust pressures. However, the Carnot cycle assumes that the exhaust gases can be released over an infinite period of time. If you want your engine to act more quickly than "never," you must take additional losses. Other limits on efficiency are imposed by friction, the mass and acceleration of the engine parts, and the timing of the ignition.

11. Practically speaking, it is not possible to create a heat engine on Earth that creates no thermal pollution, because there must be both a hot heat source and a cold heat sink. The heat engine will warm the cold heat sink and

will cool down the heat source. If either of those two events is undesirable, then there will be thermal pollution.

There are some circumstances in which the thermal pollution would be negligible. For example, suppose a satellite in space were to run a heat pump between its sunny side and its dark side. The satellite would intercept some of the heat that gathered on one side and would dump it to the dark side. Because neither of those effects would be particularly undesirable, it could be said that such a heat pump produced no thermal pollution.

It is also possible to make the heat source and sink totally self-contained. In such a case, the engine would function until the sink became as hot as the source. However, building such an engine would probably be more wasteful than simply finding an acceptable form of thermal pollution.

13. The rest of the Universe must have an entropy change of $+8.0$ J/K or more.

15. The first law is a statement of conservation of energy that says that we cannot devise a process that produces more energy than we put into it. In addition, the second law says that during the operation of a heat engine, some heat must be rejected to the environment. As a result, it is theoretically impossible to construct an engine that will work with 100% efficiency.

17. (a) The semiconductor converter operates essentially like a thermocouple, which is a pair of wires of different metals and a junction at each end. When the junctions are at different temperatures, a small voltage appears around the loop, so that the device can be used to measure temperature or to drive a small motor. (b) The second law holds that it is impossible for a cycling engine simply to extract heat from one reservoir and convert it into work. This exactly describes the first situation, in which both legs are in contact with a single reservoir and the thermocouple fails to produce work. To convert heat to work, the device must transfer heat from a hot reservoir to a cold reservoir, as in the second situation.

Review Problem
(a) 31 m/s   (b) 1/6

Problems
3. 1.1×10^4 J
5. (a) 810 J   (b) 507 J   (c) 203 J
7. (a) 6.1×10^5 J   (b) -4.6×10^5 J
9. (a) 1093 K (820°C)   (b) 6.81×10^3 J
11. $6P_0V_0$
13. (a) -152 J   (b) -248 J
15. (a) 12.0 kJ   (b) -12.0 kJ
17. (a) -180 J   (b) -188 J
19. (a) 9.12×10^{-3} J   (b) -333 J
21. (a) $W_{IAF} = 76.0$ J, $W_{IBF} = 100$ J, $W_{IF} = 88.6$ J
    (b) $Q_{IAF} = 165$ J, $Q_{IBF} = 190$ J, $Q_{IF} = 178$ J

23. (a) 560 J   (b) 350 K
25. $e = 19.7\%$
27. (a) $e = 0.294$ (b) 500 J   (c) 1.67×10^3 W
29. $\dfrac{T_c}{T_h} = \frac{1}{3}$
31. 0.48°C
33. 6.1×10^3 J/K
35. 2.70×10^3 J/K
37. (a)

Result	Possible Combinations	Total
all heads	HHHH	1
3H, 1T	THHH, HTHH, HHTH, HHHT	4
2H, 2T	TTHH, THTH, THHT, HTTH, HTHT, HHTT	6
1H, 3T	HTTT, THTT, TTHT, TTTH	4
all tails	TTTT	1

Most probable result = 2H and 2T.
(b) All H or All T   (c) 2H and 2T
39. Maximum efficiency possible with these reservoirs = 50%; claim is invalid.
41. (a) $\dfrac{-Q_h}{T_h}$ (b) $\dfrac{Q_c}{T_c}$ (c) $\dfrac{Q_h}{T_h} - \dfrac{Q_c}{T_c}$ (d) 0
43. 2.8°C
47. (a) 1.22×10^4 J   (b) -4.05×10^3 J
    (c) 8.10×10^3 J
49. 18°C
51. (a) 12.0 MJ   (b) -12.0 MJ
53. 3.27 J/K
55. (a) 4.9×10^{-2} J   (b) 1.6×10^4 J   (c) 1.6×10^4 J
57. (a) 2.49×10^3 J   (b) 1.50×10^3 J   (c) 990 J
59. (a) $2.7 \times 10^3 \dfrac{\text{metric tons}}{\text{day}}$ (b) $\dfrac{\$7.7 \times 10^6}{\text{year}}$
    (c) 4.0×10^4 kg/s

CHAPTER 13

Multiple-Choice Questions
1. c   2. d   3. b   4. b   5. d   6. a

Conceptual Questions
1. No! Because the total energy $E = \frac{1}{2}kA^2$, changing the mass while keeping A constant has no effect on the total energy. When the mass is at displacement x from equilibrium, the potential energy is $\frac{1}{2}kx^2$, independent of mass, and the kinetic energy is $E - \frac{1}{2}kx^2$. The larger mass must move slower to have the same kinetic energy. At a particular instant in time, both kinetic and potential energy would change as the mass is increased.

3. When the spring with two masses is set into oscillation in space, the coil in the exact center of the spring does not move. Thus, we can imagine clamping the center coil in

place without affecting the motion. If we do this, we have two separate oscillating systems, one on each side of the clamp. The half-spring on each side of the clamp has twice the spring constant of the full spring, as shown by the following argument. The force exerted by a spring is proportional to the separation of the coils as the spring is extended. Imagine that we extend a spring by a given distance and measure the distance between coils. We then cut the spring in half. If one of the half-springs is now extended by the same distance, the coils will be twice as far apart as they were for the complete spring. Thus, it takes twice as much force to stretch the half-spring, from which we conclude that the half-spring has a spring constant that is twice that of the complete spring. Our clamped system of masses on two half-springs, therefore, will vibrate with a frequency that is higher than f by a factor of $\sqrt{2}$.

5. The bouncing ball is not an example of simple harmonic motion. The ball does not follow a sinusoidal function for its position as a function of time. The daily movement of a student is also not simple harmonic motion, because the student stays at a fixed location—school—for a long period of time. If this motion were sinusoidal, the student would move more and more slowly as she approached her desk, and as soon as she sat down at the desk, she would start to move back toward home again.

7. If it accelerates upward, the effective "g" is greater than the free-fall acceleration, so the period decreases. If it accelerates downward, the effective "g" is less than the free-fall acceleration, so the period increases. If it moves with constant velocity, the period does not change. (If the pendulum is in free-fall, it does not oscillate.)

9. As the temperature increases, the length of the pendulum will increase, due to thermal expansion. With a longer length, the period of the pendulum will increase. Thus, it will take longer to execute each swing, so that each second according to the clock will take longer than an actual second. Thus, the clock will run slow.

11. A pulse in a long line of people is longitudinal, because the movement of people is parallel to the direction of propagation of the pulse. The speed is determined by the reaction time of the people and the speed with which they can move once a space opens up. There is also a psychological factor, in that people will not want to fill a space that opens up in front of them too quickly, so as not to intimidate the person in front of them. The "wave" at a stadium is transverse, because the fans stand up vertically as the wave sweeps past them horizontally. The speed of this pulse depends on the limits of the fans' abilities to rise and sit rapidly and on psychological factors associated with the anticipation of seeing the pulse approach the observer's location.

13. A wave on a massless string would have an infinite speed of propagation because its linear mass density is zero.

15. The kinetic energy is proportional to the square of the speed, and the potential energy is proportional to the square of the displacement. Therefore, both must be positive quantities.

17. It travels a distance of $4A$.

19. In a simple harmonic oscillator, the velocity follows the displacement by $1/4$ of a cycle, and the acceleration follows the displacement by $1/2$ of a cycle. (a) There are times when both the displacement and the velocity are positive, and therefore in the same direction. (b) There are also times when both the velocity and the acceleration are positive, and therefore in the same direction. On the other hand, (c) when the displacement is positive, the acceleration is always negative, and therefore these two always act in opposite directions.

Problems

1. (a) 24 N   (b) 60 m/s^2
3. (b) 1.8 s   (c) No, the force is not of Hooke's law form.
5. (a) 60 J   (b) 49 m/s
7. 2.94×10^3 N/m
9. 2.6 m/s
11 (a) 11 cm/s   (b) 6.3 cm/s   (c) 3.0 N
13. (a) 28 cm/s   (b) 1.81 s   (c) 26 cm/s
    (d) 3.5 cm
15. 39.2 N
17. (a) 0.628 m/s   (b) 0.500 Hz   (c) 3.14 rad/s
19. (a) 126 N/m   (b) 17.8 cm
21. 0.627 s
23. (a) at $t = 0$, $x = 0.30$ m; at $t = 0.60$ s, $x = 0.24$ m
    (b) 0.30 m   (c) $\frac{1}{6}$Hz   (d) 6.0 s
25. (a) 250 N/m   (b) 0.281 s, 3.56 Hz, 22.4 rad/s
    (c) 0.313 J   (d) 5.00 cm
    (e) $v_{max} = 1.12$ m/s, $a_{max} = 25.0$ m/s^2 (f) 1.01 cm
27. 105 complete oscillations
29. (a) gain   (b) 1.1 s
31. (a) $L_{Earth} = 25$ cm, and $L_{Mars} = 9.4$ cm,
    (b) $m_{Earth} = 0.25$ kg, $m_{Mars} = 0.25$ kg
33. 5.67 mm
35. 0.800 m/s
37. 31.9 cm
39. 219 N
41. 7.07 m/s
43. 13.5 N
45. (a) constructive interference yields $A = 0.50$ m
    (b) destructive interference yields $A = 0.10$ m
47. 0.750 J
49. (a) 588 N/m   (b) 0.700 m/s
51. 1.1 m/s
53. $\mu = 0.11$

55. (a) 0.50 m/s   (b) 8.6 cm
57. (a) 15.8 rad/s   (b) 5.23 cm
59. $F_{\text{tangential}} = -\left[(\rho_{\text{air}} - \rho_{\text{He}}) \dfrac{Vg}{L} \right] s = -ks$, $T = 1.40$ s
61. 0.0329 s

CHAPTER 14

Multiple-Choice Questions
1. e   2. b   3. e   4. c   5. d

Conceptual Questions

1. The camera is designed such that it assumes the speed of sound is 345 m/s, the speed of sound at a room temperature of 23°C. If the temperature should decrease to, say, 0°C, the speed will also decrease, and the camera will note that it takes longer for the sound to make its round trip. Thus, it will think the object is farther away than it really is.

3. Sophisticated electronic devices break the frequency range of about 60 Hz to 4000 Hz used in telephone conversations into several frequency bands and then mix them in a predetermined pattern so that they become unintelligible. The descrambler, of course, moves the bands back into their proper order.

5. Wind can change a Doppler shift but cannot cause one. Both v_0 and v_s in our equations must be interpreted as speeds of observer and source relative to the air. If source and observer are moving relative to each other, the observer will hear one shifted frequency in still air and a different shifted frequency if wind is blowing. If the distance between source and observer is constant, there will never be a Doppler shift.

7. The echo is Doppler shifted, and the shift is like both a moving source and a moving observer. The sound that leaves your horn in the forward direction is Doppler shifted to a higher frequency, because it is coming from a moving source. As the sound reflects back and comes toward you, you are a moving observer, so there is a second Doppler shift to an even higher frequency. If the sound reflects from the spacecraft coming toward you, there is another moving source shift to an even higher frequency. The reflecting surface of the spacecraft acts as a moving source.

9. At the center of the string, there is a node for the second harmonic, as well as for every even-numbered harmonic. By placing the finger at the center and plucking, the guitarist is eliminating any harmonic that does not have a node at that point, which is all the odd harmonics. The even harmonics can vibrate relatively freely with the finger at the center because they exhibit no displacement at that point. The result is a sound with a mixture of frequencies that are integer multiples of the second harmonic, which is one octave higher than the fundamental.

11. The bow string is pulled away from equilibrium and released, similar to the way that a guitar string is pulled and released when it is plucked. Thus, standing waves will be excited in the bow string. If the arrow leaves from the exact center of the string, then a series of odd harmonics will be excited. Even harmonics will not be excited because they have a node at the point where the string exhibits its maximum displacement.

13. At the instant at which there is no displacement of the string, the string is moving at maximum speed. Thus, the energy at that instant is all kinetic energy of the string.

15. A change in temperature will result in a change in the speed of sound, and therefore, in the fundamental frequency. The speed of sound for audible frequencies in the open atmosphere is only a function of temperature and does not depend on the pressure. Thus, there will be no effect on the fundamental frequency caused by variations in atmospheric pressure.

17. When the bottles are struck, standing wave vibrations are established in the glass material of the bottles. The frequencies of these vibrations are determined by the tension in the glass and the mass of the glass material. As the water level rises, there is more mass, because the glass is in contact with the water. This increased mass decreases the frequency. On the other hand, blowing into a bottle establishes a standing wave vibration in the air cavity above the water. As the water level rises, the length of this cavity decreases, and the frequency rises.

19. A vibrating string is not able to set very much air into motion when vibrated alone. Thus it will not be very loud. If it is placed on the instrument, however, the string's vibration sets the sounding board of the guitar into vibration. A vibrating piece of wood is able to move a lot of air, and the note is louder.

Problems

1. 0.20 s
3. 1.7×10^{-2} m to 17 m
5. 5.0×10^{-7} m
7. (a) 5.0×10^{-17} W   (b) 5.0×10^{-5} W
9. $I_{\text{Full Orchestra}} = 32 I_{\text{Violin}}$
11. 3.01 dB
13. (a) 130 W   (b) 96 dB
15. (a) $\dfrac{I_A}{I_B} = 2$ (b) $\dfrac{I_A}{I_C} = 5$
17. (a) 10.0 kHz   (b) 3.33 kHz
19. 595 Hz
21. 0.391 m/s
23. $\theta = 42°$
25. 800 m
27. (a) 24.0 cm   (b) 85.5 cm
29. (a) Nodes at 0, 2.67 m, 5.33 m, and 8.00 m. Antinodes at 1.33 m, 4.00 m, and 6.67 m.
    (b) 18.6 Hz

31. 8.45×10^2 Hz, 1.69×10^3 Hz, 2.54×10^3 Hz
33. (a) 4.9×10^{-3} kg/m   (b) 2 loops   (c) No standing wave exists.
35. 1.1×10^3 N/m
37. (a) 0.40 m   (b) 1.1×10^3 Hz
39. 3.0×10^3 Hz
41. (a) 0.550 m   (b) 317 Hz
43. 5.26 beats/s
45. 3.8 m/s toward station, 3.9 m/s away from station
47. 2.94 cm

49. 300 Hz. There must be relative motion between the source and observer for a Doppler shift to occur.
51. 64 dB
53. (a) 0.655 m   (b) 13.4°C
55. 4.0 beats/s
57. 7.82 m
59. 19.3 m
61. (a) 55.8 m/s   (b) 2500 Hz
63. 1.34×10^4 N
65. 1.20×10^3 Hz

Index

Page numbers in *italics* indicate illustrations; page numbers followed by "n" indicate footnotes; page numbers followed by "t" indicate tables.